Small Systems and Fundamentals of Thermodynamics

Small Systems and Fundamentals of Thermodynamics

Yu. K. Tovbin

CRC Press is an imprint of the
Taylor & Francis Group, an **informa** business

CRC Press
Taylor & Francis Group
6000 Broken Sound Parkway NW, Suite 300
Boca Raton, FL 33487-2742

First issued in paperback 2020

ISBN 13: 978-0-367-57134-4 (pbk)
ISBN 13: 978-1-138-58724-3 (hbk)

Visit the Taylor & Francis Web site at
http://www.taylorandfrancis.com

and the CRC Press Web site at
http://www.crcpress.com

Contents

Foreword

Recently, various experimental methods of research have been actively developed, which have made it possible to significantly improve the resolution of measurements of the sizes and properties of small systems. For a long time, only particles of a new emerging phase (liquid drops or microcrystals) and gas bubbles in liquids were considered small systems. For them, for the first time, methods of thermodynamic description were involved, and this state of affairs was long considered acceptable until doubts had been accumulated about their correctness.

The physical basis of these doubt is the well-known result of J.W. Gibbs (1902) that as the size of systems decreases, the role of fluctuations increases, and the thermodynamic description becomes insufficient. This general statement to the present prevails in the qualitative interpretation of numerous measurements, and the work of T.L. Hill (1963) on small systems only emphasizes the absence in the world literature of specialized books in this field. Hill's work drew attention to the need to take into account the effects of fluctuations for the description of small systems and the transition to discrete calculus. The simplest examples of such systems in different statistical ensembles were considered and the influence of fluctuations was demonstrated for them.

Recall that almost one hundred and forty years ago, J.W. Gibbs (1878) practically completed the construction of equilibrium thermodynamics, as a science on the most common thermal properties of macroscopic bodies. Thermodynamics studied the thermal form of the motion of matter – the laws of thermal equilibrium and the conversion of heat into other types of energy. In his work, Gibbs extended thermodynamics to macroscopic non-uniform systems. These provisions to date are the main reference point for many areas of knowledge: physics, chemistry, mechanics, biology, geology, etc., including all related disciplines.

Apparently, Gibbs was also the first to understand the need for a deeper understanding of thermodynamics, as an exclusively model-free field of science, which he did when developing a statistical method for studying the properties of macroscopic bodies, based on model atomic–molecular representations from the very beginning.

The foundations of thermodynamics were based on postulates that are based on experimental measurements of macroscopic systems. They were confirmed by age-old observations, so there are no strict limitations on the scope of their application in them. The shift of the interests of researchers to a small-scale region requires the same rigorous thermodynamic analysis of numerous thermophysical measurements in the submicron and nanometer ranges, as well as for macroscopic systems.

The traditional references in the traditional statistical thermodynamics to the limitations of the use of thermodynamic approaches, connected with the discreteness of the description at the molecular level of the properties of substances and phases, are sufficiently general and allow great arbitrariness in their interpretation. Until now, there was a very wide variation in the interpretation of such restrictions in this matter, since the question of the size and time limitations of the use of thermodynamics for small systems was not so acute before. These limitations exist according to the following features:

1. by the size of the region in which fluctuations are important, and in particular, what sizes of regions appear in the equations of thermodynamics (what is the magnitude of the elementary volume dV);

2. by the degree of homogeneity of the volume within the phases, and what is the minimum phase size, or how it differs from molecular associates;

3. the method of taking into account the curvature factor of curved interfaces, including the question of the applicability of the Kelvin equation;

4. The degree of non-equilibrium deviations of the non-equilibrium thermodynamics described by the equations, and how much these deviations should be small so that the equilibrium state can be considered real. In passing, there are questions about the correctness of the application of the thermodynamic approaches in kinetics, in particular, about the correctness of the use of the activity coefficient for the activated complex in it.

The monograph is devoted to an analysis of the above limitations on the use of macroscopic thermodynamic interpretations for small systems. It is based on the attraction of the methods of statistical thermodynamics which Gibbs founded and, in essence, is a comparison of Gibbs thermodynamic results with the results obtained by modern methods of his own statistical thermodynamics. The main attention is paid to the drops of a vapour–liquid system, for the description of which all existing methods of statistical physics (integral equations and stochastic Monte Carlo and molecular dynamics methods) were used.

Strict statistical approaches based on integral equations turned out to be complex for numerical realization in such a highly non-uniform system as the interface between phases with a density difference of up to two or three orders of magnitude. For the same reason, with additional complication due to strong density fluctuations, stochastic methods proved to be ineffective. The only approach that allowed to solve this problem was a discrete–continuum method based on the many-particle distribution functions introduced by N.N. Bogolyubov (1946), using discrete molecular distributions on the molecular size scale, forming cells, and with a continual description on the scale smaller than the size molecules inside the cells. This allowed us to combine known techniques in the lattice gas model, which is widely used in problems of phase transformations, taking into account all internal motions of molecules within cells, and to find a general approach to solving problems for the equilibrium distribution of molecules in strongly non-uniform systems. This method is applicable to a substance in three aggregate states, so it is the only method that provides an equal description of the three phase separation interfaces. It is an alternative method of molecular dynamics and provides three to five orders of magnitude faster calculations than the molecular dynamics method.

The key issue in solving the problem of small systems was the contradiction between the Kelvin equation (1871) and the Yang-Lee condensation theory (1952). The essence of the contradiction is related to the differences between thermodynamic and statistical constructions. Thermodynamics conditionally consists of four stages: 1) initial statements, 2) the onset of thermodynamics (the First and Second Laws of thermodynamics), 3) consequences from the beginnings of thermodynamics for macroscopic systems (these are all the formulas following from the beginnings given in all textbooks), and 4) additional hypotheses of thermodynamics outside macroscopic systems. Statistical physics differs already in the first point: instead of a small number of macroscopic parameters, the analysis proceeds with a complete ensemble of molecules (their coordinates and impulses). Further, the probability theory with its rules for averaging the distributions of the coordinates and momenta of the molecules, and obtaining the mean values, works. As a rule, the results of statistical physics are approximate, depending on the type of potential interactions and internal approximations. In statistical physics, everything is laid in the first stage, the next steps are automatically performed, and when properly operated, one should not interfere in the fourth stage. This makes it possible to verify the consequences of the thermodynamic constructions introduced for small systems in the fourth stage.

In the construction of the Kelvin equation, the relationship between the pressures in neighbouring phases is used from the condition of mechanical equilibrium, as in the mechanics of continuous media, which has nothing to do with the chemical potential (there was no such concept at the time of Laplace and Kelvin), and this leads to so-called metastable drops. These drops are absent in the Yang–Lee theory as equilibrium objects, which illustrates their contradiction. Historically, the Kelvin equation was the first equation for small systems, and it became decisive for the entire development of the direction of science. Many years of accumulated experience with the use of this equation put him in doubt, without waiting for an explanation of this contradiction. Today, the Kelvin equation is almost completely abandoned in the two most important areas of its practical application: the processes of formation of new phases and adsorption porosimetry.

The monograph deals with all the questions connected with this contradiction, as well as answers to the above questions: the concept of the 'minimum phase size', the size range of systems for which thermodynamic equations are applicable, including the notion of an 'elementary volume' dV, the notion of 'passive forces' introduced by Gibbs, and the field of applicability of the concept of 'local equilibrium' in non-equilibrium processes, and a number of other related questions.

The book consists of seven chapters. The first chapter contains examples of experimental studies of small systems and outlines the main provisions of classical thermodynamics to the extent necessary to consider the limitations of its application on the size factor and to determine the conditions for violation of local equilibrium. This refers to the formulation of the basic principles of equilibrium thermodynamics, including the beginning of thermodynamics, the problem of the Kelvin equation is formulated, and the main provisions of non-equilibrium thermodynamics are given, including the principles of constructing phenomenological equations for the transport of molecules and the rates of chemical reactions on the basis of the law of acting masses.

The multiphase systems and the Gibbs phase rule are discussed as the main object. It is the transition from macrosystems to polydisperse systems, porous bodies, and small systems, which required more detailed consideration of size effects and fluctuations unavoidable for small systems. A joint consideration of classical thermodynamics as a common system of views on equilibrium and non-equilibrium processes inevitably leads to the need to analyze the relaxation times of the thermodynamic parameters of the system.

In the second chapter, the principles of molecular description are presented: the task of this chapter is to show the difference between the molecular description of systems and the thermodynamic one. A continual description of the distribution of molecules in space is introduced (the equilibrium distribution of molecules is described by a chain of Bogolyubov–Born–Green–Kirkwood–Yvon (BBGKY) equations) and a discrete–continual distribution. In it, the description of the coordinates of molecules is divided into two scales: discrete with a characteristic size of the order of the size of the molecules (the initial presentation is constructed for the simplest case of symmetric particles), which leads to the isolation of cells, and to the continual one, which describes the distribution of the centre of mass of molecules inside the cell. The difference of this approach from different versions of previous lattice constructions is explained and its direct connection with the BBGKY chain is shown. The foundations for calculating the non-equilibrium function of molecular distributions in a discrete–continual description are introduced, and its special case is the kinetic equations of chemical reactions in dense phases (Master Equation).

The third chapter is devoted to the presentation of the traditional thermodynamic approach to the account of the interface and the molecular approach for plane and curved boundaries. The principal difference between the molecular and thermodynamic formulation of the problem of the phase boundary is shown. The molecular theory leads to the existence of a system of equations for finding the concentration profile of the density between coexisting phases. The solution of this system of equations makes it possible to calculate the surface tension between vapour and liquid and leads to the existence of equilibrium drops (2010), which can not appear in thermodynamics because of its rough description of the properties of the phase boundaries. The molecular theory makes it possible to find the size dependence of the surface tension $\sigma(R)$ on the drop radius R, which makes it possible to determine the conditions for the appearance of a new phase or the minimum phase size R_0. Traditional metastable drops are discussed in Appendix 1.

The fourth chapter quantitatively characterizes small systems and introduces for them a new method of mathematical description on the basis of a discrete calculus. The necessity of using symmetrized difference derivatives of different orders is shown, which makes it possible to control the traditional differential derivatives used in statistical physics in the search for extrema of statistical sums and the calculation of size fluctuations. Examples of the use of discrete calculus for calculating the equilibrium characteristics of small systems are given. The lower limit of the applicability of a thermodynamic description is found without taking into account the influence of the discrete nature of matter and the contribution of fluctuations. Other examples that limit the size of the regions in which thermodynamics are applicable are discussed.

The fifth chapter is devoted to non-equilibrium processes. For them, kinetic equations are introduced on the basis of the local equilibrium condition. The relaxation times allow one to assess the possibility of realizing an equilibrium process under laboratory conditions. As the degree of non-equilibrium increases, the structure of the system of kinetic equations changes, and in addition to changing the thermodynamic parameters of the state, it is necessary to take into account the relaxation times of the pair functions, which sharply increases the dimensionality of the system. This transition makes it possible to find a criterion for the realization of local equilibrium. The possibility of applying equations for highly non-equilibrium processes is discussed: turbulent flows and frozen states of solids at low temperatures. The type of equations included in the general system of the transport equation is indicated in Appendix 2, and the question of calculating the dissipative coefficients and describing the evolution of the ensemble of small systems in a two-level model is also discussed there. The introduction of relaxation times of various properties allows us to discuss the essence of passive Gibbs forces and the so-called non-equilibrium thermodynamic functions. With their help, a non-equilibrium surface tension is introduced, and the phase boundary relaxation is analyzed for different types of boundaries. In conclusion of Chapter 5 we give examples of the effect of fluctuations in the kinetics of elementary stages of the ideal and give an approach to allowance for fluctuations in non-ideal reaction systems.

In the sixth chapter, the foundations of chemical kinetics in non-ideal reaction systems are described. It is shown that these equations provide conditions for a self-consistent description of the equilibrium and kinetics of chemical reactions and transport processes for any degree of non-equilibrium. This section plays an important role in describing the continuous transition from any non-equilibrium states of systems to their equilibrium for the equations considered in Chapter 5, which allows us to relate the material on the non-equilibrium thermodynamics of Chapter 1 and the kinetic equations of Chapter 2.

The seventh chapter discusses the analysis of thermodynamic interpretations related to the Kelvin equation: an explanation is given of the contradiction between the Yang–Lee theory and the Kelvin equation. The same principle of precedence of the condition of mechanical equilibrium over the chemical equilibrium, as in the Kelvin equation, was used in Gibbs' work and in later works of the twentieth century to refine the Gibbs constructions. Today, there is relatively little that is known from the size properties of small drops, therefore, in Chapter 7, the characteristics of metastable and equilibrium drops are clarified. The discussion of the contradiction between the Yang-Lee theory and the Kelvin equation is concludes and a direct calculation of the relaxation time of the transition from metastable drops to equilibrium ones is carried out. Also, the seventh chapter questions discusses the accuracy of the molecular theories used when considering the thermodynamic characteristics, metastable states of solid-phase systems, and the use of the activity coefficients in equilibrium (Appendix 3) and in kinetics.

The conclusion gives a brief summary of the limitations on the application of thermodynamics and the current state of the molecular-kinetic theory.

The fundamentals of classical and statistical thermodynamics and their achievements for macrosystems are well known in the literature on numerous monographs. The questions of their justification are not discussed. Also, many other more complex systems (mixtures, Coulomb potentials, polymers, etc.) are not discussed. The aim of the work was not to review the numerous modifications of the discussed equations of Kelvin and Gibbs in applications for small systems (this is impossible in the past more than 140 years).

The chapters 1 and 2 present only the material on classical and statistical thermodynamics, which is necessary for the analysis of small systems. The chapters 3, 4, 5 and 7 give examples of calculations of the characteristics of small systems, and how fluctuations affect them, which, as a rule, are poorly discussed in the literature.

The monograph has an interdisciplinary character – the fundamentals of thermodynamics are expounded in many natural sciences: physics, chemistry, mechanics, biology, geology, and materials science. The material is described for specialists in the field of physical chemistry, statistical thermodynamics, physics of surface phenomena and phase transitions, kinetic theory in condensed phases and hydrodynamics, mechanics of solids and technologists engaged in the creation of new materials, as well as for students and graduate students of relevant specialties.

Symbols and Abbreviations

a_f^i –the local Henry constant of the molecule i for the site with the number f (or a_q^i of the site of type q); for a uniform surface a_i is the Henry constant of the molecule i, and a is for one substance;

a_f^{i0} –the pre-exponent of the local Henry constant of the molecule i for the site f.

A similar simplification of the notation for a transition from a non-uniform system to a uniform one and from a mixture to a single substance is retained for all characteristics.

c_i –the concentration of molecules i in the number of molecules N_i per unit volume;

$C_v(f)$ –the specific heat per particle in cell f;

$E(f)$ –the internal energy of molecules in cell f;

$E_{fg}^{iV}(\rho)$ –the activation energy of hopping of a molecule of sort i from site f by distance ρ to a free site g;

$E_{fgh}^{iVj}(\omega_r)$ –the energy contribution of the molecule j at the site h located at a distance of the r-th coordination sphere from the 'central' pair of sites fg with orientation ω_r, to the non-ideal hopping function of molecule i from site f to site g;

$d_{qp}(r)$ –the conditional probability of finding a site of type p at a distance of r-th c.s. from a site of type q;

D –the diffusion coefficient;

$D_{1,2}$ –coefficient of mutual diffusion in a binary mixture

D_{ij} –the diffusion coefficient of component i in a multicomponent mixture under the influence of the gradient of component j;

D^*_{fg} –the self-diffusion coefficient of a pure substance;

f –the site number of the non-uniform system;

F –free energy;

f_q –probability of finding a site of type q in a non-uniform system;

$f_{qp}(r)$ –the probability of finding a pair of sites of the type q and p at a distance r in the non-uniform system;

F_i^0 –the partition function of the molecule i in the gas;
F_q^i –the partition function of a particle i located at a site of type q;
F_q^{i*} –the partition function of the particle i at the site of type q in the transition state;
F_q –the fraction of sites of type q; for the monolayer q of the drop, the fraction of sites of this monolayer;
$F_i(f)$ –the component i of the external force created by the wall potential in cell f;
G –the Gibbs potential;
h –the Planck constant.
h_f –the cluster Hamiltonian;
H –the width of the neck-like pore, the number of monolayers;
i –the sort of the particle in the cell, including the vacancy;
J_i –the thermodynamic flows (Section 9) or the diffusion flux of molecules of sort i;
J_i –the partition function of a molecule of sort i; (Section 30)
k_B –the Boltzmann constant;
$K_{fg}^{iV}(\chi)$ –the rate constant for the hopping of the molecule i from the cell f to the free cell g at a distance χ;
l_i –the mean free path of component i;
m_i –the mass of the molecule i;
$m(\omega_r)$ –the set of neighbouring sites with fixed values of r and ω_r;
M –the number of cells in the system;
N_{den} –the number of densely packed particles in the same volume,
N_q^i –the number of molecules of sort i at sites of type q,
N_{qp}^{ij} –the number of pairs of molecules of sort ij at sites of type qp,
P –pressure;
P_i –the partial pressure of component i;
$P(f)$ –the pressure in cell f;
q –type of site (cells);
Q_N –the partition function of the system states;
Q_{fi} –the binding energy of a molecule of sort i, located at the site f, with an adsorbent;
Q_q^i –the binding energy of a molecule i located in a site of the type (layer) q with the walls of the pore;
$Q_{q,f}$ –the binding energy of a molecule located at a site with number f of a section of type q in a complex porous system;
r –the distance between the particles (in units of λ);

R –is the radius of the drop;
R_{lat} –the radius of the intermolecular interaction potential (in units of λ);
$P(\{p,q\},t)$ –the total distribution function N of the molecules p_1, r_1,..., p_N, r_N, where p_i is the impulse and r_i is the coordinate of the centre of the mass of the molecule i, $1 \le i \le N$, at time t;
S –entropy;
s –the number of components of the mixture;
$S_{fg}^{i}(r)$ –factor of the non-ideality function;
$S(i \to k)$ –the vacancy region of rotation of the molecule from orientation i to orientation k,
t –the number of different types of sites in the system;
T –temperature;
T_i^c –critical temperature of component i;
T_f –local temperature,
$t_{fg}^{ij}(r)$ –the conditional probability of finding the molecule j at the site g at a distance r from the molecule i at the site f;
u –flow velocity, m/s;
$\mathbf{u}(f)$ –the velocity vector in cell f with components $u_i(f)$ in the direction $i = x, y, z$;
U –internal energy;
$U(q)$ –the interaction potential of the molecule of the layer q with the wall of the pore;
$U_{fg}^{i}(\chi)$ –the probability that a molecule i jumps from cell f to a free cell g by distance χ;
$U_{\xi fg(j)}^{(i)V}(\chi)$ –the probability of a molecule jumping after collision with a molecule j at a site ξ by a distance χ;
$U(z)$ –the interaction potential of the molecule with the z coordinate with the pore wall;
V –the volume of the system, or the symbol of the vacancy;
V_{fg}^{iV} –the concentration component of the hopping rate of the molecule i from cell f to the free cell g;
$V_{\xi fg}^{(j)iV}(\chi)$ –the concentration component of the hopping rate of molecule i from cell f to the free cell g after collision with molecule j at site ξ by distance χ;
v_0 –the cell volume;
W_α –the probability of realizing the elementary stage α;
$W_i(\rho)$ –the probability of hopping of a molecule of sort i by distance ρ;
w_i –the average thermal velocity of molecules i in the gas phase;
$w_{i(j)}$ –the average relative velocity of the molecule i after its

collision with the molecule j;

w^i_{fg} –the average thermal velocity of the molecules i between the sites f and g;

x^i_f –the mole fraction of component i in site f;

z –the number of nearest neighbours of the lattice structure;

z_q –the number of nearest neighboirs in the q layer;

$z_{fg}(r)$ –the number of neighbouring sites in the layer g at a distance r from thc site in the layer f.

$z^*_{fg}(\chi)$ –the number of possible jumps between sites f and g by distance χ;

Greek characters

α –the symbol of the dense phase of the stratified system (Chapter 3);

α –the dimensionless parameter $\varepsilon^*/\varepsilon$; or the stage number;

β –the symbol of the rarefied phase of the stratified system (Chapter 3);

β –inverse value of thermal energy $(k_B T)^{-1}$;

γ^i_f –a variable describing the occupied state i of the cell (site) with the number f;

δ –the width of the monolayer;

$\delta\varepsilon^{ij}_{fh}(r)$ –the difference of the quantities $(\varepsilon^{*ij}_{fh}(r)–\varepsilon^{ij}_{fh}(r))$;

ε_i –parameter of interaction potential of adsorbate of sort i - adsorbent;

$\varepsilon^{ij}_{fg}(r)$ –the interaction parameter of the molecule i at the site f with the neighbouring molecule j at the site g at a distance r;

$\varepsilon^{*ij}_{fh}(r)$ –the interaction parameter of the activated complex of migration of molecule i from site f to free site g with neighbouring molecule j in the ground state at site h at distance r;

η_{fg} –the local shear viscosity of the mixture flow between the sites f and g;

η^q_f –the function of the correspondence between the number of the site f and its type q;

θ –the total degree of filling of the volume of the system, the dimensionless quantity, $0 \leq \theta \leq 1$;

θ^i_q –the partial degree of filling of sites of type q by molecules of sort i;

$\theta^{ij}_{fg}(r)$ –the probability of finding particles i at site f and j at site g at a distance r;

$\theta^{iV}_{fg}(\chi)$ –the probability of realizing a free trajectory from cell f to cell g of length χ.

$\theta(p_1, r_1,..., p_N, r_N)$ –the total distribution function of N molecules; p_i is the momentum and r_i is the coordinate of the centre of the mass of the molecule i, $1 \le i \le N$;

$\kappa(f)$ –the coefficient of thermal conductivity in cell f;

$\kappa_{qp}(\omega_r|\rho)$ –the number of sites with orientation $\omega_r(\rho)$ in the r-th c.s. of the central pair of particles at the sites of type q and p at a distance ρ;

λ –linear cell size;

$\lambda_{qp\xi}(\omega_r|\rho)$ –the number of sites of type ξ with orientation ω_r, located in r-th c.s. around central sites of type qp at distance ρ;

Λ^i_f –the non-ideality function of the system for molecules i at site f;

$\Lambda^i_{fg}(\chi)$ –the non-ideality function of the system for hopping of a molecule i from the site f to a free site g by a distance χ;

μ_i –the chemical potential of component i;

μ_{ij} –the reduced mass of the colliding molecules i and j;

v^i_f –a one-particle contribution to the total energy of the system by a particle of type i located at a site numbered f; v_i – for a uniform lattice structure, v - for one substance;

$\xi(f)$ –the bulk viscosity coefficient in cell f;

π –spreading pressure analog of the equation of state for a lattice system;

$\pi_r(\rho)$ –the number of orientations of the sites in the r-th c.s. around the central pair of sites at a distance ρ;

ρ –the average distance between molecules (in units of λ);

σ_{ij} –distance of the closest approach of two components i and j of the mixture;

$\phi(r) = Ar^n–Br^m$ –Mie pair potential (n, m) where the constants A, B, n, m are positive,

χ –the molecule hopping length (in units of λ);

$\omega(f)$ –the enthalpy (thermal function) of the mass unit related to a given cell f.

$\omega_r(\rho)$ –the orientation of the site h located in the r-th coordination sphere of the 'central' pair of sites fg at distance ρ; $1 \le \omega_r(\rho) \le \pi_r(\rho)$.

Brackets

Curly: $\{\eta^q_f\}$ denote a complete list of values, here, for example, for η^q_f, $1 \le f \le M$, $1 \le q \le t$;

Square: $[m_{qp}]$ fixing the numbers m_{qp} of the sites of type p in the coordination sphere of the central site of type q, $1 \le p \le t$.

Indexes

For a uniform system, the subscripts *i, j, k, l* are the sorts of particles, including the vacancy V.

For a non-uniform system, subscripts: site numbers, if the indices *f, g, h*, or the types of sites a non-uniform lattice, if the indices *q, p, ξ*; superscripts: *i, j, k* – sorts of molecules or V (vacancy).

Abbreviations used

AC –activated complex
BBGKY –Bogolyubov–Born–Green–Kirkwood–Yvon
c.s. –coordination sphere
DF –distribution function
QCA –quasi-chemical approximation
MD –molecular dynamics
LGM –lattice gas model
TARR –theory of absolute reaction rates
KE –Kelvin's equation

1

Background

1. Small systems

In thermodynamics small systems are systems that have a large surface contribution in comparison with their volume contribution to all thermophysical functions. They have sizes (radii), which can vary over a wide range from nanometers to submicron values. A traditional example of small bodies is liquid drops in a supersaturated vapour and bubbles in liquid phases [1,2].

Today in the literature there is a great deal of activity in discussing various aspects of the study of small bodies that have appeared recently [3–29], especially in connection with the development of measurements in the range of sizes from 1 to 100 nm. This interest is caused by the transition of experimental equipment to a new level of spatial resolution in the last 15–20 years. This affects both traditional fields of knowledge, studying colloidal systems and non-uniform catalysis, and many other areas of physics, chemistry and biology. In the specified size range, many physical and chemical properties change, which opens up new approaches to the study of substances. There was a separation of such concepts as nanoclusters, nanostructures, and related phenomena into a separate area of physico-chemistry. This happened mainly as a result of significant progress in obtaining and researching nanoobjects, the emergence of new nanomaterials, nanotechnologies and nanodevices. New giant nanoclusters of a number of metals, fullerenes and carbon nanotubes, many nanostructures based on them and on the basis of supramolecular hybrid organic and inorganic polymers, etc. have been synthesized. There has been remarkable progress in the methods of observing and studying the properties of nanoclusters and nanostructures associated with the development of tunnel and scanning microscopy, X-ray and optical methods using synchrotron radiation, optical laser spectroscopy, radio frequency spectroscopy, Mössbauer spectroscopy, etc. [3-29].

Small systems also include modifications of 'micro-reactor' systems as their number in a unit volume increases: aerosols, aerogels, porous and non-porous friable bodies of various structures, etc. The classification of nanoclusters and nanostructures is based on the methods of their preparation [17]. This determines their distinction into isolated nanoclusters and nanoclusters, combined in a nanostructure with weak or strong intercluster interactions or cluster–matrix interaction. The group of isolated and weakly interacting nanoclusters includes: molecular clusters, gas ligandless clusters (clusters of alkali metals, aluminum and mercury, clusters of transition metals, carbon clusters and fullerenes, van der Waals clusters), colloidal clusters. Nanoclusters and nanostructures include solid-state nanoclusters and nanostructures, matrix nanoclusters and supramolecular nanostructures, cluster crystals and fullerites, compacted nanosystems and nanocomposites, nanofilms and nanotubes.

The properties of small systems are studied by a thermodynamic approach, with the help of which one attempts to determine the patterns of their formation, growth, properties and their changes in the process of phase transitions. The generality of the thermodynamic approach allows us to consider all the listed systems from a unified position. However, the possibility of using macroscopic thermodynamics to describe the properties of small systems is limited by the very phenomenological nature of the science of the most common thermal properties of macroscopic bodies. Thermodynamics studies the thermal form of the motion of matter – the laws of thermal equilibrium and the conversion of heat into other types of energy. The main content of thermodynamics is the consideration of the laws of thermal motion in systems in thermal equilibrium, when they lack the macroscopic displacements of one part relative to the other, and also the regularities in the transition of systems to the equilibrium state. Thermodynamics considers macroscopic phenomena caused by the combined action of a huge number of continuously moving molecules or other particles that make up the surrounding bodies, without any models for their movement. Thermal motion is characterized by the temperature of the body, which is a measure of the intensity of thermal motion.

The fundamentals of thermodynamics were built on the postulates resulting from experimental measurements of macroscopic systems [30–33]. Speaking of limitations in the use of thermodynamics, we speak about the degree of correctness of application of its approaches for small systems that are not macroscopic objects, or are incorrectly considered if there is a macroscopic ensemble of small bodies. Reducing the size of small bodies increases the role of their surface. Let us recall what new factors appear and influence the properties of systems near the interfaces and on the surfaces themselves in comparison with their bulk properties. These factors increase the heterogeneity of the local properties of small systems at the molecular level.

Near-surface area. The presence of a surface requires the separation of the near-surface region from each side of the interface plane, as well as the specificity of these phases [30, 34–38]. This requires a self-consistent description of both the contacting phases themselves and the nature of the distribution of the components of these phases along the normal to the boundary. The components of volumetric phases can have a wide spectrum of mutual distributions for the given phase structures. In more complex cases, various combinations are possible between the phase transitions of stratification and ordering. For example, a system can go into a disordered or ordered phase or into two different ordered phases. The type of phase transitions and the regions of their realization are determined by the concentrations of the components, the temperature, and the interparticle interaction potentials. Similar transitions are also carried out in multicomponent solutions. The increase in the number of components increases the number of different combinations of phase transitions [39].

The nature of the distribution of particles along the normal to the surface, as a rule, differs from their distribution in the volume [40–44]. This causes local non-uniformities. In thermodynamics there is a phase separation, and all phases are treated in the same way. Although at the microlevel there are qualitative internal differences between the liquid and solid phases. Below we discuss the boundary of a solid body, for which the greatest variety of the surface properties is characteristic in comparison with liquid phases.

In solids, near-surface non-uniformities are associated with the absence or change in the symmetry of the phase in the other half-plane, which disrupts the periodicity of the crystal structure along the normal to the surface and leads to a difference in the forces acting on the atom from the surrounding other atoms of the alloy on the surface and in the volume (the atom in volume has a larger number of neighbours than the atom on the surface). The nature of the distribution of solid-state atoms in the near-surface region has a determining effect on all surface processes: the surface composition determines the properties of the surface on which adsorption of gas-phase molecules occurs and surface reactions take place, and the dissolution process of adsorbed particles and the reorganization of the region itself depend on the distribution of atoms in the near-surface region. A much more complicated situation is realized in the case of a 'frozen' strongly non-equilibrium state of the surface. In these conditions, the methods of forming surfaces are extremely important: by breaking down a part of the crystal in a vacuum, by spraying on a substrate, by precipitation from a solution, by sintering a polycrystalline powder, etc. [45, 46], since the surface preparation technique and subsequent sample processing, preceding the experiment, first of all exert their influence on the state of the surface and near-surface area.

Adsorption. Surface atoms of a solid body have a certain ability to form bonds with molecules of the gas phase, which leads to an increase in the concentration of these molecules near the surface [47–53]. At the atomic level, the surface of a monoatomic crystal is not uniform. It represents a distributed system of atoms. The different positions of the adsorbed molecule over the adsorbent atoms are energetically non-equivalent. The nature of adsorption can also depend on the state of the adsorbed particle. Most solids have on the surface regions of localization of adsorbed particles with different binding energies. The reasons for this are due to the different local chemical composition of the surface and the different geometric arrangement of solid-state atoms. Usually, both these causes are manifested simultaneously. Such surfaces are called non-uniform. The scale of non-uniformity varies widely from point imperfections and impurities to different macro-regions [54, 55].

The physical and chemical prerequisites for the existence of non-uniform adsorption centres include such well known factors as inappropriate periodicity within the bulk crystalline lattice, leading to distortion of the surface structures: Schottky and Frenkel lattice defects, non-stoichiometry of the solid, surface groups of different nature, dislocations and mechanical disturbances of the crystal during growth or under external loads, surface roughness. In addition to these factors, new types of adsorption centres form on the surface which are associated with the method of forming the surface (e.g., the angle at which cleavage takes place in the crystal) and the availability of new types of bonds between atoms of the surface and the adsorbed molecules [56].

In general, the non-uniformity of the flat surface and the crystal surface region may be associated with impaired regularity of the distribution of the surface atoms of the solid (structural non-uniformity) and the difference in the nature of surface atoms (chemical non-uniformity). Any non-uniformity leads to the fact that in the vicinity of each defect there are changes in the properties of the surrounding atoms and this leads to a modification of the properties of neighbouring nodes of the surface structure, which increases the number of types of surface nodes.

An example of chemical non-uniformity is the surface of the A_xB_{1-x} (for example, Pd–Ag), which has unlimited solubility. The structurally uniform surface of the face, let it be (100), of this alloy is non-uniform due to the different adsorption capacities of atoms A and B. The proportion of each component on the surface can vary from zero to unity, depending on the temperature and concentration of particles in the volume. As a rule, chemical and structural non-uniformities are realized simultaneously.

If the binding energy of the adsorbed particle with the surface is commensurable with the binding energy between solid-state atoms, the adsorption of molecules can cause significant changes in the state of the surface. To a large extent, this is due to the fact that the adsorbed particles compensate for the absence of neighbours in surface atoms. As a result, the state of the entire near-surface region of a solid can change. This, in turn, causes a change in the number of different adsorption centres of the surface.

In general, considerable experimental material on the effect of the gas phase on the surface state of adsorbents and catalysts has been accumulated in the literature. This fact is specially used for preliminary creation of the necessary surface composition of catalysts. At the same time, the quantitative characteristics of these changes are mainly present only for binary alloys that are in contact with the H_2, CO, O_2 molecules [57–61]. Studies of recent years show that the effect of adsorbed particles on the state of the surface of a solid body is apparently more significant than is commonly believed now. Structural transformations in surface layers can play a great role (for example, the reorganization of the surface layer of platinum upon adsorption of CO and O_2 molecules [62, 63]) and the processes of the formation of new phases in them, which are analogous to three-dimensional topochemical processes [64-66].

Absorption. Many solids absorb significant amounts of gas phase molecules (H_2, O_2, N_2, etc.) [67, 68]. Dissolved atoms occupy interstices of crystal lattices, forming interstitial alloys [69, 70]. They exert a strong influence on the state of the volume phase of a solid body, changing the short-range and long-range orders in alloys, and for large volume saturations with dissolved atoms the volume phase and the transitions to multiphase systems are possible. Similar changes are possible in the near-surface region of the solid, which change the surface states and affect surface processes.

In the presence of chemical transformations, the state of the system changes due to the migration of particles and their participation in the reactions. The relationship between reaction rates and particle mobilities can vary within very wide limits: cases where the chemical transformation stage (the traditional field of chemical kinetics [71,72]) is limited, and when particle migration (diffusion-controlled reactions [73,74]) is limited, are possible. As a whole, this leads to different regimes of processes on the surface and in the near-surface region and determines the variety of models used to describe them.

Porous bodies. Porous bodies are an example of phase inversion, when small volumes of systems relate to internal regions of solids in which there is no matter. These regions can be filled with gas and liquid molecules. The structure of porous and/or dispersed materials is extremely diverse [75–77], and it significantly affects the equilibrium distributions of molecules, the course of phase transformations and all the dynamic processes occurring inside materials (porous bodies). Here the term ‘structure’ implies the location and interconnection of the constituent elements of the system in question in space, that is, the term ‘structure’ implies a set of clearly delineated structural elements with limited autonomy. The structural elements themselves may have a crystalline structure (microcrystals) or be amorphous particles characteristic of such artificial materials as regular web weaving and fabrics, highly ordered polymer systems, etc. In general, dispersed materials can consist of structural elements of both types, for example, in porous carbon of different origin [75].

The internal processes of distribution of the mobile phase are associated with the processes of transport of adsorbed molecules (adsorbate) inside the porous structure, the processes of redistribution of molecules between different sections of the free pore volume, and the processes of phase separation (condensation) of molecules, in the case of the predominance of cooperative behaviour of molecules and the instability of uniform distribution [78, 79]. In many of these systems, it is possible to use external fields (mechanical deformation, acoustic waves, electric and magnetic fields, etc.) that cause various local reconstructions of the matrices and phase states of the systems.

Similar types of non-uniformities can be isolated in various colloidal, including micellar systems, or in other systems of 'soft materials', but the number of non-uniformities on their surfaces is much smaller. Different types of non-uniform systems can be sized both macroscopically and small. The above enumeration gives an idea of the complications in describing the surface properties in comparison with the bulk phase, which inevitably arise when considering small systems. For macroscopic non-uniform systems, thermodynamic approaches are possible. The thermodynamic description negates the differences in the properties of the systems under consideration, which can be quite strongly non-uniform in their local properties. Introducing for them the concepts of surface and/or linear tension, thermodynamics reflects the gross effects caused by the presence of a surface, but it does not allow to give a molecular interpretation of these characteristics, and their experimental determination is extremely laborious. Under these conditions, the only method of theoretical description of non-uniform systems is modelling with the help of atomic-molecular models.

2. Thermodynamic parameters of the state

In sections 2–6 we recall some of the basic theses of thermodynamics [30–33,80–93], which are necessary for analyzing its application to small bodies.

Thermodynamics is a scientific discipline that studies the transitions of energy from one form to another, from one part of the system to another, the energy effects accompanying various physical and chemical processes, their dependence on the conditions of the processes, and the possibility, direction and limits of spontaneous flow of processes in the given conditions. In other words, thermodynamics studies the laws of thermal motion in equilibrium systems and the transition of systems to equilibrium (equilibrium thermodynamics or simply thermodynamics), and also the same regularities in non-equilibrium systems (thermodynamics of irreversible processes or non-equilibrium thermodynamics).

Thermodynamic systems. We will call any material object (and even its parts) a macroscopic system or a definite set of material objects, hypothetically isolated from the environment, consisting of a large number of particles. Such systems can consist of a large number of material particles (for example, molecules, atoms, ions, etc.), or fields, for example, an electromagnetic field. The sizes of macroscopic systems are always much larger than the sizes of atoms and molecules. In thermodynamic systems we are dealing with dynamic systems possessing an extremely large number of degrees of freedom (systems with a small number of degrees of freedom are not considered by thermodynamics).

If a part of the total system is studied, then the rest will be called the surroundings, environment or thermostat, which imposes some conditions on the system under study (for example, the conditions of constancy of temperature, pressure, chemical potential, etc.).

Macroscopic parameters. The totality of all the physical and chemical properties of the system characterizes its state. This term refers to the entire system as a whole in the case of an equilibrium state, or to its locally equilibrium state in a certain region in the case of its non-equilibrium state. The considered state is uniquely determined by a set of independent macroscopic parameters that must be chosen so that they are necessary and sufficient to determine this state. Then the remaining quantities characterizing the state are functions of these variables.

The macroscopic parameters can conveniently be divided into two classes: external and internal. The values determined by the position of external bodies outside our system are called the external parameters a_i, $(i = 1, 2, ...)$, for example, the volume of the system (since their values are determined by the location of external bodies), the intensity of the force field (as it depends on the positions of the field sources – charges and currents not included in our system), etc. Consequently, the external parameters are functions of the coordinates of external bodies.

The values determined by the cumulative motion and distribution in the space of particles entering the system are called internal parameters b_j $(j = 1, 2, ...)$, for example density, pressure, energy, polarization, magnetization, etc. (since their values depend on the motion and position of the particles of the system and the charges entering into them). The spatial arrangement of the particles – atoms and molecules entering the system can depend on the location of external bodies, so the internal parameters are determined by the position and motion of these particles and by the values of the external parameters.

However, depending on the conditions in which the system is located, the same value can be either an external or an internal parameter. For example, for a fixed position of the walls of a vessel, the volume V is an external parameter, and the pressure P is an internal parameter, since it depends on the coordinates and momenta of the particles of the system. If, however, the system is in a vessel with a movable piston at constant pressure, the pressure P will be an external parameter, and the volume V is an internal parameter, since it depends on the position and motion of the particles. Therefore, in general, the difference between the external and internal parameters depends on where the boundary between the system and external bodies is drawn. This circumstance is especially important when considering mechanical contacts.

A set of independent macroscopic parameters that completely determine the state of the system at a given time and are independent of the history of the system are termed *state parameters*. The number of independent parameters describing the thermally equilibrium state is determined empirically [88]. As the state parameters, one can use any of the quantities that serve to characterize the state of the thermodynamic system (or all the physical and chemical properties of the system): temperature, pressure, volume, internal energy, entropy, concentration, polarization, magnetization, etc. The term equilibrium state means that all independent parameters of the system do not change over time (this excludes the possibility of realizing stationary flows due to the action of some external sources). Such a state of the system is called the state of thermodynamic equilibrium, and accordingly, the parameters of such a state are called the *thermodynamic* parameters.

State functions. The physical quantities that have a definite value for each thermally equilibrium state of the system are also called thermodynamic quantities or state functions. These include, for example, temperature, pressure, internal energy, enthalpy and entropy.

The internal parameters of the system are divided into intensive and extensive ones.

Intensive parameters are those parameters whose values do not depend on the mass or size of the system when it is divided into parts that do not violate the equilibrium state. If the system in a thermally equilibrium state is divided into parts by means of impermeable partitions, then each part will remain in an equilibrium state. This means that the equilibrium state of a uniform system is its intrinsic property and is determined by the thermodynamic variables that do not depend on the size of the system. Examples of intensive parameters are all the molar and specific properties, temperature, pressure, chemical potential, etc. The values of these properties are not additive.

The *extensive* parameters are those parameters whose values are proportional to the mass or size of the system when it is divided into parts that does not violate the equilibrium state, i.e. the values of the extensive parameters have the additivity property. The extensive parameters include thermodynamic potentials, entropy, volume, etc. The extensive parameters characterize the system as a whole, while intensive ones can take definite values at each point of the system. A system whose energy depends non-linearly on the number of particles is not thermodynamic, and its study by thermodynamic methods is, generally speaking, approximate or even unjustified [89].

Uniform and non-uniform systems. The whole set of thermodynamic systems is divided into two classes – uniform and non-uniform.

Uniform are such systems within which the properties change continuously when passing from one place to another within a certain region. With the uniform systems, the concept of a phase is defined, which is defined as a uniform part of the system that is uniform in composition and physical state, separated from other parts by the interface on which any properties (and the parameters corresponding to them) change discontinuously. Within the phase, the uniform system has the same physical properties in any local, arbitrarily chosen parts, equal in volume. Examples of such systems are mixtures of various gases and solutions, both liquid and solid. In these systems, reactions can occur between the constituents of the mixture, the dissociation of gas or solute, solvation, polymerization, etc. In equilibrium in such systems the reactions cease in terms of the value of the *macroscopic* parameters.

It is necessary to distinguish between aggregate states and phases. While there are only four aggregate states – solid, liquid, gaseous and plasma, the number of phases is unlimited; even the same chemically pure substance in the solid aggregate state can have several phases (rhombic and monoclinic sulphur, gray and white tin, etc.).

Non-uniform are systems that consist of several physically uniform phases or uniform bodies, so that within the systems there are discontinuities in the continuity of the variation of their properties. These systems are the aggregates or different aggregate states of the same substance (ice–water, water–vapour, etc.), or various crystalline modifications (gray and white tin, etc.), or various products of mutual dissolution (aqueous salt solution – solid salt – vapour), or products of chemical interaction of various substances (a liquid alloy and a solid chemical compound of two metals).

The non-uniform processes include aggregate transformations of individual substances, processes of evaporation, crystallization and stratification of liquid solutions, processes of dissolution and melting of solid solutions, sublimation, etc. A common feature of these processes, which makes them to a certain extent related, is that physically non-uniform substances or solutions of substances that are separated by interfaces take place in these processes. The physically non-uniform systems, delimited by interfaces, are usually called phases, and systems consisting of several phases are non-uniform systems. A feature of the non-uniform systems is the presence of interfaces, through which the interaction between phases occurs. As a result of phase processes, phase masses, their properties and composition change. It is also possible to change the number of phases. Therefore, the doctrine of non-uniform systems can be defined as the science of mutual phase conversions, as a result of which the quantitative ratio and number of phases change, as well as their compositions and properties.

3. Thermodynamic processes

The change of any thermodynamic parameter of the state of the whole system or local area is called a thermodynamic process or briefly, a process. For example, when the volume is changed, the system expands; when the characteristics of the external field change – the process of magnetization or polarization of the system take place, etc. Thermodynamics examines changes in the state of the system, occurring either on their own or under the influence of contacts with other systems.

The processes occurring at a constant temperature are called isothermal, those occurring under constant pressure are isobaric, and those occurring at a constant volume are isochoric. Adiabatic processes are those in which the system does not accept and does not give off heat, although it can be related to the environment by the work received from it and carried out over it.

Quasistatic process. This is the name for ideal processes during which the system and the environment remain in a thermally equilibrium state. Such a process is approximately realized in those cases when the changes occur rather slowly. In the extreme case of very slow changes, both processes occur along the same trajectory in opposite directions. Thus, the quasistatic process is reversible, and the system is at all times in equilibrium states. The most important quasistatic processes are: quasistatic isothermal process (the system is in contact with a thermostat having a constant temperature (thermal reservoir)) and a quasistatic adiabatic process (in which the system has no thermal (and material) contact with the environment, but under the influence of the system on the environment or, conversely, the environment on the system can work be performed).

A process is said to be non-equilibrium or non-static if the change in any parameter a occurs within a time t less than or equal to the relaxation time τ ($t < \tau$), where τ is the relaxation time of the system withdrawn from the equilibrium state and left to itself. Any process of relaxation is a non-equilibrium process.

A *thermodynamic process* is called equilibrium if during the process the system passes only through a continuous series of equilibrium states. In equilibrium thermodynamics, the concepts of the equilibrium process and reversibility coincide: the reversible process is at the same time an equilibrium process, and, conversely, every equilibrium process is reversible. With such a process at any given moment the system is in a state infinitely close to equilibrium, and it is infinitely small to change the conditions so that the process can be reversed, that is, it can flow in the opposite direction. This requirement is general, both for phase transformations and for the realization of chemical transformations. The definition of the *equilibrium* state indicated above is associated with the equilibrium process.

For stable equilibrium it is also characteristic that one can approach it in principle from two opposite directions. Thus, from a molecular point of view, a stable equilibrium is dynamic. The equilibrium is established not because of the absence or termination of the process, but because it flows simultaneously in two opposite directions at the same rate. It is the equality of the rates of the forward and reverse processes that is the cause of the conservation of the system without changing in time (with the external conditions unchanged).

Since the equilibrium process is always a process that is infinitely slow, all practically realizable processes can only approximate the equilibrium ones to some extent. This circumstance imposes certain restrictions on the difference between the concept of 'thermodynamic' characteristics and 'thermophysical characteristics', since the degree of closeness of the states of the system to the intermediate equilibrium states requires control, and, consequently, a criterion for proximity to a given equilibrium. The latter can not be done without analyzing the relaxation stages of approximation to local equilibrium: analyzing the spectrum of the relaxing characteristics (variables) and the characteristic relaxation time for each variable. We recall that in thermodynamics we consider the longest relaxation time during which an equilibrium is established for all the parameters of the given system. In this case, naturally, the question arises of the self-consistency of the description of the relaxation stages and the proper equilibrium state, which is practically never discussed in the framework of the non-equilibrium thermodynamics.

The change in the thermodynamic function in any process depends only on the initial and final states of the system and does not depend on the path of the transition. Any function of the system, the change in which in any process depends on the path of the transition, is *thermophysical*. This connects the current state of the system with the course of its evolution and brings the discussion beyond the framework of classical equilibrium thermodynamics into the region of non-equilibrium thermodynamics. Processes that can not be reversed are called irreversible.

4. Basic conditions of equilibrium thermodynamics

Classical thermodynamics is built on initial *statements* that formalize the conditions for its application, and on the *Laws of Thermodynamics* that give a mathematical notation of the law of conservation of energy for any forms of its existence. The initial statements and the Laws of Thermodynamics reflect the accumulated experience of experimental measurements.

1st statement. With time, an isolated macroscopic system comes to a state of thermodynamic equilibrium and can never leave it spontaneously. This final state is called the state of thermal equilibrium. Material particles continue their complex motion, but from a macroscopic point of view the thermally equilibrium state is a simple state, which is determined by several parameters, such as temperature and pressure. This assumption determines the scope of applicability of thermodynamics: *a*) processes that do not end with the onset of equilibrium, as well as all phenomena associated with large spontaneous deviations of the system from the equilibrium state, are excluded; *b*) systems with a small number of particles are excluded (in them the role of fluctuations is great); and *c*) the size of the systems is restricted from the above, in particular, by the volumes created by the experimenters.

2nd statement. It includes the concept of thermal equilibrium of two systems and the law of transitivity of thermal equilibrium (or the Zeroth Law of Thermodynamics).

Thermal equilibrium of two systems. If two isolated systems A and B are brought into contact with each other, then the complete system A + B eventually goes into a state of thermal equilibrium. In this case it is said that the systems A and B are in a state of thermal equilibrium with each other. Each of the systems A and B individually is also in a state of thermal equilibrium. This equilibrium will not be broken if we remove the contact between the systems, and then after a while restore it. Consequently, if the establishment of contact between the two systems A and B, which previously were isolated, does not lead to any changes, then we can assume that these systems are in thermal equilibrium with each other ($A \sim B$).

The Zeroth Law of Thermodynamics (the law of transitivity of thermal equilibrium). If the systems A and B are in thermal equilibrium and the systems B and C are in thermal equilibrium, then the systems A and C are also in thermal equilibrium with each other: $A \sim B,\ B \sim C \rightarrow A \sim C$.

5. The Laws of Thermodynamics

The First Law of Thermodynamics expresses the quantitative aspect of the law of conservation and transformation of energy in application to thermodynamic systems. The total energy of the system is divided into external and internal. Part of the energy, consisting of the energy of the motion of the system as a whole and the potential energy of the system in the field of external forces, is called external energy. The rest of the energy of the system is called internal energy.

In thermodynamics, the motion of the system as a whole and the change in its potential energy under such motion are not considered, therefore, the energy of the system is its internal energy. The energy of the position of the system in the field of external forces is included in its external energy provided that the thermodynamic state of the system does not change when it moves in the force field. If, however, the thermodynamic state changes when the system moves, then a certain part of the potential energy will be part of the internal energy. The internal energy $U = U(a_1,...,a_n; T)$ is an internal parameter and, therefore, depends in equilibrium on the external parameters a_i and temperature T.

When the thermodynamic system interacts with the surroundings, energy is exchanged. In this case, two different ways of transferring energy from the system to external bodies are possible: with changing external parameters of the system and without changing these parameters. The first way to transfer energy, associated with the change of external parameters, is called work, the second method – without changing the external parameters, but with the change of the new thermodynamic parameter (entropy) – heat, and the process of energy transfer – heat exchange.

The energy transferred by the system with changes in its external parameters is called the work W (and not the amount of work), and the energy transferred to the system without changing its external parameters is the amount of heat Q. As can be seen from the definition of heat and work, these two different methods of energy transfer studied in thermodynamics are not equivalent. Indeed, while the expended work W can directly go on to increase any kind of energy (electric, magnetic, elastic, potential energy of the system in the field, etc.), the amount of heat Q directly, i.e. without prior conversion to work, can only go to increase the internal energy of the system.

The work W and the amount of heat Q represent two different modes of energy transfer, considered in thermodynamics, and, therefore, characterize the energy exchange process between systems. For an infinitesimal equilibrium change in the parameter a, the work done by the system is $\delta W = A\, da$, where A is the generalized force, which is an equilibrium function of the external parameter a, and temperature T. When n external parameters change, the operation of the system

$$\delta W = \Sigma_i A_i da_i, \tag{5.1}$$

In the case of a non-equilibrium infinitesimal change in the parameter a, the work δW_{ne} performed by the system is also equal to $\delta W_{ne} = A_{ne}\, da_{ne}$, but in this case the generalized force A_{ne} due to the initial statements of thermodynamics is a function of the external parameters a_i, internal parameters b_j, and their derivatives in time.

The First Law of Thermodynamics establishes: the internal energy of the system is a single-valued function of its state and varies only under the effect of external influences. According to the First Law, the change in the internal energy $U_2 - U_1$ of the system upon its transition under the influence of these influences from the first state to the second is equal to the algebraic sum of Q and W, which for the final process is written in the form of equation

$$U_2 — U_1 = Q - W, \tag{5.2}$$

and for the infinitesimal (elementary) process the equation of the First Law has the form

$$\delta Q = dU + \delta W. \tag{5.3}$$

The equation of the First Law in the form (5.2) or (5.3) is valid for both equilibrium and non-equilibrium processes. The equation of the First Law allows us to determine the internal energy $U(a_1,..., a_n, T)$ in the state $(a_1,...,a_n$, T) only up to the additive constant $U(a_1^\circ, ..., a_n^\circ; T^\circ)$, which depends on the choice of the initial state $(a_1^\circ,..., a_n^\circ, T^\circ)$.

The Second Law of Thermodynamics is a law on *entropy*: entropy, like U, is a function of state, entropy exists for every equilibrium system and it does not decrease for any processes in isolated and adiabatically isolated systems. Mathematically, the Second Law of thermodynamics for equilibrium processes is written as

$$\delta Q = TdS. \tag{5.4}$$

This expression for the heat quantity element has the same form as the expression for the elementary work δW, with temperature T being the intensive heat transfer parameter (thermal generalized force), and entropy S – extensive heat transfer parameter (generalized coordinate). Entropy, by definition, is an additive quantity proportional to the number of particles in the system, i.e. the entropy of the whole system, equal to the sum of the entropy of the individual subsystems.

Basic equation of thermodynamics. For the equilibrium processes Massey [32] introduced a generalized equation for the First (5.3) and Second (5.4) Laws of thermodynamics in the form that was considered to be the basic equation of thermodynamics. Their combination gives the basic equation of thermodynamics for equilibrium processes in the form

$$TdS = dU + \sum_i A_i da_i. \tag{5.5}$$

For a simple system under the all-round pressure P, equation (5.5) has the form

$$TdS = dU + PdV. \tag{5.6}$$

Equation (5.5) is the starting point for the analysis of all equilibrium processes in thermodynamic systems with a constant number of particles.

The First Law of thermodynamics is based on the law of conservation of energy in the system under consideration (the law of conservation of mass was known even before the construction of thermodynamics). To close it, the equation of the First Law of thermodynamics must be supplemented by the equations of state (thermal $b_k = f_k(a_1,..., a_n, T)$, where b_k is the equilibrium internal parameter of the system (except for internal energy), which is conjugate to the external parameter a_i, i.e. $b_k = A_i$, and the caloric parameter $U = U(a_1,..., a_n, T)$). Equations of state can not be deduced from the principles of thermodynamics. They are established from experience or are found by methods of statistical physics, since microscopic information on the statistical behaviour of molecules is necessary for their construction. In particular, the absence in thermodynamics of a thermal equation for pressure is due to the lack of consideration of the properties of the impulse of the system.

When calculating many quantities, it is necessary to know both the thermal and the caloric equations of the state of the system. Experimentally, these equations can be obtained independently of each other. The basic equation of thermodynamics (5.5) gives the following differential equation connecting the thermal equations for a variable of type a_i and the caloric equation of state (which makes knowledge of one of them unnecessary):

$$T\left(\frac{\partial A_i}{\partial T}\right)_{a_i} = \left(\frac{\partial U}{\partial a_i}\right)_T + A_i. \tag{5.7}$$

If the caloric and thermal equations of state are known, then with the help of the principles of thermodynamics one can determine all the thermodynamic properties of the system.

Nernst's theorem (The Third Law of thermodynamics). All the energy characteristics in thermodynamics make sense for relative changes in transitions from a certain initial point (the set of parameters of the system a_i that uniquely determine its state) to any final point. In order to be able to establish a common reference point for all thermodynamic characteristics, *Nernst's heat theorem* was formulated: as the temperature approaches zero, the entropy of any equilibrium system for isothermal processes ceases to depend on any thermodynamic parameters of the state and, in the limit (T = 0 K), assumes the same universal system constant for all systems, which can be taken equal to zero by Planck's suggestion. This theorem allows us to define additive constants in expressions for entropy that can not be calculated by any other thermodynamic path.

According to the Third Law, the entropy can be found knowing only the dependence of the heat capacity on temperature and not having the thermal equation of state which is unknown for condensed bodies. The task of calculating entropy is reduced to determining only the temperature dependence of the heat capacity.

At present, the validity of the Third Law is justified for all thermodynamic equilibrium systems, but the main question of its application remains – is equilibrium always being realized during the experiment?

Characteristic functions. Characteristic are the functions of the state of the system in which all the thermodynamic properties can be expressed in the simplest and, at the same time, explicit form. In this case, the thermodynamic properties are understood as physical properties that depend only on temperature, pressure (or volume), and composition. The characteristic of a function is a consequence of the choice of independent variables (state parameters). These variables are called natural independent variables.

All the thermodynamic quantities characterizing a given system can be obtained as partial derivatives of the characteristic functions, and the so-called thermodynamic equations are the relationships between these quantities (the analytical formulation of thermodynamics). Thermodynamics can give only general information on the form of thermodynamic functions, but can not determine their specific form for each particular system. This dependence should be established empirically or by means of statistical mechanics.

We present for the uniform system the expressions of the most frequently used characteristic functions and their natural variables, as well as the total differential of the function:

internal energy U, (S, V, N_i)

$$dU = TdS - PdV + \sum_i \mu_j dN_j; \tag{5.8}$$

enthalpy $H = U + PV$, (S, P, N_i)

$$dH = TdS + VdP + \sum_j \mu_j dN_j; \tag{5.9}$$

Helmholtz free energy $F = U - TS$, (T, V, N_i)

$$dF = -SdT - PdV + \sum_j \mu_j dN_j; \tag{5.10}$$

Gibbs free energy

$$\begin{aligned} G &= F + PV = \sum_j N_j \mu_j, \ (T, P, N_i), \\ dG &= -SdT + VdP + \sum_j \mu_j dN_j. \end{aligned} \tag{5.11}$$

If the characteristic function is given as a function of natural independent variables, then the thermodynamic properties of the system are completely defined. If it is given as a function of another set of independent variables, then it is not enough to determine all the thermodynamic properties. To determine them, additional information is required to find the unknown function obtained in the integration [31–33,87]. Consequently, if the given thermodynamic function is regarded as a function of such variables for which it is not characteristic, then the other conjugate thermodynamic quantities can not be expressed in explicit form with the help of this function.

We also give the so-called Gibbs–Duhem equation

$$SdT - VdP + \sum_j N_j d\mu_j = 0. \tag{5.12}$$

Here, only the work related to the change in pressure is considered in explicit form, as in equation (5.6). Equation (5.12) plays an important role in various applications. It is the same fundamental equation as the equations for the reduced characteristic functions (5.8)–(5.11), which, because they reflect changes that are equal to work done under certain conditions, are also called thermodynamic potentials.

Gibbs called the equations (5.8)–(5.11) fundamental equations to emphasize that they express the connection between the characteristic functions and their variables and, therefore, give an exhaustive thermodynamic characteristic of the n-component mixture. The fundamental equations are equivalent to each other. Therefore, for a complete description of the thermodynamic properties of the mixture,it is necessary to have only one fundamental equation.

6. Interphase equilibrium

Phase. The concept of the phase was introduced into science by Gibbs [30]. Van der Waals [94] clarified this most important concept of the theory of non-uniform systems, in connection with the fundamentals of the derivation of the phase rule. Following these studies, we will determine that *the phase is a uniform region of a non-uniform system consisting of an individual substance or solution/mixture and bounded by an interface whose thermodynamic properties are described by a single fundamental equation and that coexists with other phases of the general non-uniform system.*

This definition bypasses the problem of the dispersity of one phase, if all its regions are macroscopic. Otherwise, this situation can be called as the Gibbs phase approximation for describing a macroscopic non-uniform system. In this definition, the extension of this concept to metastable states, proposed in various papers, in particular, in [87], is excluded. This phase definition is used below for small systems.

The question of how many phases can exist simultaneously in a system under given external conditions and how many state variables can be changed simultaneously is answered by the phase rule [87–90]. To use the phase rule, it is necessary to have an idea of the number of phases and components, as well as the idea of the possibility of chemical reactions in the system. The number of degrees of freedom of the thermodynamic system f consisting of n components participating in k independent reactions within the r phases is $f = n - k - r + b$, where the number $b = 2$ (these numbers correspond to the parameters of the system T and P) in the absence of external fields, and the given number b increases accordingly by the number of external fields affecting the states of the molecules of the system.

Conditions for thermodynamic equilibrium and stability. The theory of thermodynamic equilibrium was developed by Gibbs on the basis of Lagrange's approach to mechanical statics, that is, by generalizing and extending the principle of virtual displacements to thermodynamic systems. It is known from mechanics that a mechanical system with perfect connections is in equilibrium if the sum of the work of all the given forces for any virtual displacement of the system is zero (the principle of virtual displacements). Writing analytically this principle (the general equilibrium condition) in the form of an equation and solving it together with the equations determining virtual displacements, it is possible to find concrete equilibrium conditions for the mechanical system in each given problem.

The equations that are satisfied by virtual displacements and the equation of the principle of virtual displacements are written as follows. Let the state of the mechanical system be determined by the coordinates $q_1,\ldots,q_n$, and the constraints imposed on the system are expressed by the conditions $f_s\ (q_1,\ldots,q_n) = 0$, $s = 1,2, \ldots, k \le n$. Then the displacements $\delta q_1, \ldots, \delta q_n$, permitted by these links and called virtual or possible displacements, obviously satisfy the equations

$$\sum_{i=1}^{n} \frac{\partial f_s}{\partial q_i} \delta q_i = 0. \tag{6.1}$$

If Q_i is a generalized force conjugate to the coordinate q_i, then the principle of virtual displacements has the form

$$\sum_{i=1}^{n} Q_i \delta q_i = 0. \tag{6.2}$$

Solving jointly equations (6.1) and (6.2) by the method of uncertain Lagrange multipliers, one can find the equilibrium conditions for a given mechanical system.

This method of determining the equilibrium conditions was extended by Gibbs to thermodynamic systems. The equilibrium state of a thermodynamic system is determined by the temperature T and external parameters $a_1 \ldots . a_n$, which characterize the ratio of the system to external bodies.

According to the second initial statement of thermodynamics, at equilibrium all internal parameters are functions of external parameters and temperature, and therefore, when a_i and T are given, they are not needed to determine the state of the equilibrium system. If the system is deviated from the equilibrium state, then the internal parameters are no longer functions of only external parameters and temperature. Therefore, the non-equilibrium state must be characterized by additional independent parameters. It is believed that this makes it possible to treat a non-equilibrium system as an equilibrium system, but with a larger number of parameters and corresponding generalized forces that 'hold' the system in equilibrium than in an equilibrium state [84, 89]. In this case, the thermodynamic functions of the system in the non-equilibrium state should be considered equal to the values of these functions for an equilibrium system with additional 'retaining' forces (their role is played by external fields and adiabatic partitions).

For non-equilibrium processes, the basic inequality of thermodynamics that is a sequel of equality (5.5), which expresses the First and Second Laws of thermodynamics, must be rewritten in the form

$$TdS \geq dU + \Sigma_i A_i da_i \tag{6.3}$$

where the equality sign refers to equilibrium, and inequalities to non-equilibrium elementary processes. On the basis of such a representation, considering the output of a system from an equilibrium state as a result of virtual deviations of internal parameters from their equilibrium values, one can, using the basic inequality of thermodynamics (6.3) for non-static processes, obtain general conditions for thermodynamic equilibrium and stability.

Since now the state of thermodynamic systems is determined not only by mechanical parameters, but also by special thermodynamic (temperature, entropy, etc.) and other parameters, instead of one general equilibrium condition for mechanical systems (6.2) for thermodynamic systems there will be several depending on the ratio of the system to external bodies (adiabatic system, isothermal system, etc.).

The general conditions for the stable equilibrium of thermodynamic systems in various cases are determined by the extremal values of the corresponding thermodynamic potentials. These conditions are not only sufficient but also necessary if all other conditions for establishing equilibrium are ensured (since the conditions we have found are not the only ones for the possibility of processes) – Gibbs associated these constraints on the realization of the equilibrium with 'passive forces' (see below). Thermodynamic potentials can have several extrema (for example, entropy has several maxima). The states corresponding to the largest (entropy) or the smallest

(Helmholtz energy, etc.) of them are called stable (absolutely stable equilibrium states), others are metastable (semi-stable). In the presence of large fluctuations, the system can go from a metastable state to a stable state.

Two-phase system. As the simplest example of determining specific equilibrium conditions, starting from the general equilibrium conditions established above, let us consider an isolated two-phase system of the same substance. This example is given in all textbooks (taking into account the surface contribution, it will be considered in section 45). If S_α and S_β are the entropy of the first and second phases, respectively, then the entropy of the entire system $S = S_\alpha + S_\beta$, its general equilibrium condition $\delta S = \delta S_\alpha + \delta S_\beta = 0$.

Each of the phases is a one-component system with a variable number of particles, and the basic equations of thermodynamics for them will accordingly be:

$$T_\alpha \delta S_\alpha = \delta U_\alpha + P_\alpha \delta V_\alpha - \mu_\alpha \delta N_\alpha, \ T_\beta \delta S_\beta = \delta U_\beta + P_\beta \delta V_\beta - \mu_\beta \delta N_\beta.$$

Determining by these formulas the expressions for δS_α and δS_β, we obtain the general equilibrium condition $\delta S = 0$ in the form

$$\frac{\delta \mathrm{U}_\alpha + \mathrm{P}_\alpha \delta \mathrm{V}_\alpha - \mu_\alpha \delta \mathrm{N}_\alpha}{T_\alpha} + \frac{\delta \mathrm{U}_\beta + \mathrm{P}_\beta \delta \mathrm{V}_\beta - \mu_\beta \delta \mathrm{N}_\beta}{T_\beta} = 0. \tag{6.4}$$

Since the system is isolated, its extensive parameters obey the following coupling equations: $U_\alpha + U_\beta = U =$ const (internal energy of the system), $V_\alpha + V_\beta = V =$ const (volume of the whole system), $N_\alpha + N_\beta = N =$ const (total number of particles). As independent parameters of the system, we choose U_α, V_α, T_α; as dependent – U_β, V_β, T_β. According to these relationships, virtual changes in system parameters are expressed as

$$\delta U_\beta = -\delta U_\alpha, \ \delta V_\beta = -\delta V_\alpha, \ \delta N_\beta = -\delta N_\alpha.$$

Solving jointly the equation of the general equilibrium condition (6.4) with the equations for virtual changes of the indicated extensive parameters of the system, we find that

$$\left(\frac{1}{T_\alpha} - \frac{1}{T_\beta}\right)\delta U_\alpha + \left(\frac{P_\alpha}{T_\alpha} - \frac{P_\beta}{T_\beta}\right)\delta V_\alpha - \left(\frac{\mu_\alpha}{T_\alpha} - \frac{\mu_\beta}{T_\beta}\right)\delta N_\alpha = 0. \tag{6.5}$$

Because of the independence of the variations δU_α, δV_α, δN_α, the following particular phase equilibrium conditions for a two-phase single-component system are obtained, as the equality of three particular phase equilibrium conditions

$$P_\alpha = P_\beta,\ T_\alpha = T_\beta,\ \mu_\alpha = \mu_\beta. \tag{6.6}$$

This cntry is a consequence of the analog of the principle of mechanical equilibrium for thermodynamic parameters – the equality of thermodynamic forces in both phases.

Instead of three equalities for particular equilibria, one can write down one condition of complete equilibrium

$$\mu_\alpha(P,T) = \mu_\beta(P,T), \tag{6.7}$$

which shows that when two phases of the same substance are in equilibrium, the pressure is a function of temperature, that is, the parameters T and P cease to be *independent*.

These phase equilibrium conditions (6.6) or (6.7) are valid only for uniform phases, that is, in the absence of a field of external forces. If, on the other hand, the phases are in an external field (for example, in a gravitational field), then in equilibrium only the temperatures are equal in both phases, the pressure and chemical potential in each phase are functions of coordinates – this is a consequence of the fact that the parameter independent of the coordinates is not the chemical potential but it is the total chemical potential, which includes the potential energy of a particle in external fields.

Diagrams of state. The ultimate goal of analyzing the equilibrium of phases in non-uniform systems is to establish strict relationships between the parameters characterizing the state of the system. Knowing the dependence between the state parameters, it is possible not only to determine the equilibrium state of the non-uniform system, but also to predict the nature of the phase transformations with varying temperature, pressure, and concentration in a certain direction.

In the thermodynamics of non-uniform systems, they share analytical and geometric methods [30]. The analytical method is based on partial differential equations. The problem is to establish the regularities of the flow of the equilibrium heterogeneous processes on the basis of differential equations [94].

Graphical interpretation of the change in the characteristic thermodynamic functions as a function of the state parameters and the establishment on this basis of regular graphic relationships between them underlies the theory of the diagrams of the state of non-uniform systems [87,95–97]. A phase equilibrium diagram, or a state diagram, is a graphical representation of the relationships between the state parameters. Each point on the state diagram, called a figurative point, specifies the numerical values of the parameters characterizing the given state of the system. The straight lines connecting the figurative points of two phases in equilibrium are called nodes. All the details of the chemical interaction process, for example, the appearance of new phases and certain compounds, the formation of liquid and solid solutions, find an exact and definite reflection in that geometrical complex of surfaces, lines and points that forms a chemical diagram [97,98]. The state diagram gives an answer to the question of how many and exactly which phases form a system at given values of the state parameters. In accordance with the choice of temperature, pressure and concentration as the main parameters of the state of the non-uniform system, we unambiguously arrive at the choice of the free Gibbs energy as the main thermodynamic function that characterizes it.

To formulate these representations, different physical properties of systems are measured: the temperature of phase transitions, thermal properties (thermal conductivity, heat capacity, thermal expansion), electrical (electrical conductivity, dielectric constant), optical (refractive index, rotation of the polarization plane of light), density, viscosity, hardness and others, and also the dependence of the rates occurring in the system of transformations on its composition [95–99]. 'Self' thermodynamics does not determine the number of phases in the system, their aggregate states, the minimum concentrations of components, etc., which must be taken into account in deciphering the experimental data. Gibbs' introduction of the presence of macroscopic non-uniform phases is the key in practical work on phase transitions, and serves as a tool for the self-consistency of data on the thermodynamic functions in non-uniform systems (as a phase approximation).

When working with nanometer dispersed phases, a more detailed interpretation of the measured characteristics is required due to the description of the properties of the interface and their surface tension, which is currently relatively rare. The listed possibilities on interrelations of various experimental methods are necessary at researches of conditions of real polydisperse systems.

However, the most important questions for small bodies: the minimum phase size and the minimum size at which macroscopic connections of thermodynamics are correctly used, until recently remained open.

7. The problem of the Kelvin equation

The small systems mentioned above include both isolated small bodies and their ensembles (macroscopic and finite) – polydisperse systems, polycrystalline structures, porous bodies, etc. Today, when moving from macrosystems to small systems, the theoretical interpretation of the latter systems is based on the Young–Laplace [100,101] and Kelvin [102] equations.

The Kelvin equation is the first thermodynamic work on phase equilibrium for small systems. It was the basis of all subsequent work. Very often the derivation of the formula for the saturated vapour pressure over the curved drop surface is carried out as follows [34,103]: first, the conditions of mechanical equilibrium on the curved surface of the vapour–liquid interface are sought, and then the chemical potentials in the vapour and liquid are equated.

The Young–Laplace equation. Consider a small element of a curved surface between two phases α (liquid) and β (vapour) having two radii of curvature R_1 and R_2. Let the surface be displaced by a small distance dz under the influence of a tensile pressure equal to $P_\alpha - P_\beta$. The change in its area is $dA = (x + dx)(y + dy) - xy \approx xdy + ydx$. For a system in equilibrium, the total work of a small displacement must be zero. It consists of work on the displacement of the surface by dz (equal to $\sigma dA = \sigma(xdy + ydx)$), where σ is the surface tension, and from the expansion work of the vapour under the action of excess pressure $P_\alpha - P_\beta$ (equal to $(P_\alpha - P_\beta)\, xydz$). That is, $\sigma(xdy + ydx) = (P_\alpha - P_\beta)\, xydz$. Taking into account that in the given geometry $(x + dx) / (R_1 + dz) = x/R_1$, then $dx = xdz/R_1$ and $dy = ydz/R_2$, we obtain the Young–Laplace equation

$$P_\alpha - P_\beta = 2\sigma / R,\ 1/R = (1/R_1 + 1/R_2), \tag{7.1}$$

where $1/R = (1/R_1 + 1/R_2)$ is the mean value of the radius of curvature.

The Kelvin equation is obtained from the condition that the chemical potentials of the liquid and the vapour are equal to $\mu_\alpha = \mu_\beta$ under the condition of mechanical equilibrium according to the Young–Laplace equation. If we go from one equilibrium state to another at a constant temperature, then $dP_\alpha - dP_\beta = d(2\mu/R)$ and $d\mu_\alpha = d\mu_\beta$. For each phase, we can apply the Gibbs–Duhem equation $s_\alpha dT + V_\alpha dP_\alpha + d\mu_\alpha = 0$, where s_α and V_α are the molar entropies and molar volumes of the phase α (similarly for phase β). At T = const, these equations give the relation $V_\alpha dP_\alpha = V_\beta dP_\beta$. Excluding dP_α, this allows us to rewrite the changes in the Young–Laplace equation as $d(2\sigma/R) = [(V_\alpha - V_\beta)/V_\alpha]\, dP_\beta$. The molar volume of liquid V_α is very small in comparison with the molar volume of steam V_β, and if the vapour behaves like an ideal gas ($V_\beta = k_BT/P_\beta$), then the last equation can be transformed as $d(2\sigma/R) = -[k_BT/(P_\beta V_\alpha)]dP_\beta$. Integration from (R, P_β) and (∞, P_s) gives the Kelvin equation (KE)

$$P_\beta(R)/P_s = \exp\{2\sigma V_\alpha/(k_BTR)\}, \tag{7.2}$$

where P_s is the saturated vapour pressure corresponding to $R = \infty$ at a given temperature T (it is assumed that the fluid is not compressible and the quantity V_α is independent of pressure).

It is believed that the KE shows how the vapour pressure around small metastable drops exceeds the analogous vapour pressure around larger metastable drops, the count is taken from the pressure $P_s(T)$ in a system with a planar interface. It follows from the KE that the saturated vapour pressure above the convex (concave) surface of the meniscus is greater (lower) than the saturated vapour pressure above the flat interface. The Kelvin equation and its analogues (Thomson's equation for the phase transition temperature shift and the Ostwald-Frendlich equation for changing the solubility of particles from the curvature of small particles) [35] were actively used for all questions of analysis of the new phase [1] and in problems of adsorption porosimetry [103]. For adsorption in pores, this means that if the meniscus of the adsorbed liquid is concave (convex), then the filling of the pores occurs at pressures lower than (greater than) the saturated vapour pressure in the bulk phase.

The Kelvin equation leads to an unlimited increase in the pressure of the metastable vapour at small droplet radii, which has no physical meaning due to the absence of this concept in molecular associates, and this equation does not allow us to explain the phenomenon of limiting supersaturation for small particles (Oswald ripening).

The application of the Kelvin equation was 'supported' using coarse models for the equations of state, which lead to loops of the van der Waals type in the intermediate region of the parameters of the binodal curve. The van der Waals loop on isotherms exists in many equations of state, and it is well known that its origin is the result of the approximate nature of the equations used. The Maxwell rule [104] corrects this inaccuracy [104], which makes it possible to determine the densities of the coexisting phases of vapour and liquid at a given temperature, which corresponds to the saturated vapour pressure $P_s(T)$. The answer to the question of how to describe metastable states in the framework of molecular theories, statistical physics, speaks of the need to change the type of equations and the transition from equilibrium to non-equilibrium bonds. The use of the concept of metastable states is associated with the convenience of interpreting hysteresis phenomena, when instead of the kinetic theory very simple equations of thermodynamics are applied.

The same attitude is maintained today, when the fundamental results of statistical physics are ignored, such as the Yang–Lee theory [104–107]: all the state parameters inside the binodal curve are non-equilibrium and for them there is no equilibrium equation of state. According to the Yang–Lee theory, the thermodynamic Kelvin equation is erroneous (see section 18).

In order to show the incorrectness of the Kelvin equation, it is necessary to recall the following thermodynamic considerations for macrosystems: 1) in thermodynamics there are concepts of intense and extensive properties that do not depend on the amount of matter in the system; 2) the true state of equilibrium does not depend on the path to the final state; 3) the complete equilibrium of the non-uniform system is the consequence of three kinds of mechanical, thermal and chemical equilibria [30], therefore, in complete equilibrium, it is not important which order of equilibrium is first to be postulated.

Recall that the original statements and the Laws of thermodynamics were not derived from any principles, but were formulated on the basis of experience. Thermodynamics works with statements or hypotheses (and not with proofs like the molecular theory) and their consequences according to the rules for constructing equilibrium conditions and their stability, taken from mechanics and generalized to a greater number of degrees of freedom. Therefore, working with hypotheses applies to the same degree to the conditions for mechanical, thermal and chemical equilibrium of phases in a non-uniform system, as corresponding to the condition of complete equilibrium of the system, and then the consequences of them arc looked. This was noted in the Preface as three stages of thermodynamics for macrosystems which are tested by experience.

Despite the explicit presence in the Kelvin equation of the radius of a small system R (a drop or another new phase), it is not indicated anywhere that this equation contradicts thermodynamics. How to understand this obvious mismatch of concepts?

The size of the drop R in the Kelvin equation appeared during the consideration of mechanical equilibrium. For mechanics, the size is a natural characteristic both at the microlevel (when looking at individual atoms/molecules), and for an ensemble of individual particulates – stresses are defined at their boundaries. But in the transition to the mechanics of continuous media, as in thermodynamics, the concept of the particle size disappears, there is only the size of the system that determines the boundary conditions. In the thermodynamics of non-uniform systems, there are concepts of the weight or volume fraction of the phase, depending on the choice of the type of variables, but there is no linear size of the individual particles (for more details, see [33]).

In terms of its meaning, the parameter (drop radius) R in the Kelvin equation can not be an extensive quantity proportional to the number of particles of any subsystem of the size not less than the value of the elementary volume dV (see Section 2). The intensive parameters also formally refer to the subsystem not smaller than the size of the elementary volume dV, i.e. the concepts of the intensive and extensive parameters for macrosystems have nothing to do with the 'size' of the subsystem. In other words, in thermodynamics, the independence of the thermodynamic properties from the sizes of the subsystem is assumed [31–33]. The size of small bodies is transferred to thermodynamics from mechanics, and this transfer refers to the fourth stage of the construction of thermodynamics. The correctness of such a transfer must be verified by the molecular theory.

We should note the difference between the items 2 and 3. Item 2 refers to the general system which before Gibbs was associated with a uniform system. The conditions for the realization of particular equilibria are not discussed in the initial propositions and postulates of thermodynamics (these are its first two stages). They are a consequence of the introduced generalized rules on the equilibrium states in the heterophase system (without allowance for surface contributions). The principle of introducing the description of phase states in the non-uniform systems and the establishment of the phase equilibrium conditions go beyond the framework of a purely technical technique, since the essence of the method of describing the state of the system changes in it – the number of parameters varies (this is the first stage of thermodynamics). Instead of the property of one uniform system, a set of properties of different systems appeared. And these properties had to be differentiated, because just the concept of 'complete equilibrium' of the system became vague. As soon as the properties became different, the very notion of partial equilibria arose (see [82]). After that, they can be established in any order, and this is not the same thing as complete equilibrium. The presence of partial equilibria allows the possibility that these equilibria can be realized or not.

The Kelvin equation is derived on the basis of a consistent application of the postulates of thermodynamics about two equilibria (mechanical and chemical) between a small subsystem and a macrosystem. First, the question of pressures in neighbouring phases in the presence of a curved surface is considered, and then, using the fixed values of these pressures, the condition of equality of chemical potentials is considered (i.e. postulated), or the extreme properties of the Gibbs energy of the complete system are considered, which lead to the same results [33,108] (see Section 44), since an implicit connection between these particular equilibria is used in the derivation.

If we change the order of introduction of equilibrium conditions: first chemical, and then mechanical, the answer will be different. The first condition gives the equality of chemical potentials $\mu_\alpha = \mu_\beta$, and from it for T = const follows the condition that the pressures in both phases are equal $\mu_\alpha (P_\alpha, T) = \mu_\beta (P_\beta, T) \rightarrow P_\alpha = P_\beta$, as for a flat boundary. As a result, the droplet size is excluded from consideration due to the rules of operation of thermodynamics with macroscopic objects. This path contradicted the old knowledge of the Laplace equation for curved surfaces in the case of mechanical equilibrium, although they were introduced *apriori* in the absence of the notion of chemical equilibrium. The way out of this contradiction (with the retention of the drop size) would be the possibility of adjusting the mechanical equilibrium under the chemical equilibrium of the system. But the conditions of interphase equilibrium are determined by the postulates of thermodynamics for macrosystems independently of each other (as in the general interpretation [33,108]), and therefore they can not in principle provide a relationship. It turns out that the complete equilibrium of the system depends on the order of the establishment of equilibria by the types of subsystems, which contradicts the basic postulate of complete equilibrium of the system (item 2).

For equilibrium functions, the influence of the path to the final result implicitly indicates the dependence of the process on the time characteristics. It is shown below that the analysis of this equation leads to the need to take into account temporal characteristics, therefore, for a complete analysis of small systems, we will consider the main points of non-equilibrium thermodynamics.

8. Fundamentals of non-equilibrium thermodynamics

The thermodynamics of non-equilibrium processes is a general theory describing macroscopic thermodynamically non-equilibrium processes [89,109–116]. Classical equilibrium thermodynamics for non-equilibrium processes establishes only inequalities that indicate the direction of the processes. The problem of non-equilibrium thermodynamics is to give a quantitative description of non-equilibrium processes depending on the initial and/or external conditions for states that do not differ much from the equilibrium states. Taking into account the model-free nature of equilibrium thermodynamics, in non-equilibrium thermodynamics all processes are also considered as in the mechanics of continuous media, i.e. in the continuum description, and their state parameters are considered as variables from the continuous coordinates and time.

In thermodynamically equilibrium systems, as is well known, the pressure P, the temperature T, and the chemical potential μ_i are constant along the entire system: grad $P = 0$, grad $T = 0$, grad $\mu_i = 0$. Under non-equilibrium conditions, pressure P, temperature T and chemical potential μ_i (grad $P \neq 0$, grad $T \neq 0$, grad $\mu_i \neq 0$) are not constant, therefore irreversible processes of impulse, energy, mass transfer, etc., appear in the system.

When generalizing the classical thermodynamics to non-equilibrium processes, one starts with the idea of local equilibrium. It is generally accepted that the relaxation time increases with increasing system size, so that individual *macroscopically* small parts of the system come to themselves in an equilibrium state much earlier than the equilibrium between these parts is established. Non-equilibrium thermodynamics assumes that, although in general the state of the system is non-equilibrium, its individual small parts are in equilibrium (more precisely, quasi-equilibrium), but they have thermodynamic parameters that vary slowly with time and from point to point.

The sizes of these physically small equilibrium parts of the non-equilibrium system and the times of the change in the thermodynamic parameters in them are determined experimentally in thermodynamics. It is usually assumed that the physical elementary volume L^3 on the one hand contains a large number of particles ($v_0 \ll L^3$, $v_0 = \gamma_s\lambda^3$ – volume per particle, γ_s – shape factor, λ - average distance between particles), and on the other hand, the non-uniformities of the macroscopic parameters $a_i(r)$ over the length L are small in comparison with the value of these parameters ($|\partial a_i/\partial x|L \ll a_i$), i.e.,

$$v_0^{1/3} \ll L \ll \left|\frac{1}{a_i}\frac{\partial a_i}{\partial x}\right|^{-1}. \tag{8.1}$$

Time τ of the changes in the thermodynamic parameters in physically small equilibrium parts is much longer than the relaxation time τ_l within them and much less than time τ_L, during which equilibrium is established throughout the system:

$$\tau_l \ll \tau \ll \tau_L. \tag{8.2}$$

The properties of a non-equilibrium system are determined by local thermodynamic potentials, which depend on spatial coordinates and time only through the characteristic thermodynamic parameters for which the equations of thermodynamics are valid. Thus, if the local internal energy density $u(\mathbf{r}, t)$, the specific volume $v(\mathbf{r}, t)$ ($v = \rho^{-1}$, ρ is the local mass density of the medium) and the local concentrations $c_i(\mathbf{r}, t)$ are selected as the characteristic variables, the state of a physically elementary volume in a neighbourhood of the point r at time t is described by the local entropy $s = s[u(\mathbf{r}, t), v(\mathbf{r}, t), c_1(\mathbf{r}, t), \ldots, c_n(\mathbf{r}, t)]$, determined by the basic equation of thermodynamics

$$Tds = du + Pdv - \textstyle\sum_i \mu_i dc_i. \tag{8.3}$$

This equation (8.3) for specific (by mass) local magnitudes is also the basic equation of non-equilibrium thermodynamics. With its help, all expressions for the mass, impulse and energy fluxes (in a more general case, charges) in the dynamics of non-equilibrium processes are obtained.

The general evolution of entropy is expressed in terms of two contributions as

$$\frac{dS}{dt} = \frac{d_eS}{dt} + \frac{d_iS}{dt}, \tag{8.4}$$

where the first contribution means the flow of entropy, and the second – the production of entropy.

The local entropy s (mass units or ρs – volume units) depends on the thermodynamic parameters $a_i(\mathbf{r}, t)$ like at full equilibrium, so for an irreversible process in the adiabatic system the entropy rate per unit volume (entropy production) is equal to

$$\frac{d_iS}{dt} = \frac{d(\rho s)}{dt} = \sum_i \frac{d(\rho s)}{da_i}\frac{da_i}{dt} = \sum_i J_i X_i. \tag{8.5}$$

where the following notations are introduced: quantities $d(\rho s)/da_i \equiv X_i$ are called thermodynamic forces, and the quantities $da_i/dt \equiv J_i$, determining the rate of change of the parameters $a_i(\mathbf{r}, t)$ are called thermodynamic flows. These names are related to the fact that an increase in entropy is the 'cause' of an irreversible process when the local macroscopic parameters $a_i(\mathbf{r}, t)$ change under adiabatic conditions.

The entropy of the entire non-equilibrium system is composed additively from the entropy of its individual parts

$$S = \int_V \rho s dV. \tag{8.6}$$

9. Equations of non-equilibrium thermodynamics

The so-called equation of the entropy balance (8.4) plays a central role in non-equilibrium thermodynamics. This equation expresses the fact that the entropy of a certain volume element varies with time for two reasons. First, it changes due to the presence of some entropy flow into a given volume element; secondly, due to the presence of some source of entropy, the existence of which is due to irreversible phenomena within the volume element. We always deal with a positive source of entropy, since entropy can only arise, but not be destroyed. With reversible processes, there are no sources of entropy. This is the local formulation of the Second Law of thermodynamics.

To determine the thermodynamics of the non-equilibrium system of entropy production and the change in time of all its other thermodynamic functions with the help of the basic equation (8.3), it is necessary to add to this equation the balance equations for a number of quantities (mass, internal energy, etc.), as well as equations connecting the fluxes J_i with the thermodynamic forces X_i. The conservation laws include such quantities as the diffusion flow and flows of heat ad tensor pressure which characterise accordingly the transfer of mass, energy and an impulse.

Any extensive quantity $B(x, y, z, t)$ of the system is subject to the balance equation

$$\partial B/dt = -\text{div}\mathbf{J}_B + I_B \tag{9.1}$$

where $\mathbf{J}_B$ is the density of the total flux of the quantity $B = \rho b$ (the density of matter, b is the value of B referenced to the mass), I_B is the change in B due to its sources, referred to volume and time. Equation (9.1), in which I_B vanishes, expresses the law of conservation of B. Thus, the law of conservation of mass has the form of the hydrodynamic equation of continuity

$$\partial\rho/\partial t = -\text{div}(\rho\mathbf{u}) \tag{9.2}$$

where $\mathbf{u}$ is the mass velocity at a given point x, y, z at time t.

The density of the total flux $\mathbf{J}_B$, generally speaking, does not reduce to the convective flux $B\mathbf{u}$, i.e., to the transfer of the quantity B to the flow of matter, but also contains the terms J_B of another non-convective portion of the flow (heat flux, diffusion flux, etc.):

$$\mathbf{J}_B = B\mathbf{u} + I_B \tag{9.3}$$

Thus, the balance equation (9.1) of the additive quantity can be written in the form

$$\partial(\rho b)/\partial t = -\text{div}(\rho b\mathbf{u} + J_B) + I_B \tag{9.4}$$

where the partial derivative on the left $\partial(\rho b)$ determines the change in the value $B = \rho b$ at a given fixed point in space. This derivative can be expressed in terms of the total (substantial) derivative of B, which refers to a 'particle' of substance moving in space (as a continuous medium). The change dB in the value B of the substance particle is made up of two parts: from a change in B at a given point in space and from a change in B in transition from a given point to a point distant from it by a distance $d\mathbf{r}$ traversed by the particle under consideration during the time dt. The first of these parts is equal to

$\dfrac{\partial B}{\partial t}dt$, and the second part is equal to

Consequently

$$\frac{\partial B}{dt}=\frac{\partial B}{\partial t}+(u,\nabla)B. \qquad (9.5)$$

Therefore, the law of conservation of mass (9.2) and the balance equation for the quantity B (9.4) can be written accordingly in the form

$$\frac{d\rho}{dt}=-\rho\operatorname{div}\mathbf{u} \text{ and } \rho\frac{db}{dt}=-\operatorname{div}\mathbf{J}_B+I_B. \qquad (9.6)$$

In accordance with the general formula (9.6), the equation of the entropy balance will be

$$\rho\frac{ds}{dt}=-\operatorname{div}\mathbf{J}_s+d(\rho s)/dt, \qquad (9.7)$$

where $\mathbf{J}_s$ is the entropy flux density, and the local rate of entropy generation (8.5).

To find the explicit form of $\mathbf{J}_s$ and $d(\rho s)/dt$, formula (9.7) is compared with the expression for $\rho ds/dt$, obtained from equation (8.3).

$$\rho\frac{ds}{dt}=\frac{\rho}{T}\frac{du}{dt}+\frac{\rho p}{T}\frac{dv}{dt}-\sum_i\frac{\rho\mu_i}{T}\frac{dc_i}{dt} \qquad (9.8)$$

in which the expressions for the derivatives with respect to time and entropy production (8.5) are substituted.

The system of conservation laws, together with the equation of the entropy balance and the equations of state, is non-closed: it must be supplemented by a set of phenomenological equations connecting the reversible fluxes and the thermodynamic forces entering into the expression for the intensity of the source of entropy. In a first approximation, the fluxes are linear functions of the thermodynamic forces. The Fick's law of diffusion, the law of Fourier thermal conductivity and the Ohm electrical conductivity law belong to the class of linear phenomenological laws. These linear laws should, generally speaking, also reflect possible cross effects, since each flux can in principle be a linear function of all the thermodynamic forces that are necessary to characterize the intensity of the source of entropy in the system. In the general case of irreversible processes, the production of entropy is caused both by the phenomena of transport (energy, electric charge, etc.), and by internal transformations in the system (chemical reactions, relaxation phenomena) [112].

Let us give an expression for the production of entropy in chemical reactions in a uniform system. Here, the reactions are not related to the mass transfer processes and the flows to the equilibrium state occur in the coordinates of the composition of the system N_i (the number of particles of type i), but not in spatial coordinates. Let the uniform system consist of n substances i ($i = 1, 2,..., n$), between which r chemical reactions j ($j = 1, 2, ..., r$) can occur. If N_i is the number of particles of type i, v_{ij} is the stoichiometric coefficient of substance i in reaction j, then the change in the number d_jN_i of particles of type i over the time interval dt in reaction j is equal to

$$d_jN_i = v_{ij}I_jdt,\ I_j = \frac{1}{v_{ij}}\frac{d_jN_i}{dt} = \frac{d\xi_j}{dt}, \tag{9.9}$$

where I_j is the reaction j rate, and the differential $d\xi_j = d_jN_i/v_{ij}$ determines the 'affinity of the reaction' and has the same meaning and sign for all substances participating in the given reaction. For this reason, ξ_j is taken as the internal parameter of the system and is called the degree of completeness of the reaction j.

The change in the number of particles of the i-th sort for all reactions in a closed system is equal to $dN_i = \sum_{j=1}^{r} d_jN_i = \sum_{j=1}^{r} v_{ij}d\xi_j$. Therefore, the basic equation (8.3) in the absence of transfer processes and for the constant volume of the system, takes the form

$$TdS = -\sum_{i=1}^{n}\sum_{j=1}^{r}\mu_i v_{ij}d\xi_j = \sum_{j=1}^{r}A_jd\xi_j, \tag{9.10}$$

where A_j is the chemical affinity of the reaction j ($j = 1, 2,..., r$):

$$A_j = -\sum_{i=1}^{n}\mu_i v_{ij}. \tag{9.11}$$

The entropy production in the considered case of chemical reactions in a uniform multicomponent system, according to formulas (9.11) and (9.9), is equal to

$$d(\rho s)/dt = \sum_{j=1}^{r}I_jA_j/T. \tag{9.12}$$

In the case of a separate reaction in a closed system $d(\rho s)/dt = IA/T$.

For further consideration, it is necessary to know the relationship of the fluxes $\mathbf{J}_j$ and/or I_j to the forces X_j which are known functions of concentration. When applying the theory in specific cases, one proceeds as follows: make up the balance of entropy, find from it a definite expression for the dissipative function, then determine the fluxes and forces in such a way that all these quantities are independent and become equal to zero in equilibrium, formulate phenomenological equations, and finally, apply the Onsager reciprocity relations.

Non-equilibrium thermodynamics has found many diverse applications in physics and chemistry. To classify these applications, the different irreversible phenomena were grouped according to their 'tensor properties' [112]. Our task does not include an account of known conditions for non-equilibrium thermodynamics. Below we will consider only on its limitations related to the description of the properties of small systems. To this end, we discuss the question of a self-consistent description of non-equilibrium and equilibrium processes and the concept of Gibbs' 'passive forces'. We also recall that all the results of non-equilibrium thermodynamics are obtained under the condition of local equilibrium realization, and the question of its correspondence with the strongly non-equilibrium states of the system remains open.

10. Self-consistency of equilibrium and dynamics

Irreversible processes take place until either a stationary state or equilibrium is established (excluding the possibility of realization of periodic processes). If several irreversible processes are superimposed and the final state attained corresponds to equilibrium, then in certain cases it is possible to obtain general conditions for the coefficients that describe irreversible processes, without the application of the thermodynamics of irreversible processes.

As an illustrative example, consider the chemical reaction $L \leftrightarrow M$ [109], which can flow in both directions in a uniform volume. If the reaction occurs in an ideal gas phase or in an ideally dilute solution, then it is known from the experiment that the reaction rate w is determined by expression

$$w = \kappa c_L - \kappa' c_M, \tag{10.1}$$

where c_L and c_M are the molar volume concentrations of particles of the form L and M, respectively; κ and κ' are the rate constants for the reactions from left to right and from right to left. Chemical equilibrium occurs when the velocities of both reactions become equal, i.e., when the rate of the entire reaction w becomes zero:

$$w = 0. \tag{10.2}$$

Now, however, the equilibrium conditions of classical thermodynamics in the above case mean:

$$c_M / c_L = K, \tag{10.3}$$

where K is the equilibrium constant. The quantity K, as κ and κ', does not depend on c_L and c_M. It follows immediately from (10.1)–(10.3) that

$$\kappa / \kappa' = K. \tag{10.4}$$

Thus, from equation (10.1), which follows directly from the experiment, the obvious condition (10.2) and relation (10.3), taken from classical thermodynamics, we obtain the dependence between the coefficients κ and κ', which describe irreversible processes, and the equilibrium constant K.

This simplest example indicates that the ratio between the rate constants of the reaction in the forward and backward directions must necessarily give an equilibrium constant, or the equilibrium constant can be determined in different ways: either from equilibrium or from kinetic measurements. In this sense, we can say that the empirical regularities for describing the reaction and equilibrium rates in the system under consideration give a self-consistent description of this process at any time intervals including finite deviations of the equilibrium state, as well as the limiting equilibrium case itself.

The law of mass action. In the general case of more complex elementary stages of chemical reactions in the gas phase, the law of mass action, which was empirically established by Guldberg and Waage (1867), is used to describe reaction rates. For reversible reactions of a general form we can write $\sum_i \nu_i [A_i] \underset{k_2}{\overset{k_1}{\rightleftarrows}} \sum_j \nu_j [A_j]$ where the symbols A_i and A_j in parentheses denote different reacting particles, the values of ν_j and ν_i are equal to the negative and positive values of the stoichiometric coefficient (the sign of the coefficient is determined by their location: on the left or right side of the equation). The constants k_1 and k_2 are the reaction rate constants in the forward and backward directions. Numerically, they are equal to the reaction rate at single values of the concentration of each of their reagents in the forward direction.

The rate of the considered reaction within the framework of the law of mass action [71, 72] will be written as

$$w = k_1 \prod_i c_i^{\nu_i} - k_2 \prod_j c_j^{\nu_j}. \tag{10.5}$$

In the equilibrium state, the rate is zero $w = 0$, and it follows from (10.5) that the rate constants in the forward and backward directions are related to each other in the form

$$k_1 / k_2 = \prod_j c_j^{\nu_j} / \prod_i c_i^{\nu_i} = K \tag{10.6}$$

where $K = k_1/k_2$ is the equilibrium constant of this stage.

The law of mass action is justified for an ideal gas mixture and dilute solutions for which the following expression for the chemical potential can be written $\mu_i = \mu_i^0 + k_B T \ln(c_i)$ where μ_i^0 and c_i are the chemical potential of the standard state and the molar volume concentration of component i.

For non-ideal systems, the concept of activity coefficients is used, expressing the chemical potential as

$$\mu_i = \mu_i^0 + k_B T \ln(a_i) = \mu_i^0 + k_B T \ln(\alpha_i c_i) = \mu_i^{id} + k_B T \ln(\alpha_i). \tag{10.7}$$

The activity coefficient is defined as the ratio $a_i = a_i/c_i$, here a_i is the activity of component i in solution, depending on all concentrations of the solution components (provided the total concentration is determined $c = \sum_i c_i$) and on all their molecular properties, including intermolecular interaction energies.

The intermolecular interactions in non-ideal reaction systems, according to the theory of absolute reaction rates (TARR) [117, 118], are taken into account by using thermodynamic bonds known from the theory of non-ideal solutions [82]. This approach is related to the preservation in the expression for the reaction rates of the concentration factor used in the form of the law of mass action (through the product of reagent concentrations), as in formula (10.5), and with the change in the rate constants of reactions in the form

$$K_i(ef) = K_i^0 \frac{\alpha_i}{\alpha_i^*} \exp(-E_i / k_B T) = K_i \alpha_i / \alpha_i^*,$$
$$K_{ij}(ef) = K_{ij}^0 \frac{\alpha_i \alpha_j}{\alpha_{ij}^*} \exp(-E_{ij} / k_B T) = K_{ij} \alpha_i \alpha_j / \alpha_{ij}^*, \tag{10.8}$$

where α_i is the activity coefficient of molecules of type i, α_i^* and α_{ij}^* is the activity coefficient of the activated complex of the mono- and bimolecular stage; $K_i = K_i^0 \exp(-E_i/k_BT)$ and $K_{ij} = K_{ij}^0 \exp(-E_{ij}/k_BT)$ are the rate constants of mono- and bimolecular stages, K_i^0 and K_{ij}^0 are the pre-exponentials, and E_i, E_{ij} are the activation energies of these stages.

Continuous system. Another example of the connection between the equilibrium and kinetic characteristics relates to the problem of the relationship between diffusion and sedimentation [109]. In a continuous system with two independently moving substances, in the presence of concentration gradients and the action of stationary external force fields (gravitational or centrifugal), the simultaneous occurrence of diffusion processes (the transfer of matter due to the concentration gradient) and sedimentation (the motion of matter under the action of external fields) in the volume element under normal conditions in accordance with experience can be described by the following equation:

$${}_w\mathbf{J}_2 = -D \operatorname{grad} c_2 + c_2 s_2 \mathbf{g}, \tag{10.9}$$

where ${}_w\mathbf{J}_2$ is the appropriately determined diffusion flux (diffusion flux density vector) of substance 2; c_2 is the molar volume concentration of substance 2; $\mathbf{g}$ is the vector of gravitational or centrifugal acceleration; D is the diffusion coefficient; s_2 is the sedimentation coefficient. The last two quantities do not depend on grad c_2 and g. At equilibrium (sedimentation equilibrium), the diffusion and sedimentation in each volume element must obviously be compensated. Consequently, we have

$${}_w\mathbf{J}_2 = 0. \tag{10.10}$$

However, now the equilibrium condition for classical thermodynamics for such a case has the form [109]

$$(M_2 - \rho V_2)\mathbf{g} = \left(\frac{\partial \mu_2}{\partial c_2}\right)_{T,P} \operatorname{grad} c_2. \tag{10.11}$$

In this equation, M_2 is the molar mass (molecular weight), V_2 is the partial molar volume, μ_2 is the chemical potential of substance 2, ρ is the density, T is temperature, and P is the pressure in the volume element. Comparison of the equations (10.9) and (10.11) gives

$$\frac{D}{s_2} = \frac{c_2}{M_2 - \rho V_2}\left(\frac{\partial \mu_2}{\partial c_2}\right)_{T,P} \tag{10.12}$$

The equation (10.12) is the relationship between the two transport coefficients D and s_2 and the equilibrium values obtained from the empirical equation (10.9), the obvious condition (10.10) and the relation (10.11), which follows from classical thermodynamics. A similar relationship is extended to the system with any number of substances and in the same way a general relationship between filtration and osmosis for binary systems with a membrane was obtained [109].

Determining the general criterion for obtaining such links in the above way for transport processes, it can be established that when a number of transport processes are superimposed in the system in which the final state of the system corresponds to equilibrium in an non-uniform system, then by means of classical thermodynamics the general relationship between the various transport coefficients is established. This general situation facilitates the derivation of dependences between transport values, since in certain cases the application of more complicated methods of thermodynamics of irreversible processes becomes superfluous. However, this approach is not applicable if the imposition of several transport processes leads to equilibrium in a uniform system.

11. Passive Gibbs forces

Discussing the question of relaxation times, we should note the path that was introduced by Gibbs when considering slow or 'frozen' processes. The criteria he introduced in discussing equilibrium and stability are formulated with respect to only possible changes. In the general case, processes that are 'essentially unrealistic' can be proposed (see pp. 63–64 [30]).

In Appendix 3 in [30] the following comment is given: "Here we are talking about states that, although they are not truly equilibrium, remain unchanged during the observation time. For example, in a mixture of hydrogen and oxygen under normal conditions, no changes are observed, although the state of such a system is not an equilibrium state. In such a system, a weak external action, such as an electric spark, can produce a noticeable change in state, and the degree of this change is in no way related to the intensity of the external action (with the intensity of the spark current or the intensity of the sunlight). In such situations, it is considered that the transition of the system to a state of equilibrium is prevented by some 'passive forces of resistance'."

The mechanical analogue of passive resistances can be the friction force at rest which keeps the body on an inclined plane. Gibbs distinguishes between the equilibrium caused by passive resistances and the equilibrium that is provided by the 'balance of active tendencies acting in the system' (p. 64, [30]). Forces of resistance, similar to viscosity or sliding friction, which only slow the transitions, but do not prevent them, are not enough to neglect the corresponding changes in state.

Thus, the term 'passive forces' means powerful enough force actions, the nature of which is not disclosed, and the introduction of the very notion of passive forces is associated with the time interval that the experimenter has in carrying out own measurements. The main issue of the correctness of the application of this method is the interpretation of the situation, depending on the accuracy and reliability of the measurements. For example, geological rocks can not be considered equilibrium due to their genesis, but the application of thermodynamic methods to them is a common practice and one of the necessary methods for their analysis [33], although the treatment of measurements performed, obviously non-equilibrium systems as thermodynamic rather than thermophysical ones, remains the responsibility of the authors. The same applies to macroscopic samples of metals and alloys consisting of grains of different sizes [10,24]. Another example of a wide range of opinions about relaxation times is the processes of adsorption deformation in which a rigid body is either inert [119] or deformed, but for some reason it is assumed that any initial state of the sample is equilibrium [85,120].

The need to take into account different relaxation times of components in complex systems is explicitly mentioned in the separation of components into mobile and frozen components [30]. The same principle is actively used in considering adsorption and absorption processes, when fixed components (solid-state atoms) create local fields (cells) that are filled with mobile (ad- or absorbed) components.

As the size of the system decreases, the relaxation times should decrease and this aspect remains often controversial due to the limited time of experimental measurements. In real systems, the times of relaxation processes always correlate with the sizes of small bodies and with the nature of the polydispersity of the material.

12. Necessity of taking into account relaxation times

The order of introduction of equilibria (first mechanical, and then chemical) in the derivation of the Kelvin equation is strictly fixed, and they can not be rearranged. The essence of the contradiction lies in the fact that different relaxation times are implicitly compared here: 1st for relaxation of pressure, and 2nd for relaxation of the chemical potential. In physical sense, the 1st relaxation time for pressure is much less than the second relaxation time for the chemical potential. In thermodynamics this remains behind the scenes; it is meant that we are talking about complete equilibrium or about many times longer than the relaxation time for the chemical potential.

Size and time. The concept of 'time' is also excluded from classical equilibrium thermodynamics, like 'size'. The size of the small body was introduced in equilibrium with the help of the Kelvin equation and, as shown above, it also automatically introduces the concept of the relaxation time. Thus, if we take a more general point of view than only the concept of complete equilibrium (without time) and take into account the differences in the relaxation times of different properties of the system, then the incorrectness of the Kelvin equation follows without the results of statistical physics. To do this, it is necessary to include the notion of different relaxation times for mechanical and chemical equilibria.

This is a fairly natural course of describing the properties of real systems, because it is directly connected with the concept of quasistatic transitions between the initial and final states of the system. More precisely, in a real experiment, there should be a consideration of the relaxation times for various properties that characterize the rate of convergence of the current state of the system with the equilibrium one. The specificity of all real processes requires differentiation of the conditions for reaching full equilibrium through three types of equilibria for specific properties of the system: mechanical, thermal and chemical. By virtue of the differences in properties, there is a clear system of relations between the relaxation times of $\tau_{mech} \leq \tau_{therm} \leq \tau_{chem}$ related to the transfer of impulse, energy and mass, which also follows from numerous experiments [82,92].

The appearance of the relaxation time of the properties due to the appearance of 'size' in thermodynamics naturally raises the question of the quasistatic process itself, and on what sizes it is performed. The smaller the body size, the less the relaxation time and the faster the equilibrium is established. Therefore, for the *macroscopic* sizes of thermodynamic objects, the question of the reality of attainability of equilibrium is not abstract. It requires experimental proof (and not just agreement with such an assertion), i.e., the applicability of the postulate of equilibrium must be proved experimentally, only after this it is possible to apply thermodynamic equations.

The question of the time factor plays a particularly important role in solid-phase systems. Real polydisperse materials, which are used in experiments with solids, consist of grains. The smaller the grain size, the faster the redistribution of atoms is achieved in it and the chemical equilibrium is established (but at the same time the role of the size factor and the inapplicability of macroscopic bonds for its recording increase). The larger the grain size, the slower the chemical equilibrium is established in it. As a result, the question of real equilibrium in most polydisperse solid materials often remains open. As examples, one should consider Elliot's data [123–125] and the overwhelming number of other experimental systems with complex structures of numerous phases [95–99]. All in all, this requires the introduction of a number of criteria on the correctness of the application of macroscopic laws of thermodynamics (in micro-non-uniform systems) for small particles and their ensembles.

General view of classical thermodynamics. An analysis of the problem of the Kelvin equation points to the need for a more general view of the conditions for the application of the postulates of thermodynamics, taking into account the inclusion of relaxation times of various properties of the system into consideration, as a necessary element for discussing the progress of any equilibrium establishment processes, which is actually the main task of thermodynamics.

In the transition to non-equilibrium thermodynamics, it is usually assumed [89,109–116] that the formulation of the combined equation of thermodynamics is sufficient for any local region dV. To justify this assumption, it is necessary to indicate the criteria that distinguish the region of locally equilibrium states. This will be done below, but it is prudent to formulate two refinements which are needed in the joint consideration of equilibrium and non-equilibrium processes. In part, these updates have already been introduced earlier, and here they are given for a more precise exposition of the part of the material that has not previously been accented.

For equilibrium Massey systems, a generalized equation was introduced for the First and Second Laws of thermodynamics (5.6) in the form $dU = TdS - PdV$, which became the basic equation of equilibrium thermodynamics. In developing the foundations of non-equilibrium thermodynamics, the condition of local equilibrium for elementary macroscopic volumes of macrosystems was introduced and this equation is written as the inequality $dU \leq TdS-PdV$ [89,109–116]. This automatically transferred all the initial positions of equilibrium thermodynamics to the region of non-equilibrium processes. Today, when discussing the state of classical thermodynamics, it is advisable to consider both types of processes together from a single point of view.

Refinement 1. On the need for a self-consistent description of the dynamics and equilibrium state of the system. Refinement 1 is already implied in the Second Law of thermodynamics when discussing the possibility of an exit and return of the system from an equilibrium state. It is a prerequisite for introducing the concept of 'affinity' in chemical thermodynamics, working both in equilibrium and in non-equilibrium [82,125]. It also serves as a justification for the introduction of the concept of thermodynamic functions in non-equilibrium states [109], since these functions automatically change to their equilibrium functions during the transition of the system to the equilibrium state.

Refinement 1 is necessary to introduce a single elementary volume dV in equilibrium and the dynamics to be found, as well as a criterion for the correctness of the construction of molecular models of kinetic stages and elementary stages. When the rates of reversible reactions are equal in the forward and backward directions, these models should ensure the same expressions for equilibrium constants as the equilibrium constants constructed in the framework of only equilibrium distributions. For ideal systems, this was discussed in section 10.

Refinement 2. On the need to take into account the differences in the characteristic relaxation times τ for the transfer of various properties: impulse (τ_{imp}), energy (τ_{ener}) and mass (τ_{mass}). These mechanical invariants retain their meaning also for macroscopic ensembles, both in equilibrium and in non-equilibrium thermodynamics and in statistical mechanics. Refinement 2 reflects the experience of studying dynamic processes and is necessary as a reflection of real experimental measurements. The relaxation time characterizes the process of establishing thermodynamic equilibrium in the macroscopic system under study. Because of the unequal physical parameters of the thermodynamic system, they tend to their equilibrium at different rates. This fact is reflected as the presence of 'partial equilibrium' when introducing the concept of affinity for non-equilibrium processes with chemical reactions occurring at constant pressures and temperatures (τ_{imp}, $\tau_{ener} \ll \tau_{mass}$) [82,125]. In this case, partial equilibrium with respect to the impulse means that there is no convective flow in the system, and partial equilibrium with respect to temperature means that there is no flow of heat. The complete equilibrium of the system corresponds to the times when the whole system achieves its equilibrium in all its parameters. Or this is achieved at times τ exceeding the relaxation time of the slowest thermodynamic parameter (τ_{imp}, τ_{ener}, $\tau_{mass} \ll \tau$).

Thus, the problem of describing the properties of small bodies leads to the necessity of simultaneously taking into account the basic positions of not only equilibrium but also non-equilibrium thermodynamics.

References

1. Frenkel J., Kinetic theory of liquids, Academy of Sciences of the USSR, 1945.
2. Skripov V.P., Metastable liquid. Moscow, Nauka, 1972.
3. Nanomaterials: Synthesis, Properties and Applications. Edited by A.S. Edelstein, R.S. Commarata. Bristol: Institute of Physical Publishing. Bristol and Philadelphia. 1996..
4. Gusev A.I., Rempel' A.A., Nanocrystalline materials. Moscow, Fizmatlit, 2001.
5. Uvarov N.F., Boldyrev V.V., Usp. khimii. 2001. V. 70. No. 4. P. 307. [Russ. Chem. Rev. 70, 265 (2001)]
6. Petrii O.A., Tsirlina G.A., Ibid. 2001. V. 70. No. 4. P. 330. [Russ. Chem. Rev. 70, 285 (2001)]
7. Haruta M., Date M., Appl Catal. A: General. 2001. V. 222. P. 427.
8. New materials. Team of authors, ed. Yu.S. Karabasov. Moscow, MISIS. 2002.
9. Daniel M.-C., Austric D., Chem. Rev. 2004. V. 104. P. 293.
10. Haruta M., Gold Bull. 2004. V. 37. No. 1–2. P. 27.
11. Chuvil'deev V.N., Non-equilibrium grain boundaries in metals. Theory and applications. Moscow. Fizmatlit. 2004.
12. Smirnov V.V., Lanin S. N., Vasil'kov A. Yu., Nikolaev S. A., Murav'eva G. P., Ty-

urina L. A., Vlasenko E. V., Ross. nakotekhnologii. 2007. V. 2. No. 1–2. P. 47. [Russ. Chem. Bull. 54, 2286 (2006)]
13. Rostovshchikova T. N., Smirnov V. V., Kozhevin V. M., et al., Ross. Nanotekhnol. 2 (1–2), 47 (2007)
14. Ozin G.A., Arsenalt A.C. Nanochemistry. A chemical approach to nanomaterials. Cambridge: The Royal Society of Chemistry. 2005.
15. Rambidi N., Berezkin A., Physical and chemical fundamentals of nanotechnologies. Moscow, Fizmatlit, 2008.
16. Fisher T.S., Lessons from nanoscience: A lecture notes series: Thermal energy at the nanoscale. V. 3, World Scientific, 2014.
17. Suzdalev I.P., Nanotechnology: physical chemistry of nanoclusters, nanostructures and nanomaterials. KomKniga, Moscow. 2006.
18. Cao G., Nanostructures & Nanomaterials. Synthesis, Properties & Applications. University of Washington: Imperial College Press. 2006.
19. Gemming S., Schreiber M., Suck J.B. Materials for Tomorrow: Theory, Experiments @ Modeling. Berlin. Heidelberg: Springer-Verlag. 2007.
20. Suzdalev I.P.. Electric and magnetic transitions in nanoclusters and nanostructures. Moscow. Krasand, 2011.
21. Handbook Springer of Nanotechnology, Bharat Bhushan (Ed.) 2nd revised and extended edition. Berlin - Heidelberg - New York: Springer. Science + Business Media Inc., 2007.
22. Gusev A.I. Nanomaterials, nanostructures, nanotechnologies. Moscow, Fizmatlit, 2007.
23. Ryzhonkov D.I., et al., Nanomaterials. Moscow, Binom, Laboratory of knowledge. 2008.
24. Zhilyaev A.P., Pshenichnyuk A.I., Superplasticity and grain boundaries in ultrafine-grained materials. Moscow. Fizmatlit. 2008.
25. Poole C., Owens F., Nanotechnology. Moscow, Tekhnosfera. 2009.
26. Nanostructured materials. Ed. R. Hannink, F. Hill. Moscow, Tekhnosfera. 2009.
27. Fakhl'man B., Chemistry of New Materials and Nanotechnologies. Dolgoprudnyi, Intellekt, 2011.
28. Eliseev A.A., Lukashin A.V., Functional nanomaterials. Moscow, Fizmatlit. 2010.
29. Vorotyntsev V.M., Nanoparticles in two-phase systems. Moscow, Izvestiya, 2010.
30. Gibbs J.W., Thermodynamics. Statistical mechanics. Moscow, Nauka. 1982.
31. Putilov K.A., Thermodynamics. Moscow, Nauka, 1971.
32. Krichevsky I.R., Concepts and the basics of thermodynamics. Moscow, Khimiya, 1970.
33. Voronin G.F., Fundamentals of thermodynamics. Moscow, Publ. MGU, 1987.
34. Adamson A. W., The Physical Chemistry of Surfaces,Mir, Moscow, 1979. [Wiley, New York, 1976]
35. Rusanov A.I., Phase equilibrium and surface phenomena. Leningrad, Khimiya, 1967.
36. Ono S., Kondo S., Molecular theory of surface tension. Moscow, IL, 1963. [Handbuch der Physik, Vol X (Springer) 1960]
37. Rowlinson, J., Widom B., Molecular theory of capillarity. Moscow, Mir, 1986. p. [Oxford: ClarendonPress, 1982]
38. Tovbin Yu.K., Theory of physico-chemical processes at the gas-solid interface, Moscow, Nauka, 1990. [CRC, Boca Raton, Florida, 1991].
39. Gufan Yu.M., Structural phase transitions. Moscow, Nauka, 1982.
40. Santen R.A. von, Sachtler W.M.H., J. Catal. 1974. V. 37. P. 202.

41. Williams F.L., Nason D., Surface Sci. 1974. V. 45. P. 377–408.
42. Kymar F., et al., Phys. Rev. B. Solid State. 1979. V. 19. P. 1954–1962.
43. Moran-Lopez J.L., Falicov L.M., Ibid. 1978. V. 18. P. 2542–2548.
44. Moran-Lopez J.L., Falicov L.M., Ibid. P. 2549–2554.
45. Farncuort H.E. Interphase gas–solid boundary, Ed. E. Flad, trans. from English. Ed. A.V. Kiselev. Moscow, Mir, 1970, PP. 359–370.
46. Lodiz R., Parker R. Growth of single crystals. Ed.A. A. Chernov, A.N. Lobachev. Moscow, Mir, 1974.
47. De Boer J., Dynamic nature of adsorption. Ed. B. M. Gryaznov. Moscow, IL, 1962. 290 p.
48. Crowell A., Interphase gas–solid boundary, Ed. E. Flad, trans. from English. Ed. A.V. Kiselev. Moscow, Mir, 1970. P. 150–174.
49. Vol'kenshtein, F.F., Physicochemistry of the surface of Semiconductors, Moscow, Nauka, 1973.
50. Kiselev V.F., Krylov O.V., Adsorption processes on the surface of semiconductors and dielectrics. Moscow, Nauka, 1978.
51. Steele W.V. Interphase gas–solid boundary, Ed. E. Flad, trans. from English. Ed. A.V. Kiselev. Moscow, Mir, 1970, P. 260.
52. Morrison S., Chemical physics of the surface of a solid body. Ed. F.F. Vol'kenstein. Moscow, Mir, 1980.
53. Krylov O.V., Kiselev V.F., Adsorption and catalysis on transition metals and their oxides. Moscow, Khimiya, 1981.
54. Roginsky S.Z., Adsorption and catalysis on non-uniform surfaces. Moscow and Leningrad, Publishing House of the USSR Academy of Sciences, 1948.
55. Dunning V.V., Interphase gas–solid boundary, Ed. E. Flad, trans. from English. Ed. A.V. Kiselev. Moscow, Mir, 1970. P. 230.
56. Jaycock M. Parfitt J., Chemistry of interfaces.– Moscow: Mir, 1984. – 270 p. [Wiley, New York, 1981].
57. Leygraf C, Hultquist G., Surface Sci. 1976. V. 61. P. 60.
58. Franken P.E.C., Ponec V., J. Catal. 1974. V. 35. P. 417.
59. Kugler E.L., Boudart M., Ibid. 1979. V. 59. P. 201.
60. Wang T., Schmidt L.D., Ibid. 1981. V. 71. P. 411.
61. Sedlacek J., Hilaire L., Legare P. et al., Surface Sci. 1982. V. 115. P. 541.
62. Ertl G., Norton P.R., Rustig J., Phys. Rev. Lett. 1982. V. 49. P. 177.
63. Griffits K., et al., Surface Sci. 1984. V. 138. P. 113.
64. Delmon B., The kinetics of non-uniform reactions. Ed. V.V. Boldyrev, Moscow, Mir, 1972.
65. Rozovskii A.Ya., Kinetics of topochemical reactions. Moscow, Khimiya, 1974.
66. Barre P., Kinetics of non-uniform processes. Ed. V.V. Boldyrev. Moscow, Mir, 1976.
67. Fast J.D., Interaction of metals with gases. Ed. L. A. Shvartsman. Moscow, Metallurgiya, 1975. V. 2.
68. Gel'd L.V., Ryabov P.A., Mokhracheeva L.P., Hydrogen and physical properties of metals and alloys: Hydrides of transition metals. Moscow, Nauka, 1985.
69. Andrievskii R.A, Umaiskii Ya.S., Interstitial phases. Moscow, Nauka, 1977.
70. Smirnov A.A., The theory of interstitial alloys. Moscow, Nauka, 1979.
71. Emanuel J.M., Knorre D.T., Course of chemical kinetics. 4th ed., Moscow, Vysshaya shkola, 1984.
72. Eremin E.Ya., Fundamentals of chemical kinetics. Moscow, Vysshaya shkola, 1976..
73. Frank-Kamenetsky D.A., Diffusion and heat transfer in chemical kinetics. Moscow, Publishing House of the Academy of Sciences of the USSR, 1967.

74. Ovchinnikov A.A., Timashev S.F., Belyi A.A., Kinetics of diffusion-controlled chemical processes. Moscow, Khimiya. 1986.
75. Fenelonov V.B. Porous carbon. Novosibirsk, Institute of Catalysis, 1995.
76. Modeling of porous materials. Novosibirsk: Siberian Branch of the USSR Academy of Sciences, 1976.
77. Kheifets L.I., Neimark A.V., Multiphase processes in porous bodies. Moscow, Khimiya, 1982.
78. Frevel L.K., Kressey L.J., Annal. Chem. 1963. V. 35. P. 1492.
79. Mayer R.P., Stowe R.A., J. Coll. Sci. 1965. V. 20. P. 893.
80. Plank M., Treatise on thermodynamics. New York, Dover Pub. 1945.
81. Guggenheim E.A., Modern thermodynamics, described by the method of J.W. Gibbs. Leningrad and Moscow, State. Nauch.-Tekh. Izdat. Khim. Lit., 1941.
82. Prigogine I., Defay R., Chemical thermodynamics. Novosibirsk, Nauka, 1966.
83. Sommerfeld A., Thermodynamics and statistical physics. Moscow, IL, 1955.
84. Leontovich M.A., Introduction to thermodynamics. Statistical physics. Moscow, Nauka, 1983.
85. Guggenheim E.A., Thermodynamics. An advanced treatment for chemists and physics, 5th edition. Amsterdam: North-Holland, 1967. PP. 166-169.
86. Semenchenko V.K., Selected chapters of theoretical physics. Moscow, Prosveshchenie, 1966.
87. Storonkin A.V., Thermodynamics of non-uniform systems. Leningrad, LGU, 1967.
88. Kubo R., Thermodynamics. Moscow, Mir, 1970. 304 p.
89. Bazarov I.P., Thermodynamics. Moscow, Vysshaya shkola, 1991.
90. Kireev V.A., Course of physical chemistry. Moscow, Khimiya, 1975. 773 p.
91. Kvasnikov I.A., Thermodynamics and statistical physics. V. 1: Theory of equilibrium systems: Thermodynamics. Moscow, Editorial URSS, 2002.
92. Landau L.D., Livshits E.M., Theoretical physics. V. 5. Statistical physics. Moscow, Nauka, 1964.
93. Van der Waals I.D., Kohnstamm Ph., Thermostatics course. Moscow, ONTI, 1936.
94. Sychev V.V., Differential equations of thermodynamics. Moscow, Vysshaya shkola, 19910.
95. Kurnakov N.S., Introduction to physical and chemical analysis. 4 ed., Moscow and Leningrad, 1940.
96. Anosov V.Ya., Pogodin S.A., The main principles of physical and chemical analysis. Moscow and Leningrad, 1947.
97. Mikheeva V.I. Method of physical and chemical analysis in inorganic synthesis. Moscow, Khimiya, 1975. 272 p.
98. Glazov VM, Pavlova L.M. Chemical thermodynamics and phase equilibria. Moscow, Khimiya. 1981.
99. West A.R., Solid state chemistry and its applications, Chichester, etc, John Wiley & Sons, 1984.
100. Young T., Philos. Trans. R. Soc. London. 1805. V. 95. P. 65.
101. Laplace P.S., Traite de Mecanique Celest; Supplement au dixeme livre, Sur l'Action Capillaire, Courcier, Paris. 1805. V. 4.
102. Thomson W.T., Phil. Mag. 1971, V. 42. P. 448.
103. Greg S., Sing K. Absorption, specific surface, porosity. Moscow, Mir, 1984.
104. Huang K., Statistical mechanics. Moscow, Mir, 1966.
105. Yang C.N., Lee T.D., Phys. Rev. 1952. V. 87. P. 404.
106. Lee T.D., Yang C.N., Phys. Rev. 1952. V. 87. R. 410.
107. Hill T.L., Statistical Mechanics. Principles and Selected Applications. Moscow: Izd.

Inostr. lit., 1960. – 486 p. [N.Y.: McGraw–Hill Book Comp. Inc., 1956]
108. Kvasnikov I.A., ref. [92], pages 113-116.
109. Haase R., Thermodynamics of irreversible processes. Moscow, Mir, 1967. [Dr. Dietrich Steinkopff, Darmstadt, 1963]
110. Onsager L., Phys. Review. 1931. V. 37. P. 405.
111. Prigogine I., Introduction to the thermodynamics of irreversible processes. Izhevsk, Regular and chaotic dynamics, 2001. [Charles C Thomas Sprinfield, Illinois, U.S.A., 1955]
112. de Groot S., Mazur P., Non-equilibrium thermodynamics. Moscow, Mir. 1964. [Amsterdam. North – Holland Publ. Company. 1962]
113. Gyarmati, I., Non-equilibrium thermodynamics. Field theory and variational principles. Moscow, Mir, 1974. [Springer, Berlin, Heidelberg, New York, 1970]
114. Landau, L.D., Lifshitz E.M., Theoretical physics. VI. Hydrodynamics. Moscow, Nauka, 1986.
115. Bird R.B., Stuart W.E., Lightfoot E.T. Transport phenomena. Moscow, Khimiya, 1974. [Wiley, New York, London, 1965]
116. Lykov A.V. Heat and mass transfer. Moscow, Energiya, 1978.
117. Eyring H. J., Chem. Phys. 1935. V. 3. P. 107.
118. Glasston S., Laidler K.J., Eyring H., Theory of absolute reaction rates. Moscow, IL, 1948 [Princeton Univ. Press, New York, London, 1941].
119. Hill T., Catalysis, questions of theory and methods of investigation, Moscow, IL, 1955.
120. Bakaev V.A., Izv. AN SSSR, Ser. khim. 1971. No. 12. P. 2648.
121. Rice O.K., J. Phys. Chem. 1927. V. 31. P.207.
122. Prigogine I., Defay R., J. Chim. Phys. 1949. V. 46. P. 367.
123. Elliott G.R.B., Lemons J.F., J. Phys. Chem. 1960. V. 64. P. 137.
124. Elliott G.R.B., Lemons, J.F., Advanced in chemistry series, No. 39. R.F. Gloud, Ed., Amer. Chem. Soc., Washington, D.C., 1964. P. 144, 153.
125. Roof R.B. Jr., Elliott G.R.B., Inorganic Chemistry. 1965. V. 4. P. 691.

2

Fundamentals of molecular theory

In development to the thermodynamic method, J.W. Gibbs (1902) [1, 2] also proposed a statistical method for studying the properties of macroscopic bodies, from the very beginning based on model atomic–molecular representations. The main task of statistical physics is to establish the laws of behaviour of a macroscopic quantity of matter, knowing the laws of the behaviour of particles from which the system is constructed (atoms, molecules, ions, electrons, photons, etc.). Accordingly, the conclusions of statistical physics are valid only to the extent that the assumptions made about the behaviour of the particles of the system are valid. From the very beginning, this method of solving the problem traces the atomic–molecular mechanism of phenomena. The statistical method makes it possible, on the one hand, to give a rigorous justification for the laws of thermodynamics, and on the other hand, to establish the limits of their applicability, and also to predict the conditions for violation of the laws of classical thermodynamics due to fluctuations and to estimate their scale.

All subsequent works of statistical physics [3–15] continue to use the method developed by Gibbs for the averaging procedures for ensembles of particles with respect to velocities and coordinates [2]. Statistical physics gave statistical interpretations of internal energy and entropy, previously conducted in thermodynamics. It is necessary to single out the general approach in static mechanics proposed by N.N. Bogolyubov [6,16], from a unified position considering equilibrium and non-equilibrium processes in the gas phase. Subsequently, these approaches were transferred to dense phases [17–30]. At present, they make it possible from a unified point of view to consider three aggregate states of a substance and their interface, both for equilibrium and non-equilibrium processes by means of a discrete–continual description of the particle distributions (based on the 'lattice gas model' (LGM)) [31–35].

13. Microscopic states of molecules and their description

The microscopic specification of the N body system includes: a) specifying the thermodynamic parameters of the system that determine its macroscopic state, or the specification of the external conditions into which the system under consideration is placed; b) the actual task of the system based on the atomistic approach: we know all the microscopic characteristics of the system, that is, the masses and structure of its constituent molecules, the charges and spins of the particles, the potentials of their interaction with external fields and with each other, etc.

The task of the system is carried out by specifying the total energy of the system (the Hamiltonian), which includes the kinetic energy T, the potential energy in the external field E_{ex}, and the internal energy of the interparticle interaction E_{in}. If the masses m_i of the particles of the system and their number N are given, then $T = T(p_1 ... p_N) = \sum_{1 \le i \le N} p_i^2 / 2m_i$, where p_i is the impulse of the i-th particle.

The potential energy of interaction with external fields, which are given as external parameters, is defined as the sum of the potential energies E_i of each of the N particles: $E_{ex} = \sum_{1 \le i \le N} E_i$. So for particles in a uniform vertical field of gravity $E_i = m_i g Z_i$, Z_i is the coordinate along the Z axis, etc. The potential E_i includes the potential of the walls of the vessel, fixing the external macroscopic parameter V, the volume of the system. The simplest and most convenient model for E_w is the impermeable wall model (a three-dimensional potential well bounded by an infinitely high vertical barrier): $E_w = 0$ if the point **r** is inside the vessel, or $E_w = \infty$ if the point **r** is outside the vessel [36]. In the case of finding a point **r** on the surface of a vessel, it is required to detail the potential type, or its calculation, for example, based on the atom–atom wall potential, depending on its structure and composition [31].

In determining the energy E_{in}, taking into account the interaction of the particles with each other, we will assume that this quantity is expressed as the total sum of all pair contributions, regardless of whether there are other particles around them or not. Such a potential of pair forces can be determined either experimentally when investigating data on the scattering of two isolated particles of the required sort by each other, or with the help of theoretical calculations.

$$E_{in} = \sum_{1\le i<j\le N} E_{ij}, \tag{13.1}$$

where in the general case the potential E_{ij} depends on the arrangement of the particles (i.e., on $\mathbf{r}_i$ and $\mathbf{r}_j$), their orientations and the values of their spins, electric moments, if any, etc. Such interaction forces are called non-central, or tensor. When developing a number of specific problems, it is often enough to assume that the interaction of particles is central and isotropic, i.e. what $E_{ij} = E(r_i - r_j) = E(|r_i - r_j|)$. Such an approximation, when the interaction depends only on the modulus of the distance between the particles, from the physical point of view corresponds to the description of systems such as a gas or a liquid. Of course, it is not satisfactory when studying questions related, for example, to crystallization and the appearance of ordered spatial particle configurations.

In the simplest case, for a system N of spinless identical particles with a central interaction, the Hamiltonian is written in the form

$$H = \sum_{1\le i\le N} \frac{\mathbf{p}_i^2}{2m_i} + \sum_{1\le i\le N} E(\mathbf{r}_i) + \sum_{1\le i<j\le N} E(|\mathbf{r}_i - \mathbf{r}_j|). \tag{13.2}$$

If necessary, terms with a more complex dynamic structure involving multiparticle interactions can be included in H.

The task of the statistical theory based on the microscopic task of the system is: (1) to determine all the thermodynamic equilibrium properties and characteristics of the system (the problem of statistical thermodynamics); (2) to evaluate the nature and magnitude of deviations of a given parameter of the system from its average value, that is, to construct a theory of fluctuations; (3) to investigate the simplest types of non-equilibrium processes occurring in statistical systems, their temporal evolution, i.e., to construct a kinetic theory. These tasks are implemented below under the LGM.

14. Continuous functions of molecular distributions

A natural approach to the study of many-particle systems is the use of probabilistic formalism. The microscopic idea of the initial discreteness of the system when using the probabilistic way of describing it is not preserved. The efforts of classical mechanics are aimed at finding the trajectories of the motion of each of the N particles of the system, which always preserves the discrete nature of the system. In statistical theory, the main attention is paid to the consideration of the probability density $w_N(\mathbf{r}_1, \ldots, \mathbf{r}_N, t)$ continuous in the space of coordinates and momenta of the field, which describes only the probability of detecting a system in a microscopic state corresponding to a set of its arguments. In the equilibrium theory, the time argument t is absent, and in the mechanical formulation of the problem it always remains a dynamic value.

The problem of constructing a statistical theory of equilibrium systems is the establishment of general expressions for statistical distributions, i.e., such distributions when the averages calculated with their help correspond to the observed macroscopic quantities that appear in thermodynamic relationships. Starting with [2], the calculations were constructed by direct averaging of the characteristics using the Gibbs distribution function $w_N(\mathbf{r}_1, \ldots, \mathbf{r}_N)$. This path is convenient for weakly interacting particles. More general and fruitful was the approach to the theory of non-ideal statistical systems, developed by N.N. Bogolyubov [6,16]. It is based on the idea of investigating not the integral value of the partition function $Q = Q(T, V, N)$, but the correlation properties of the particles of the system expressed in terms of the corresponding correlation functions. This makes it possible not to calculate the infinite–dimensional integral Q in the forehead, but to solve a system of several integro-differential equations for the correlation functions. The idea has acquired such a general significance in statistical mechanics that it embraced not only the theory of non-ideal equilibrium systems, but also the kinetic theory.

Correlation functions. In the equilibrium theory, the initial moment for constructing the correlation functions is the probability density of the particle momentum distributions in $6N$-dimensional phase space of particles and their coordinates $(p, r) = (p_1, \ldots, p_N, r_1, \ldots, r_N)$ the canonical Gibbs distribution $w(p,r) = \text{const} \exp\{-\beta H(p,r)\}$, where $H(p, r)$ is defined in (13.2), $\beta = (k_B T)^{-1}$, k_B is the Boltzmann constant, and T is the temperature. The value $w_N(r_1, \ldots, r_N)dr_1 \ldots dr_N = \frac{1}{Q_N}\exp[-\beta H(p,r)]\frac{dr_1 \ldots dr_N}{V^N}$ is the probability of detecting a state of the system when its particles are located in differentially small volumes $(r_1, r_1 + dr_1), \ldots, (r_N, r_N + dr_N) = \{r, r + dr\}$

(the curly brackets denote the complete set of quantities), Q_N is the statistical sum of a system of N particles

$$Q_N = \frac{1}{N!h^{3N}} \int\limits_V \cdots \int\limits_V dp_1 dr_1 \cdots dp_N dr_N \exp(-\beta H), \tag{14.1}$$

The function $w_N(r)$ itself is a general symbol, which is convenient for the formulation of connections between correlation functions of different nature and for connection with thermodynamic functions. Using this definition, we can introduce distribution functions that depend on the arguments of a small number of particles. Thus, the single-particle distribution function $F_1(r_1) = V \int_V w_N(r) dr_2 ... dr_N$ determines the probability of detecting a particle of the system in the volume (r_1, $r_1 + dr_1$). Since all the N particles of the system are the same and the function $w_N(r)$ does not change when the particle indices are rearranged, instead of r_1 in this expression, any of the r_j, where $1 \leq j \leq N$.

The two-particle distribution function (often called the pair correlation function) $F_2(r_1, r_2) = V^2 \int_V w_N(r) dr_3 ... dr_N$ determines the probability of detecting one particle of the system in the volume (r_1, $r_1 + dr_1$), and the other in the volume (r_2, $r_2 + dr_2$). In the general case, the s-partial (s = 1, 2, 3, ...) distribution function

$$F_s(r_1, ..., r_s) = V^s \int_V w_N(r) dr_{s+1} ... dr_N \tag{14.2}$$

determines a similar probability for s selected particles of the system.

From the general properties of the introduced correlation functions, the conditions for their normalization and the relationship between functions of different orders are important, which are written as

$$\int_V \frac{F_s(r_1, ..., r_s)}{V^s} dr_{s+1} ... dr_N = 1, \quad \int_V \frac{F_s(r_1, ..., r_s)}{V} dr_s = F_{s-1}(r_1, ..., r_{s-1}), \tag{14.3}$$

and also the condition for weakening correlations with increasing distances between molecules: so for the simplest pair correlation function for $F_2(r_1, r_2)_{|r_1 - r_2| \to \infty} \to F_1(r_1) F_2(r_2)$ $|r_1 - r_2| \to \infty$, which means that the molecules become statistically independent of each other.

A similar expression for weakening the correlation can also be written not only between individual molecules, but also between their groups, when the distance between them exceeds a certain minimum distance, the so-called correlation radius, less than which the molecules can not be regarded as independent. First of all, it is determined by the radius of the interaction potential R_{lat}, which directly affects the behaviour of a pair of molecules, as well as their indirect influence through neighbouring molecules.

If the system is isotropic, then the pair function $F_2(r_1,r_2) = g(|r_1 - r_2|) \equiv g(r)$ depends only on one argument – the distance between the particles $r = |r_1 - r_2|$. In this case the probability of detecting a particle at a distance from r to $r + dr$ from some other particle $dw(r) = g(r)\frac{4\pi r^2}{V}dr$ is equal, therefore the function $g(r)$is called the radial distribution function of a gas or liquid.

The introduced correlation functions make it possible to express the thermodynamic properties of the system. In particular, the internal energy, the equation of state, etc. are expressed through the function $g(r)$. The total internal energy U, referred to the volume V, is written as follows

$$\frac{U}{V} = \frac{3}{2}\rho kT + \frac{4\pi\rho^2}{2}\int g_2(r)E(r)r^2 dr, \tag{14.4}$$

where ρ is the density of matter, $E(r)$ is the pair interaction potential, and the pressure P is expressed as

$$P = \rho kT - \frac{4\pi\rho^2}{6}\int g_2(r)\frac{\partial E(r)}{\partial r}r^3 dr. \tag{14.5}$$

15. Equations on the continuous functions of distributions

To circumvent the problems of direct calculation of various averages and corresponding statistical sums, it was proposed to construct systems of coupled equations for the distribution functions of different dimensions. This system of equations is called the BBGKY (Bogolyubov–Born–Green–Kirkwood–Yvon) chain of equations. It is obtained by differentiating the expression for the total partition function of the system from the coordinates of one of the particles.

If we denote the total distribution function by $F(\{N\})$, in the form [13]

$$F(\{N\}) = \frac{\lambda^N}{N!}\exp[F - \beta E(\{N\})], \tag{15.1}$$

where $F = -kT\ln(QN)$ is the free Helmholtz energy of a system of N particles, $\lambda = (2\pi mkT)^{3/2}$, then the s-partial correlation function $F(\{s\})$ can be expressed in terms of $F(\{N\})$ $F(\{s\}) = \frac{\lambda^N}{(N-s)!}\int\dots\int \exp[F - \beta E(\{N\})]d\{N-s\}$ as $F(\{s\})$ where is the symmetrized function (14.2), the curly brackets mean the abbreviated entry for integrals whose number is $(N-s)$.

Denote by $D(\{N\})$ the quantity $\exp[F-\beta E(\{N\})]$ and write the identity: $D(\{N\}) = \exp[F-\beta E(\{N\})]$, which is differentiated by the coordinate of the i-th particle r_i, which gives

$$\frac{\partial D(\{N\})}{\partial r_i} + \beta \frac{\partial E(\{N\})}{\partial r_i} D(\{N\}) = 0. \tag{15.2}$$

Multiply expression (15.2) on the right and left by a factor $\lambda^N/(N-s)!$ and integrate over the coordinates $r_{s+1},\dots,r_N$. Taking into account the equality for the interparticle interactions (13.1), so that $\frac{\partial E_{in}(\{N\})}{\partial r_i} = \frac{\partial}{\partial r_i}\sum_{1\le i<j\le N} E_{ij}(r_i,r_j) = \frac{\partial E_{in}(\{s\})}{\partial r_i} + \frac{\partial}{\partial r_i}\sum_{s+1\le j\le N} E_{ij}(r_i,r_j)$ we write

$$\begin{aligned}&\frac{\partial F(\{s\})}{\partial r_i} + \beta \frac{\partial E_{in}(\{s\})}{\partial r_i} F(\{s\}) + \\ &+\beta \sum_{s+1\le j\le N} \int\dots\int \frac{\partial}{\partial r_i} E_{ij}(r_i,r_j) \frac{\lambda^N}{(N-s)!} D(\{N\}) dr_{s+1}\dots dr_N = 0\end{aligned} \tag{15.3}$$

Integrating over all coordinates $r_{s+1},\dots,r_N$ except one r_k ($k = s+1, \dots, N$) in the last term of the left side of the equation, we obtain the function $F_{s+1}(r_1,\dots,r_s,r_k)/(N-s)$. In this case, the factor $(N-s)!$ is replaced by $(N-s-1)!$, so that as a result of the summation the factor $(N-s)$ disappears. After this, the function in the integrand is defined by definition, $F(\{s+1\})$which gives the desired expression

$$\begin{aligned}&\frac{\partial F(\{s\})}{\partial r_i} + \beta \frac{\partial E_{in}(\{s\})}{\partial r_i} F(\{s\}) + \\ &+\beta \int \frac{\partial}{\partial r_i} E_{is+1}(r_i,r_{s+1}) F(\{s+1\})\, dr_{s+1} = 0.\end{aligned} \tag{15.4}$$

This structure of equations has the general property of non-closed: the equations for the correlation functions form a chain of successively entangled equations. Each equation for F_s contains in the integral term the function F_{s+1}, so that the solution of these equations must be preceded by the procedure for disengaging the chain, so that the remaining group of equations would be closed. The closure operation is performed in different ways, depending on the type of system under consideration and the physical conditions in which it is located. This question is examined in detail in various specific situations [9,15–29]. In principle, this universal approach allows solving almost any problem.

16. Discrete functions of molecular distributions

For dense phases (liquid, solid in bulks and for adsorbed phases on the surfaces of condensed phases), lattice models in which the states of molecules were characterized by a small number of parameters: the coordinate of the centre of the cell (or the lattice site) and its energy of interaction with neighbouring molecules were widely used. In addition to this set of state parameters, the lattice models differ in the nature of the site's occupancy states.

States of occupation of cells. The experimental data that appeared in the thirties testifying to the proximity of many properties of a liquid and a crystal: comparatively small relative changes in volume and energy during melting, close values of the heat capacity of substances in the liquid and solid states, short-range ordering in a liquid found in X-ray diffraction study, etc. [37], served as a basis for the transfer of the idea of lattice models for describing the liquid state. A model of non-spherical molecules was proposed in [38], and a theory describing all three aggregate states (gas, liquid, and solid) was developed in a series of papers [39, 40].

Lattice models allow us to reflect the most important part of potential functions in the vicinity of the minimum of the potential curve and all the numerous variants of these models are to some extent related to the properties of the intermolecular potential in the vicinity of this minimum. This is most clearly manifested in the fact that all lattice models operate with the number of nearest neighbours z, which is directly related to the local structure of the liquid. If for the adsorbed particles the number z is determined by the potential relief of the substrate, then for a liquid this number is the average statistical characteristic that is associated with the cooperative behaviour of a large ensemble of molecules and the properties of intermolecular interactions.

The central feature of lattice models is that the total volume of the system V is divided into some elementary volumes v_0, which form a periodic lattice. The statistical problem of the spatial distribution of molecules is greatly simplified by introducing the assumption of spatial regularity of the distribution of molecules. This replaces the continual description of the distribution of molecules to a discrete one. In addition, having introduced a number of assumptions, it is possible to simplify so much the problem of calculating the configuration integral of the system with allowance for intermolecular interaction, which is possible for its analytical solution.

Various variants of cell filling with different number of particles were proposed. There have been attempts to bring cellular models closer to real liquids, assuming that the cells in the system have different sizes and are placed irregularly [41,42]. Models with irregular cell structures are used mainly to calculate the radial distribution function; good results are also obtained for thermodynamic functions. It was also assumed that cells can contain one or more molecules, but they do not have the concept of a free cell.

The most important is the situation where each cell is allowed to contain one particle or to be empty – this is the so-called lattice gas model (LGM) that 'came' from the theory of adsorption: the well-known Langmuir model [43] is in fact the first formulation of the LGM, about filled and free sites. It should be clearly divided the use of LGMs from so-called lattice models. The account of vacancies allows to take into account the irregularity of the structure and does not require the artificial introduction of the concept of collective entropy [44]. These models are applicable not only to liquids, but also to gas; they are able to give an equation of state describing both phases simultaneously and transferring a liquid–vapour phase transition. Note that the LGMs and models of irregular cell structures [45] allow us to describe not only the liquid–vapour phase transition (this requires attraction between molecules), but also the liquid–solid phase transition, i.e., the liquid–solid phase transition or a phase transition of the melting type [46,47].

The LGM corresponds to three situations [9,27]: adsorption and absorption of gas-phase molecules, as well as solid solutions (the well-known analogs of the LGM are the binary alloy model without allowance for vacancies, and the Ising model with two directions of spin along and against the field [48]). If we confine ourselves to spherical molecules of the same kind, then the cell size v_0 is directly related to the size of the molecule, which is given by the parameter σ in the potential Lennard-Jones function (LJ). This parameter characterizes a solid incompressible sphere of a molecule whose nature follows their exchange interaction of the electron shells of two molecules as they approach each other, leading to repulsion of the molecules. Such a choice of the quantity $v_0 = \gamma_s\lambda^3$ imposes the condition of a single filling of the cell volume when a molecule is placed in it (γ_s is the cell shape factor, λ is the average distance between the particles). Taking into account the explicit form of the LJ potential, we have $\lambda = 2^{1/2}\sigma \approx 1.12\ \sigma$. The ratio $M = V/v_0$ determines the number of cells in the system under consideration.

Fixation of the molecule in the centre of the cell corresponds to its occupation state. Mathematically, this event is described by the value γ_f^i, where f is the cell number, $1 \le f \le M$, the index i denotes the occupation state of the cell with the number f. If there is a particle A in the site f, then $\gamma_f^A = 1$. If cell f is free, then there is a vacancy in it, therefore $\gamma_f^A = 0$, and $\gamma_f^V = 1$. For two types of occupation states $s = 2$ of any lattice structure site, corresponding to a one-component system for which i = A or V are vacancies, the random variables γ_f^i obey the following relations:

$$\sum_{i=1}^{s} \gamma_f^i = 1 \text{ and } \gamma_f^i \gamma_f^j = \Delta_{ij}\gamma_f^i, \tag{16.1}$$

where Δ_{ij} is the Kronecker symbol, which means that any site is necessarily occupied by some, but only one, particle. These conditions mean the absence of multiple filling of any cell. In particular, the equality $(\gamma_f{}^i)^k = \gamma_f{}^i$ of any integer $k > 1$ holds. In the general case, equations (16.1) hold for any value $s > 2$.

Particle configurations. The number of different states of the system is equal to s^M, where it increases sharply both with increasing number of occupied states of the site s and with increasing number of sites of the system M. In this connection, it is impossible to use the detailed description of real systems with the help of complete sets $\{\gamma_f^i\}$ and have go to their abbreviated description. (For example, detailed description is impossible at sufficiently small values of s and M, for example, for $s = 3$ and $M = 15$.) For a brief description of the state of the system, discrete many-particle distribution functions or correlation functions (simpler, correlators) are defined as follows

$$\theta_{f_1 \ldots f_m}^{i_1 \ldots i_m} = \sum_{k_1=1}^{s} \cdots \sum_{k_M=1}^{s} \prod_{n=1}^{m} \gamma_{f_n}^{k_n} P\left(\{\gamma_f^k\}, \tau\right), \text{ where } \gamma_{f_n}^{k_n} = \begin{bmatrix} 0, & k_n \ne i_n \\ 1, & k_n = i_n \end{bmatrix}. \tag{16.2}$$

where the sums are taken over all sites of the lattice (the types of particles in them are numbered by the symbols k_1, ..., k_M from 1 to s, i_f is the particle sort at site f; m is the order or dimension of the correlator characterizing the probability of the following particle configuration at time τ: in the site f_1 there is a particle i_1, in the site f_2 – a particle of the sort i_2, etc. up to the m-th particle inclusive. The state of occupation of the remaining (M–m) lattice sites is of no interest to us but averaging is carried out over them.

In the LGM, elementary events are described by the quantities γ_f^i (16.1), (16.2), and their various products are complex events. The probabilities of these complex events characterize the probabilities of the corresponding local particle configurations. For $m = M$, the correlator (16.2) is the total M-particle distribution function through which any correlator is expressed as

$$\theta_{f_1 \ldots f_m}^{i_1 \ldots i_m}(\tau) = \sum_{i_{m+1}=1}^{s} \ldots \sum_{k_M=1}^{s} \theta_{f_1 \ldots f_M}^{i_1 \ldots i_M}(\tau). \qquad (16.3)$$

This connects correlators of the dimension m and m–1 with each other as

$$\theta_{f_1 \ldots f_{m-1}}^{i_1 \ldots i_{m-1}}(\tau) = \sum_{i_{m+1}=1}^{s} \theta_{f_1 \ldots f_m}^{i_1 \ldots i_m}(\tau). \qquad (16.4)$$

Averaging over the states of occupation of cells determines the degree of their filling with particles of type i: $\theta_i = \langle \gamma_f^i \rangle$. This is the simplest single-particle (unary) correlator, which characterizes the probability of finding a particle i at the instant of time τ at the site with the number f – the local density of the component i at the point f at time τ (expressed in mole fractions). The numerical density θ_i is related to the concentration c_i as $c_i = \theta_i / v_0$.

The second-dimensional correlators are pair correlators

$$\theta_{fg}^{i_1 i_2} = < \gamma_f^{i_1} \gamma_g^{i_2} >_\tau \qquad (16.5)$$

which characterize the probability of finding a particle i_1 at site f and particle i_2 at site g at time t. By virtue of the projection properties of the quantities γ_f^i (16.1) here $f \neq g$. Similarly, we can consider higher-dimensional correlators: the third, fourth, etc. order [27].

In the case of an equilibrium particle distribution, the normalized distribution function is written as [27]

$$P(\{\gamma_f^i\}) = \frac{1}{Q} \exp[-\beta H(\{\gamma_f^i\})], \qquad (16.6)$$

where Q is a normalizing factor, usually called the partition function or the sum over the states of the system; $H(\{\gamma_f^i\})$ is the effective Hamiltonian (or Hamiltonian) function of the lattice system:

$$H(\{\gamma_f^i\}) = \tilde{H}(\{\gamma_f^i\}) - \sum_{f,i} \mu_i \gamma_f^i,\ \tilde{H}(\{\gamma_f^i\}) = H_{kin}(\{\gamma_f^i\}) + H_{pot}(\{\gamma_f^i\}). \quad (16.7)$$

In the formula (16.7) $\tilde{H}(\{\gamma_f^i\})$ is the Hamiltonian function, which characterizes the total energy of the system – the sum of the kinetic $H_{kin}(\{\gamma_f^i\})$ and potential $H_{pot}(\{\gamma_f^i\})$ energies; μ_i – the chemical potential of particles i – the distribution (16.6) is written for a large canonical ensemble. The expression for the partition function of the lattice system has the form

$$Q = \sum_{i_1=1}^{s} \dots \sum_{i_M=1}^{s} \exp[-\beta H(\{\gamma_f^i\})]. \quad (16.8)$$

The concrete form of writing the Hamiltonian H is uniquely related to the physical model. The change in the physical model requires the construction of a new Hamiltonian. Changing the interaction potential, we will change the form $H_{pot}(\{\gamma_f^i\})$. In old lattice models, the particles are assumed to be fixed at the lattice sites. Their kinetic energy does not contain translational degrees of freedom, but consists of vibrational and, if the particle has a complex structure, from rotational degrees of freedom. Previously, it was believed that: 1) the motion of each particle independently of the motion of other particles is an analogue of the Einstein oscillator model [11]; 2) the expressions for $H_{kin}(\{\gamma_f^i\})$ are determined for the pure component (or for ideal models) and they do not change when interparticle interaction is taken into account.

To calculate the thermodynamic characteristics, averaging over all configurations is required. The possibility of solving exact problems even for discrete distribution functions, as for any other system of many interacting bodies, is quite unique. It is for such discrete productions of problems that few exact solutions have been obtained within the framework of non-one-dimensional structures [49]. One-dimensional structures always admit exact solutions [48,50]. This was first discovered in the work of Ising [48], after which such model systems became known as the Ising model. Nevertheless, in a number of situations, one-dimensional models are also useful in many problems of adsorption of molecules on the edges of faces, at the base of steps of stepped surfaces, on linear polymers, etc.

The exact solution was first obtained for the two-dimensional Ising model only in the absence of an external field. Onsager [51] studied a rectangular lattice; later, solutions were found for other types of two-dimensional lattices [49]. For a two-dimensional Ising model in a nonzero field and a three-dimensional model, the results, which can be called exact, are available only in the form of expansions in density and temperature [9, 52]. The presence of exact solutions for the two-dimensional case and practically exact solutions for three-dimensional systems makes the Ising model (and 'Ising' lattice gas) the most important model basis for studying the regularities of phase transitions and critical phenomena. Significant advances in the theory of phase transitions are associated with the consideration of the generalized Ising model, in which the variable characterizing the state of the site takes not two values, but more (in particular, when there is a continuous series of values).

Approximate methods. In the overwhelming number of situations where there are no exact solutions to the problem, it is necessary to use approximate solutions, which are also important for understanding the behaviour of cooperative systems. The question of the nature and accuracy of approximate calculations occupies the central part of the theory of condensed phases [27].

In the theory of discrete distribution functions, different ways of approximating the probabilities of many-particle configurations through the probabilities of configurations of smaller dimension are used. For simplicity, we consider a one-component system whose sites are occupied by particles A (we will assume that these are adsorbed particles A) or V (vacancies). The complete spectrum of the particle A configuration in the first coordination sphere around the central (not shown) particle on the lattices $z = 4$ and 6; adjacent particles A – blackened circles, the remaining sites of the first coordination sphere are occupied free (or occupied by the particle V). Let us denote by the symbol $\theta_i\ (n\sigma)$ the probability that in the first coordination sphere of the central particle of type i, i = A and V, there are n particles A with an arrangement of σ (Fig. 16.1).

Examples of such approximations are shown in Fig. 16.2 for planar lattices with $z = 4$ (*a*) and 6 (*b*). The first variant of the approximation corresponds to the mean-field approximation when the probability of the multiparticle configuration is expressed in terms of the product of the probabilities of the sites to be occupied by their particles θ_i of type i (or through unary distribution functions), $\theta_i\ (n\sigma) = \theta_i \theta_A^n \theta_V^{z-n}$ which are related to each other by the normalization condition

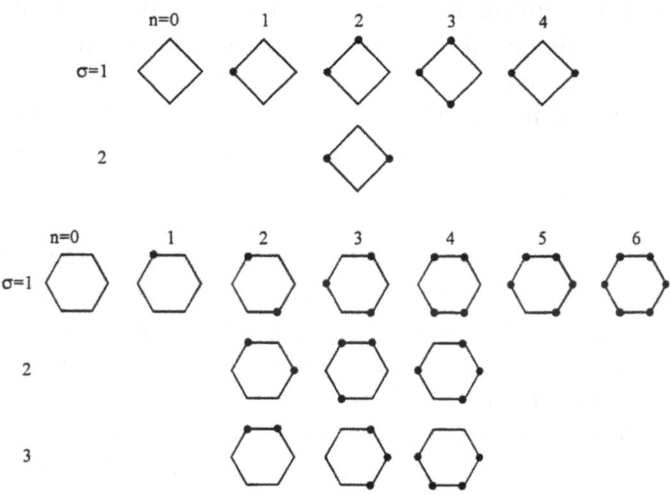

Fig. 16.1. The configurations of particles A in the first coordination sphere around the central particle on the sieves $z = 4$ and $z = 6$; blackened circles – adsorbed particles A; n is the number of particles A, $\sigma(n)$ is the type of configuration for a fixed value of n, $R = 1$ [27].

$\theta_A + \theta_V = 1$. Obviously, the method of arranging the particles, fixed by the symbol σ, does not matter. With this approximation, there are no correlations.

The second variant of approximation is also one-particle – in it the many-particle configuration of neighbours around the central particle is approximated through the product of unary distribution functions (polynomial approximation). The difference from the first way of disengaging is that the correlation itself between the central particle and its nearest neighbour is preserved.

The third variant of the approximation is that the probabilities of the many-particle particle configurations $\theta_i(n\sigma)$ are approximated in terms of the local probabilities θ_i for the detection of particles i (unary distribution functions) and in the pair distribution functions $\theta_{ij} = \left\langle \gamma_f^{\,i} \gamma_g^{\,i} \right\rangle$. This is the so-called Guggenheim's quasi-chemical approximation (QCA) [53] or the Bethe–Peierls approximation [54,55]. The correlation effect is explicitly taken into account via the pair distribution function θ_{ij}, which characterizes the probability of finding a pair of particles ij, ij = A, V at neighbouring sites. In this approximation, we have $\theta_i(n\sigma) = \theta_i t_{iA}^{\,n} t_{iV}^{\,z-n}$, where $t_{ij} = \theta_{ij}/\theta_i$ is the conditional probability of finding the particle j near the particle i; and $\theta_{iA} + \theta_{iV} = \theta_i$ or $t_{iA} + t_{iV} = 1$. Note that in QCA the weights of configurations with different values of σ (for n = const)

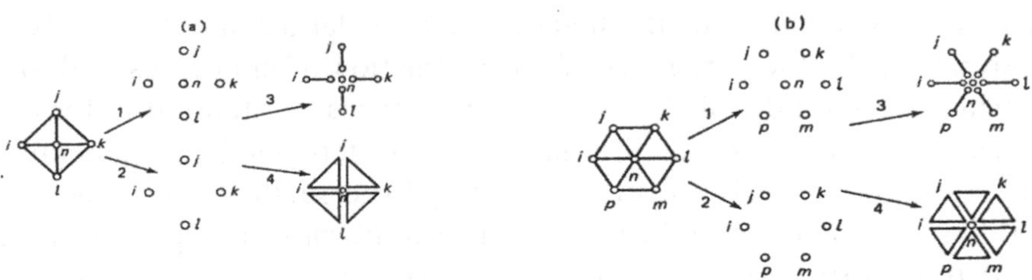

Fig. 16.2. Schemes of approximate calculations of many-particle configurations of neighbouring molecules for two-dimensional lattices $z = 4$ (a) and 6 (b): 1 - mean-field approximations and chaotic (through the concentrations of all molecules); 2 - polynomial approximation (through the concentrations of neighbouring molecules); 3 - quasi-chemical approximation (through the probabilities of pairs of molecules); 4 - allowance for indirect correlations in the quasi-chemical approximation and allowance for triple correlations (through the probabilities of triples of molecules) [27].

are equiprobable, or the symbol σ does not affect the calculation of $\theta_i(n\sigma)$.

The fourth variant of approximation is the simplest example of taking into account the correlation effects of three molecules. This approximation is a discrete version of the so-called Kirkwood superposition approximation [56], in which for three particles located at any distances an approximation is introduced of the probabilities of simultaneous realization of triples of particles θ_{ijk} of type *ijk* in the form $\theta_{ijk} = \theta_{ij}\,\theta_{kj}\,\theta_{ki}/(\theta_j\theta_j\theta_k)$ in terms of the probabilities of pairwise and unary distribution functions. Within the framework of such an approximation, the symbol σ determines contributions from pairs of particles at different distances. When using the superposition approximation, the problem is also closed through equations for unary and pairwise distribution functions.

The structure of the fourth approximation allows us to go beyond just the pair approximation, if we assume that additional equations will be built on the probabilities of triple configurations. However, such a path sharply increases the dimensionality of the system of equations, which must be solved in order to find the equilibrium particle distribution [57].

Adsorption systems. The widest application of lattice models is associated with the development of the possibility of simultaneously taking into account intermolecular interactions and heterogeneity of lattice structures. This is a typical case of adsorption systems, which are discussed in detail in monographs [27, 31, 58–62]. Here the lattice models are rigorously justified by the existence of a surface potential, creating regular or irregular regions of localization of adsorbed particles. The nature of the heterogeneity of lattice structure cells is determined by the potentials of the interaction of molecules with the surface of a solid and the chemical properties of atoms that form a surface and their structural organization. Heterogeneity is generated at the supramolecular level by the polydisperse structure of porous bodies, and at the atomic level by non-uniformity of open surfaces and pore wall surfaces: surface potential type, lattice defects near surfaces, the presence of different surface groups on the surface, dislocations and surface roughness, surface amorphization, etc. All these features of adsorption systems are conveniently described by the LGM.

Description of non-uniform systems. We will call the system non-uniform if, under the conditions of the equilibrium distribution of particles, the occupation states of different sites are different. This definition combines two different factors that lead to different occupation states by particles of the same kind of different sites. The first factor is the non-uniformity associated with the difference in interactions with the surrounding particles, which do not change their state at the characteristic relaxation times of the process. Examples are different adsorption surface centres and different absorption centres in the volume of the solid, which differ in binding energies with molecules of the gas phase. The second factor is the non-uniformity, which is related to the nature of the particle distribution due to the interaction between them. Examples are two- and multiphase systems and interfaces of stratifying phases, including the subsurface region of the gas-solid interface, as well as the ordering of particles on an uniform surface or in the volume of a solid. In the second case, the non-uniformity is due to the cooperative properties of the system, leading to the realization of phase transitions of stratification and (or) ordering. We note that the simultaneous 'action' of various factors that cause the non-uniformity of the system also leads to its non-uniformity. As an example, we point out the case of ordering of adsorbed particles (the 2nd factor) on the stepped surface (1st factor) [27].

Non-uniform lattices reflect a wide range of real situations, so in the following we formulate a general approach regardless of the physical nature of the non-uniformity. We introduce the quantities η_f^q characterizing the type q of the lattice site with the number f, $1 \leq f \leq M$, $1 \leq q \leq t$, t is the number of site types [27]. The quantities η_f^q are analogues of the quantities γ_f^i. That is, $\eta_f^q = 1$ if site f is a site of type q, and $\eta_f^q = 0$ otherwise. For them analogous relations

$$\sum_{q=1}^{t} \eta_f^q = 1, \ \eta_f^q \eta_f^p = \Delta_{qp} \eta_f^q \tag{16.9}$$

The set of values $\{\eta_f^q\} = \eta_1^q, ..., \eta_M^q$ uniquely determines the non-uniformity of the lattice sites. However, γ_f^i and η_f^q there is a fundamental difference between the quantities: η_f^q the set is assumed to be fixed and unchanged, whereas over the set γ_f^i averaging is carried out. As in the case of quantities γ_f^i, it is difficult to work with a full set of values η_f^q and a transition to the distribution functions of non-uniform sites is necessary. Their dimension, as will follow from the following, depends on the interaction potential, on the approximation used for taking into account the interaction, on the type of the elementary process, and on the dimensionality of the particle configurations.

The state of occupation of sites of a non-uniform lattice, as above, is characterized by random variables γ_f^i, but the site itself is further characterized by its intrinsic property (type) through a given value η_f^q. In the general form, the correlators on a non-uniform lattice have the form analogous to (16.2):

$$\theta_{f_1 \dots f_m}^{i_1 \dots i_m}(q_1 \dots q_m \mid \tau) = < \gamma_{f_1}^{i_1} \dots \gamma_{f_m}^{i_m} \mid \eta_{f_1}^{q_1} \dots \gamma_{f_m}^{q_m} >_{\tau} \tag{16.10}$$

In the formula (16.10), the types of sites on which the correlation function of dimension m is defined are additionally indicated. The same addition is present in all the above expressions (16.3)–(16.8), so there is no need to rewrite them for non-uniform lattices. The condition of statistical independence for correlators is written in the following form:

$$\theta_{f_1 \dots f_m}^{i_1 \dots i_m}(q_1 \dots q_m \mid \tau) = \prod_{n=1}^{m} \theta_{f_n}^{i_n}(q_n \mid \tau) \tag{16.11}$$

It means that the filling of each site does not depend on the occupation state of other sites, that is, there are no correlation effects. Strictly speaking, such a situation is possible only in the absence of interaction between the particles. A special case of using condition (16.11) is the recording of the law of mass action (surfaces) for calculating the rate of chemical transformations in the kinetic regime.

Knowing the local correlators, it is easy to construct macroscopic characteristics of the particle distribution over the lattice sites. In the equilibrium on non-uniform lattices (surfaces), the nature of the filling of sites of each type is of great importance. To this end, in the expressions (16.10) there is an indication of the types of sites. A more 'active' role is played by the values η_f^q in the expressions for the local energy of the lattice systems, they reflect the influence of the site type on the energy characteristics of the bond and the activation energies.

17. Functions of molecular distributions in a discrete-continual description

The transition to discrete-continual models was connected with the solution of the problem of substantiating lattice models of a liquid [31,63,64] and in connection with the refinement of the description of the thermal motion of molecules inside cells [32]. The question of the validation of lattice models of a liquid was solved using the cluster approach [27, 65].

The molecular–statistical theory of equilibrium systems is based on the Gibbs distribution for microscopic states with a fixed selected set of macroscopic variables (defining the ensemble type) describing the observed state of the system [4–7]. For the canonical ensemble belonging to N particles of one kind that are in the volume V at temperature T, this distribution is indicated above in Section 14. The essence of the cluster approach consists in replacing the calculation of the partition function of the system under study by solving a system of equations for the cluster distribution functions that characterize the probabilities of realization different local configurations of molecules $\theta(1, \ldots, N)$ with coordinates $r_1, \ldots r_N$. This idea is similar to the idea of constructing equations for the correlation functions in Section 15.

To construct this system of equations, we choose m "central" sites in the centre of the region N. The width of the remaining region containing $(N-m)$ sites should be equal to the radius of the interaction potential between the particles R. In this case, particles in the central sites do not interact with particles in the sites outside the considered region. We denote such distribution functions as $\theta(\{m\}|N)$, where the symbol $\{m\}$ stands for the list γ_f^i, $1 \leq f \leq m$ and $1 \leq i \leq s$, the occupation states m of the central sites. If we fix the types of particles in the central sites and consider the relations between the functions $\theta(\{m\}|N)$ with different occupation states of the central sites, then, for example, for $m = 1$, we obtain relations of the type $\theta(i|N)/\theta(j|N)$ with particles of sorts i and j. These relations are very simply expressed in terms of cluster Hamiltonians (or the total energy of the central particles) $h(i|N)$ for a region containing N sites and a particle i at the centre, as

$$\theta(i\,|\,N)\ /\,\theta(j\,|\,N)\ = \exp\{\beta[h(j\,|\,N)\ - h(i\,|\,N)]\}, \qquad (17.1)$$

Using the principle of inclusion–elimination of probabilities and normalization relations, the functions $\theta(i|N)$ can be expressed in terms of a sequence of correlators of lower dimension (16.2)–(16.4).

The discrete distribution functions are in many respects similar to the functions of the continuum distributions (14.2) and (14.3), except for the fact that it is more convenient to introduce their normalization not into the entire volume of the system, but into the volumes of local sites – i.e. each local correlator is defined on the ensemble of copies of identical lattices. This allows us to introduce the concept of a fully distributed model in which each site can be considered as a separate sort of site, and any size group of sites of the same type can be formed from a set of identical sites. Their contributions will be taken into account by the weights of the corresponding functions of the site distributions for the non-uniform lattice (for details, see [27,31,66]).

Accounting for the interaction of nearest neighbours. Below is given a concrete example of constructing cluster equations in the case of interactions between the nearest neighbours. We shall assume that the internal degrees of freedom of the particles are weakly dependent on the change in the energy of the interparticle interaction of neighbouring particles when a particle is replaced one kind by another. Consider a uniform lattice system, any site of which can be occupied by a particle of sort i, $1 \le i \le s$, s is the number of components. If we denote the parameters of pair interaction of neighbouring particles of sort i and j, then the potential energy of the system will be composed of all possible pairs of interacting particles and the effective Hamiltonian will be written as [27,65]

$$H = \sum_{f,i} v_i \gamma_f^i - \frac{1}{2}\sum_{f,g}\sum_{i,j} \varepsilon_{ij}\gamma_f^i\gamma_g^j, \qquad (17.2)$$

where the sum over f is taken over all sites of the lattice, and the sum over g with respect to all z neighbours of site f; the sums over i and j are taken over all the states of occupation of the lattice sites (the factor 1/2 takes into account that each pair is calculated twice), v_i is the one-particle contribution of component i to the energy of the system.

In the cluster approach, the original lattice system is presented as a set of clusters. We confine ourselves to clusters with one central site. The initial lattice is presented as clusters consisting of $z + 1$ sites: the central site and its z neighbours. For each cluster, we can introduce the quantities h_f, characterizing the contributions to the total energy of the system of central particles (cluster Hamiltonians). When the interaction of the nearest neighbours is taken into account, the cluster Hamiltonian for a cluster with one central site has the form

$$h_f = \sum_{i=1}^{s} h_f^i, h_f^i = \left(v_i - \sum_g \sum_{j=1}^{s} \varepsilon_{ij} \gamma_g^j \right) \gamma_f^i \tag{17.3}$$

We define two types of cluster correlators. Correlators of the first type can be represented as follows:

$$\begin{gathered} \theta_{fg_1 \dots g_z}^{ij_1 \dots j_z} = \frac{1}{Q} \sum_{i_1=1}^{s} \dots \sum_{i_M=1}^{s} \gamma_f^i \prod_{g=1}^{z} \gamma_g^j \exp(-\beta H) = \\ \frac{1}{Q} \sum_{i_1=1}^{s} \dots \sum_{i_M=1}^{s} \gamma_f^i \prod_{j=1}^{s} (\gamma_g^j)^{n_j} \exp[-\beta(h_f + G_f)] \equiv \theta_{\{[n]\}}^i, \end{gathered} \tag{17.4}$$

where $G_f = H - h_f$. The second equality determines the transition from the indication of the sort of particle located at the site g, where g is any of the sites of the coordination sphere, to the number of energy bonds of the central particle i with neighbouring particles j in the coordination sphere of the cluster. On a uniform lattice, the energy of the central particle does not depend on the number of sites on which the particle and its neighbours are located, therefore, for simplicity, the subscripts of the sites f, g_1 ... g_z are omitted. Direct brackets in the functions $\theta_{\{[n]\}}^i$, where $\{[n]\} = [n_1 \dots n_s]$, mean that there are exactly n_1 particles of sort 1 in the coordination sphere of the cluster, exactly n_2 of particles of sort 2, etc. For unambiguous assignment of the occupation state of the sites of the coordination sphere it is sufficient to set $(s - 1)$ the value of n_j, since

$$n_s = z - \sum_{j=1}^{s-1} n_j, \quad 0 \le n_j \le z, \tag{17.5}$$

Thus, particles of sort s can be considered as an addition to the complete filling of the sites of the coordination sphere of the cluster. The brackets mean the full set of values of n_j. Correlators of the first type (17.4) characterize the probability of finding particle i in the centre of the cluster, and n_j particles of the sort j, $1 \le j \le s$, at the sites of its coordination sphere.

The expression for G_f in (17.4) does not depend on the occupation state of site f, but depends on the states of occupation of the sites of its coordination sphere; therefore, for any particular set of particles $\{[n]\}$, formula (17.4) can be rewritten as

$$\theta^i_{\{[n]\}} = \Lambda \exp[\beta(-\nu_i + \sum_{j=1}^{s} \varepsilon_{ij} n_j)], \ \Lambda = \mathrm{const}(\{[n]\})/Q, \tag{17.6}$$

where the unknown constant const($\{[n]\}$) takes into account the energy contributions from all the lattice sites, except for the central site of the cluster with particle i.

Correlators of the second sort are obtained from the correlators (17.4) when averaging over the occupation states of a part of the coordination sphere sites. Using the numbers of energy bonds of the central particle with its neighbours, these correlators will be written in the form

$$\theta^i_{\{n\}} = \left\langle \gamma^i_f \prod_{j=1}^{s-1} (\gamma^j_g)^{n_j} \right\rangle, \ 0 \le n_j \le z, \tag{17.7}$$

The functions (17.7) characterize the probabilities that there is a particle i in the centre of the cluster, and in the coordination sphere of the cluster there are at least n_1 particles of sort 1, n_2 of particles of sort 2, etc., not less than n_{s-1} particles of the sort s–1. In this case, (17.5) does not hold, and in the notation (17.7), straight brackets are omitted.

All the thermodynamic characteristics of the system are expressed through the correlators (17.7). Formula (17.7) is a special case of formula (16.2), which characterizes the probability of finding an arbitrary configuration of particles on the lattice: there is a particle of sort i_1 at site f_1, a particle of sort i_2 is located at site f_2, and so on, up to the m-th particle inclusive.

So $\theta^i_{\{0\}} = \theta^i_{0\ldots0} = \langle \gamma^i_f \rangle = \theta_i$ represents the probability of finding a particle i at any lattice site; fixation of the value of θ_i in the sorption isotherms (adsorption or absorption) determines the external pressure of the molecules of the gas phase or vice versa.

The functions $\theta^i_{0...j...0} = \langle \gamma^i_f \gamma_g{}^j \rangle = \theta_{ij}$ (with j-th index equal to one) are the probabilities of finding two particles i and j at neighbouring sites of the uniform lattice (pair functions). Knowing these averages, one can find the heats of sorption or mixing, heat capacity, etc.

The introduction of correlators of two sorts makes it possible to reduce the entire procedure for constructing systems of equations for correlators only to work with the probability relations between them. Let's consider the steps of this work for a binary system.

We construct equations describing the local distribution of the particles A and B of the binary solution in the equilibrium state [27, 65]. For clarity, we confine ourselves to taking into account the pair interactions between the central particle and its neighbours, according to Eq. (17.2). In the first stage, we use expression (17.6), and consider the ratio of the correlators of the first sort, which differ in the sort of the central particle, but have the same state of occupation of the sites of the coordination sphere:

$$\begin{aligned} &\theta^B_{[n]} = M_n \theta^A_{[n]},\ M_n = \exp[-\beta(\nu + n\omega)], \\ &\nu = \nu_B - \nu_A + z(\varepsilon_{AB} - \varepsilon_{BB}),\ \omega = \varepsilon_{AA} + \varepsilon_{BB} - 2\varepsilon_{AB}, \end{aligned} \tag{17.8}$$

where n is the number of particles A in the coordination sphere of the cluster. The ratio of the functions (17.6) eliminates the unknowns Q and const([n]}).

In the second stage, we consider the connection of the correlators of the first and second types. By definition of correlators of the second type, containing not less than n particles A in the coordination sphere of the cluster, we represent

$$\theta^A_n = \sum_{k=0}^{z-n} C^k_{z-n} \theta^A_{[n+k]},\ 0 \le n \le z, \tag{17.9}$$

The coefficients C^k_z of (17.9) are easily obtained from the following arguments. Adding one more particle to the n particles, we pass to the configuration containing n + 1 particles A and characterized by the correlator $\theta^A{}_{[n+1]}$. Such a transition is possible in (z–n) ways. A similar transition to (n+2) particles is possible with the addition of two particles A (z–n) (z–n–1)/2 ways, etc. The coefficients C^k_m are the number of combinations of m elements in k: $C^k_m = m!/(k!(m-k)!)$.

Turning the linear system of equations (17.9) with respect to correlators of the first sort, we find

$$\theta_{[n]}^{A} = \sum_{k=0}^{z-n} (-1)^k C_{z-n}^{k} \theta_{n+k}^{A}. \tag{17.10}$$

A similar expression holds for correlators of the first type with a central particle B.

In the third stage, we substitute the expressions (17.10) for the central particles A and B into relations (17.6):

$$\sum_{k=0}^{z-n} (-1)^k C_{z-n}^{k} \theta_{n+k}^{B} = M_n \sum_{k=0}^{z-n} (-1)^k C_{z-n}^{k} \theta_{n+k}^{A} \tag{17.11}$$

This linear system of equations connects the unknown functions θ_n^A and θ_n^B with each other. It is solved sequentially, beginning with $n = z$. Its solution has the form

$$\theta_n^B = \sum_{k=0}^{z-n} C_{z-n}^{k} M_{n+k} \sum_{r=0}^{z-n-k} C_{z-n-k}^{r} \theta_{n+k+r}^{A} = \exp[-\beta(\nu + n\omega)] \sum_{k=0}^{z-n} C_{z-n}^{k} x^k \theta_{n+k}^{A} \tag{17.12}$$

where $x = \exp(-\beta\omega) - 1$. In the second equality, the terms are rearranged so that the first sum changes the dimension of the correlators, and the energy factors are in the internal sum, and the explicit form M_n (17.10) is taken into account.

The system of equations (17.12) is the initial one for the calculation of cluster correlators. With a suitable sequential search of all lattice sites, the system (17.12) allows in principle to obtain an exact solution. This is done for a one-dimensional lattice [67]. For two- and three-dimensional lattices, the number of equations of the system (17.12) is less than the number of unknowns θ_n^A and θ_n^B (taking into account the normalization condition and the invariance condition for the filling of the central and neighbouring cluster sites), so it must be closed. Depending on the method of closing the system of equations (17.12) by approximating the higher correlators through correlators of lower dimensionality, we obtain different approximations differing in the accuracy of taking into account the correlation effects. The structure of the equations obtained is such that it is necessary to formulate a general rule for approximating all higher correlators through the lower ones, which determines the functional dependence of the closed system for uncoupled correlators. After finding the uncoupled correlators, this, in turn, allows one to self-consistently calculate all higher correlators, both of the first and second sorts in the approximation under consideration.

Continuous distribution of molecules. To describe the distribution of molecules, we use the continuous analog of the cluster approach [63, 64, 68]. For a group with the same number of molecules N, but differing in their coordinates, we consider the ratio $F(\{s\})$ of the functions defined in Section 15 below (15.1) and replaced by the symbol $\theta(\{r_j\})$ (as above, $\{r_j\}$ is the complete set of coordinates particles)

$$\theta(r_1,\ldots,r_N) = \theta(r_1^*,\ldots,r_N^*)\xi\exp\left\{\beta\sum_{1\le i<j\le N}[\varepsilon(|r_i^*-r_j^*|-\varepsilon(|r_i-r_j|)]\right\},$$

$$\xi=\Lambda/\Lambda^*,\ \Lambda=\int_V\cdots\int_V dr_{N+1}\cdots dr_M\exp\left\{-\beta\left[\sum_{1\le i\le N;N+1\le j\le M}\varepsilon(|r_i-r_j|)+\sum_{N+1\le i<j\le M}\varepsilon(|r_i-r_j|)\right]\right\}$$

(17.13)

where the coordinates of the molecules in different groups are denoted by r_i and r_i^*, respectively (the coordinates with an asterisk correspond to Λ^*).

Let us select from N molecules a group containing m molecules in a sphere with volume ω, which we will consider to be central. The volume ω is surrounded by a sphere ω_R whose radius is not less than $\omega^{1/3}+R$, where R is the radius of the interaction potential between the molecules. In the volume $(\omega_R-\omega)$ there are the remaining $(N-m)$ molecules. If their coordinates are fixed, then independently of the position m of the central molecules $\xi=1$ and formula (17.13) is rewritten as

$$\theta(r_1,\ldots,r_m\,|\,r_{m+1},\ldots,r_N) = \theta(r_1^*,\ldots,r_m^*\,|\,r_{m+1},\ldots,r_N)\exp\{\beta\times$$

$$\times\left(\sum_{1\le i<j\le m}[\varepsilon(|r_i^*-r_j^*|)-\varepsilon(|r_i-r_j|)]+\sum_{1\le i\le m;m+1\le j\le N}[\varepsilon(|r_i^*-r_j|)-\varepsilon(|r_i-r_j|)]\right)\}$$

(17.14)

Expressions (17.14) are a set of relationships that differ in the different locations of both the central molecules and molecules in the surrounding region $(\omega_R-\omega)$.

In the case of a discrete distribution of molecules, these relationships were used in the cluster approach for lattice structures at $m = 1$ and 2 [23, 25]. The same relationships can be used to describe systems with a continuous distribution of molecules. To prove the latter, we need to consider the system (17.13), in which

N is replaced by m, and the value of M by N, and put $r_i = r_i^*$, $2 \le i \le m$, $r_1^* = r_1 + dr_1$. Expanding the expression in (17.13) in terms of dr_1, we obtain

$$\frac{\partial}{\partial r_1}\ln\theta(r_1,\cdots,r_m)+\beta\frac{\partial}{\partial r_1}E_m+\frac{\partial}{\partial r_1}\xi=0, \qquad (17.15)$$

where E_m is the energy of the group of m molecules, defined in (13.1), and the derivative with respect to ξ (according to (13.1), definition $F(\{s\})$ (14.2) and (17.13)) leads to the following expression

$$\frac{\partial}{\partial r_1}\xi=\frac{\beta(N-m)}{V\theta(r_1,\cdots,r_m)}\int_V dr_{m+1}\theta(r_1,\cdots,r_{m+1})\frac{\partial}{\partial r_1}E(|r_1-r_{m+1}|). \qquad (17.16)$$

Thus, equations (17.15) and (17.16) are a system of integro-differential BBGKY equations [6] (they can be transformed in (15.4)). Equations (17.13) determine the relationship of different configurations of groups consisting of the same number of molecules, which differ in their coordinates. Differential changes of these probabilities with a coordinate change in the positions of *each* of the molecules of the group obey the system of BBGKY equations.

Equations (17.14) are one of the specific methods for locating molecules that obey the integral relations (17.13). The procedure for enumerating molecular arrangements in the regions ω and $(\omega_R - \omega)$ can be organized arbitrarily. This question is discussed in [64, 68] by dividing the volume of the system into a fine grid, smaller than the diameter of the molecules. It is useful to note that for the first time the idea of introducing a shallow 'grid' in the description of continual quantities was introduced by Boltzmann [69] in calculating the entropy for the velocity distribution of molecules in an ideal gas, which led to the well-known results on the increase of entropy under equilibrium conditions (see also [70]).

In the case of a fine grid, the difference system of equations (17.4) for discrete distribution functions gives a description of the structure of the liquid state, as in theories of integro-differential or integral liquid equations. This explains that in [71,72] a procedure was given for constructing the average values of the effective parameters of the interparticle interaction with respect to the continual potential curves in the mean-field approximation. This procedure, which reveals the meaning of effective parameters, retains its value even with the new procedure for constructing successive approximations for different lattice parameters λ_n. Lattice fluid models have a rigorous statistical justification and are directly related to the functions of molecular distributions, which are described by the BBGKY equations. Traditional lattice models are the first step in the sequence of partitioning of space into elementary volumes, which in the limit approach a continual description. This procedure provides an arbitrarily close approach to the continual description.

The first approximation is sufficient to obtain the thermodynamic properties of simple liquids and their mixtures, but in order to describe the structural characteristics, subsequent approximations are required that allow us to take into account not only the average density of the system but also the inhomogeneities of the local density. Approximation to the continual description can be achieved both by increasing the number of places occupied by a molecule in the lattice, which is equivalent to a finer division into cells for a molecule of a fixed size, and also taking into account the displacement of the centre of mass of molecules from the centre of the cell. For lattice systems with molecules that block several cells, the general procedure for their statistical analysis based on the cluster approach is given in [65,73]. The breakdown of the volume into cells in the LGM does not impose any restrictions on the motion of the particles; the latter can move around the entire volume. They take into account not only the 'standard' positions of the molecules in the centre of the cell, but also the displacements of the particles from the indicated positions, which allows obtaining the structural characteristics of the liquid.

The static justification of the LGM is key to the possibility of a unified description of all three phases. The initial setting of the LGM strictly bound to the crystalline lattice of a solid limited its use to solids and their surfaces. The removal of this restriction provides a general approach for liquid and vapour, as well as for all their interfaces.

18. Connection between thermodynamic functions and correlation functions

A system of equations for the pair distribution functions in the QCA, in which the pair functions are related to each other as

$$^*\theta_{fg}^{AA}(r)^*\theta_{fg}^{VV}(r) = {}^*\theta_{fg}^{AV}(r)^*\theta_{fg}^{VA}(r), \quad {}^*\theta_{fg}^{ij}(r) = \theta_{fg}^{ij}(r)\exp[-\beta\varepsilon_{fg}^{ij}(r)]. \qquad (18.1)$$

and this system allows us to express the thermodynamic functions $\sum_{\lambda=1}^{\Phi}\theta_{fg}^{i\lambda}(r)=\theta_f^i$, $\sum_{l=1}^{\Phi}\theta_{fg}^{i\lambda}(r)=\theta_g^{\lambda}$ for a uniform volume with allowance for lateral interactions inside R_{lat} c.s. in the traditional form for a discrete lattice.

The summary energy of a one-component system is written as the sum of the contributions $F = F_{lat} + F_{vib} + F_{tr}$ from the lattice system $F_{lat} = E_{lat} - TS_{lat}$ and the vibrational (F_{vib}) and translational (F_{tr}) motions of the particles. In particular, $E_{lat} = -\langle H\rangle$ – internal energy, and S_{lat} – entropy, are written as

$$E_{lat}=-\sum_f \theta_f^A v_f^A+\frac{1}{2}\sum_r\sum_{f,g}\varepsilon_{fg}^{AA}(r)\theta_{fg}^{AA}(r) \tag{18.2}$$

$$S_{lat}/k_B=\sum_{f,i=A,V}\theta_f^i\ln\theta_f^i+\frac{1}{2}\sum_r\sum_{f,g}\sum_{i,j=A,V}\left[\theta_{fg}^{ij}(r)\ln\theta_{fg}^{ij}(r)-\theta_f^i\theta_g^j\ln\theta_f^i\theta_g^j\right] \tag{18.3}$$

In the expression for E_{lat}, there are only functions θ_f^A and $\theta_{fg}^{AA}(r)$ for particles A, whereas in the expression for S_{lat} there are all sorts of occupation states of sites (particles and vacancies). The expressions for F_{vib} and F_{tr} are concretized in the approximations used [31].

The isothermal relationship between the pressure in the thermostat and the density inside the system (the expression for the chemical potential) has the form

$$\begin{gathered}\theta_f^s=\theta_f^i\exp\beta(v_f^i-v_f^s)\prod_{r=1}^{R_{lat}}\prod_{g\in z_f(r)}S_{fg}^i(r),\ \sum_{j=1}^{s}\theta_f^j=1,\\ S_{fg}^i(r)=1+\sum_{j=1}^{s-1}t_{fg}^{ij}(r)x_{fg}^{ij}(r),\ x_{fg}^{ij}(r)=\exp[-\beta\varepsilon_{fg}^{ij}(r)]-1.\end{gathered} \tag{18.4}$$

The equation of state (pressure inside the system)

$$\begin{gathered}P=\frac{kT}{\mathrm{v}_0}\theta-\frac{1}{2d\mathrm{v}_0}\sum_{\chi=1}^{R_{lat}}\sum_{f,g}\sum_{i,j=1}^{s-1}\int_{\mathrm{v}(f)}\int_{\mathrm{v}(g)}\theta_{ij}(r_{fg}\,|\,\chi)r_{ij}\times\\ \times(\partial\varepsilon_{ij}(r_{fg}\,|\,\chi)/\partial r_{ij})dr_idr_j,\end{gathered} \tag{18.5}$$

where d is the dimension of the lattice.

Here it is assumed that the internal degrees of freedom (translational and oscillatory motion) are separated from the configuration states. These expressions are common for three aggregate states of matter. Their generalization in the framework of a discrete–continual description is given in [73, 74].

The essence of the QCA was discussed in Section 16. In the existence of two phases, it leads to the so-called van der Waals loop, and to the necessity of using the Maxwell rule to determine the densities of coexisting phases of gas and liquid [9,11,27]. The meaning of Maxwell's construction and the essence of metastable states that are inside the stratification curve (or the binodal curve) in the statistical theory is provided by the Yang–Lee condensation theory [9,11,105,106].

The Yang–Lee condensation theory and isotherm loops. The Yang–Lee theory [105,106] is based on the study of the number of roots of a large partition function of the system

$$\Xi(z,V)=\sum\nolimits_{N=0}^{\infty}z^N Q_N(V),\ \text{where}\ Q(V)=\frac{1}{N!\lambda^{3N}}\int d^{3N}r\exp[-\beta E(r_1,\dots,r_N)],$$

$$E(r_1,\dots,r_N)=\sum_{i<j}u(r_i,r_j),\quad u(r_i,r_j)=u(|r_i-r_j|),$$

for a fairly general sort of potential short-range interactions between molecules. Let the interaction potential be given in the form of a solid sphere ($u(r) = \infty$, $r \le a$) with an attractive potential of radius r_0 and maximum depth $-\varepsilon$($u(r) = -\varepsilon$, $a< r <r_0$), and $u(r) = 0$, $r \ge r_0$. The equation of state is obtained by eliminating the value of z from the two parametric equations $\beta P=\frac{1}{V}\ln\Xi(z,V)$, and $\frac{1}{v}=\frac{1}{V}z\frac{\partial}{\partial z}\ln\Xi(z,V)$, where the quantity v is the specific volume, i.e. a parameter that does not depend on the total volume V; z is activity, defined as $z = \exp(\beta\mu)/\Lambda^3$, μ – chemical potential, $\Lambda^2 = h^2/(2\pi mkT)$, h – Planck's constant.

The idea of the theory is to investigate the singularities of the behaviour of the limiting relations $\beta P=\lim_{V\to\infty}\left[\frac{1}{V}\ln\Xi(z,V)\right]$ and $\frac{1}{v}=\lim_{V\to\infty}\left[\frac{1}{V}z\frac{\partial}{\partial z}\ln\Xi(z,V)\right]$, without making an explicit calculation of $P(v)$, and in the limiting case $V\to\infty$, then the detection of the singularity of the equation of state can be interpreted as a phase transition.

For small particle numbers N, the large partition function $\Xi(z,V)=1+zQ_1(V)+z^2Q_2(V)+\dots+z^{N_m}Q_{N_m}(V)$ is a polynomial of degree N with positive coefficients (here N_m is the maximum number of spheres that can be placed inside a given volume V), and this polynomial can not have real roots. Such roots can appear only as a result of the limiting transition $N\to\infty$, under the condition $V\to\infty$, so that N/V = const. The passage to the limit $V\to\infty$ must be understood in a strict mathematical sense. In particular, the operations of passing to the limit *lim* with $V \to \infty$ and taking the derivatives $z(d/dz)$ may not be permutable.

The Yang–Lee theory showed that there really are situations when such singularities exist [75]. In the particular case of the two-dimensional Ising model, for which the partition function can be calculated accurately, the correctness of this study [76], independent of a particular type of potential, was shown, and not related to the specific type of closure of higher DFs (distribution functions) through the lower ones. The phase transition is completely characterized by the distribution function of the roots. The Ising model undergoes a first-order phase transition, and mathematical analysis provides answers to many questions of the theory of phase transitions. Omitting the proof of this theory, we formulate its main corollaries for first-order phase transitions.

This theory justified the correct application of Maxwell's rule, to search for coexisting densities and saturated vapour pressure using the secant so-called van der Waals loop, which is obtained in all approximate methods of calculating the equations of state (and / or isotherms). This proof allows completely to refuse any information about the properties of the approximate approach used, since in it there are no intermediate points relating to the two-phase region of the system or points within the stratification curve (or binodal). Note that the Maxwell rule itself uses the isotherms of approximate equations, so it was important to show that the properties of the limit points themselves – the coexisting vapour and liquid densities – do not depend on the type of approximate isotherms (although the type of approximation used determines their concrete position, as well as the saturated vapour pressure P_s).

Figure 18.1 *a* shows the curve (1) of the isothermal relationship between the density and the chemical potential, calculated by the approximate method (according to equation (18.4)), at which a van der Waals type loop appears. The position of the secant Maxwell, determined from the condition of equality of areas above and below the secant, gives the saturated vapour pressure P_s and the density of the coexisting phases of the fluid θ_f and the pair θ_v.

We recall that the Maxwell rule can be applied in the 'pressure-volume' coordinates for a fixed number of molecules, i.e. 'pressure-specific volume', or in the coordinates 'chemical potential–density' (Here, numerical densities are used instead of the usual specific volumes, defined as $v = 1/\theta$ [11]). But Maxwell's rule does not hold in the coordinates 'pressure–density' and 'chemical potential–specific volume' coordinates [9,77]. Figure 18.1 *b* shows shaded areas on an isotherm loop of equal area; v^+ corresponds to the specific volume of the liquid phase, v^- to the gaseous phase (the vertical curve $v = 1$ corresponds to the limiting value of volume on a rigid lattice in the vapour–liquid system, the value of v for a compressible lattice may be less than unity [31]).

The results of the Yang–Lee theory are obtained on a complete set of molecular configurations, which rejects any treatment of the so-called metastable states as equilibrium states. They indicate that the binodal curve is a set of singular points for which one-sided limits are defined in the direction of the nearest point of the binodal, but there are no limits for any directions from the interior points to the binodal. Or, that only two-phase solutions correspond to strictly equilibrium solutions below the critical point. Only such solutions satisfy the thermodynamic conditions for the phase equilibrium of the phases α and γ: $T_\alpha = T_\gamma$, $P_\alpha = P_\gamma$, $\mu^i_\alpha = \mu^i_\gamma$. All other solutions within the region of two-phase solutions are not equilibrium. Therefore, this theorem uniquely eliminates any metastable state from the number of equilibrium states.

The formal continuation of single-phase solutions to the two-phase region is possible for any phases. If the phases have one symmetry, as in the case of vapour and liquid, then the continuation of the solutions leads to the van der Waals loop, which is the consequence of using the canonical distribution. This distribution follows from the full statistical sum, if only its maximum term is used, instead of completely taking into account all the terms. It is this simplification, which is purely technical in the calculation of isotherms, that is responsible for the appearance of loops. When going over to the full spectrum of particle configurations, this loop disappears even in approximate approaches [9], which completely agrees with the result of Li-Yang's work [76], since all the results in the LGM (in this book) belong to the class of quasi-Ising models for which it was performed, including, if we use the most rough approximation of correlation effects, the mean-field approximation [9]. If crystals with different symmetries participate in the phase transition, as is the case in solid-state transitions of the second kind, then the equations of state and expressions for the thermodynamic functions for each phase are different, and as stated above, the Maxwell rule is not applicable. A similar situation exists for the solid–liquid transition [12, 78].

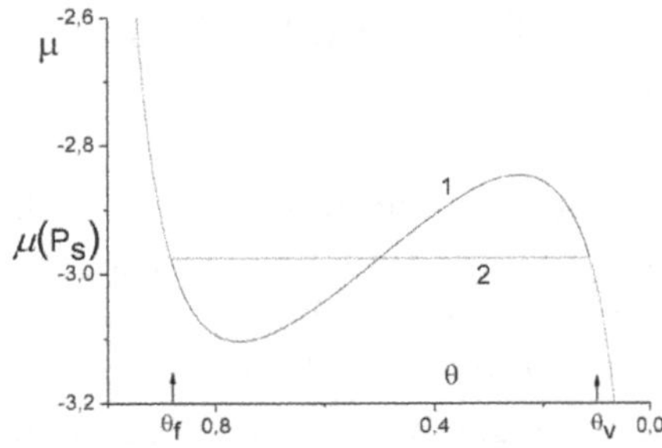

Fig. 18.1a. Isotherm with van der Waals loop (1) and Maxwell cross section (2) in the 'chemical potential–density' coordinates.

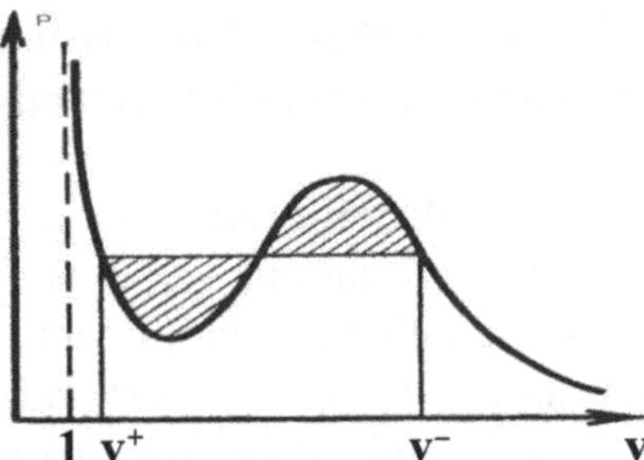

Fig. 18.1b, Isotherm with van der Waals loop and Maxwell cross-section in the 'pressure–specific volume' coordinates.

At the same time, the Yang–Lee theory also answers the question about the nature of metastable states: the existence of any metastable state under experimental conditions is a consequence of natural inhibitions that do not allow the realization of the *complete spectrum of molecular configurations* and, accordingly, the equilibrium of the system.

The exact result of the Yang–Lee theory is important as a result obtained without introducing approximate connections between the distribution functions of all dimensions and preserving exact relations between them. If there are changed relationships between the weights of different configurations, then their change is possible only in time, and this change must be described only by kinetic equations. That is why the exclusion of metastable states from equilibrium concepts opens a real opportunity to apply kinetic approaches and to select the real dynamic variables for which the kinetic equations are constructed. In the absence of dynamic variables, kinetic equations can not be formulated.

In the general case for solid-phase systems in the absence of equilibrium, the number of non-equilibrium states is not limited by anything – it can be anything, because is determined by the mechanical stability and mechanism of the process and the course of evolution of the system starting with some initial conditions. The absence of complete mixing in solids leads to the realization of many intermediate states with different in duration characteristic life times, which are usually called metastable. The evolution of such 'frozen' states should be described only by kinetic equations. This situation is reflected in various kinds of relaxation theories that extend the thermodynamic approach to weakly non-equilibrium states [79–83].

19. Fundamentals of the calculation of non-equilibrium functions of molecular distributions

Continuous distribution. In a macroscopic volume of a gas or liquid, the Liouville equations serve as the starting position for constructing transport equations, which operate with the total distribution function of the molecules of the system. Six variables describing its spatial coordinates and velocities are used to describe the state of each molecule [6,15,21,22,84–86]. In describing the dynamics of the system, we must pass to the generalized chain of coupled Bogolyubov equations [6] for time (non-equilibrium) distribution functions. They are written as follows: the expression for the function $\theta_{(s)}$ can be written as

$$L^0_{x_1,\dots,x_s}\theta_{(s)} - \sum_{1\le i\le s} B_{ij}\theta_{(s)} = n\sum_{1\le i\le s}\int B_{i,s+1}\theta_{(s+1)}dx_{s+1}, \tag{19.1}$$

where: t is the time $L^0_{x_1,\dots,x_s} = \frac{\partial}{\partial t} + \sum_{1\le i\le s}(\mathbf{v}_i\frac{\partial}{\partial r_i} + F_0\frac{\partial}{\partial p_i})$, $F_4 = -\partial u(\mathbf{r}_i)\,/\,\partial \mathbf{r}_i$ is the external force, $u(\mathbf{r}_i)$ is the potential energy of the molecule i in the external field (Section 13), $x_i = (\mathbf{r}_i, \mathbf{v}_i)$, $\mathbf{v}_i$ is the velocity of the molecule i, $\mathbf{p}_i$ is its momentum; $B_{ij} = \frac{\partial E_{ij}}{\partial r_i}\frac{\partial}{\partial p_i} + \frac{\partial E_{ij}}{\partial r_j}\frac{\partial}{\partial p_j}$, E_{ij} is the intermolecular interaction potential between molecules i and j (13.1).

Because of the extreme complexity of the chain of equations for the sequence of distribution functions $\theta_{(1)}$, $\theta_{(2)}$, ... it is natural to seek approximate closed equations for the simplest distribution functions, which is achieved by a more crude description of processes in the system under consideration.

Discrete description of distributions. The kinetic equation for the non-equilibrium discrete distribution functions is constructed, as for the continual description, in terms of the reduced functions $\theta_{(s)}(\{i,f,\mathbf{r}^i_f,\mathbf{v}^i_f\},t)$, here s is the order of the distribution function, $1 \le s \le N$. The state of the system at time t is described by the total distribution function $\theta_{(N)}(\{i,f,\mathbf{r}^i_f,\mathbf{v}^i_f\},t)$, which characterizes the probability of being in the cell with the number f, $1 \le f \le N$ (the complete list of cells is denoted by curly brackets { }), particles of sort i (the state of occupation of the site f), at the point with the coordinate $\mathbf{r}^i_f$, and having the velocity $\mathbf{v}^i_f$. This designation corresponds to the total distribution function $P(\{\gamma_f^i\},t)$ (16.6), for which the values of the velocities $\mathbf{v}_f^i$ were not determined.

The local distribution functions in the space $\theta_{(N)}(\{i,f,\mathbf{r}_f^i,\mathbf{v}_f^i\},t)$ used here differ from the continual distribution functions by their normalization to the cell volume (rather than the volume of the system). They are not defined outside the cells under consideration: $\theta_{(N)}\left(\{i,f,\mathbf{r}_f^{\ i},\mathbf{v}_f^{\ i}\},t\right)_{r_{f,\alpha}\notin\omega_f}=0$, where $1 \le f \le N$, $\alpha = x, y, z$; $\mathbf{r}_f \in \omega_f$ (here the symbol ω_f is used for the volume of the cell with the number f instead of $\mathbf{v}_f$, in order to distinguish this symbol from the velocity of the particle). The boundary conditions in the velocity space coincide with the usual conditions: these functions vanish when the velocity modulus tends to infinity $\theta_{(N)}\left(\{i,f,\mathbf{r}_f^i,\mathbf{v}_f^i\},t\right)_{v_{f,\alpha}=\pm\infty}=0$.

We will use QCA accounting for intermolecular interactions, which allows the probability of any configuration of molecules, i.e. $\theta_{(1)}(\mathbf{r}_f^i,\mathbf{v}_f^i,t)$ and paired $\theta_{(2)}(\mathbf{r}_f^i,\mathbf{v}_f^i,\mathbf{r}_g^j,\mathbf{v}_g^j,t)$ distribution functions, $1 \le f,\ g \le N$, in the following form [115–117]

$$\theta_{(N)}\left(\{i,f,\mathbf{r}_f^{\ i},\mathbf{v}_f^{\ i}\},t\right)=\prod_{f=1}^{N}\theta_{(1)}(\mathbf{r}_f^{\ i},\mathbf{v}_f^{\ i},t)\times \times\prod_{g\in z_f}\left\{\xi_{fg}^{ij}\left(\mathbf{r}_f^{\ i},\mathbf{v}_f^{\ i},\mathbf{r}_g^{\ j},\mathbf{v}_g^{\ j}t\right)\right\}^{1/2} \tag{19.2}$$

where the exponent 1/2 takes into account that the pairs of cells are listed twice; the index g runs through all z neighbours of the site f; $\xi_{fg}^{ij}(\mathbf{r}_f^i,\mathbf{v}_f^i,\mathbf{r}_g^j,\mathbf{v}_g^j,t)=\theta_{(2)}(\mathbf{r}_f^i,\mathbf{v}_f^i,\mathbf{r}_g^j,\mathbf{v}_g^j,t)/[\theta_{(1)}(\mathbf{r}_f^i,\mathbf{v}_f^i,t)\theta_{(1)}(\mathbf{r}_g^j,\mathbf{v}_g^j,t)]$ is a pair correlation function (we shall omit the argument of time t for simplicity below).

Kinetic equations are constructed for the complete set of these local distribution functions. Taking into account the normalization relations for a single-component substance, it is sufficient to construct the kinetic equations for the local unary $\theta_{(1)}(x_f)$ and the pair $\theta_{(2)}(x_f, x_g)$ distribution functions, which have the 'usual' form [6,15,21,22, 84–86].

The kinetic equations for unary functions are written as [87–89]

$$\left(\frac{\partial}{\partial t}+\mathbf{v}_f\frac{\partial}{\partial \mathbf{r}_f}+\frac{F(f)}{m}\frac{\partial}{\partial v_f}-\sum_h\int\frac{\partial\varepsilon_{fh}}{\partial \mathbf{r}_f}\frac{\partial t_{fh}(x_f,x_h)}{m\partial \mathbf{v}_f}dx_h\right)\theta_{(1)}(x_f)= \\ =I=\int\frac{\partial\varepsilon_{fg}}{\partial r_f}\frac{\partial\theta_{(2)}\left(x_f,x_g\right)}{m\partial v_f}dx_g \qquad (19.3)$$

where ε_{fg} is the potential interaction function of molecules in cells f and g; m is the mass of the molecule; $\mathbf{F}(f)$ is the vector of the external conservative force in the cell f (in the narrow pores, the main contribution is made by the wall potential, and the gravitational field can be neglected). Here we use the numerical density θ instead of the traditional mass density ρ ($\rho = m\theta/v_0$).

The sum over h is taken over all neighbours of the site f, it describes the terms created by the interactions of neighbouring molecules in the neighbourhood of the site f with which the molecule at the site f does not collide (the so-called Vlasov contributions). The magnitude of this time interval is determined by the time variation in the left-hand side of the kinetic equation in the derivative $\partial/\partial t$. In their structure, these terms are completely similar to the collision integral, which is on the right. The intermolecular interaction potential ε_{fg} corresponds to the molecular arrangements at the sites h at a distance up to R_{lat} c.s. For rarefied gases this term is absent, then the formula (19.3) goes into the Boltzmann equation. The presence of neighbours sharply complicates the form of the kinetic equation and requires knowledge of the pair distribution functions. Usually in the theory of a solid and plasma [19, 86] these terms are considered in the mean-field approximation, in which the closure occurs at the level of unary distribution functions, i.e. Instead of functions $\partial\theta_{(2)}(x_f,x_g)/\partial\mathbf{v}_f$, derivatives $\partial[\theta_{(1)}(x_f)\,\theta_{(1)}(x_g)]/\partial\mathbf{v}_f$ are considered. If we neglect the contribution of the collision integral I, we obtain the Vlasov equation [15, 90]. In this case, the short-range LJ potential is considered for the vapour–liquid system and therefore it is necessary to preserve both types of terms. As the density increases, the role of Vlasov's contributions increases. However, as shown in Chapter 6, unary distribution functions do not provide a self-consistent description of systems with a wide range of density variations, so it is necessary to preserve the pair distribution functions.

The kinetic equations for the pair distribution functions are written out in a similar manner [87–89]

$$\left(\frac{\partial}{\partial t}+\mathbf{v}_f\frac{\partial}{\partial q_f}+\frac{F(f)}{m}\frac{\partial}{\partial \mathbf{v}_f}+\mathbf{v}_g\frac{\partial}{\partial \mathbf{r}_g}+\frac{F(g)}{m}\frac{\partial}{\partial \mathbf{v}_g}-\frac{\partial \varepsilon_{fg}}{\partial r_f}\frac{\partial}{m\partial v_f}-\frac{\partial \varepsilon_{fg}}{\partial r_g}\frac{\partial}{m\partial v_g}\right)\theta_{(2)}(x_f,x_g)-$$
$$-\sum_{\xi}\int\left\{\frac{\partial \varepsilon_{f\xi}}{\partial \mathbf{r}_f}\frac{\partial}{m\partial \mathbf{v}_f}+\frac{\partial \varepsilon_{g\xi}}{\partial \mathbf{r}_g}\frac{\partial}{m\partial \mathbf{v}_g}\right\}\theta_{(3)}\left(x_f,x_g,x_\xi\right)dx_\xi = \tag{19.4}$$
$$=\int\left\{\frac{\partial \varepsilon_{fh}}{\partial \mathbf{r}_f}\frac{\partial}{m\partial \mathbf{v}_f}+\frac{\partial \varepsilon_{gh}}{\partial \mathbf{r}_g}\frac{\partial}{m\partial \mathbf{v}_g}\right\}\theta_{(3)}\left(x_f,x_g,x_h\right)dx_h$$

The potential function ε_{fg} refers to a pair of molecules considered at the sites f and g. At the sites ξ there are neighbouring molecules that interact (simultaneously or separately depending on the distance) with molecules at the sites of f and g, but do not collide with them in the considered time range. As above, as the density of molecules increases, the role of Vlasov's contributions increases. The system of equations (19.3) and (19.4) is closed with the aid of superposition or QCA at the level of paired functions and describes the dynamics of the non-uniform system.

20. Kinetic equations in dense phases

In dense phases, the main role is played by interparticle interactions, and the contribution of translational motion is small. Under these conditions, the kinetic processes of migration and chemical transformations of particles in space are described with the help of the basic kinetic equation (Master Equation) for the total distribution function $P(\{\gamma_f^i\},\tau)$ (for spin systems these are the so-called Glauber type equations [91, 92]. They are obtained under the condition that the contribution of the translational motion of particles is small: migration in a solid, in a liquid at rest, on surfaces of solid and liquid phases, and also in the absence of the effect of translational motion of particles on the rate of their reactions (local equilibrium conditions).

These equations are simpler than the equations of the previous section. Movements of molecules in space are carried out through elementary particle jumps to neighbouring free sites (vacancies) – the so-called migration stage in a multistage description of physicochemical processes. To calculate the thermal migration rate of molecules, a transition state model is used that treats displacements as an activation process to overcome a barrier created by neighbouring particles. This model was proposed for gases by Eyring [93, 94] and later transferred to condensed phases in [27,95–105].

At the present time, the kinetic theory at the atomic–molecular level within the framework of the LGM [27,100–105] can be used practically throughout the entire time range, starting from the characteristic times of atomic vibrations to macroscopic ones, including the times of reaching equilibrium states. The theory considers the complete set of elementary processes of molecular displacements and their chemical transformations taking place in a system of non-equivalent lattice sites. To construct the general structure of the kinetic equations of the lattice model, we shall assume that the lattice sites are non-equivalent. The nature of the heterogeneity of the lattice sites is assumed to be known and unchanged in time. From a physical point of view, the non-uniformities in the distribution of particles are caused both by interactions between the particles of the system itself and by the possible additional influence of external fields or interactions (for example, the field of the substrate potential). This formulation of the problem makes it possible from a unified point of view to cover a wide range of questions (see Section 1) related to the spatial distribution of particles and taking into account chemical transformations in the course of chemical reactions.

According to Section 16, $\{\gamma_f^i\} = \gamma_1^i,\ \gamma_2^i, \ldots, \gamma_N^n$ is the complete collection (or a complete list) of the values of the values γ_f^i of all lattice sites that uniquely determine the complete configuration of particle arrangements on the lattice at time τ. For brevity, we denote this state as $\{\mathrm{I}\} \equiv \{\gamma_f^i\}$. Let the general process under consideration consist of a set of stages and let α denote the step number of the elementary process. The basic kinetic equation (Master Equation) for the evolution of the total distribution function of a system in the state {I}, due to the realization of elementary processes α in condensed phases, has the form

$$\frac{d}{d\tau} P(\{\mathrm{I}\},\tau) = \sum_{\alpha,\{\mathrm{II}\}} \Big[W_\alpha(\{\mathrm{II}\} \to \{\mathrm{I}\}) \times \\ \times P(\{\mathrm{II}\},\tau) - W_\alpha(\{\mathrm{I}\} \to \{\mathrm{II}\}) P(\{\mathrm{I}\},\tau) \Big], \tag{20.1}$$

where $W_\alpha(\{\mathrm{I}\} \to \{\mathrm{II}\})$ is the probability of the realization of the elementary process α (the probability of a transition through the channel α), as a result of which at the time moment τ the system from the initial state {I} goes to the final state {II}. In formula (20.1), the sum is taken over various types of direct processes (index α) and over all inverse processes {II}, in which the state of occupation of each of the sites of the system changes.

If the elementary process runs on the same site, the lists of occupation states of the sites of the system {I} and {II} differ only for this site. One-site processes are processes associated with a change in the internal degrees of freedom of a particle, with adsorption and desorption of undissociated molecules, with a reaction by a collision mechanism. If the elementary process proceeds at two neighbouring lattice sites, then the lists of states {I} and {II} differ by the occupation states of these two sites. Two-site processes – exchange reactions, adsorption and desorption of dissociating molecules, migration processes by the vacancy and exchange mechanisms, etc. The sum over the states {II} corresponds to a change in the occupation states of all lattice sites. The interrelation of the states {I} and {II} depends on the mechanism of the process, which determines the set of elementary stages α.

Equation (20.1) is written in the Markov approximation, for which it is assumed that the relaxation processes of the internal degrees of freedom of all particles proceed faster than the processes of changing the states of occupation of different sites of the lattice system.

The transition probabilities W_α are subject to the condition of detailed balancing

$$\begin{aligned} &W_\alpha(\{\mathrm{I}\}\rightarrow\{\mathrm{II}\})\exp(-\beta H(\{\mathrm{I}\}\rightarrow\{\mathrm{II}\}))= \\ &=W_\alpha(\{\mathrm{II}\}\rightarrow\{\mathrm{I}\})\exp(-\beta H(\{\mathrm{II}\})), \end{aligned} \tag{20.2}$$

where $H(\{\mathrm{I}\})$ is the total energy of the lattice system in the state {I}. In the equilibrium state, $P\left(\{\gamma_f^i\},\tau\rightarrow\infty\right)=\exp\left(-\beta H\left(\{\gamma_f^i\}\right)\right)/Q$ where Q is the partition function of the system, the system (20.1) in QCA transforms to equations (18.4).

Expressions for $W_\alpha(\{\mathrm{I}\}\rightarrow\{\mathrm{II}\})$ are constructed taking into account all molecular features of the system: 1) each site f is characterized by a definite set of number of particle sort s_f that can be in it; 2) the internal degrees of freedom F_j^i of a particle of some sort i depend on the number of the lattice site; 3) the interaction parameters $\varepsilon_{fg}^{ij}(r)$ of the particles i and j, located at the sites with the numbers f and g at a distance r from each other depend on the site numbers.

The large dimensionality of the system (20.1) does not allow us to use it to study the dynamics of macroscopic systems by direct integration; therefore, kinetic equations are constructed with respect to the functions of lower-order distributions through which higher-order distribution functions are closed. To this end, instead of the total distribution function $P(\{\gamma_f^i\},\tau)$, the abbreviated method of its assignment through time distribution functions (correlators), defined in (16.2), is used to describe the evolution of the system,

The introduced local time functions (16.2) imply that at each instant of time averaging is performed over the complete ensemble of copies of the non-uniform lattice system under consideration over all of its realizable states. This definition is analogous to the determination of the averages in the equilibrium Gibbs statistical theory and the non-equilibrium theory of gases and liquids. The difference from the non-equilibrium theory of gases and liquids is that the local inhomogeneities under consideration are realized on a microscopic atomic scale (instead of small elementary volumes containing a macroscopic amount of matter in the theory of gases and liquids).

The kinetic equations for the correlators are obtained by multiplying expression (20.1) by $\prod_{n=1}^{m}\gamma_{f_n}^{i_n}$ and averaging over all states of the system. The sorts of particles at the sites f_n, leading to nonzero contributions to the change in the correlators (16.2), depend on the direction of the elementary process and are determined by the initial {I} or final {II} states. This leads to the following kinetic equations:

$$\frac{d}{d\tau}\theta_{f_1\ldots f_m}^{i_1\ldots i_m}(\tau)=$$
$$\sum_{\alpha,\{II\}}\left[\left\langle\prod_{n=1}^{m}\gamma_{f_n}^{i_n^*}W_\alpha(\{\mathrm{II}\}\to\{\mathrm{I}\})\right\rangle-\left\langle\prod_{n=1}^{m}\gamma_{f_n}^{i_n}W_\alpha(\{I\}\to\{II\})\right\rangle\right], \tag{20.3}$$

where the sorts of particles i_n^* in the first term on the right correspond to the states of occupation of the lattice sites in the state {II}. Equations (20.3) are the starting points for obtaining the kinetic equations of the processes occurring in the condensed phases.

The construction of the kinetic equation for the local concentration of molecules ($m = 1$) leads to a large increase in the order of the unknown correlation functions in the right-hand side of the equation. This dimension is equal to the number of all neighbouring molecules that affect the rate of this stage. We recall that in the BBGKY kinetic chain the dimension of the correlators on the right-hand side of the kinetic equations is successively increased by one. This structure of the kinetic equations of the LGM differs from the traditional form of the BBGKY kinetic chains, and accordingly the kinetic theory raises the problem of calculating high-dimensional correlators. In a discrete version of the theory based on the LGM, the system of kinetic equations also has a high dimensionality and, in practice, tends to be limited to a minimal dimension that preserves the effects of correlation between molecules.

The necessity of considering high-order correlation functions is due to the fact that the interaction of activated complexes with surrounding particles not only quantitatively changes the dynamic characteristics of the transient regimes, but can qualitatively change the evolution of the system, for example, the number of stationary states of the system increases [106]. In addition, it was shown [107] that the type of transition probabilities affect the calculated values of the critical dynamic index for the correlation length, which contradicts the dynamic hypothesis of universality [108, 109]. A consistent introduction of the theory of absolute reaction rates into the theory of kinetic equations for an arbitrary character of particle mobility is given in [27,100–105] when considering processes on smooth and rough surfaces.

A closed system of equations for the first $\left(\theta_f^i = \left\langle \gamma_f^i \right\rangle\right)$ and the second $\left(\theta_{fg}^{ij}(r) = \left\langle \gamma_f^i \gamma_g^i \right\rangle\right)$ correlators in general form is written as

$$\frac{d}{dt}\theta_f^i = I_f^i = \sum_\alpha \left[U_f^b(\alpha) - U_f^i(\alpha)\right] + \sum_r \sum_h \sum_j \sum_\alpha \left[U_{fh}^{bd}(r|\alpha) - U_{fh}^{ij}(r|\alpha)\right] \tag{20.4}$$

$$\begin{aligned} &\frac{d}{dt}\theta_{fg}^{ij}(r) = I_{fg}^{ij}(r) = \sum_\alpha \left[U_{fg}^{bd}(r|\alpha) - U_{fg}^{ij}(r|\alpha)\right] + P_{fg}^{ij}(r) + P_{gf}^{ji}(r) \\ &P_{fg}^{ij}(r) = \sum_\alpha \left[U_{fg}^{(b)j}(r|\alpha) - U_{fg}^{(i)j}(r|\alpha)\right] + \\ &\sum_h \sum_m \sum_\alpha \left[U_{hfg}^{(cb)j}(r|\alpha) - U_{hfg}^{(mi)j}(r|\alpha)\right], \end{aligned} \tag{20.5}$$

where $U_f^i(\alpha)$ are the rates of elementary single-site processes $i \leftrightarrow b$ (here $h \in z_f$), $U_{fg}^{ij}(r|\alpha)$ are the elementary two-site processes $i + j_\alpha \leftrightarrow b + d_\alpha$ $(h \in z(r))$ at a distance r; the second term in $P_{fg}^{ij}(r)$

describes the stage $i + m \leftrightarrow b + c$ at neighbouring sites f and h at a distance r. All the velocities of the elementary stages $U_f^i(\alpha)$ and $U_{fg}^{ij}(r|\alpha)$ are calculated within the framework of the theory of absolute reaction rates for non-ideal reaction systems written out in the QCA for accounting for interparticle interaction (see below in Sections 49 and 51).

Equations (20.4) and (20.5) satisfy the normalizing relations (16.3), which are satisfied at any time.

Of great importance is the question of the separation of dynamic variables describing the state of molecules into fast and slow ones. This question is solved depending on the chosen characteristic time scale determined by the values of $W_\alpha(\{I\}\rightarrow\{II\})$. Slow variables are described by kinetic equations, and fast variables are described by algebraic equations, since they usually refer to particles having an equilibrium distribution. In the general case, fast particles (described by fast variables) form a tuning subsystem that exerts its influence on the energy of slow elementary processes in kinetic equations. In turn, slow particles (described by slow variables) determine the character of the distribution of fast particles. They form a spatial region in which fast processes are realized (excluding their common spaces, regions occupied by slow particles), and influence their potential on the character of the distribution of fast particles.

Usually, the spatial coordinates of molecules or quantum numbers related to their electron terms are usually referred to slow variables, and to quantum numbers referring to the rotational and vibrational states of molecules [109]. The latter does not exclude the possibility of considering the population of vibrational levels of molecules in the gaseous phase as slow variables. On the other hand, the stages of migration of molecules can also be fast. For example, at a low activation energy compared to the desorption energy, surface migration is a fast step compared to the desorption stage.

These kinetic equations are used in Chapter 5 to construct a local equilibrium criterion, and in Chapter 6 to discuss the issue of self-consistency of equilibrium and kinetics. Chapter 7 discusses the use of thermodynamic interpretations in kinetics.

References

1. Gibbs J.W., Elementary principles in statistical mechanics, developed with especial references to the rational foundations. New York, 1902.
2. 2. Gibbs J.W., Thermodynamics. Statistical mechanics. Moscow, Nauka, 1982. 584 .
3. 3. Mayer J.E., Goeppert-Mayer, M., Statistical mechanics. Moscow, Mir, 1980. [Wi-

ley, N.Y.-Sydney-Toronto, 1977]
4. Fowler R.H.. Statistical Mechanics. Cambridge. Cambridge Univer. Press, 1936.
5. Fowler R.H., Guggenheim E.A.. Statistical Thermodynamics. Cambridge. Cam¬bridge Univer. Press, 1939.
6. Bogolyubov N.N.. Problems of dynamic theory in statistical physics. Moscow, Gostekhizdat, 1946. [Interscience, New York, 1962]
7. Sommerfeld A.. Thermodynamics and statistical physics. Moscow, IL, 1955. [Academic, New York, 1964].
8. Leontovich M.A., Introduction to thermodynamics. Statistical physics. Moscow, Nauka, 1983.
9. Hill T.L., Statistical Mechanics. Principles and Selected Applications. – Moscow: Izd. Inostr. lit., 1960. – 486 p. [N.Y.: McGraw–Hill Book Comp. Inc., 1956].
10. Hill T.L., Thermodynamics of Small Systems. Part 1. New York Amsterdam: W. A. Benjamin, Inc., Publ., 1963. Part 2. 1964.
11. Huang K., Statistical mechanics. Moscow, Mir, 1966.
12. Landau L.D., Livshits E.M., Theoretical physics. V. 5. Statistical physics. Moscow, Nauka, 1964.
13. Kubo R. Statistical mechanics. Moscow, Mir, 1967.
14. Uhlenbeck J., Ford J. Lectures on static mechanics. Moscow, Mir, 1965.
15. Vlasov A.A., Statistical functions of distributions. Moscow, Nauka, 1966.
16. Gurov K.P., Foundations of the kinetic theory. Moscow, Nauka, 1967.
17. Fischer I.Z., Statistical theory of liquids. Moscow, Fizmatgiz, 1961.
18. Zubarev D.N., Non-equilibrium statistical thermodynamics. Moscow, Nauka, 1971.
19. Bazarov I.P., Statistical theory of the crystalline state. Moscow, Izd-vo MGU, 1972.
20. Statistical physics and quantum field theory. Ed. N.N. Bogolyubov. Moscow, Nauka, 1973.
21. Balescu, R. Equilibrium and non-equilibrium statistical mechanics. V.2. Moscow, Mir, 1978..
22. Croxton K., Physics of the liquid state. Moscow, Mir, 1979.
23. Bazarov I.P., Nikolaev P.N., Correlation theory of a crystal. Moscow, Publishing House of Moscow State University, 1981.
24. Gurov K.P., Kartashkin B.A., Ugaste Yu.E., Mutual theory in multiphase metal systems. Moscow, Nauka, 1981.
25. 25. Bazarov I.P., Gevorkyan E.V., Statistical theory of solid and liquid crystals, Moscow State University Publishing House, Moscow, 1983.
26. Klimontovich Yu.L., Statistical physics. Moscow, Nauka, 1982.
27. Tovbin Yu.K., Theory of physico-chemical processes at the gas–solid interface, Moscow, Nauka, 1990. [CRC, Boca Raton, Florida, 1991].
28. Martunov G.A., Fundamental Theory of Liquids: Method of Distribution Functions. Bristol, A. Hilger, 1992.
29. 29. Martunov G.A., Classical Statistical Mechanics (Fundamental Theories of Physics, V.89). Dordrecht: Kluwer Acad. Publ., 1997.
30. Zubarev D.N., et al., Static mechanics of non-equilibrium processes, V. 1 and 2, Moscow, Fizmatlit, 2002.
31. Tovbin Yu.K., The Molecular Theory of Adsorption in Porous Solids, Moscow, Fizmatlit, 2012. [CRC Press, Taylor@Francis Group, 2017]
32. Tovbin Yu.K., Zh. fiz. khimii. 1995. V. 69. No. 1. P.118. [Russ. J. Phys. Chem. 1995. V. 69. No. 1. P. 105]
33. Tovbin Yu.K., Zh. fiz. khimii. 1998, V.72. №5, P .775 [Russ. J. Phys. Chem. , 1998 V. 72, No. 5, P. 675]

34. Tovbin Yu.K., Senyavin M.M., Zhidkova L.K. Zh. fiz. khimii.1999, V. 73. № 2. P. 304. [Russ. J. Phys. Chem. 1999. V. 73. № 2. P. 245]
35. Tovbin Yu.K., Zh. fiz. khimii. 2017. 91. No. 3. P. 381. [J. Phys. Chem. A 91, 403 (2017)].
36. Kvasnikov I.A., Thermodynamics and statistical physics. V. 2: The Theory of Equi¬librium Systems: Statistical Physics. Moscow, Editorial URSS, 2002.
37. Frenkel Ya.I., Kinetic Theory of Liquids. Moscow, Academy of Sciences of the USSR, 1945.
38. Fowler R.H., Rushbrooke G.S., Trans. Far. Soc. 1937. V. 33. P. 1272.
39. Lennard-Jones J.E., Devonshire A.F., Proc. Roy. Soc., 1937. V. 163A. P. 53.
40. Lennard-Jones J.E., Devonshire A.F., Proc. Roy. Soc., 1939. V. 169A. P. 317.
41. Collins R., Phase transitions and critical phenomena. Ed. C. Domb, M.S. Green. London – New York, Academic Press, 1972. P. 271.
42. Vortler N.L., et al., Physica, A. 1979. V. 99. P. 217.
43. Langmuir I., J. Amer. Chem. Soc. 1916. V. 38. P. 2217.
44. Hirschfelder J. O., Curtiss C. F., Bird R.B., Molecular theory of gases and liquids. Moscow, IL, 1961. [Wiley, New York, 1954].
45. Runnels L.K., Phase transitions and critica1 phenomena, Ed. C. Domb, MS Green. London–New York, Academic Press, 1972. V. 2. P. 305.
46. O'Reilly D.E., Phys. Rev. A. 1977. V. 15. P. 1198.
47. Shinomoto S.-G., Progr. Theor. Phys. 1983. V. 70. P. 687.
48. Ising E., Zs. Phys. B. 1925. 31. S. 253.
49. Baxter R., Exactly solvable models in statistical mechanics. Moscow, Mir, 1985. [Academic Press, London, 1982].
50. Kac M., Probability and related topics in physical sciences. Moscow, Mir, 1965. [Proceeding of the summer seminar Boulder, Colorado, 1957, Interscience Pub. Ltd. London].
51. Onsager L., Phys. Rev. 1944. V. 65. P. 117.
52. Domb C., Phase transitions and script phenomena, Ed. C. Domb, M.S. Green. London–New York, Academic Press, 1974. V. 3. P. 356-484.
53. Guggenheim E.A., Proc. Roy. Soc. London. A. 1935. V. 148. P. 304
54. Bethe H.A., Proc. Roy. Soc. London. A. 1935. V. 150. P. 552.
55. Peierls R., Proc. Camb. Phil. Soc. 1936. V. 32. P. 471.
56. Kirkwood J.G., Monroe E., J. Chem. Phys. 1942. V. 10. P. 395.
57. Tovbin Yu.K., Progress in Surface Science. 1990. V. 34, No. 1-4, P. 1-236.
58. Dunning W., Interphase boundary gas-solid. Moscow, Mir, 1970. P. 230.
59. Roberts M.W., McKee C.S., Chemistry of the metal-gas interface, Clarendon Press, Oxford, 1978.
60. Somorjai G. A., Chemistry in two-dimension surface, Cornell Univ. Press L., N.Y., Ithaca, 1981.
61. New in the study of the surface of a solid, Ed. T. Jayadevay, R.M. Vanselov, Moscow, Mir, 1977. Issue 2. [CRC Press, Inc., Cleveland, 1974]
62. Jaycock M. Parfitt J., Chemistry of interfaces.– Moscow: Mir, 1984. – 270 p. [Wiley, New York, 1981].
63. Tovbin Yu.K., Doct. Sci. Thesis, Moscow, NIFHI L.Ya. Karpova. 1985.
64. Tovbin Yu.K., Theoretical methods for describing the properties of solutions. Interuniversity collection of scientific works. Ivanovo. 1987. P. 44.
65. Tovbin Yu.K., Zh. fiz. khimii. 1981. V. 55. No. 2. P. 273.
66. Tovbin Yu.K., Zh. fiz. khimii. 1990. P. 64. No. 4. P. 865. [Russ. J. Phys. Chem. 1990. V. 64. № 4. P. 461]

67. Fedyanin V.K., Statistical physics and quantum field theory. Moscow, Mir. 1973.
68. Tovbin Yu.K., Zh. fiz. khimii. 2005. V. 79. No. 12. C. 2140. [Russ. J. Phys. Chem. , 2005 V. 79, No. 12, P. 1903]
69. Boltzman L., Selected Works. Moscow, Nauka, 1984..
70. Rumer Yu.B., Ryvkin M.Sh., Thermodynamics, statistical physics and kinetics. Moscow, Nauka, 1971.
71. Kirkwood J.G., J. Chem. Phys. 1950. V. 18. P. 380.
72. Salsburg Z.W., Kirkwood J.G., J. Chem. Phys. 1952. V. 20. P. 1538.
73. Tovbin Yu.K., Zh. fiz. khimii. 2006. V. 80. No. 10. C. 1753. [Russ. J. Phys. Chem. 2006. V. 80. № 10. P. 1554]
74. Tovbin Yu.K., Zh. fiz. khimii. 2015. V. 89. No. 11. P. 1704. [Russ. J. Phys. Chem. A. 2006. V. 89. № 11. P. 1971]
75. Yang C.N., Lee T.D., Phys. Rev. 1952. V. 87. P. 404.
76. Lee T.D., Yang C.N., Phys. Rev. 1952. V. 87. P. 410.
77. Kvasnikov I.A., Thermodynamics and statistical physics. V. 1: Theory of equilibrium systems: Thermodynamics. Moscow, Editorial URSS, 2002.
78. Landau L.D., Zh. Eksp. Teor. Fiz. 1937. V. 5. P.627.
79. Rostiashvili V. G., Irzhak V. I., Rozenberg B. A., Glass Transitions in Polymers, Leningrad, Khimiya,1987.
80. Leontovich M.A., Zh. Eksp. Teor. Fiz. 1936. V. 6. P. 561.
81. Mandelstam L.I., Leontovich M.A., Zh. Eksp. Teor. Fiz. 1937. V. 7. No. 7. P. 438.
82. Mikhailov I.G., et al., Fundamentals of molecular acoustics. Moscow, Nauka, 1964..
83. Haase R., Thermodynamics of irreversible processes. Moscow, Mir, 1967. [Dr. Dietrich Steinkopff, Darmstadt, 1963]
84. Chapman S., Kauling T., Mathematical theory of non-uniform gases. Moscow, IL, 1960.
85. Ferziger J., Kaper G., Mathematical theory of transport processes in gases. Moscow, Mir, 1976.
86. Klimontovich Yu.L., Kinetic theory of non-ideal gas and non-ideal plasma. Moscow, Nauka, 1975.
87. Tovbin Yu.K., Modern chemical physics, Moscow, Izd-vo MGU, 1998.
88. Tovbin Yu.K., Khim. Fizika. 2002. V. 21. No. 1. P.83.
89. Tovbin Yu.K., Zh. fiz.khimii. 2002. V. 76. No. 1. P.76. [Russ. J. Phys. Chem. 2002. V. 76. № 1. P. 64]
90. Vlasov A.A., Zh. Eksp. Teor. Fiz. 1938. V. 8. P. 291.
91. Glauber J., J. Math. Phys. 1963. V. 46. P.541.
92. Stanley G., Phase transitions and critical phenomena. Moscow, Mir. 1973.
93. Eyring H. J. Chem. Phys. 1935. V. 3. P. 107.
94. Glasston S., Laidler K.J., Eyring H., Theory of absolute reaction rates. Moscow, IL, 1948 [Princeton Univ. Press, New York, London, 1941].
95. Tovbin Yu.K., Fedyanin V.K., Kinetika i kataliz. 1978. Vol. 19. No. 4. P. 989.
96. Tovbin Yu.K., Fedyanin V.K., Kinetika i kataliz. 1978. V. 19. No. 5. P. 1202.
97. Tovbin Yu.K., Fedyanin V.K., Fiz. Tverd. Tela. 1980. V. 22. No. 5. P. 1599.
98. Tovbin Yu.K., Fedyanin V.K., Zh. fiz. khimii. 1980. V. 54. No. 12. P. 3127, 3132.
99. Tovbin Yu.K., Zh. fiz. khimii. 1981. Vol. 55. No. 2. P. 284.
100. Tovbin Yu.K., Dokl. AN SSSR. 1982. V. 267. No. 6. P. 1415.
101. Tovbin Yu.K., Dokl. AN SSSR. 1984. P. 277. No. 4. P. 917.
102. Tovbin Yu.K., Progress in Surface Science. 1990. V. 34, No. 1-4. P. 1-236.
103. Tovbin, Yu.K., Poverkhnost. Fizika. Khimiya. Mekhanika. 1989. No. 5. P. 5.
104. Tovbin Yu.K., Dynamics of Gas Adsorption on heterogeneous Solid Surfaces. Eds.

W. Rudzinski, W.A. Steele, G. Zgrablich. Elsevier: Amsterdam, 1996. P. 240-325.
105. Tovbin Yu.K., Thin Films and Nanostructures. Vol. 34. Physico-Chemical Phenomena in Thin Films and at Solid Surface. Eds. L. I. Trakhtenberg, S. H. H. Lin, and O. J. Ilegbusi, Elsevier, Amsterdam, 2007. P. 347.
106. Tovbin Yu.K., Cherkasov, A. N., Teor. Eksp. Khimiya. 1984. V. 20. No. 4. P. 507.
107. Pandit R., Forgacs G., Rujan P., Phys. Rev. V. 1982. V. 25. P. 1860.
108. Ma Sh., Modern theory of critical phenomena. Moscow, Mir, 1980.
109. Nikitin E.E., Theory of elementary atomic-molecular processes in gases. Moscow, Khimiya. 1970.

3

Phase separation boundary

21. Thermodynamic values of the surface layer

Interfaces between coexisting phases are the simplest example of local non-uniformities in a system. They inevitably arise within the framework of the phase approximation of describing the properties of non-uniform systems by Gibbs as *heterogeneous* systems, when the contributions from the phase boundaries can not be neglected in thermodynamic functions. Our task is to compare the hypotheses of the thermodynamic approach with the results obtained on the basis of the molecular theory, using the same hypotheses. Below, the thermodynamic introduction to the problem is limited to reminding the basic assumptions about the concepts of the dividing surface. Determination of the surface of tension and methods for determining the surface tension based on thermodynamic and hydrostatic approaches in metastable spherical drops, as well as other issues of comparing the continuum description of surface tension with the molecular discrete description in the LGM (lattice gas model), are discussed in Appendix 1 and Chapter 7.

Dividing surface. Consider a two-phase system with a phase boundary that can be flat or curved. Imagine a mathematical surface called the dividing surface, which is introduced in such a way as to accurately divide the two phases, denoted as α or β [1,2]. In general, the position of the dividing surface is chosen in such a way that it is normal to the density gradient in the transition zone and, therefore, in the case of a flat boundary, it must be a plane. The use of a dividing surface has the advantage that it allows us to consider the interphase layer without indicating its thickness.

Any quantity related to one of the phases will be denoted by the lower index of the phase α or β. Let the volume of the system be divided into two volumes V_α and V_β. If the dividing surface is selected lying in the transition layer, then V_α contains the bulk phase α together with a part of the transition zone material, and V_β contains the bulk phase β and the rest of the transition zone material.

Let us imagine a hypothetical system built of two volume phases α and β, which remain strictly uniform up to the above dividing surface and therefore have volumes V_α and V_β, respectively [1,2]. The superscript i will be used to denote the i-th molecular component. Let N_α^i and N_β^i be the numbers of molecules of the i-th sort in the above-mentioned hypothetical phases α and β. Specifically, N_α^i is equal to n_α^i – the number of molecules of the i-th sort per unit volume of the phase α multiplied by the volume occupied by this phase V_α, and N_β^i is, respectively, $n_\beta^i V_\beta$.

In the general case, N_i – the total number of molecules of the i-th sort in the real system – is not necessarily equal to the number $N_\alpha^i + N_\beta^i$ for the above-mentioned hypothetical system; However, it can be represented in the form

$$N^i = N_\alpha^i + N_\beta^i + N_b^i \tag{21.1}$$

where N_b^i can be considered as an additive associated with the existence in the real system of the interface or interphase region. The number N_b^i and any other similar quantity will be called the surface quantity, and mark with the subscript b.

Figure 21.1 shows the schemes of two-phase systems with a flat phase interface located in a vessel, having the shape of a rectangular parallelepiped and arranged so that one of its edges is parallel to the direction of gravity. These schemes sequentially detail the interface region.

The scheme in Fig. 21.1 *a* shows the mathematical surface or the dividing surface between the phases α and β [1,2]. The scheme in 21.1 *b* reflects the finite width of the transition region between the phases α and β, following Guggenheim [3,4]. This region is regarded as a separate layer of matter of finite thickness, bounded by the planes *K'H'* and *K"H"* parallel to the dividing surface *KH*. It is assumed that the parts of the system above the *K'H'* plane and below the *K"H"* plane are completely uniform, so that the spatial variation of the thermodynamic quantities is limited by the surface phase between *K'H'* and *K"H"* and having a thickness of $\Delta = \Delta_\alpha + \Delta_\beta$.

In molecular approaches, the same system will be considered as a system consisting of a set of monomolecular layers parallel to the

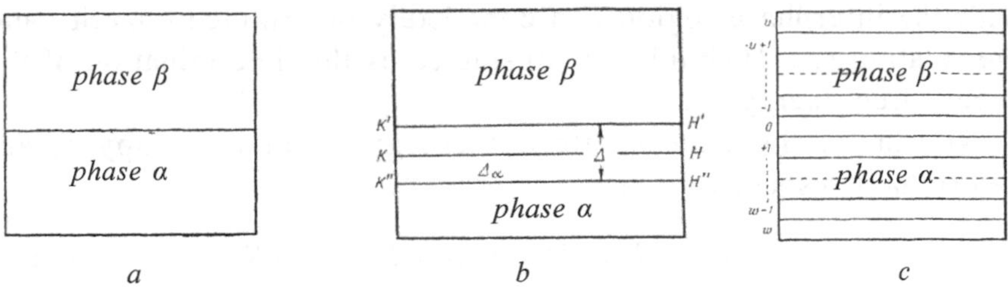

Fig. 21.1. (*a*) Two-phase system with a flat interface. (*b*) The phase separation boundary is regarded as a separate layer of a material of finite thickness bounded by the planes *K'H'* and *K"H"* parallel to the dividing surface *KH*. (*c*) The phase boundary is considered as a set of monomolecular layers parallel to the phase interface.

phase interface (Fig. 21.1 *c*) and perpendicular to the direction of gravity [2]. Plane bounding monolayers will be called equipotential. We denote the width (thickness) of the monolayer by λ; all of them have the same area A. For brevity, we shall assume that one of the equipotential planes is a zero fiber and assign numbers –1, –2,..., $-u$ to successive layers lying above zero, and numbers 1, 2,..., ω to layers lying below zero (Fig. 21.1 *c*). The way in which the zero layer is selected depends on the method for determining the surface tension. One of the most commonly used definitions is an equimolecular surface. The quantities related to the monolayer q will be denoted by the subscript q, for example, the number of molecules in the q-th layer will be denoted by N_q.

The contribution of the interphase layer to thermodynamic quantities. Any extensive thermodynamic property of the whole system, like the number of molecules, can be considered with respect to an arbitrarily chosen dividing surface in the form of a sum of three contributions: the contribution of the phase α, the contribution of the phase β, and the contribution of the interphase region. For example, the free Helmholtz energy for the whole system can be written as

$$F = F_\alpha + F_\beta + F_b. \tag{21.2}$$

where F_α is the free energy of the volume phase α in the case when the latter remains uniform up to the dividing surface, F_β is an analogous value for the volume phase β. In general, $F_\alpha + F_\beta$ is the free energy of a hypothetical system in which intense physical properties jump discontinuously on the dividing surface from the values they have in the bulk phase α to the values they take in the bulk phase β and F_b is the contribution of the transition layer to real free energy. The total free energy F of the two-phase system together with the interphase region is a completely determined quantity and the relation (21.2) should be considered as the determination of the excess free energy F_b.

Similarly, the excess internal energy U_b and entropy S_b are determined respectively as

$$U = U_\alpha + U_\beta + U_b \text{ and } S = S_\alpha + S_\beta + S_b. \tag{21.3}$$

Using the relation $F = U - TS$ and the definitions given above, we obtain

$$F_b = U_b - TS_b, \tag{21.4}$$

The absolute temperature T is assumed constant in all parts of the system.

22. The planar interface of macroscopic phases

Determination of surface tension. Let the area of the boundary A (Fig. 21.1 *a*) increase by a value dA by means of a reversible isothermal displacement of the side walls of the vessel. If the system is in hydrostatic equilibrium at a pressure P, the work done by it in the described process can be divided into two parts: the work PdV spent on increasing the volume dV, and the excess work associated with the increase in the area of the boundary by dA, which we denote $-\sigma dA$. Then the total work done by the system in this process is $PdV-\sigma dA$. Now let's shift the upper and lower covers of the vessel so that we return the system to its original volume. The work done by the system in this process is $-PdV$. At the end of these two processes, the system will have the same pressure, composition and temperature as it originally had.

The only change is that the area of the border has increased by dA. This means that the work done by the system in order to cause an increase in the area of the boundary by dA for constant values of volume, pressure and temperature is σdA. Therefore, σ represents the work done by the system with a single increase in the area of the interface. It is called surface tension. The elementary work dW produced by the system with a change in its volume and boundary area, respectively, on dV and dA should be written as

$$dW = PdV - \sigma dA. \tag{22.1}$$

This expression can be regarded as the definition of the surface tension σ. It can be seen from (22.1) that in the case of a plane boundary the surface tension does not depend on the position of the dividing surface, since in this case the changes of the latter do not affect the area of the boundary in any way.

If the surface tension σ determined by the described method were negative, then the work done by the system with increasing interface area would be positive and, consequently, the interface area could grow spontaneously. This means that a flat interface is stable only for positive σ. Therefore, the surface tension value defined above can not be negative, just as the pressure in the liquid, determined in the usual way, can not be negative.

The internal energy U of the two-phase system also varies with the area of the interphase boundary. If the amount of heat obtained by the system in the process of an infinitely small increase in the area of the boundary dA is denoted by dQ, then the First Law of thermodynamics is written in the form

$$dQ = dU + PdV - \sigma dA. \quad (22.2)$$

Equality (22.2) is also valid for non-equilibrium systems as long as they are in mechanical equilibrium at constant pressure.

Surface free energy. If the system is in thermodynamic equilibrium, then the amount of heat dQ absorbed during the reversible change in state, according to the Second Law, is equal to the increase in the entropy of the system dS, multiplied by the temperature T. Combining this with the first law (22.2), we obtain

$$dU = TdS - PdV + \sigma dA. \quad (22.3)$$

This important relation can be represented in a more convenient form, using the expression for the Helmholtz free energy:

$$dF = -PdV - SdT + \sigma dA. \quad (22.4)$$

In an open system that can exchange with matter and energy with the environment, the number of molecules can vary and therefore the total differential of free energy given by expression (22.4) should have an additional term associated with this change:

$$dF = -PdV - SdT + \sigma dA + \sum_{i=1}^{s-1} \mu_i dN_i, \quad (22.5)$$

where $(s-1)$ is the number of components (the contribution of vacancies is omitted), μ_i is the chemical potential calculated for the molecule of the *i*-th component, which should be constant throughout the system if it is in chemical equilibrium. The fundamental Gibbs equation (22.5) for a two-phase system can be regarded as an equation that determines the surface tension σ in an open system.

If the free energy F is known as a function of volume, temperature, interface area and number of molecules, then σ can be easily calculated from relation

$$\sigma = \left(\frac{\partial F}{\partial A}\right)_{T,V,\mathbf{N}} \tag{22.6}$$

where the symbol $\mathbf{N}$ represents the set of numbers N_1, N_2,...,N_s. We will increase the volume of the two-phase system shown in Fig. 1 *a*, while limiting the thermodynamically equilibrium case and keeping the temperature, pressure, composition and height of the vessel unchanged. Then the quantities F, V, $\mathbf{N}$, A vary in the same proportion. Thus, F is a uniform function of the first power of the variables V, N, and A, and therefore, by Euler's theorem, we obtain from (22.5)

$$F = \sum_{i=1}^{s-1} \mu_i N_i - PV + \sigma A. \tag{22.7}$$

For the volume phases α and β, the free energy expressions are written as,

$$F_\alpha = \sum_{i=1}^{s-1} \mu_i N_\alpha^i - PV_\alpha,\ F_\beta = \sum_{i=1}^{s-1} \mu_i N_\beta^i - PV_\beta, \tag{22.8}$$

then subtracting (22.8) from (22.7) and using the equalities (21.1), (21.2) and the volume additivity condition, we obtain

$$F_b = \sum_{i=1}^{s-1} \mu_i N_b^i + \sigma A. \tag{22.9}$$

In contrast to the surface tension σ, excess values such as F_b, U_b and N_b^i depend on the position of the dividing surface. If we choose the dividing surface in such a way that the sum $\sum_{i=1}^{s-1} \mu_i N_b^i = 0$, then (22.9) reduces to the relation

$$F_b = \sigma A. \tag{22.10}$$

In the case of a single-component system, this particular dividing surface plays an important role in the theory of surface tension and is called an equimolecular dividing surface. We emphasize that N_b^i

can not always be called the number of molecules of sort i adsorbed in the interphase region, since N_b^i depends on the position of the artificially introduced dividing surface. It is seen from (22.10) that the surface tension becomes equal to the surface density of free energy only with such a special choice of the dividing surface [5,6]. It should be emphasized that surface tension is neither internal energy nor potential energy per unit surface area.

Thermodynamic relationships for a plane boundary. The fundamental Gibbs equations for the volume phases α and β are written as follows:

$$dF_\alpha = -PdV_\alpha - S_\alpha dT + \sum_{i=1}^{s-1} \mu_i N_\alpha^i, \tag{22.11}$$

$$dF_\beta = -PdV_\beta - S_\beta dT + \sum_{i=1}^{s-1} \mu_i N_\beta^i, \tag{22.12}$$

Subtracting (5.1) and (5.2) from (4.3) and using (1.1), (2.1) and (2.3), we obtain the relation

$$dF_b = -S_b dT + \sigma A + \sum_{i=1}^{s-1} \mu_i N_b^i. \tag{22.13}$$

which can be regarded as the fundamental Gibbs equation for the interphase layer. Thus, we see that the interphase thermodynamic quantities are related to each other by the same relation as in the case of the usual volume phase, and, consequently, they can be treated as if they determined a certain third phase. However, this analogy of the resulting expressions for some third phase has a purely formal meaning. Such equations can be used to construct thermodynamic bonds only between surplus surface characteristics, as defined above. But the boundary itself is not an autonomous phase, and under no circumstances can the boundary be considered a phase. This fact has a fundamental limitation on thermodynamic constructions for the total content of molecules in the transition region.

On the basis of equation (22.13), all thermodynamics is constructed for a planar interphase layer [2]. For example, differentiating (22.9) and using (22.13), we arrive at the general form of the Gibbs adsorption equation

$$d\sigma + s_b dT + \sum_{i=1}^{s-1} \Gamma_i d\mu_i = 0, \tag{22.14}$$

where the surface density of the entropy is usually called the surface entropy $s_b = S_b/A$, and the value $\Gamma_i = N_b^i/A$ will be called the surface density of the number of molecules of sort i.

The contribution of σA also appears in all other thermodynamic potentials when it is necessary to take into account the surface properties of the system. Thus, if for a volume phase the Gibbs potential G is written as $G_v = \Sigma_i \mu_i m_i = U - TS + PV$, then for a plane boundary this potential has the form $G_b = \Sigma_i \mu_i m_i = U - TS + PV - \sigma A$, designations for internal energy U, entropy S, temperature T, pressure P, volume V, surface tension σ; μ_i and m_i are the chemical potential and mass of component i. In thermodynamics, the concept of a phase interface is formulated for macroscopic systems, for which the inclusion of surface contributions is only necessary if the surface area A is so developed that the contribution of the term $\sigma A = G_b - G_v$ to all thermodynamic potentials becomes commensurable with the contribution from bulk phases: $\sigma A \sim G_v$ and it can not be neglected. If $G_v \gg \sigma A$, or $G_b - G_v \approx 0$, then the surface contribution to the thermodynamic functions of the systems under consideration is omitted.

Mechanical determination of surface tension for a plane interface. The physical interface of the phases is not a geometric surface of zero thickness, but a transition layer whose thickness is finite (Fig. 21.1 *b*). Since the density in the transition zone varies noticeably in the direction normal to the interface, within this region the pressure tensor, defined in the hydrostatic sense, must also change, referring to the isotropic and constant hydrostatic pressure within each phase.

We introduce a rectangular coordinate system (x, y, z) with the z axis directed along the normal to the flat interface from the phase α to the phase β and the plane xy parallel to the plane of the boundary [2]. In any uniform phase, the pressure is isotropic, i.e., the force acting on the unit area is normal to it and is the same for all unit area orientations. Consequently, within each of the uniform phases α or β, the pressure tensor $\mathbf{P}$ is reduced to the hydrostatic pressure P multiplied by the unit tensor $\mathbf{I}$: $\mathbf{P} = P\mathbf{I}$, or

$$P_{xx} = P_{yy} = P_{zz} = P, P_{xy} = P_{yz} = P_{zx} = P_{yx} = P_{zy} = P_{xz} = 0. \quad (22.15)$$

Inside the interphase region, the force acting along the normal to the unit area is different in different directions. However, from the symmetry requirements it is easy to see that in the planar case, even in the transition region, $P_{xy} = P_{yz} = P_{zx}$ should disappear, and P_{xx} and P_{yy} must be equal to each other and not depend on either x

or y. For the remaining non-zero components (or components) of the pressure tensor, it is convenient to introduce the notation $P_{zz} = P_N(z)$ and $P_{xx} = P_{yy} = P_T(z)$, which we will call the normal and tangential components of the pressure tensor **P** at the point z.

The pressure tensor on a planar vapour–liquid interface is written as

$$\mathbf{P} = P_T(z)(\mathbf{e}_x\mathbf{e}_x + \mathbf{e}_y\mathbf{e}_y) + P_N(z)\mathbf{e}_z\mathbf{e}_z, \tag{22.16}$$

where $\mathbf{e}_x$, $\mathbf{e}_y$, $\mathbf{e}_z$ are unit vectors directed along the Cartesian axes $\alpha = x, y, z$; P is the isotropic pressure in bulk coexisting phases. In general, the tangential component of $P_T(z)$ can vary along z in a complex manner, whereas the values of the normal component $P_N(z)$, according to the hydrostatic equilibrium condition between the coexisting phases, must equal P even in the transition zone: $P_N(z) = P$.

We choose a unit area in the form of a strip having a unit width in the y direction and extending from $-\ell/2$ to $\ell/2$ in the z direction. It is obvious that the total stress $\Delta\Sigma_x$ acting in the direction of the x axis through this unit area can be written as

$$\Delta\Sigma_x = -\int_{-\ell/2}^{\ell/2} P_T(z)dz. \tag{22.17}$$

If the value of $\Delta\Sigma_x$ did not contain a contribution from the transition zone, it would simply be $-P\ell$. For a planar boundary, the excess stress due to the presence of an interphase boundary (here the unit area is chosen as the strip shape as specified above) is [2.7]

$$\sigma = -\int_{-\ell/2}^{\ell/2} P_T(z)dz + P\ell = \int_{-\ell/2}^{\ell/2} [P - P_T(z)]dz \tag{22.18}$$

where the condition that the normal component of the pressure tensor of the plane boundary and pressure in the bulk phase is equal ($P_N(z) = P$).

This equation shows that a real system with a planar boundary between phases can be treated as if it consisted of two uniform phases separated by a plane membrane of zero thickness carrying the tension σ defined by (22.18). Thus, equation (22.18) is a definition of surface tension, and we will take it as the mechanical definition of σ.

It is also obvious that the surface tension determined by (22.18) is independent of ℓ provided that ℓ has a macroscopic length such that for $z = \ell/2$ and $z = -\ell/2$ the value of $P_T(z)$ becomes equal to the hydrostatic pressure P From this consideration, we can assume that ℓ is infinitely large, and rewrite (22.18) in the conventional form:

$$\sigma = \int_{-\infty}^{\infty} (P - P_T)\, dz. \tag{22.19}$$

Using the fact that P_N does not depend on z and is equal to P, it is possible to obtain Becker's equation [7] from (22.19)

$$\sigma = \int_{-\infty}^{\infty} (P_N - P_T)\, dz. \tag{22.20}$$

In ordinary hydrodynamics, the physical quantities at the lengths of the order of the range of intermolecular forces are assumed to be constant, therefore the value of $P_T(z)$ introduced above can not be directly identified with pressure in the hydrodynamic sense. At the microscopic level, the question of the correspondence between $P_T(z)$ and $P_N(z)$ with the local components of the pressure tensor and hydrodynamic concepts was developed much later in the framework of microscopic hydrodynamics [8] (see also Section 37).

Thermodynamics introduced the concept of surface tension, suggesting its definition from experimental data. To calculate its value, it is necessary to use molecular models.

23. Molecular description of a planar interface

The most consistent way in solving this problem for pure substances was associated with the use of molecular models. For this purpose, all the existing methods of statistical thermodynamics [2,9–25] are involved. These include the theory of van der Waals capillarity [9,11,12], the method of integral equations [2,9,10,13,14] and its simplified version without correlation effects – the density functional method [15–19], the method molecular dynamics [20,21] (and Monte Carlo [13]) and the lattice gas model (LGM) [22–25].

Continual description [2]. Consider a system in a vessel that has the shape of a rectangular parallelepiped with edges of length l_1, l_2, l_3, directed along the axes of the rectangular coordinate system x, y, z, respectively; the z axis is directed vertically upwards – opposite to the direction of gravity. The volume of the system is $V = l_1 l_2 l_3$ and the interface is $A = l_1 l_2$ (see Fig. 23.1).

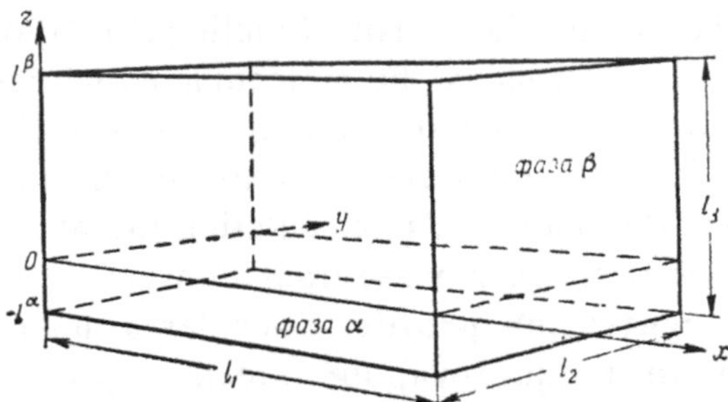

Fig. 23.1. The flat interface of the phases [2].

The free Helmholtz energy of a two-phase system depends on A and V, $\left(\frac{\partial F}{\partial l_1}\right)_{l_2,l_3,T,N} = l_2\left(\frac{\partial F}{\partial l_1}\right)_{V,T,N} + l_2 l_3\left(\frac{\partial F}{\partial l_1}\right)_{A,T,N} = l_2\sigma - l_2 l_3 P$ so from which $\left(\partial F / \partial l_1\right)_{A,T,N} = \left(\partial F / \partial l_3\right)_{l_1,l_2,T,N} / (l_1 l_2)$, using the relationship, we obtain the following relation $\sigma = \frac{1}{l_1 l_2}\left[l_1\left(\frac{\partial F}{\partial l_1}\right) - l_3\left(\frac{\partial F}{\partial l_3}\right)\right]$.

In order to differentiate the partition function Z (because $F = -kT\ln Z$), a device specially developed by Bogolyubov should be used in terms of volume, which allows us to relate the expression for the surface tension for a planar interface through the distribution functions

$$\sigma = \frac{1}{2}\int_{-\infty}^{\infty} dz_1 \int \theta(z_1, \mathbf{r}_{12}) \frac{\partial \varepsilon(r_{12})}{\partial r_{12}} \frac{(x_{12}^{\ 2} - z_{12}^{\ 2})}{r_{12}} d\mathbf{r}_{12}$$

where $\theta(z_1,\mathbf{r}_{12})$ is paired distribution function (DF), characterizing the probability of finding two particles at points with coordinates z_1 and $z_1 + \mathbf{r}_{12}$; $\varepsilon(r_{12})$ is the paired interaction potential of the specified particles, $x_{12} = x_1 - x_2$ is the component of the distance vector between two particles $\mathbf{r}_{12} = \mathbf{r}_1 - \mathbf{r}_2$ along the x axis, z_{12} – similarly for the z axis.

The dependence of the surface tension on the distribution functions formally solves the problem. For computations, the values of unary and paired DFs are necessary inside a vapour–liquid interface that is non-uniform in density. The equations for these DFs are given in [2,9,10]. Because of the large computational problems with the use of integral equations for unary and paired DFs, the van der Waals theory of capillarity and the density functional method are widely used. The difficulties in using the molecular dynamics method for calculating the drop characteristics are discussed in [21]. However, these methods give rather different results, so the thermodynamic approach remains the main one in calculating the nucleation energy for the kinetics of first-order phase transitions.

Discrete concentration profile. Consider a macroscopic two-phase system with a liquid–vapour interface (Fig. 21.1 *c*). The simplest version of the LGM, which has the simplest equations for the concentration profile, is discussed. Recall that LGM was the first molecular approach to the theory of surface phenomena [2,9,26–29], which was applied both to simple gases and liquids, as well as to metals and alloys – see the survey papers [29–31]. It makes it possible to calculate the surface tension, considering the parameters of the model either as determined from the data of a direct experiment, or as found from other bulk properties, for example, on the stratification curves. The equations for the volume phase are given in monographs [8, 29–33] (see also Section 18).

The volume of the system is divided into individual elementary cells or volume sites $v_0 = \gamma_s \lambda^3$ (see Section 16). We combine the cells into monolayers of width λ. All sizes in the system (drop radius, width of the transition region, etc.) will be measured in units of λ. The sites located in one monolayer q are assigned the type q. A site of type q is characterized by the numbers of its nearest neighbours z_{qp}, $1 \leq q, p \leq \kappa$, both in the same monolayer $p = q$ and in neighboring monolayers $p = q \pm 1$, for any site in the monolayer q, and $\sum_{p=q-1}^{q+1} z_{qp} = z$. Here κ is the number of monolayers in the flat transition region between the vapour and the liquid, and z is the coordination number of the lattice. For a planar lattice or bulk layer, the number of cells in all monolayers is the same and the numbers $z_{q,p}$ are constant: $z_{q,q-1} = z_{q,q+1} \equiv z_1$, $z_{q,q} = z-2z_1$. For example, for a flat lattice with a coordination number $z = 6$, we have $z_{q,q\pm1} = z_1 = 1$, $z_{q,q} = 4$.

For a given temperature T, according to the Maxwell rule [8, 32, 33], in the bulk phase, we determine the equilibrium pressure P_0 and the numerical densities of the coexisting phases $\theta^{(L)}$ and $\theta^{(V)}$, liquid and vapour, respectively. We will include monolayers from each of the coexisting phases in the number: by the construction of the system of equations, the index $q = 1$ corresponds to the dense phase ($\theta_1 = \theta^{(L)}$), and the index $q = \kappa$ corresponds to the vapour phase ($\theta_\kappa = \theta^{(V)}$).

The equations for the concentration profile $\{\theta_q\}$, $1 \le q \le \kappa$, have the form

$$a_q P = \frac{\theta}{1-\theta}\Lambda_q,\ \Lambda_q = \prod_{p=q-1}^{q+1}\left(S_{qp}\right)^{z_{qp}}, \tag{23.1}$$

here the equations take into account the energy non-uniformity of the lattice sites along the normal to the surface and the interaction between nearest neighbours ($R_{lat} = 1$); Λ_q is a function of non-ideality, depending on the type of approximation. Henry's constants $a_q = \beta F_q \exp(\beta\varepsilon_q)/F_0$ reflect internal motions of molecules in different regions with variable local density. In the formula (23.1), $s = 2$ are particles A and vacancies. All local partial fillings θ_q are functions of pressure P. The non-ideality functions Λ_q take into account only direct correlations between interacting particles in the QCA. In the case of QCA, the functions $S_{qp} = 1 + xt_{qp}$, $t_{qp} = \theta_{qp} / \theta_q \equiv \theta_{qp}^{AA} / \theta_q^A$ where they are defined by the following formulas (here $x = x_{qp}^{AA} = \exp(\beta\varepsilon)-1$ from eq, (18.4))

$$\theta_{qp}^{AA} = \frac{2\theta_q^A\theta_p^A}{\delta_{qp}+b_{qp}}, \delta_{qp} = 1 + x_{qp}^{AA}(1-\theta_q^A-\theta_p^A),$$
$$b_{qp} = \left\{\left[\delta_{qp}\right]^2 + 4x_{qp}^{AA}\theta_q^A\theta_p^A\right\}^{1/2}, \tag{23.2}$$

$$\sum_{i=1}^{s}\theta_{qp}^{ij} = \theta_p,\ \sum_{j=1}^{s}\theta_{qp}^{ij} = \theta_q,\ \sum_{i=1}^{s}\theta_q^i = 1. \tag{23.3}$$

The system of equations (23.1) with allowance for the normalizations to the functions θ_f^i and θ_{fg}^{ij} is closed. The concentration profile of the transition region $\{\theta_q\}$ is found by iterating by the Newton method from the solution of the system of equations (23.1), giving $P_q = P_0$ for a given value of the width of the transition region $\{\theta_q\}_{q=1}^{q=\kappa}$ and the initial approximation to the density profile of the transition region that varies monotonically from $\theta_1 = \theta^{(L)}$ to $\theta_\kappa = \theta^{(V)}$ The accuracy of the solution of this system is not less than 0.1%.

Knowing the concentration profile, we can calculate the thermodynamic characteristics of the interface. The expression for the free energy of the system F for the distributed model of the system under consideration with $R_{lat} = 1$, according to (18.2), (18.3), will be written as [34]

$$F = \sum_q \sum_i \theta_f^i M_f^i = \sum_q \sum_i \left\{ \theta_q^i \left(v_q^i + kT \ln \theta_q^i \right) + \frac{kT}{2} \times \right.$$
$$\left. \times \sum_{p=q-1}^{q+1} z_{qp} \sum_{i=1} \left[\theta_{qp}^{ij} \ln {}^* \theta_{qp}^{ij} - \theta_q^i \theta_p^i \ln \left(\theta_q^i \theta_p^1 \right) \right] \right\}, \tag{23.4}$$

where $^*\theta_{fg}^{ik} = \theta_{fg}^{ik} \exp(-\beta\varepsilon_{fg}^{ik})$, and

$$M_f^i = v_f^i + kT \ln \theta_f^i + \frac{kT}{2} \times$$
$$\times \sum_{p=q=1}^{q+1} z_{qp} \ln \left[{}^*\theta_{fg}^{ii} \; {}^*\theta_{fg}^{ik} / \left(\theta_f^i \right)^2 {}^*\theta_{fg}^{ki} \right], \tag{23.5}$$

where the symbol k refers to one of the components of the mixture – a reference sort of particles, $1 \leq k \leq s$, $k \neq i$. The choice of k is determined by the convenience of calculation.

Calculation of the width of the interphase region. The quantity κ is found by varying the upper limit of the system of equations (23.1)–(23.3) in order to obtain a constant molecular density profile over the layers for given values of the temperature T and the chemical potential μ (or external pressure P). The procedure for finding the width of the transition region and calculating the distributions of molecules at the vapour–liquid interface is demonstrated using the example of Fig. 23.2. The length of the transition region κ is the maximum number of monolayers at which the local densities θ_q change monotonically from the values of $\theta^{(V)}$ corresponding to the vapour phase to the values of $\theta^{(L)}$ corresponding to the liquid phase for all q, $1 \leq q \leq \kappa$. Figure 23.2 shows the results of calculation of the transition region in the volume and shows the effect of the number of monolayers on the value of the free energy F. The value of the free energy is calculated from the found concentration profile (23.1).

Figure 23.2 *a* shows the change in the numerical density θ in the transition from the vapour phase to the liquid phase. The abscissa is the number of the layer (in the volume it is just the site number) of the transition area. The ordinate is the density θ_q at the site of the layer q, $1 \leq q \leq \kappa$, $\kappa = 12$ for $\tau = T/T_c = 0.8$. Dotted lines reflect the discrete nature of the transition region, consisting of κ

sites. Figure 23.2 *b* shows the free energy values of the transition region F calculated from the formulas of Ref. [34] for the values of κ obtained during the iterative determination of the length of the transition region. Figure 23.2 *c* shows the nature of the change in the increment $\Delta F = F(\kappa) - F(\kappa - 1)$ with increasing κ. In view of the discreteness of the change in κ, we have $\Delta\kappa = 1$ and the increment ΔF is an analog of the derivative $dF/d\kappa$.

The curve in Fig. 23.2 *c* shows that for the values of $\kappa \geq 11$ the derivative of free energy along the length of the transition region is practically constant. A further increase in the length of the region leads to a certain limiting value of the specific free energy of the system, the concentration profile ceasing to depend on the length of the transition region, which determines the size of the transition region at a given temperature.

The initial section θ_q for $q < 5$ on the profile of Fig. 23.2 *a* remains practically constant. The real density difference between the liquid and the vapour is realized at a length of only l = 5–6 monolayers, that is $l \sim \kappa/2$. A further increase in κ can lead to a non-monotonicity of the profiles of local densities in the transition region. This is due to the possibility of formation of fluid structures having a more developed surface between vapour and liquid than the monotonic structure of the transition region at the vapour–liquid interface.

Figure 23.3 shows the density distribution profiles in the transition region between liquid and vapour for a plane boundary for a wide range of reduced temperatures. The ordinate in this figure shows

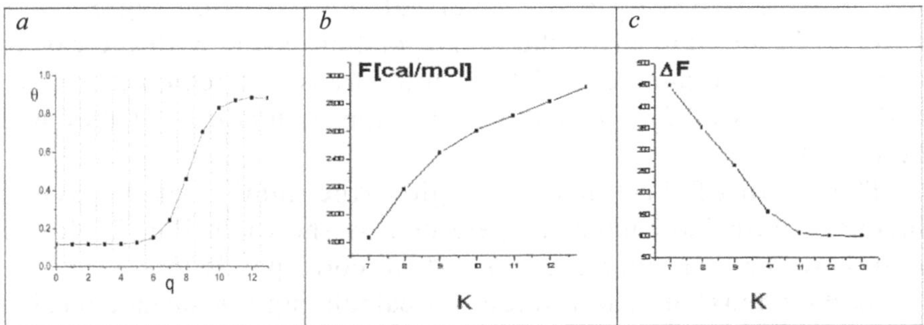

Fig. 23.2. Characteristics of the transition region of argon in the volume at $\tau = 0.8$. (*a*) The density profile θ_q in the transition region of the vapour–liquid interface in the volume for $\kappa = 12$; on the axis of the abscissas, q is the number of the layer in the transition region. (*b*) The free energy of the transition region $F(\kappa)$ as a function of the length of the transition region κ. (*c*) The free energy increment of the transition region ΔF as a function of the length of the transition region κ [8].

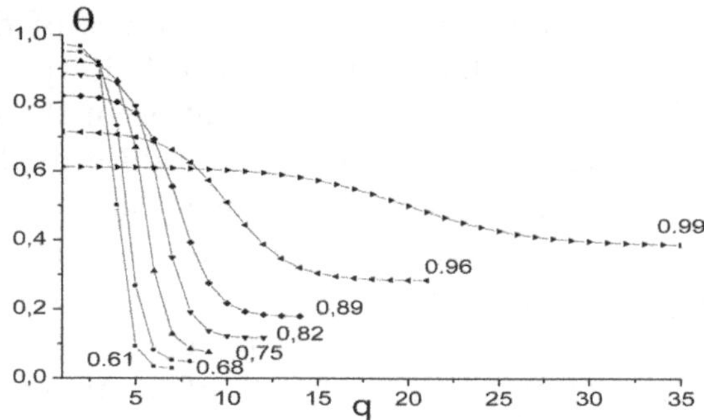

Fig. 23.3. Profiles of the density of the transition region from liquid to vapour at different temperatures for a planar lattice. The values of $\tau = T/T_c$ are indicated on the curves. The abscissa is the number of the monolayer of the transition region q, and the local densities θ_q along the ordinate.

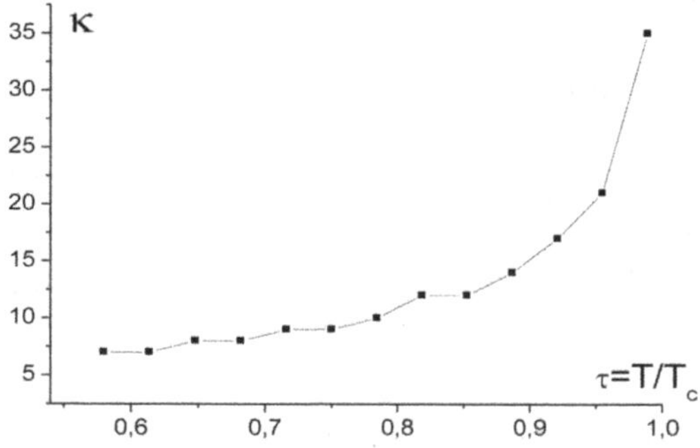

Fig. 23.4. Dependence of the width of the transition region κ on the reduced temperature τ.

the local densities θ_q in the layers of the transition region. The abscissa is the number of the corresponding layer. With increasing temperature, the number of layers in the transition region increases, and the values of $\theta_1 = \theta^{(L)}$ and $\theta_\kappa = \theta^{(V)}$ approach (for $\tau = 1$ they will be equal).

The width of the transition region (the number of its layers) increases with increasing temperature, as shown in Fig. 23.4. For $\tau \to 1$, we have an increase in κ, which corresponds to $\kappa \to \infty$.

Surface tension. The formula for calculating the surface tension for a plane boundary by the thermodynamic definition in terms of excess free energy is written, according to [2,9,28,29,34], as

$$\sigma A = \sum\nolimits_{q=1}^{\kappa} (M_q^V - \mu_V), \tag{23.6}$$

where the functions M_q^V and have the form

$$M_q^V = kT\ln\theta_q^V + \frac{kT}{2}\sum_{p=q-1}^{q+1} z_{qp}\ln\left[\theta_{qp}^{VV}\theta_{qp}^{Vk} / \left(\theta_q^V\right)^2 \theta_{qp}^{kV}\right],$$

$$\mu_V = kT\ln\theta_V + \frac{1}{2}kTz\ln\left[\theta_{VV} / (\theta_V)^2\right] \qquad (23.7)$$

We note that the functions M_q^i defined by formula (23.5) in expression (23.6) can refer to any sort of particle i. The formula (23.6) follows from the direct thermodynamic definition as the excess of free energy in the transition region (22.10) $\sigma A = \sum_{i=1}^{s} \sum_{q=1}^{\kappa} (M_q^i - \mu_i)\theta_q^i$ due to the fulfillment of the local equilibrium equations (23.1).

The internal pressure π in the lattice system corresponds to its connection with the chemical potential of vacancies in the form $\pi v_0 = -\mu_V$. For non-uniform regions this relation has the form $\pi_q v_q^0 = -M_q^V$. Then the local pressure π_q in the non-uniform region q with respect to the layered structure of the interface is written as

$$\beta\pi_q v_q^0 = -\ln\theta_q^V - 1/2\sum_{p=q-1}^{q+1} z_{qp}\ln(\theta_{qp}^V / (\theta_q^V\theta_p^V) \qquad (23.8)$$

Formula (23.8) corresponds to the average pressure in a non-uniform system. In the uniform phase (23.8), the isotropic pressure P is described. At the interface, the non-uniform distribution of molecules leads to a change in the isotropic properties of the phase, including the violation of the isotropy of the pressure at the interface – the latter differs from the pressure in the bulk phase.

Along with the thermodynamic determination of surface tension, hydrostatic determination of surface tension is actively used [2.9] through the mechanical work of creating the surface. On the basis of expression (23.8), we can select the components of the pressure tensor π_q^α as follows

$$\beta\pi_q^\alpha v_q^0 = -\ln\theta_q^V - 3/2\sum_{p=q-1}^{q+1} z_{qp}\cos^2(qp,\alpha)\ln\left(\theta_{qp}^{VV} / \theta_q^V\theta_p^V\right), \qquad (23.9)$$

where the subscript α denotes the direction of the design axis in space, $\alpha = N$ and T. The symbol (qp, α) denotes the angle between the direction of α and the direction of communication between the sites in the layers q and p. For a planar boundary, N is the direction of the normal to the interface of the vapour–liquid interface; T is one of the two tangential directions, so $\cos^2(qp,N) + 2\cos^2(qp,T) = 1$.

The above structural characteristics z_{qp} allow us to have the identical structure of expression (23.9) for interfaces with different symmetries. Thus, for $z = 6$, the normal components of the pressure tensor are written in the form

$$\beta\pi_q^N v_q^0 = -\ln\theta_q^V - 3/2[z_{qq-1}\ln(\theta_{qq-1}^{VV}/\theta_q^V\theta_{q-1}^V) + z_{qq+1}\ln(\theta_{qq+1}^{VV}/\theta_q^V\theta_{q+1}^V)], \tag{23.10}$$

and the tangential components of the pressure tensor differ in the coefficients of the second term:

$$\beta\pi_q^T v_q^0 = -\ln\theta_q^V - 3z_{qq}\ln(\theta_{qq}^{VV}/\theta_q^V\theta_q^V)\ /\ 4. \tag{23.11}$$

According to the hydrostatic definition of surface tension, instead of formula (23.6)

$$\sigma A = \sum_{q=1}^{\kappa} (\pi_{1,\kappa} - \pi_q^T), \tag{23.12}$$

which is identical to the Becker equation (22.20), since because of the equality of pressure in the coexisting phases $\pi=\pi_1=\pi_\kappa=\pi_q^N$.

This model was used to describe various experimental data [28-30,35]. For example, calculations of the temperature dependences of the surface tension for substances whose properties are close to the so-called law of the corresponding states are presented [36]. Details of calculations are indicated in [35], they are performed for 19 substances. Some of them are shown in Fig. 23.5, and the parameters are indicated in Table 23.1. The results of the calculation are plotted with solid lines and are marked with the corresponding sequence number from Table 23.1 (field *a*: 1–6, field *b*: 7–11), the experimental points are marked with symbols.

24. Molecular description of the curved interface

A rigorous statistical theory describing the equilibrium distributions of molecules leads to flat boundaries [2,9,10]. The presence of curved surfaces, from the point of view of thermodynamics [1,40–42], indicates a metastable state of the liquid.

The layer structure of spherical monolayers [24, 25]. The most detailed (distributed) model in the LGM was determined in [33, 34]. With its help, you can form any type of non-uniform systems by combining sites of the same sort into a common group. In particular, by combining sites into layers, the layered model of a flat boundary is formulated above. A similar principle was used in [24] to describe spherical phase interfaces. It considers a layer model of the transition region of a spherical drop, which is divided into monomolecular layers of width λ that are uniform in their properties. These layers are numbered by the index q, where q is the site number pertaining to the monolayer in question, $1 \le q \le \kappa$, here κ is the width of the transition region. Each monolayer is characterized by its density θ_q. As above, we restrict ourselves to the interaction of only the nearest neighbours.

The structural characteristics of z_{qp} must be related to the average numbers of bonds between sites in different monolayers q and p, and the number of sites in each monolayer of radius R must be related

Table 23.1. The table data of the substances examined and the model parameters applicable to them

No.		ε_0/k, K [36]	σ_0, Å [36]	T_{cr}, K [37]	τ	T_{melt}/T_{cr} [37]	R_L	$\delta\sigma_0$ %	$\delta\varepsilon$ %	σ_{exp}, mN/m	σ_{theor}, mN/m	δ, %
1	CH_4	148.2	3.817	190.55	0.48	0.47	1	0	−10	18.00 [38]	18.03	0.17
2	N_2	95.05	3.698	126.25	0.51	0.5	1	0	−6.8	11.77 [39]	11.91	1.2
3	O_2	117.5	3.58	154.77	0.42	0.35	1	0	−4.3	19.40 [39]	19.67	1.4
4	C_2H_6	230	4.418	305.3	0.44	0.30	1	0	−4.3	24.62 [38]	24.57	0.20
5	CO	100.2	3.763	132.91	0.53	0.52	1	0	−3.2	12.11 [37]	12.15	0.33
6	n-C_4H_{10}	297	4.971	426.2	0.40	0.32	1	0	−2.2	27.2 [38]	26.91	1.1
7	C_6H_6	440	5.270	562	0.50	0.5	2	0	−10	30.24 [38]	30.25	0.033
8	F_2	112	3.653	144	0.48	0.37	2	0	−4.2	17.90 [38]	17.91	0.056
9	C_{12}	257	4.4	417.15	0.48	0.41	3	0	−8.2	33.00 [37]	33.06	0.18
10	C_2H_4	199.2	4.523	282	0.55	0.37	4	+10	−6.2	19.05 [38]	19.13	0.42
11	C_3H_8	254	5.061	368.8	0.38	0.23	4	+10	−5.1	27.80 [38]	27.62	0.65

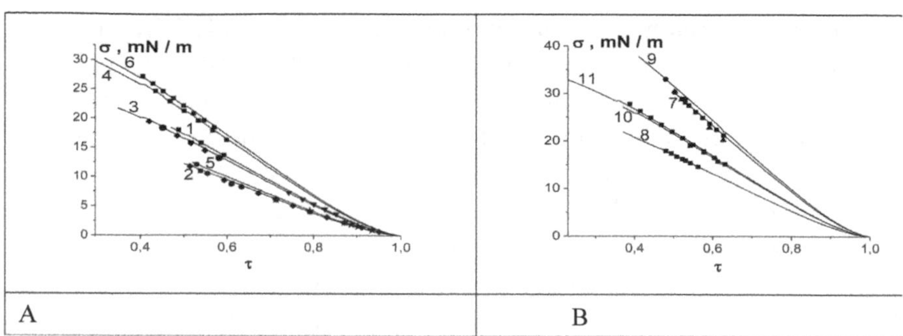

Fig. 23.5. Dependence of surface tension σ on reduced temperature τ for substances from Table 23.1.

to its curvature. As a result, the structure of the entire transition region is given by the average numbers $z_{qp}(R)$, denoting the number of neighboring sites of the layer p around any site of the layer q, where $1 \leq q \leq \kappa$. Each layer q of the transition region has a radius of curvature $q = R$, $z_{qp}(R) \equiv z_{Rp}$. For a planar interface (for $R \to \infty$), this formulation of the problem automatically goes over to the model of Section 23 described earlier. The construction of equations describing the distribution of molecules in the transition region of the interface of coexisting phases reduces to constructing the numbers $z_{qp}(R)$.

We assume that the drop has a sufficiently large diameter, then the deviations of the lattice structure in the drop from the strictly regular rectangular lattice in the bulk phase will be small. In the 'curved' spherical drop layers one can retain the same layer-by-layer method for describing the number of nearest neighbours $z_{qp}(R)$, which should depend on the number of the monolayer and the radius of the drop R [24, 25]. The count of the number of monolayers comes from the centre of the drop. For simplicity, Fig. 24.1 *a* shows a lattice with $z = 6$ nearest neighbours (see Fig. 24.1 – field *b*). The closest neighbours of the filled cell located in the monolayer q are 2 cells located in the monolayers $p = q-1$ and $p = q+1$ (in the figure – left and right), as well as 4 cells located in the same monolayer $p = q$. Two of them are shown in Fig. 24.1 *b* (top and bottom), and the other two are located in front of the filled cell and behind it along the normal to the plane of the drawing.

Figure 24.1 shows a fragment of a drop of radius R with centre at point O and a transition region consisting of κ monolayers from the radius R to $R+\kappa$. By radius R we mean the radius of the liquid part of the drop (internal, shaded part of the drop (region 1)), it corresponds to the index $q = 1$ of the transition region. The outer region of the 3-vapour, the region of the vapour corresponds to the radius $R + \kappa$.

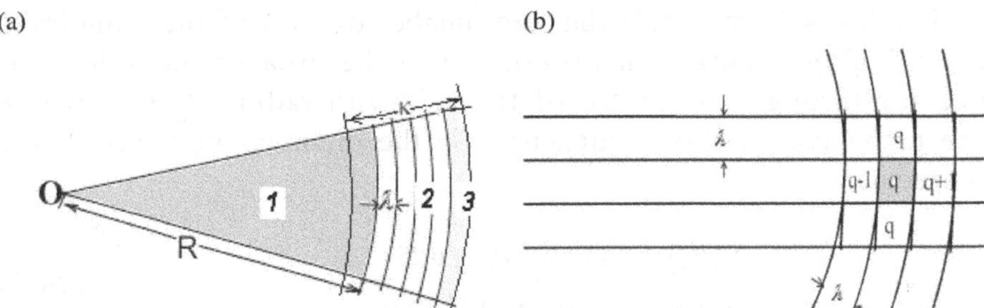

Fig. 24.1. (a) A schematic fragment of a drop of radius R, the inner part of which is filled with liquid, the outer part (3) – with vapour, and the transition region from liquid to vapour consists of κ spherical monolayers of width λ, including one monolayer of liquid and one monolayer of vapour. (b) scheme of the nearest neighboring sites in the 'curved' lattice $z = 6$.

The transition region 2 represents κ spherical monolayers, including one monolayer of liquid and one monolayer of vapour (Fig. 21.1 *a*). In the curved spherical drop layers, the layer-by-layer method for describing the number of nearest neighbours $z_{qp}(R)$ (Fig. 21.1 *b*) is preserved for a plane boundary. Accordingly, all functions become dependent on R.

The overall balance of the sites of bonds between the surrounding molecules is preserved in the form

$$\sum_{p=q-1}^{q+1} z_{qp}(R) = z. \tag{24.1}$$

Each monolayer $R_q=R+q-1$ contains $N(R_q)$ of elementary cells v_0 (sites). The relation between $N(R_q)$ and the radius of the monolayer R_q is determined from a simple geometric relationship:

$$\begin{aligned} N\left(R_q\right) &= \left[V\left(R_q+0.5\right)-V\left(R_q-0.5\right)\right]/v_0, \\ V\left(R_q+0.5\right) &= 4\pi/3\left[\lambda\left(R_q+0.5\right)\right]^3, \end{aligned} \tag{24.2}$$

where $V(R_q+0.5)$ is the volume of a sphere with a radius $(R_q+0.5)$.

The difference in square brackets determines the volume of the spherical monolayer. The centre of the drop corresponds to the value $R = 0$ in the expression (24.2). The expression (24.2) gives rough estimates of $N(R_q)$ for $R = 1 \div 4$, therefore $N(R_q) = 5$ as the minimum value of $N(R_q) = 5$. The question on the small size of drops should be considered within the framework of a discrete description and here it is not discussed.

It follows from (24.2) that the number of sites of the monolayer R_q is highly accurate in proportion to the area of the sphere S_R passing through the centre of the cell with radius R_q+0.5, where the monolayer radius is measured in the diameter numbers of the molecule λ

$$N\left(R_q\right) = 4\pi/3\left[3R_q^{\ 2}+0.25\right] = 4\pi\left(R_q^{\ 2}+1/12\right) \approx 4\pi R_q^{\ 2} = S(R_q). \qquad (24.3)$$

Each monolayer q, $1 \leq q \leq \kappa$ is assigned its weight $F_q(R)$ – the fraction of sites of type q in the interphase layer. The contribution of each monolayer in the weight function of the sites of the transition region is expressed as

$$F_q(R) = N_q(R)\ /N(R),\ 1 \leq q \leq \kappa,\ N\left(R\right) = \sum_{q=R+1}^{R+\kappa-1} N_q(R), \qquad (24.4)$$

For a plane boundary, the values of $F_q = \kappa^{-1}$.

From the condition of the balance of bonds between the nearest sites in adjacent layers $q = R$ and R+1, the equation $N(R_q)$

$$N(R_q)z_{qq+1}(R) = N(R_{q+1})z_{q+1q}(R), \qquad (24.5)$$

which connects the number of neighbours between different neighbouring layers in the forward $z_{qq+1}(R)$ and the inverse $z_{q+1q}(R)$ directions from the centre of the drop.

Formulas (24.1) and (24.5) hold for each value of R. Two equations are not sufficient to uniquely determine the three values $z_{qp}(R)$ ($p = q$, $q \pm 1$) for any R. A strict procedure for determining $z_{qp}(R)$ from the analysis of successive addition of surface atoms and minimization of the surface area of the drop. Such a procedure is an independent problem, so we confine ourselves to determining one of the quantities $z_{qp}(R)$, starting from physical assumptions.

We represent the structure numbers for the curved lattice $z_{qq\pm1}(R)$ in terms of analogous numbers for the planar lattice $z_{qq\pm1}$ in the form

$$z_{qq-1}(R) = z_{qq\pm1}(1 - \alpha / R), \qquad (24.6a)$$

then from equations (24.1) and (24.5) we obtain

$$z_{qq+1}(R) = z_{qq\pm1}\left[1+ (2-\alpha)/R + (1-\alpha)/R^2\right], \qquad (24.6b)$$

$$z_{qq}(R) = z_{qq} - 2z_{qq\pm1}\ (1-\alpha)\ (1+1/(2R))/R$$

Equations (24.6) can be interpreted as approximation expressions for two cases having a clear physical picture.

In the first case, for $\alpha = 0$, the relation (24.6a) means that for any site of the curved monolayer R_q, the number of bonds to the centre of the drop remains the same as it was in the bulk phase for a plane boundary. This assumption follows from a consideration of the microcrystallite at low temperatures [1.42], for which there are many faces of different orientation of the single crystals. Each face has a flat (three-dimensional) structure, and their docking, forming regions with projected surface atoms between the faces, are regions with an increased number of broken bonds, whereas all the connections deep into the crystal for these surface atoms remain. In this case, for a curved boundary, the number of bonds between neighboring layers oriented from the centre of the drop is larger than for a plane boundary ($z_{qq}+1(R) > z_{qq\pm1}$), and the values of the number of bonds within one layer are smaller ($z_{qq}(R) < z_{qq}$). In the asymptotic limit, the values of $z_{qq+1}(R)$ and $z_{qq}(R)$ tend to their limits $z_{qq\pm1}$ and z_{qq}.

In the second case with $\alpha = 1$, the numbers $z_{qq\pm1}(R)$ are related to the surface area of this monolayer R with the areas of adjacent monolayers $p = q \pm 1$:

$$z_{qq\pm1}\left(R\right) = z_{q\pm1q}S(R_{q\pm1}) \;/\; S(R_q), \tag{24.7}$$

where the plus sign refers to the farther layer from the centre of the drop, and the minus sign to the layer closer to the centre of the drop. This leads to an increase in the number of connections of the drops directed from the centre $z_{qq+1}(R) \approx z_{qq-1}(1+2/R)$ due to a decrease in the fraction of the bonds directed to the centre of the drop $z_{qq-1}(R) \approx z_{qq-1}(1-2/R)$. At the same time, the number of neighbours in the same layer does not change $z_{qq}(R) = \text{const} = z_{qq}$. For $R \to \infty$, the ratio $S(R_{q\pm1})/S(R_q) \to 1$, and this relation is consistent with the volume values z_{qp}. This case refers to large drop radii and high temperatures when the spherical surface is slightly curved.

In the general case, the parameter α is a function of the temperature and size of the drop. When it deviates from the limiting cases 0 and 1, formula (24.6) takes into account the change in all bond numbers $z_{qp}(R)$ from the curvature of the monolayer in accordance with the law λ/R, but with different numerical coefficients.

Structural characteristics [43]. The behaviour of the structural characteristics of a drop as a function of its radius R, varying from 2^2 to 2^{10}, is shown in Figs. 24.2–24.4 for a transition region of width

κ = 8 monolayers and a reference dividing surface located on the average of the 4th monolayer.

The ratio of the number of sites $\eta_\delta = \sum_{q=4-\delta+1}^{4} N(R_q) / \sum_{q=5}^{5+\delta-1} N(R_q)$ in δ monolayers before and after the dividing surface, for different numbers of monolayers δ = 1, 2, 3 and 4 is shown in Fig. 24.2. This characteristic shows the effect of the drop size on the relative contribution of the internal and external monolayers to the energy of the drop, provided that the dividing surface is in the middle. All curves tend to monotone. But for δ = 1 this is achieved at $R \sim 2^8$, and for δ = 4 the ratio close to unity is reached at $R \sim 2^{10}$, purely geometric deviations from unity with respect to the volumes of internal and external groups of monolayers that play an important role in calculating the values of free energy and surface tension are noticeable for sufficiently large drop sizes.

The values of the weight coefficients F_q for different monolayers of the transition region with κ = 8 are given in Fig. 24.3. All the curves tend to the limit $1/\kappa$ = 1/8. For monolayers that are internal with respect to the dividing surface, $F_q \rightarrow 1/\kappa$ from below, and for monolayers external to the dividing surface, $F_q \rightarrow 1/\kappa$ from above. In the case of κ = 8, the ratios F_q differ by more than 7 times for small drops, and become commensurable (the differences between them do not exceed 15%) only at $R > 2^6$.

In addition to the influence of the number of sites in different monolayers, structural characteristics play a role in the distribution of molecules through the values of the numbers $z_{qp}(R)$ in the formulas for local isotherms. Figure 24.4 shows the ratios $z_{4p}(R)/z_{qp}$ for the two variants $z_{qp}(R)$ calculated by formulas (24.6) for α = 0 and 1. Both variants give the same dependence on the drop size for $z_{45}(R)$,

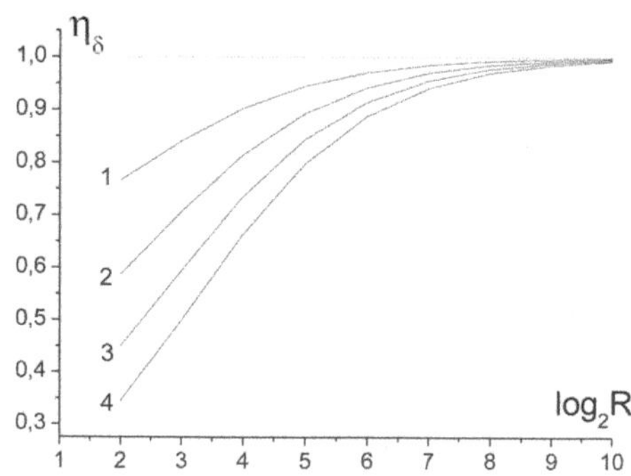

Fig. 24.2. Dependences of structural characteristics $\eta_\delta = \sum_{q=4-\delta+1}^{4} N(R_q) / \sum_{q=5}^{5+\delta-1} N(R_q)$ δ = 1, 2, 3,4 on the radius of the drop R.

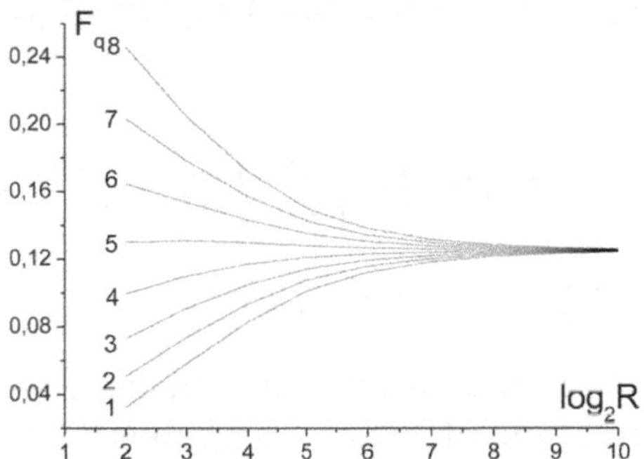

Fig. 24.3. Dependence of the weighting coefficients F_q, $1 \leq q \leq \kappa$ on the radius of the drop R.

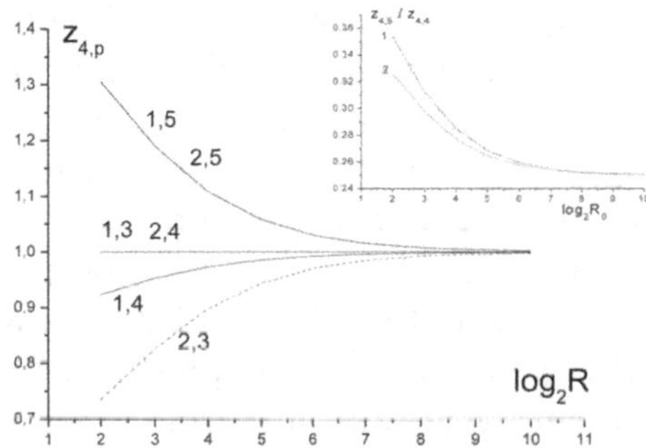

Fig. 24.4. Dependences of the structural characteristics $z_{4p}(R)/z_{4p}$ on the drop radius R for a transition region consisting of $\kappa = 8$ monolayers calculated in two ways. The notation on the curves is b, p, where $b = 1$ for calculations with $\alpha = 0$ and $b = 2 - \alpha = 1$. The insertion gives the ratios $z_{45}(R)/z_{44}(R)$ for the same two variants of constructing the numbers $z_{qp}(R)$.

and the difference between them reduces to redistribution between the numbers of the $z_{44}(R)$ and $z_{43}(R)$ bonds. The insertion gives the relations $z_{45}(R)/z_{44}(R)$ for the same two variants of constructing the numbers z_{qp}. Taking into account that the given numbers $z_{qp}(R)$, like the numbers z_{qp} in (23.1), are exponents in local isotherms, their change with increasing drop radius should cause noticeable differences in the dependences of local densities on the drop size.

Local isotherms. The transition region between phases is a non-uniform system, which exists due to intermolecular interaction. The equation of the local isotherm for the monolayer q relates the probability of filling the cell θ_q and the chemical potential of the molecule μ_q in the layer q, $2 \leq q \leq \kappa-1$. According to [24, 25], we obtain the following system of equations for local isotherms for cells of different types in the transition region of vapour–liquid at a fixed temperature T

$$\beta v_0 a_q P = \exp\left(\beta\mu_q\right) = \frac{\theta_q}{1-\theta_q}\prod_p \left(S_{qp}\right)^{z_{qp}(R)}, \qquad (24.8)$$
$$S_{qp} = 1 + t_{q,p}^{AA}x, \; x = \exp(-\beta\varepsilon) - 1,$$

where $\beta = (k_B T)^{-1}$, k_B is the Boltzmann constant; ε – parameter of intermolecular interaction of nearest neighbours; t_{qp}^{AA} is defined in (23.2). As for a plane boundary above, the normalization relations (23.3) are satisfied.

The system of equations (24.8) with respect to the local densities θ_q has the dimension equal to the number of layers (κ–2) of the transition region between the vapour and the liquid. It is solved by Newton's iteration method for given values of the vapour density (θ_v) for q = 1 and liquid (θ_f) for $q = \kappa$, as for plane boundaries. In the absence of the Laplace equation, the pressure inside the drop and in the surrounding pair are the same – such drops will be called equilibrium drops. For them, the chemical potential in the drop and vapour is also the same. In this chapter, attention is paid to equilibrium drops, which are absent in thermodynamics and appear only in molecular theories.

The difference from the equations for a plane lattice is associated only with the quantities $z_{qp}(R)$, which depend on the radius of the drop. Accordingly, the formulas for free energy and for local pressures also change. On the basis of the local distributions of molecules $\{\theta_q\}$ found, the expression for the free energy of the transition region is written in the form [25] $F(R) = \sum_{q=2}^{\kappa-1}\sum_{i=A,V} F_q(R) M_q^i(R)\theta_q^i$, where

$$M_q^i(R) = \Delta_{i,A}\left(v_q^i + \sum\nolimits_{p=q-1}^{q+1} z_{qp}(R)\varepsilon_{qp}^{ii}/2\right) +$$
$$+\beta^{-1}\left[\ln\theta_q^i + \frac{1}{2}\sum\nolimits_{p=q-1}^{q+1} z_{qp}(R)\ln\left(\frac{\theta_{qp}^{ii}}{\theta_q^i\theta_p^i}\right)\right], \qquad (24.9)$$

where F_q is defined in (24.4). The contribution of a particle of sort i that fills the site q to free energy is denoted by M_q^i, where Δ_{iA} is the Kronecker symbol, $i = A$, V is the sort of particle filling the site q; this contribution is absent for vacancies. The expression (24.9) for the free energy is normalized to one site of the system. If all p coincide with q, then formula (24.9) determines the chemical potential of particle i in the bulk phase q.

The average value of the local pressure π_q in the layer q with respect to the layer structure of the spherical interface is expressed as

$$\beta\pi_q(R)v_q^0 = -\ln\theta_q^V - 1/2\, z_{qp}(R)\ln(\theta_{qp}^{VV}/(\theta_q^V\theta_p^V)). \qquad (24.10)$$

Formula (24.10) corresponds to the average pressure in the layer q of the non-uniform transition region.

The thermodynamic definition for the surface tension σ can be written in terms of the mean pressures $\pi_q(R)$, which is a direct analog of expression (23.6) for a plane boundary.

$$\begin{aligned}\sigma A &= \frac{1}{F_\rho}\left(\sum_{q\le\rho} F_q(R)(\pi_1 - \pi_q(R)) + \sum_{q>\rho} F_q(R)(\pi_\kappa - \pi_q(R))\right) \\ &= \frac{1}{F_\rho}\sum_{q=1}^{\kappa} F_q(R)(\pi - \pi_q(R))\end{aligned} \qquad (24.11)$$

where the first equation is written in the traditional way for metastable drops, the symbol ρ refers to the radius of the reference dividing surface. (Metastable drops are discussed in Appendix 1 and Chapter 7.) The second equality holds because the equilibrium drop exists on both sides of the interface $\mu_\kappa^V = \mu_1^V = \mu_V = -\pi v_0$. Here the functions $M_q^V(R) = -\pi_q(R)v_0$ have the form (24.9). The rest is like for a flat border.

Macroscopic symmetry of a drop. The drop symmetry requires that, in the absence of an external field, the normal and two tangential components of the pressure tensor P [2.9] are isolated, as in the planar case,

$$P = P_T(r)(\mathbf{e}_\theta\mathbf{e}_\theta + \mathbf{e}_\varphi\mathbf{e}_\varphi) + P_N(r)\mathbf{e}_R\mathbf{e}_R, \qquad (24.12)$$

where $\mathbf{e}_R$, $\mathbf{e}_\theta$, $\mathbf{e}_\varphi$ are orthogonal unit vectors corresponding to the coordinates R, θ, φ; $P_N(R)$ and $P_T(R)$ are the normal and tangential pressures. The values of $P_N(R)$ and $P_T(R)$, according to spherical symmetry, do not depend on the angular variables θ and φ. In the transition to uniform macrophases of vapour and liquid, both quantities coincide with the pressures in both phases P_α and P_β, respectively. In the general case, when the spherical system $P_N(R)$ is mechanically examined, it turns out that it is no longer a constant but a function of R, which can be determined from the hydrostatic equilibrium condition. The functions $P_N(R)$ and $P_T(R)$ depend on the size of the drop. The dependence of P_N on R is determined from the condition of mechanical (hydrostatic) equilibrium. In the absence of external fields, the hydrostatic equilibrium equation has the form

$$\nabla \mathbf{P} = 0. \tag{24.13}$$

It should be noted that Eq. (24.13) is an equation for the mechanics of continuous media – they imply a practically constant density, or the presence of macroscopic discontinuities [44]. Both variants do not correspond to the interface of the vapour-liquid phases with variable density. This question arose long ago in connection with the equation for the plane boundary of Becker [7]. The mechanical equilibrium must be considered at the microlevel of the discrete model. For equilibrium pressure drops inside and outside the drop, the same $P_\alpha = P_\beta$, therefore $P_N(R) = \text{const}$.

Mechanical definition of σ. For the components of the pressure tensor π_q^α, $\alpha = N$ and T, formula (23.9) is rewritten for the curved boundary as follows

$$\beta\pi_q^a(R)\, v_q^0 = -\ln\theta_q^V - 3/2\, z_{qp} (R)\cos^2(qp,\alpha)\ln(\theta_{qp}^{VV}/\theta_q^V\theta_p^V), \tag{24.14}$$

As above, the index α denotes the direction of the design axis in space, and the symbol (qp,α) denotes the angle between the direction of α and the direction of communication between the sites in the layers q and p. For a spherical boundary, N is the direction of the normal to the interface of the vapour–liquid boundary; T is one of the two tangential directions, so $\cos^2(qp,N)+2\cos^2(qp,T) = 1$. It follows from (24.13) that the ordinary coupling $\pi_q(R) = (\pi_q^N(R)+2\pi_q^T(R))/3$.

According to the hydrostatic definition of surface tension, we have

$$\sigma = \frac{1}{F_\rho(R)}\left[\sum_{q\le\rho} F_q(R)(\pi_1 - \pi_q^T(R)) + \sum_{q>\rho} F_q(R)(\pi_\kappa - \pi_q^T(R)\right], \tag{24.15}$$

where the functions $\pi_q^T(R)$ are the tangential components of the pressure tensor (24.14).

The boundary value of this function for $q = 1$ in the liquid phase drops is calculated by the formula (24.10) (index of the fluid f is replaced by the index 1). For another boundary value of the function M_κ^V for $q = \kappa$, we have the same formula with the replacement of the subscript 1 by the index κ. Equilibrium drops characterized by the condition that the chemical potential ($\mu_\kappa = \mu_1$) within the drops in the vapour phase at $P_0(T)$ vapour pressure and a given temperature T. This means that the internal pressure in the vapour and liquid in the same and equal π_q^N. These expressions are determined by the formulas indicated in (24.10).

The imposition of an additional condition: the equality of the normal components of local pressures in all layers of the interphase region of equilibrium drops means $M_q^{V(N)}(R) = M_1^V = M_\kappa^V = \mu_V$. It follows that $\pi_q^T(R) = [3\pi_q(R) - \pi_q^N]/2$, therefore, (24.15) follows

$$\sigma = \frac{3}{2F_\rho(R)} \sum_{q=1}^{\kappa} F_q(R)[\pi - \pi_q^T(R)]. \tag{24.16}$$

The additional consideration of the mechanical condition for the equality of the normal components in all layers between the vapour and the liquid leads to an expression that differs from the purely thermodynamic determination of the surface tension (24.11) by a constant coefficient of 3/2 [45]. It is important to note the difference in the two mechanical definitions of surface tension by formulas (24.15) and (24.16), using geometric constraints of the discrete model and the macroscopic symmetry condition at the microlevel.

These differences are due to the fact that in Sections 23 and 24 the simplest way of calculating σ on a rigid lattice is discussed, therefore, the description of the mechanical properties (in particular, of the internal pressure) of the system, which is related to lattice deformability, is approximate [46]. The analysis showed that expression (24.16) gives higher values of σ than formula (24.15) for any R, but this difference has a systematic character and it is of the order of up to 10%. Most importantly, the dimensionless ratio $\sigma(R)/\sigma_{bulk}$, where σ_{bulk} is the value of the surface tension for a planar lattice (Section 23), which is below the goal of our analysis, does not depend on additional accounting for mechanical equilibrium. Thus, the absolute differences in magnitude do not affect any of the size characteristics. The same applies to the ratio $\sigma(R)/\sigma_{bulk}$, calculated by the formula (24.11).

The radius ρ_e of an equimolecular surface for equilibrium drops is determined in the usual way

$$\sum_{q=1}^{\rho_e} F_q(R)(\theta_q - \theta_1) + \sum_{q=\rho_e+1}^{\kappa} F_q(R)(\theta_q - \theta_\kappa) = 0, \tag{24.17}$$

where the index $q = 1$ refers to the drop, the index $q = \kappa$ refers to the pair. For $q \leq \rho$ there are layers with increased density, with $q > \rho$ – layers with reduced density. This surface is a dividing surface that divides the intermediate region into two subregions, related to the liquid and the vapour, respectively.

25. Properties of equilibrium drops [45]

Molecular models make it possible to calculate the concentration density profile in the interphase region. For curved boundaries, the procedure for calculating the concentration profile is identical to the procedure for a flat boundary. The radius of the liquid part of the drop R is an additional parameter of the state of the 'drop–vapour' system. The temperatures below are given in the dimensionless form $\tau = T/T_c$, where T_c is the critical temperature in the bulk phase and is defined as the disappearance temperature of two-phase regions (stratification).

Figures 25.1 and 25.2 show curves showing the influence of the drop radius and temperature on the properties of the interphase region. From the comparison of Figures 25.1 *a* and 25.2 *a*, on which the concentration profiles of the interphase region are given for three drop sizes $R = 30$ (1), 120 (2), 480 (3), it is evident that the width of the interphase region increases with increasing temperature. For the low temperature ($\tau = 0.55$, Fig. 25.1 *a*), the profiles with $R =$ 120 (2) and 480 (3) practically coincide with the volume profile, whereas at high temperatures ($\tau = 0.82$, Fig. 25.2 *a*) the profile even for $R = 480$ (3) differs from the volume profile.

Figures 25.1 *b* and 25.2 *b* show the local values $(\pi_q - \pi_0)$ used in the formulas for calculating the surface tension. The finite values of σ depend on their distribution with respect to q. These values are differently distributed in the transition region for different drop radii R. With increasing temperature and, respectively, the width of the interphase region, their maximum value sharply decreases and the curves become smoother. The curves in these figures, starting from zero ($\pi_1 - \pi_0 = 0$), increase sharply and reach a maximum, almost as sharply decrease to a negative value, after which they slowly increase to zero ($\pi_\kappa - \pi_0 = 0$).

Figures 25.1 *c* and 25.2 *c* show the values of σ/σ_b for different positions of the reference surface $q_r = q$ within the interphase region. The values of $\sigma(q)$ decrease monotonically with increasing number of the layer q, $1 \leq q \leq \kappa$, because the monolayer q determines the

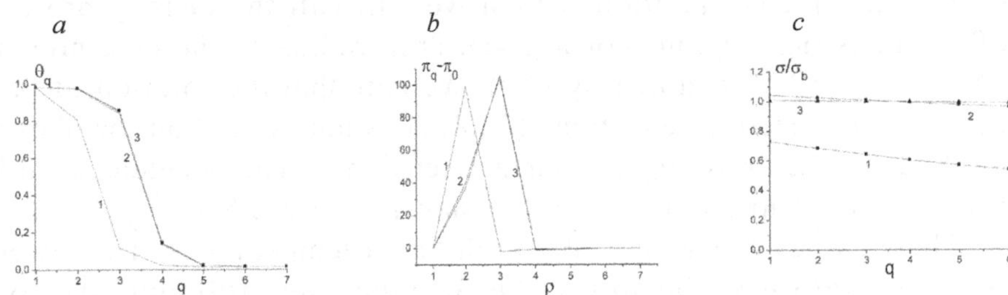

Fig. 25.1. Change of the layer number q in the interphase region to the local concentration value θ_q, (*a*), the value $(\pi_q - \pi_0)$ (*b*) and the surface tension value σ (*c*) for drops of size $R = 30$ (1), 120 (2), 480 (3) at $\tau = 0.55$.

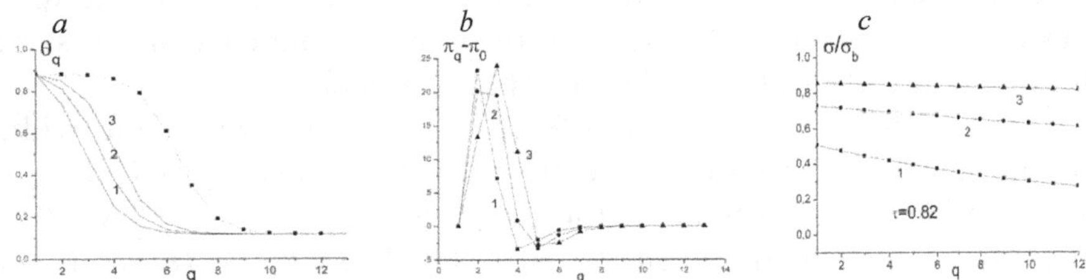

Fig. 25.2. The influence of the layer number q in the interphase region on the value of the local concentration θ_q, (*a*), the values $(\pi_q - \pi_0)$ (*b*) and the surface tension σ (*c*) for drops of size $R = 30$ (1), 120 (2), 480 (3) at $\tau = 0.82$. Points on 3, and correspond to a flat boundary.

position of the reference surface ρ_r, the denominator in formula (24.15) increases with increasing q, and the numerator remains constant. Consequently, the maximum of $\sigma(q)$ is reached for $q = 1$. Hence the main conclusion is that for the equilibrium drops, one can not use the traditional thermodynamic concept of the *surface of tension*, as a reference surface located inside the transition region, in which the quantity σ has an extremum. This conclusion agrees with the fact that equilibrium drops in the transition region lack a dividing surface on which a pressure jump occurs according to the Laplace equation. The position of the equimolecular dividing surface $\rho = \rho_r = \rho_e$ is uniquely determined by the concentration profile.

Width of the transition area. Above, a method has been described for determining the width of the interphase region $\kappa(T)$ and the concentration profile of the density of the transition region of a plane boundary. It is completely preserved for the spherical boundary.

The value of κ for a drop depends on the radius R and temperature. As the width of the transition region we select the smallest value of κ for which there exists a monotone solution θ_q, $1 \leq q \leq \kappa$ of the system under consideration, which we will call the density profile of the transition region. For a given drop radius R, the determined value of κ will be denoted by $\kappa(R)$. We note that the solution of the equilibrium system of equations (24.8) does not depend on the above methods for determining the surface tension σ. The dependences of $\kappa(R)$ for different temperatures are shown in Fig. 25.3.

These curves differ markedly at different temperatures. The lower the temperature τ, the lower the R value, the minimum size of the transition region is established. The stepwise course of the curves is due to a discrete change in the number of monolayers. The main conclusion is that with increasing drop radius, the quantity $\kappa(R)$ increases, and the width of the interphase region of the drop $\kappa(R)$ does not exceed the width κ for a plane boundary at the same temperature: $\kappa(R) \rightarrow \kappa(bulk)$ and $\kappa(R) \leq \kappa(bulk)$.

Figure 25.4 illustrates the effect of the drop size on the width of the interfacial region. For different temperatures, $\gamma = V_\kappa/V_{drop}$ is the fraction of the volume of the interphase layer $V_\kappa = \Sigma_{q=2}^{\kappa-1} N_q(R)$ in the full volume of the drop $V_{drop} = V_{liq} + V_\kappa$. Here $V_{liq} = 4\pi R^3/3$ is the volume of the liquid part of the drop.

With increasing drop size, the share of the interphase region in the total volume of the drop sharply decreases. With increasing temperature, the width of the interphase region κ, and, consequently, the value $\gamma = V_\kappa/V_{drop}$ increase. At R = 500, V_κ is 3.5% of the total volume of the drop for τ = 0.62 and 6% of the total volume of the drop for τ = 0.82. For small drop sizes, the fraction of the interphase layer in the total volume of the drop sharply increases. Near R = 50, this fraction γ reaches 40–50%, and at R = 10, the share of the interphase region in the total volume of the drop prevails. The inset in Fig. 25.4 shows the range of variation of ρ_e for drops of radius

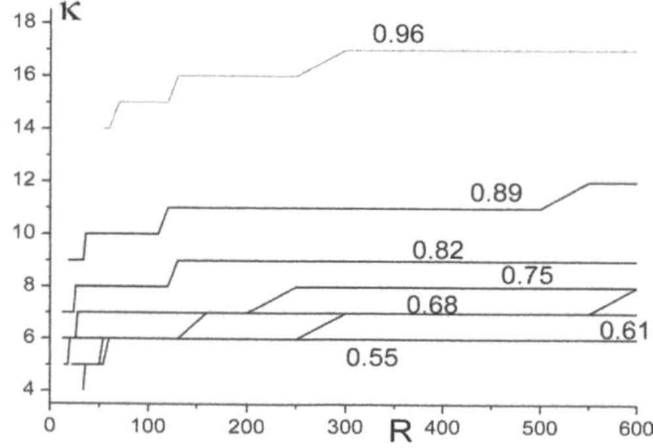

Fig. 25.3. Dependence of the width of the transition region of the spherical drop $\kappa(R)$ on its radius R. The numbers at the curves denote reduced temperature τ.

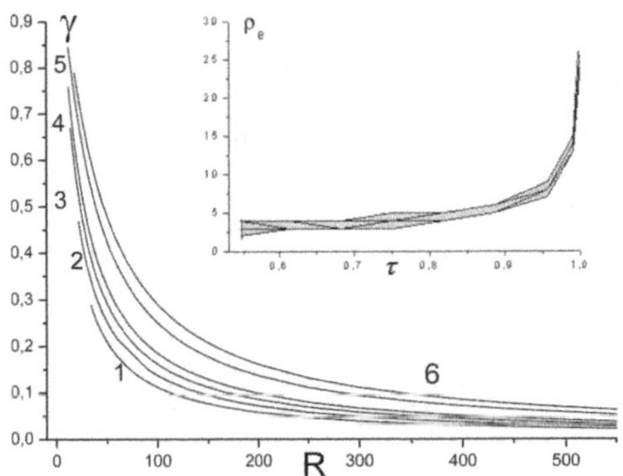

Fig. 25.4. The dependence of the quantity $\gamma = V_\kappa/V_{drop}$ on the drop size R for the temperatures $\tau = 0.55$ (1), 0.61 (2), 0.68 (3), 0.75 (4), 0.82 (5), 0.89 (6), here $\alpha = 0.5$. The inset shows the dependence of the intervals of variation of ρ_e for drops of size from $R = 8$ to $R = 1000$ on the temperature.

from $R = 8$ to 10^3 as a function of temperature. The calculation was carried out for 7 radii (with a practically constant step on a logarithmic scale). It can be seen that the spread of the values of ρ_e is only about 2 monolayers. The behaviour of the curves in Fig. 25.4 coincides with the course of the behaviour of the curve ρ_e in Fig. 23.4, constructed for a plane lattice, for which we can conditionally assume that $\rho_e \approx \kappa/2$.

Size dependence of surface tension. Figure 25.5 shows the calculated temperature dependence of σ on the drop size normalized to the surface tension of the flat lattice σ/σ_b (at the same temperature). The calculations were carried out with two limiting values with the parameter $\alpha = 1$ and 0. It is seen that the surface tension increases monotonically with increasing drop radius, tending to its volumetric value $\sigma(R) \to \sigma_{bulk}$. For relatively low temperatures $\tau < 0.75$, an

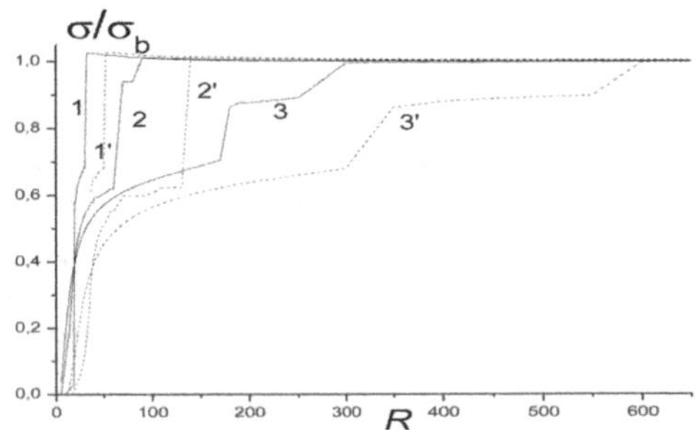

Fig. 25.5. Dependence of σ/σ_b on the drop size; (a): $\tau = 0.55$ for $\alpha = 1$ (1), $\tau = 0.55$ for $\alpha = 0$ (1') $\tau = 0.61$ for $\alpha = 1$ (2), $\tau = 0.61$ for $\alpha = 0$ (2') $\tau = 0.68$ for $\alpha = 1$ (3), $\tau = 0.68$ for $\alpha = 0$ (3').

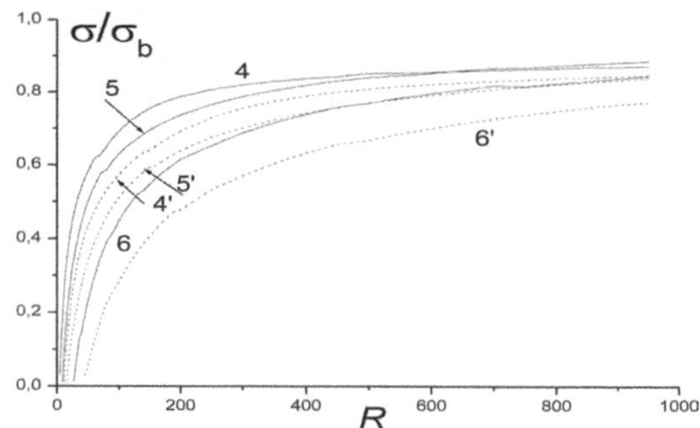

(b): $\tau = 0.75$ for $\alpha = 1$ (4), $\tau = 0.75$ for $\alpha = 0$ (4′) $\tau = 0.89$ for $\alpha = 1$ (5), $\tau = 0.89$ for $\alpha = 0$ (5′) $\tau = 0.96$ for $\alpha = 1$ (6), $\tau = 0.96$ for $\alpha = 0$ (6′).

abrupt change in surface tension is possible. These jumps occur at small values of the radius R in connection, which raises the question of the proportionality of the width of the transition region and the size of the molecules. In traditional thermodynamic constructions it is considered that the size of a molecule can be neglected. With increasing temperature, the number of monolayers κ increases, and the relative change of κ with a growth of R plays a smaller role.

Figure 25.5 shows that the theory allows us to consider a very wide range of drop sizes. This served as a basis for analyzing the characteristic scales of drop sizes: from nanometers to the submicron range (see the next section 26).

Temperature dependence of surface tension. Figure 25.6 gives the temperature dependences of the surface tension of drops $\sigma(T|R = \text{const})$ for a number of fixed values of radii from 16 to 10^3. For each temperature, the normalization was carried out to its volume value σ_b.

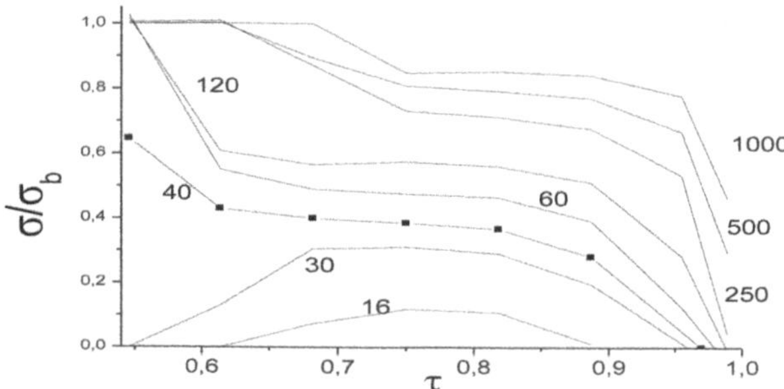

Fig. 25.6. Dependences of σ/σ_b for drops of different sizes R from 16 to 1000 on temperature τ, $\alpha = 1$. The values of the drop radius R are shown on the curves.

There are also two types of curves. The first type of curves refers to sufficiently large radii ($R \geq 40$), for which the surface tension is positive, starting from low temperatures. The general trend of these curves is a decrease in σ with increasing temperature, although a small non-monotonicity of the curves is possible in the intermediate temperature range. Near the critical temperature, all curves have a negative value of $\sigma(R)$. This indicates the instability of this drop size at high temperatures.

For the second type of curves, the dependences $\sigma(T|R = \text{const})$ increase with increasing temperature from negative values of $\sigma(R)$ to a certain maximum and after its passage decrease to negative values, as in the first case. In the case of a positive maximum, there is a range of temperatures for the specified drop sizes, in which the equilibrium drops are stable. These drops can not exist at low and high temperatures because of the negative value of the surface tension. In particular, this refers to $R = 16$ and 30. At the same time, the curve for $R < 10$ has a maximum at negative values of σ which indicates its instability at any temperatures. The limiting value of R separating both types of curves is $R \sim$ from 35 to 40. The reversal of the surface tension value to zero corresponds to the loss of stability of the drop as a liquid phase. Or the condition $\sigma = 0$ means that the molecules of the system under consideration are not split into two phases. Recall that the positive value of surface tension indicates the impossibility of spontaneous unlimited increase in the area of the interface [2,9].

This circumstance indicates a natural (in physical sense) criterion of formation (small in size), since by construction within the drop there is necessarily a region with constant properties, which is necessary by the definition of the phase, which was introduced by Gibbs [1]. For more details, see below in Section 27.

26. Three characteristic drop size scales [46]

A drop is an object of an intermediate size: from molecular associates to macrodrops of a new phase. The physical basis for changing concepts, in fact, is absent. The nature of the intermolecular potential acting in the system of molecular associates implies the absence of saturation and, as a consequence, the absence of a limit for the degree of association [40]. When describing the equilibrium drop there is no Laplace equation, which greatly simplifies mathematical analysis. Let us consider the behaviour of the surface tension on the drop size $\sigma(R)$. Three characteristic sizes can be distinguished which correspond to: 1) the value of the radius R_b, the greater of which the surface tension values are close to the volume value, with small curvatures of the surface of large drops; 2) the value of the radius R_t above which the thermodynamic description of the surface tension of the drop is justified, when both the discrete nature of the matter and the contributions of spontaneous fluctuations can be neglected; and 3) the critical value R_0 of the beginning of the formation of the phase associated with the appearance of the surface tension of the dense phase.

Large radii. For the analysis of large drops, it is convenient to count the drop properties from the equilibrium profile of the concentrations of the plane boundary (we denote its characteristic by index ∞). The deviation of the surface tension of the curved boundary $\sigma(R)$ from σ_∞ can be written as

$$\delta\sigma(R) = \sigma_\infty - \sigma(R) = \frac{1}{F_\rho(R)} \sum_{q=1}^{\kappa} F_q(R)[\pi_q(\infty) - \pi_q(R)]. \tag{26.1}$$

If, as a first approximation, we use the assumption that the density profiles on the curved $\{\theta_q\}_R$ and the flat $\{\theta_q\}_\infty$ separation boundaries are close, then the functions are also close to analogous functions in the volume, and the problem reduces to bulk densities and structural characteristics. Expression (6) gives

$$\delta\sigma(R) = \frac{1}{2N_\rho} \sum_{q=1}^{\kappa} N_q \sum_{p=q-1}^{q+1} \delta z_{qp}(R) \ln[\theta_{qp}^{VV}(R)/(\theta_q^V \theta_p^V)]_\infty, \tag{26.2}$$

here $\delta z_{qp}(R) = z_{qp}(R) - z_{qp}$, where $\Sigma_p \delta z_{qp}(R) = 0$, according to the formulas (24.1). The presence of the logarithmic factor, taking into account expressions (24.6) for the numbers $z_{qp}(R)$, which lead to members of type $1/R$, results in the non-zero value of the second sum over the index p. The ratio under the logarithm sign in (26.2) depends on the temperature; nevertheless, the coefficient before $(1/R)$ in the second sum can be qualitatively estimated as one. Then $\delta\sigma(\mathrm{R}) = \sum_{q=1}^{\kappa} N_q/(2RN_\rho) \approx \kappa/(2R)$ therefore, as the size of the drop decreases, σ decreases.

The main result (26.2) is that the quantity $\delta\sigma(R)$ depends on the radius as $1/R$, i.e. *any* curvature of the surface leads to a change in the surface tension, or the range of the drop size values, within which the surface tension depends on the curvature of the dimension, is large. If we assume that the volumetric value of σ_∞ for a plane boundary can be used as a reference for the quantity $\sigma_b = 0.99\sigma_\infty$, then this condition is reached at $R_b/\lambda > 10^2\kappa/2$. Here, R_b denotes the size of large drops, for which the values of σ differ little from their volume value for the flat interface σ_∞. For all values of $R > R_b$, this difference can be neglected. As the size of the drops decreases, the formula (26.2) ceases to work. In addition to changing the structural characteristics, it is necessary to take into account the change in the $\{\theta_q\}_R$ profile. However, the sign of $\delta\sigma(R)$ is preserved, i.e. the smaller the radius R, the smaller $\sigma(R)$.

The width of the interphase region κ depends strongly on the temperature: from ~3λ (near the triple point) to 40λ (near the critical point). It follows that even for low temperatures, the size range of drops having surface tension values close to volume values can reach about 10^2 molecular diameters, i.e. a value of the order of 40 nm for argon atoms, and at high temperatures reaches a micron scale. For atoms of inert gases and simple molecules ranging in size from 0.3 to 0.6 nm, this occurs at a drop radius up to the micron size. This fact was not paid attention before, and, as a rule, the bulk value of the surface tension is traditionally attributed to relatively small drops.

Small radii ~R_0. We have shown above that for small drops there exists a radius $R_0 = R(\sigma = 0)$ at which the surface tension vanishes, since a small drop, in fact, becomes a molecular associate or cluster without surface tension [40]. The size of the drops R_0 corresponds to the critical size of the associate, which determines the onset of the formation of the phase and the appearance of the surface tension of the dense phase. For values of $R < R_0$, there is a negative value of the surface tension σ, at the boundary of the drop, which is considered unstable according to thermodynamic concepts. The condition $\sigma = 0$ defines the minimum drop size as a cooperative property of the system, by analogy with the *stable* two-phase equilibrium, in contrast to the condition on the unstable state of the system in thermodynamics [40]. This definition in principle differs from the critical size in thermodynamics, which is introduced through the condition on the maximum of the free Gibbs energy for the *unstable* state of the embryo [40] (and also because thermodynamics does not determine this size, but it is established a priori [40]). The same critical radius R_0^* ($\sigma = 0$) exists for metastable drops [43]. At the point R_0^* the pressure inside the drop also becomes the total pressure

$P_f = P_v = P_s$. The values of R_0^* and R_0 coincide (in more detail in Section 60).

The mathematical analysis of the condition $\sigma = 0$ consists in solving the system of equations for the concentration profile (24.1). In [46], a qualitative estimate was obtained for $R_0 = 7\lambda$ on the basis of the simplest model consisting of two monolayers and a degree of asymmetry of the profile of the order of ~1.3 (from the data of numerical calculations). Thus, it turns out that the critical value of the equilibrium nucleus R_0 must be at least 7 monolayers. This radius corresponds to sufficiently large relative fluctuations in the number of molecules in the whole drop $\eta_V = 2.6\%$ or only on its surface $\eta_s = 4.0\%$ (for more details see Section 34).

Intermediate sizes R_t. The sizes R_0 and R_b cover a wide range of sizes, from the field of existence of molecular associates to the two-phase states of a system with a practically flat boundary. Between them is the intermediate size of the drops R_t, at which it is necessary to take into account, on the one hand, the discreteness of the substance, and, on the other hand, the requirement of absence of spontaneous (thermal) fluctuations in the substance [47]. These two interrelated concepts define the lower limit of the applicability of the thermodynamic approach.

Any thermodynamic potential contains a surface contribution only if it gives a sufficiently appreciable effect on the calculated characteristics of the system. Otherwise, the contribution of the surface can be neglected – it is sufficient to consider the bulk coexisting phases. The presence of a surface contribution means the development of the surface, and the large contribution of the interface between the phases in the thermodynamic characteristics of the system. For the calculation of the latter, the values of the first and second derivatives of the thermodynamic potentials (by temperature for heat capacities and by pressure for the compressibility coefficients) are necessary. Non-planar interfaces are characterized by the curvature of the local surface area, which is also expressed in terms of the first and second derivatives of the equation of the surface $z(x,y)$ along the x and y coordinates. In order for these derivatives to be meaningful, all the thermodynamic functions must be defined at the points of space in question, as well as for plane boundaries [24]. From this it follows that there is some minimal characteristic length that determines the size required for the calculation of spatial derivatives.

From the analysis of the properties of the surface tension of spherical drops, it is possible to distinguish three characteristic sizes of a drop $R_0 < R_t < R_b$. The first characteristic size, equal to $R_0 \sim 10\lambda$, corresponds to the onset of the formation of a dense phase and to the appearance of a surface tension of the drop. For $R < R_0$, we have molecular associates. The second characteristic size, equal to R_t (it is estimated in Section 34 from 40λ to 90λ), separates the range of applicability of the thermodynamic description of the surface tension of a drop. For $R > R_t$, the discreteness of the matter and the contributions of spontaneous fluctuations can be completely neglected. The third characteristic size, equal to $R_b \sim 10^2 \div 10^3\lambda$ for different temperatures, separates the size range of large drops, in which the surface tension values are close to the bulk value.

27. Criterion for the minimum phase size

Figure 25.6 indicates the existence of solutions of the system of equations (24.8) for which the values of σ are negative. This occurs at critical drop sizes. The condition $\sigma = 0$ can be achieved both by raising the temperature near the critical temperature, so by decreasing the size of the drop. In the first case, the system is in the region of large thermal fluctuations and the selected radii are smaller than the average size of the fluctuations at a given temperature. In the second case, the 'phase' inside the drop becomes unstable due to the large fraction of the interphase region in the total volume of the drop (see Fig. 25.4).

Calculation of the surface tension of the drop $\sigma(R)$ of an arbitrary radius R allows one to answer the question posed about the minimum phase size R_0. The condition $\sigma > 0$ means that there is no spontaneous increase in the surface, which is a sign of the thermodynamic stability of the drop as an independent phase. The value of the drop radius $R_0 = R(\sigma = 0)$, at which the equality $\sigma = 0$ is satisfied, is taken as the lower boundary of existence of the liquid phase. With the drop size $R > R_0$, we have a two-phase system; therefore, in fact, the value of R_0 is limited by the possibility of introducing the concept and calculating the single-phase chemical potential on the traditional $\mu(T)$ curves for two coexisting phases. This follows from the fact that direct contact of these phases is an indispensable condition for the two-phase coexistence of molecules. In turn, molecular associates from dimer to drops of size $R < R_0$ can exist for a long time as a result of the dynamic process of formation and decomposition of associates, but they can not be attributed to the properties of the equilibrium phase and equilibrium surface tension. Thus, the concepts of cluster and drop have distinct differences related to the thermodynamic behaviour of these objects – with the value $R_0 = R(\sigma = 0)$.

A simplified analytical estimate gave the smallest value of R_0 equal to 7λ. The calculations for the full model showed slightly larger values of R_0/λ from 8 to 12, depending on the structure of the interphase boundary, i.e. $R_0 \sim 10\lambda$. According to the parameters of the intermolecular interaction, found from the description of the experimental data for plane boundaries, the size dependences of the surface tension for 19 substances were calculated, some of which are given in Section 23. Figure 27.1 shows these dependences of the surface tension $\sigma(R)$ normalized to the surface tension of the flat lattice σ_{bulk} on the radius of the drop R normalized to the width of the monolayer λ, and in Fig. 27.2 values of R_0/λ as a function of reduced temperature τ [35].

These calculations of the values of R_0 should be refined: they refer to the values of the structural parameter $\alpha = 1$, which correspond to macroscopic values of the radii (refinement of α should increase R_0). This question was considered in more detail (the effect of the structural parameter α) in [45]. The calculations also did not take into account the scaling corrections near the critical temperature [8]. Nevertheless, these results show that there is a general universality of the behaviour of 19 substances, and it reflects the characteristic features of the behaviour of the cooperative system at small sizes.

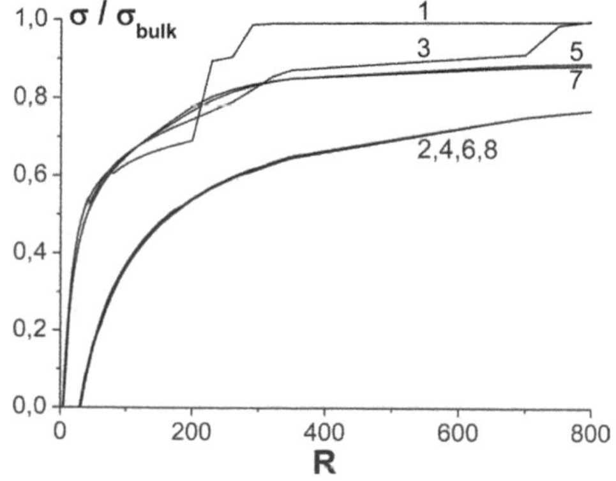

Fig. 27.1. The normalized surface tension $\sigma(R)$ for two reduced temperatures $\tau = 0.6$ (1, 3, 5, 7) and 0.95 (2, 4, 6, 8) with interatomic interaction within the limits of the c.s. $R_{lat} = 1$ (1, 2), 2 (3, 4), 3 (5, 6), 4 (7, 8) (see Table 23.1). [35].

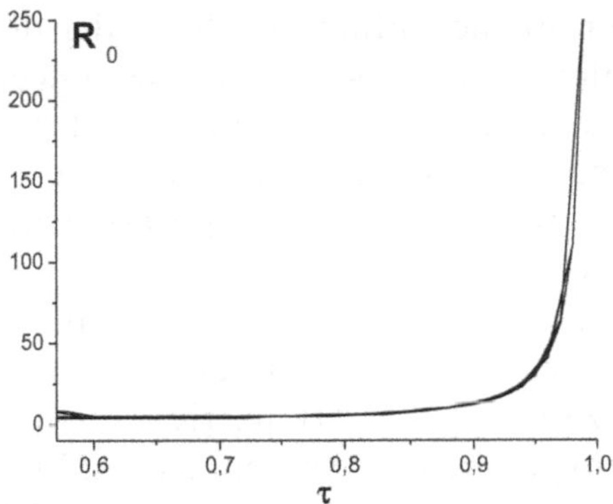

Fig. 27.2. The dependence of the minimum radius of the drop R_0, corresponding to the formation of a new phase, on reduced temperature τ [35].

The universality of estimating the minimum size of the equilibrium phase [48]. The question of the lower boundary of the sizes of the existence of equilibrium phases is important in the description and interpretation of experimental data at the microlevel, since it determines the type of equations that should be used in calculating thermodynamic characteristics (recall that the very concept of 'phase' is thermodynamic [1]). There are three types of experimental data in which the question of the minimum phase size plays a key role: 1) the formation of a new phase in the process of vapour condensation, in which the concept of a critical nucleus is introduced [1,40,49]; 2) capillary condensation in narrow pores, in which adsorption–desorption hysteresis disappears when the width of the pore width decreases [50–52]; and 3) spontaneous magnetization of particles of small size, in which the magnetization hysteresis disappears as the crystallite size decreases [53]. In [48], these systems are discussed from a single point of view of comparing the values of the minimum particle size that have the properties of the equilibrium phase. The systems under consideration have one common property: cooperative behaviour of molecules or spins.

A free drop exists only through intermolecular interaction – this is a purely cooperative property. For isolated drops, the critical size is discussed above.

Fluid in the pores. For porous systems, a systematic analysis of phase transitions in small systems is given in [8, 54]. The exfoliating simple (low-molecular) liquid, which is in a limited volume of the pore, decomposes into two phases of vapour and liquid. A characteristic feature of such systems is that the surface potential rapidly decreases along the normal from the pore wall to its central part. The restriction of the volume to the walls of the pores only affects the wall region in which several dense molecular monolayers can form before the capillary condensation of the molecules in the central part of the pore. Such a potential can not cause condensation in the central part, if there is no cooperative behaviour of the molecules themselves.

The conditions for stratification of fluid in narrow slit-like, cylindrical, globular and other types of pores over a wide temperature range were analyzed in [55]. Of particular importance is the description of the experimental data on the capillary condensation of a number of adsorbed molecules in long cylindrical channels of MCM-41 [56], which have quasi-one-dimensional properties for sizes $R_c^* < 6.5–7\lambda$. Similar results were obtained for channels of other cross sections [54].

The knowledge of R_c^* allows us to find the radius of the spherical pore R_s^* containing the same number of molecules [54], which is in a limited section of the channel with radius R_c^* and length L^* if we assume that the cooperative properties of molecules of an infinite channel are approximated by this section. Approximation is achieved under the condition $2/L < \delta_1^*$, where δ_1^* is a numerical criterion corresponding to the performance of this approximation. In terms of its meaning and size, the quantity δ_1^* corresponds to a value of $1/R_c^*$ [54]. As a result, it was obtained that the minimum critical channel length $L^* = 2R_c^*$. Whence follows the minimum volume of the spherical pore R_s^* equal to $R_0 = R_s^* \sim 10\lambda$. Consequently, the beginning of the formation of an equilibrium liquid phase in a vapour and the stratification of the fluid in bounded volumes of porous systems of different geometry occur at the same smallest size of a new phase: $R_0 \sim 10\lambda$ (or about 4 nm for Ar atoms).

Magnetization of microcrystals. From the point of view of the statistical theory [32,57–59] of the first-order phase transitions, there is a one-to-one correspondence between the phase behaviour of liquid stratifying systems and the spontaneous magnetization of magnetic materials having spin 1/2. In particular, the theory of phase transitions [57–59] ascribes to them the same critical exponents in the vicinity of the critical point. Small solid particles exist due to the interaction between its atoms. Spin–spin interactions are realized for a fixed distribution of atoms inside the particles. They depend on the size and shape of the atomic subsystem, including the dependence on the property of the boundary of a small particle. The number of atoms in a particle must be large enough to ensure the existence of a spontaneous magnetization of the spin subsystem. If the particle is uniform in its central part, then the behaviour of the spin subsystem is equivalent to the behaviour of molecules inside the central part of the pore. The data on the magnetization of small crystals of magnetic substances (Fe, Ni, and Co metals) are given in Ref. [48], indicating the possibility of the appearance of spontaneous magnetization [60-64]. An indication of the spontaneous magnetization is the existence of a hysteresis loop when an external magnetic field is applied. The loop disappears as the temperature rises and/or the size of the microcrystal changes to a paramagnetic state, so the picture is completely analogous to adsorption hysteresis.

In the work [60] nanowires of Fe, Co, and Ni ferromagnetic metals were obtained in porous alumina matrixes with channel diameters from 5 to 40 nm. The course of the hysteresis curves of the magnetization depends on the diameter of the channel. The resulting nanowires were polycrystalline. The smallest crystal sizes were several nanometers. Hysteresis is observed on large crystals. If the wires were uniform in length, then such systems would be analogous to capillary condensation in MCM-41 or for a system of larger pores. It is noted [60] that with a decrease in the pore size, fluctuations in the system are markedly increased.

The study of Fe powders led to an estimate of the particle size in which a magnetization hysteresis appeared exceeding 5 nm [53], which corresponds to the condition $R_0 \sim 10\lambda$, where $\lambda = 0.26$ nm is the crystal lattice size of Fe. For the Co powder used as a catalyst, particle size estimates were obtained for the transition from the paramagnetic state to the ferromagnetic state from 7 to 8 nm [61, 62], which corresponds to values R_0 from 14λ to 16λ at $\lambda = 0.25$ nm. In [63], with clusters of Ni in very narrow carbon channels not exceeding 2.5 nm, only paramagnetic states of the clusters were detected. These examples are in good agreement with the results of capillary condensation in narrow-porous systems.

However, in general, more complex situations are possible in magnetic systems, when magnetism arises due to the joint properties of embedded metal clusters and the original matrix, as is the case in the Pd clusters system in the carbon matrix [64]. Such situations in capillary condensation are not realized.

Thus, by comparing the minimum size of thermodynamically stable liquid drops and the pore sizes in which the fluid stratification into two coexisting phases occurs, as well as data on the hysteresis of the spontaneous magnetization of microcrystals, one can conclude that there exists a universal size of the volume of atoms / molecules that corresponds to the condition for the realization of the equilibrium phase. Estimates [48] indicate a qualitative correspondence between the three types of experimental data and require more detailed analysis by refining the factors affecting the numerical values of the minimum sizes for each of the experiments. The maximum value of R_0 converted to the volume of the sphere does not exceed 15λ; exactly such an increase in R_0 gives the above estimates for magnetic systems for which these deviations are due to the inaccuracy of the polydispersity of the powders, the oxidation state of metals, the non-spherical shape, etc. Of course, the thermodynamic approaches can not be used for this kind of refinement of the properties of small bodies, especially when considering the sizes of small particles smaller than the minimum estimates for the existence of equilibrium phases $R_0 \sim 10\lambda$.

28. Equilibrium drops and phase rule

The molecular theory leads to the existence of a new object of small systems – equilibrium drops. Such drops are impossible in thermodynamics – the problem of the Kelvin equation was discussed in Section 7.

The solutions to the concentration profile of the equilibrium drops completely correspond to the result of the Yang–Lee theory [65], since they do not contain any information about the states of the system inside the binodal curve. The boundary conditions on the concentration profile correspond to the macroscopically stable states of the stratifying phases of vapour and liquid, as for plane interfaces. For them, the conditions for mechanical and chemical equilibrium of the bulk phases are satisfied. As the drop size increases, the concentration profile becomes the profile of the planar interface. Accordingly, the surface tension $\sigma(R,T)$ of the drop is transformed into the surface tension of the plane macroscopic boundary $\sigma(T)$.

Equilibrium drops are an intermediate state of matter in the transition from small associates to macroscopic phases, as their size increases, under the same fixed external conditions T, P. The existence of equilibrium drops indicates the need for a change in the recording of the phase rule for curved boundaries: practically in all publications, and even in the Commentaries on page 506 of [1] that surface tension σ can increase the number of degrees of thermodynamic freedom per unit due to the presence of a curved boundary even in the absence of the external field. Obviously, the drop type should not influence the phase rule for two reasons.

Firstly (in terms of physical meaning), any boundary is a 'non-autonomous phase'. The influence of the boundary extends only to a certain near-surface region. Thus, the surface does not affect the internal properties of the phases, unlike the external fields acting on all molecules within a given volume. In other words, the properties of non-autonomous phases can not be parameters of the state of the system. They necessarily functionally depend on the parameters of the state of the bulk coexisting phases and must obey the same conditions of thermal, mechanical and chemical equilibrium at each point of the transition region, as do the coexisting phases themselves.

Secondly (from a formal point of view), despite the presence of two pressures P_α and P_β on the curved boundary of the metastable drop (i.e., the number of parameters of such a system is increased by one, as it contains more parameters), but for the phase rule this it does not matter – it should take into account all the relationships between the parameters of the system. In the case of a metastable drop, there is an additional connection according to the Laplace equation, which connects these two pressures through $\sigma_{met}(R,T)$. Therefore, the total number of degrees of thermodynamic freedom for a curved surface does not change with respect to such a number for a plane boundary (and equilibrium drops again emphasize this). This conclusion fully corresponds to the physical meaning of the 'phase' as a macroscopic object, for which the state of the surface is secondary, but the surface necessarily exists and limits its size.

The existence of equilibrium drops allowed us to pursue a strict concept of the minimum phase size, and this concept turns out to be the same for both equilibrium and metastable drops, because it satisfies the Laplace equation (Section 60). This fact indicates that the use of rigorous molecular theories makes it possible to extend the thermodynamic concept of 'phase' to small systems, down to values less than 10 nm. This is actually two orders of magnitude smaller than the traditional concept of the phase, as a macroscopic formation of the order of the micron size.

Note that in recent years, compared with the references [11–23], little has changed. As an example, we cite almost randomly selected papers [66–69] from the set of analogous papers. They also almost arbitrarily refer to the concepts of the minimum phase size, the regions of phase transformations and, accordingly, to surface tension. In each of these works, these concepts are interpreted in their own way, depending on the type of system and methods of carrying out calculations without any justification for their ideas and their connection with thermodynamics.

References

1. Gibbs J.W., Thermodynamics. Statistical mechanics. Moscow, Nauka, 1982. 584 c.
2. Ono S., Kondo C., Molecular theory of surface tension. Moscow, IL, 1963. [Handbuch der Physik, Vol X (Springer) 1960]
3. Guggenheim E.A., Trans. Faraday Soc. 1940. V. 37. P. 397.
4. Guggenheim E.A., Thermodynamics, Amsterdam, 1950.
5. Fower R.H., Proc Roy. Soc. A. 1937. V. 159. P. 229.
6. Fower R.H., Physica. 1938. V. 5. P. 39.
7. Bakker G., Kapillaritat und Oberflachenspannung, Handbuch der Experimental physik, Bd. VI, Leipzig, 1928.
8. Tovbin Yu.K., Molecular theory of adsorption in Porous Solids. Moscow: Fizmatlit, 2012. [CRC, Boca Raton, Florida, 2017].
9. Rowlinson, J., Widom B., Molecular theory of capillarity. Moscow, Mir, 1986. [Oxford: ClarendonPress, 1982]
10. Croxton K., Liquid State Physics. A Statistical mechanical Introduction. Moscow, Mir, 1978. 400 p. [Cambridge Univer. Press, 1974]
11. Iwamatsu M., J. Phys.: Condens. Matter. 1994. V. 6. L173.
12. Baidakov V.G., Boltachev G.Sh., Zh. fiz. khim. 1995. V. 69. P. 515.
13. Moody M.P., Attard P., J. Chem. Phys. 2002. V. 117. P. 6705.
14. He S., Attard P., Phys. Chem. Chem. Phys., 2005, V. 7. P. 2928.
15. Oxtoby D.W., Evans R., J. Chem. Phys. 1988. V. 89. P. 7521.
16. Bykov T.V., Shchekin A.K., Neorg. mater.. 1999. V. 35. No. 6. P. 759.
17. Bykov T.V., Shchekin A.K., Kolloid. zh. 1999. V.61. No. 2. P. 164.
18. Bykov T.V., Zeng H.S., J. Chem. Phys. 1999. V. 111. P. 3705.
19. Bykov T.V., Zeng H.S., J. Chem. Phys. 1999. V. 111. P. 10602.
20. Thompson S.M., Gubbins, K.E., Walton, J.P. R., et al., J. Chem. Phys. 1984. V. 81. P. 530.
21. Zhukhovitsky D.I., Kolloid zh. 2003. V. 65. No. 4. P. 480.
22. Appert C., Pot V., Zaleski S., Fields Institute Communications. 1996. V. 6. P. 1.
23. Ebihara K., Watanabe T., Eur. Phys. J. B 2000, V. 18. P. 319.
24. Tovbin Yu.K., Zh. fiz. khim. 2010. V. 84. No. 2. P. 231. [Russ. J. Phys. Chem. A 84, 180 (2010)]
25. Tovbin Yu.K., Zh. fiz. khim. 2010. V. 84. No. 10. P. 1882. [Russ. J. Phys. Chem. A 84, 1717 (2010)]
26. Prigogine I.R., Molecular theory of solutions. Moscow, Metallurgiya.
27. Tovbin Yu.K., Kolloid zh. 1983. V. 45. No. 4. P. 707.
28. Okunev B.N., Kaminsky V.A., Tovbin Yu.K., Kolloid zh. 1985. V. 47. No. 6. P. 1110.
29. Smirnova N.A., Molecular theory of solutions. Leningrad, Khimiya, 1987.
30. Morachevsky A.G., et al., Thermodynamics of liquid–vapour equilibrium. Ed. Morachevsky A.G., Lenigrad, Khimiya, 1989.
31. Prausnitz J.M., Lichtenthaler R.N., de Azevedo E.G., Molecular thermodynamics of fluid-phase equilibria. Second ed., Prentice-Hall Inc., Englewood Cliffs, New Jersey, 1986.
32. Hill T.L., Statistical Mechanics. Principles and Selected Applications. – Moscow:

Izd. Inostr. lit., 1960. – 486 p. [N.Y.: McGraw–Hill Book Comp. Inc., 1956]
33. Tovbin Yu.K., Theory of physico-chemical processes at the gas-solid interface, Moscow, Nauka, 1990. [CRC, Boca Raton, Florida, 1991].
34. Tovbin Yu.K., Zh. fiz. khim. 1992. V. 64. No. 5. P. 1395. [Russ. J. Phys. Chem., 1992 V. 66, No. 5, P. 741]
35. Tovbin Yu.K., Zaitseva E.S., Rabinovich A.B., Zh. fiz. khim. 2017. V. 91. No. 10. P. 1734. Rus. J. Phys. Chem. A, 2017, V. 91, No. 10, P. 1957]
36. Hirschfelder J. O., Curtiss Ch. F., Bird R. B., Molecular theory of gases and liquids. – Moscow: Inostr. Lit., 1961. – 929 p. [Wiley, New York, 1954].
37. Tables of physical quantities. Directory. Ed. I.K. Kikoin. Moscow, Atomizdat, 1976.
38. Vargaftik N.B., Handbook of thermophysical properties of gases and liquids, Nauka, Moscow, 1972.
39. Abramzon A.A., Surface-active substances: properties and applications. 2nd ed. Leningrad, Khimiya, 1981.
40. Frenkel J., Kinetic theory of liquids. Moscow, Academy of Sciences of the USSR, 1945.
41. Kubo R., Thermodynamics. Moscow, Mir, 1970. [Amsterdam: North-Holland Publ. Comp., 1968.]
42. Bazarov, I.P., Thermodynamics. Moscow, Vysshaya shkola. 1991.
43. Tovbin Yu.K., Rabinovich A.B., Izv. AN. ser. khim. 2009. No. 11. P. 2127. [Russ. Chem. Bull. 58, 2193 (2009)]
44. Sedov L.I. Continuum mechanics. V. 1. Moscow, Nauka, 1970.
45. Tovbin, Yu.K., Rabinovich A.B., Izv. AN. ser. khim. 2010. No. 4. P. 663. [Russ. Chem. Bull. 59, 677 (2010).]
46. Tovbin Yu.K., Zh. fiz. khim. 2010. V. 84, No. 4. P. 797. [Russ. J. Phys. Chem. A 84, 705 (2010)]
47. Tovbin Yu.K., Zh. fiz. khim. 2012. V. 86. No. 9. P. 1461. [Russ. J. Phys. Chem. A 86, 1356 (2012)]
48. Tovbin Yu.K., Zh. fiz. khim. 2010. V. 84. No. 9. P. 1795. [Russ. J. Phys. Chem. A 84, 1640 (2010)]
49. Laudise R. A., Parker R. L., The growth of mono-crystals. – Moscow: Mir. 1974. – 540. [Prentice-Hall, Inc., Englewood Cliffs, New Jersy, 1970; Solid State Physics, V. 25, 1970].
50. Dubinin M.M., Usp. khimii. 1955. Vol. 24. P. 3.
51. Gregg, S.J. Sing, K.G.W., Adsorption, Surface Area and Porosity. – Moscow: Mir, 1984. [Academic Press, London, 1982].
52. Karnaukhov A.P., Adsorption. Texture of dispersed porous materials. Novosibirsk, Nauka, Siberian Branch of the Russian Academy of Sciences, 1999.
53. Petrov Yu.I., Physics of small particles. Moscow, Nauka, 1982.
54. Tovbin Yu.K., Petukhov A.G., Izv. AN. ser. khim. 2008. No. 1. P. 18. [Russ. Chem. Bull. 2008. V. 57. No. 1. P. 18]
55. Tovbin Yu.K., Zh. fiz. khim. V. 82. No. 10. P. 1805. [Russ. J. Phys. Chem. A. 2008. V. 82. No. 10. P. 1611]
56. Beck J.S., et al., J. Am. Chem. Soc. 1992. V. 114. P. 10834.
57. Fischer M., Nature of the critical state. Moscow, Mir, 1968..
58. Stanley G., Phase transitions and critical phenomena. Moscow, Mir.
59. Ma Sh.. Modern theory of critical phenomena. Moscow, Mir, 1980.
60. Zeng H., et al., Phys. Rev. B, 2002. V. 65. P. 134426.
61. Chernavsky P.A.. Ross. khim. zh. 2002. V.46. No. 3. P. 19.
62. Chernavskii P.A., Khodakov A.Y., Pankina G.V., Girardon J.-S., Quinet E., Applied

Catalysis A: 2006. V. 306. P. 108.
63. Fedosyuk V.M., et al., Fiz. Tverd. Tela. 2003. V. 45. No. 9. P. 1667.
64. Shanina B.D., et al., Zh. Eksper. Teor. Fiz. 2009. V. 136. No. 4. P. 711.
65. Yang C.N., Lee T.D., Phys. Rev. 1952. V. 87. P. 404.
66. Dolgusheva E.B., Trubitsyn V.Yu., Fiz. Tverd. Tela. 2010. V. 52. No. 6. P. 1163.
67. Factorovich M.H., Molinero V., Scherlis D.A., J. Am. Chem. Soc. 2014. V. 136. P. 4508.
68. Lau G.L.V., et al., J. J. Chem. Phys. 2015. V. 142. P. 114701.
69. Belashchenko D. K., Zh. fiz. khim. 2015. V. 89. No. 3. P. 517. [Russ. J. Phys. Chem. A. 2015. V. 89. No. 3. P. 513]

4

Small systems and size fluctuations

29. Small system fluctuations

Macroscopic thermodynamics, which refers to a large number of small systems, deals with strictly defined thermodynamic functions [1–3]. The current trend, consisting in the transfer of the thermodynamic approach to smaller particles, is connected with the hope of preserving the same thermodynamic functions and their interrelations in a wide range of system sizes. The question of the possibility of such an extension of the domain of the thermodynamic description remains open. To solve it, it is necessary to use the methods of statistical thermodynamics. With a decrease in the volume of the system, equilibrium fluctuations increase for all the thermodynamic characteristics [1,4]. The existence of fluctuations is a direct consequence of the discreteness of the structure of matter at the atomic–molecular level of any system and the thermal motion of molecules [5–7].

As the size of the substance decreases, the proportion of surface particles increases in comparison with their total number, and spontaneous density fluctuations increase. The problem of the role of equilibrium fluctuations was discussed by Hill [8–10], who laid the foundations of the thermodynamics of small systems, and in [11] in which the role of macroscopic fluctuations on non-uniform surfaces was discussed in order to improve the accuracy of the description of adsorption isotherms. However, due to the macroscopic size of the systems [11], the contribution of fluctuations did not allow obtaining fundamentally new information, and this topic has been forgotten for many years.

When formulating the question which systems are considered small, it was assumed that the error from ignoring the surface contribution is 1% [9]. Then for a surface property in a three-dimensional system we have $M^{2/3}/M = 0.01$, or $M = 10^6$. That is, the system is considered 'small' when $M < 10^6$. For the boundary effect in a two-dimensional system: $M^{1/2}/M = 0.01$; or $M < 10^4$. For a logarithmic term, the condition $\ln M/M = 0.01$ implies $M < 600$. For a term of the order of unity (the boundary effect in unit dimension, or the 'net' effect): $1/M = 0.01$; $M < 100$.

If we go over to analyzing the value of the mean-square fluctuation η, then its choice is not arbitrary. Since the problem is posed of comparing the calculated characteristics in classical and statistical thermodynamics, the choice of criterion η refers to the comparison of two different (continuum and discrete–molecular) points of view, and this criterion should not depend on the accuracy of existing experimental techniques. The thermodynamic description of the curved boundary is characterized by the curvature of the mathematical surface to which the magnitude of the surface tension refers. To comply with this situation in molecular theories, it is necessary to abandon all the features of the distribution of molecules at the interface between phases, which depend on the temperature and type of the model.

In these conditions, this choice of one percent is rather crude. In the case of interest to us, Table 29.1 shows the relationship between the number of molecules M and the value of the corresponding radii of spherical drops related to the indicated values $\eta = 100/M^{1/2}$ % with respect to the fluctuations in the number of surface particles (R_s) and the volume (R_v) of the sphere, where the radii are measured in units of λ (λ is the average distance between the molecules in the liquid, equal to $\lambda = 1.12\sigma$, σ is the size of the solid sphere of the spherical molecule, which appears in the Lennard-Jones potential).

From Table 29.1 it follows that the density fluctuations on the surface do not exceed 1%, the particle size R_s should be in the range from 28 to 89 λ.

Volume or surface. The question of what characteristics of a small system should be compared, related to its volume or surface, plays an important role [12]. From Table 29.10 follows an obvious fact: for $M > 2 \cdot 10^2$ we have $R_s > R_v$, i.e. for a given accuracy of describing the properties of the volume of the drop, the radius of the drop R_v is considerably smaller than the radius R_s in the case of describing the surface properties. We note that

Table 29.1. The scale of the particle sizes containing different numbers of drop molecules M related to their surface (R_s) and bulk (R_v) regions at a given level of the relative magnitude of the mean-quadratic fluctuations of the number of molecules $\eta = 100/M^{1/2}$ %

$M =$	10^8	10^6	10^5	10^4	10^3	10^2	49
η, %	0.01	0.10	0.32	1.0	3.2	10	14
R_s	2800	280	90	28	9	2.8	2.0
R_v	275	62	27.5	13.5	6.2	2.9	2.3

the concept of the interface between phases in thermodynamics refers to macrosystems. Allowance for the surface contributions is necessary if the surface area A is so developed that the contribution of the surface contributions to the thermodynamic potentials becomes commensurable with the contribution of the bulk phases. If the Gibbs potential G for a volume phase is written as $G = \Sigma_i \mu_i m_i = U - TS + PV$, then for a plane boundary and an equilibrium drop this potential has the form $G = \Sigma_i \mu_i m_i = U - TS + PV - \sigma A$, and for the curved boundary of the metastable drop $G = \Sigma_i \mu_i m_i = U - TS + P_\alpha V_\alpha + P_\beta V_\beta - \sigma A$, where the usual notations are: internal energy U, entropy S, temperature T, pressure P, volume V, μ_i and m_i is the chemical potential and the mass of component i. The contribution σA, where σ is the surface tension, also appears in all other thermodynamic potentials, when it is necessary to take into account the surface properties of the system. In the case of metastable drops, the phases are in equilibrium at different pressures P_α and P_β.

For an isolated small system, all terms of the thermodynamic potential must be described with equal accuracy, and the equations of thermodynamics treat both the volume and surface terms from the continuum point of view – without taking into account the effects of fluctuations. If the volume of a small system is chosen as a condition for choosing the accuracy of the description, then its surface characteristics will include fluctuation effects, which changes the accuracy of the description of the entire system. Therefore, when comparing the continuum and discrete–molecular points of view for small systems, it is necessary to choose surface characteristics. The most convenient for this purpose is the 'surface tension' itself, which naturally appears in disperse systems. Below we consider the values of the fluctuations η on the surface of tension and not inside the volume of the drop. This is necessary to have not only the volume, but also the surface not experiencing fluctuations, i.e. corresponding to the requirements of thermodynamics. The question of the fluctuations of the internal volume of the phases is considered in Section 34.

30. The discreteness of matter

The discreteness of matter is the only natural limitation on the possibility of using thermodynamic approaches. The effects of the discreteness of matter are most clearly seen in the case of adsorption, which, in the absence of fluctuation effects, fairly well describes the main factors of adsorption systems: surface non-uniformity and lateral interactions between adsorbed molecules [13]. But in the transition to ultradispersed particles, the existing theory should be supplemented by taking into account the equilibrium fluctuations and analyzing their influence on all the molecular characteristics of real surfaces. Traditionally in thermodynamics solid adsorbents are considered as homogenized objects. For this reason, all the results of [8–10] refer to small uniform systems. The actual non-uniformity of solid particles is well known [13] (see Section 1).

We will consider the distribution of molecules in a large canonical ensemble ($\{\mu\}$, V, T) or ($\{\mu\}$, M, T), where the symbol $\{\}$ is the set of values of μ_i chemical potentials of molecules of sort i. The symbol M is used to denote the total number of sites of a system of volume V, since for a rigid lattice structure $V = v_0 M$, where v_0 is the volume of the site. The non-uniform surface consists of sections of sites of different types whose number is equal to M_q, $1 \leq q \leq t$, t is the number of site types; $M = \sum_{q=1}^{t} M_q$.

The number of occupied states of any lattice site, including vacant sites, is s (the index of sort $i = s \equiv V$ refers to vacancies). Each molecule of sort i (except for vacancies) has an internal statistical sum J_q^i, depending on the type of site q occupied by a given particle. Denote by N_q^i the number of molecules of sort i that are located at sites of type q. The number of free sites of type q is denoted by $N_q^V = M_q - \sum_{i=1}^{s-1} N_q^i$.

The total energy of the non-uniform ideal lattice system in the grand canonical ensemble is written as [11]

$$H = \sum_{f=1}^{M} \sum_{i=1}^{s} v_q^i \gamma_f^i \eta_f^q, \quad v_q^i = -\beta^{-1} \ln J_f^i - \varepsilon_f^i - \mu_f^i \tag{30.1}$$

where J^i_f is the partition function of a particle of sort i located at a site with number f of type q; $1 \le f \le M$, $1 \le i \le s$, ε^i_f is the binding energy of a particle i with a site of type q having the number f on the surface of the adsorbent, μ^i_f is the chemical potential of particle i at site f of type q, which in equilibrium equals the chemical potential of the particle i in the thermostat μ_i outside the lattice system, $\mu_i = \mu^i_f$, $\mu_i = kT\ln(P_i / J_i)$, J_i is the partition function of a particle of sort i in the gas phase, P_i is its partial pressure. The variable γ^i_f and the quantity η^q_f are defined in Section 16.

δ-shaped distributions. In the absence of lateral interactions, each centre is filled independently, so the probability of its filling is expressed as:

$$\theta_f{}^i = N^i_q / M_q = a^i_f P_i / (1 + a^i_f P_i), \quad a^i_f = J^i_f / (kTJ_i)\exp(\beta\varepsilon^i_f). \tag{30.2}$$

where a^i_f is the local Henry coefficient characterizing the degree of retention by the centre f of a particle of sort i.

This result is easily obtained in the grand canonical ensemble for a non-uniform surface in which the large partition function Ξ for an ideal non-uniform systems is expressed in terms of the 'local' partition functions Ξ_q for the sites q as $\Xi = \prod_{q=1}^{t} \Xi_q$. In turn, each local partition function Ξ_q is connected with the partition function in the canonical ensemble $Q_q(\{N_q{}^i\})$, which is expressed in terms of the statistical weight of the given distribution $\Omega_q(\{N^i_q\}) = C^{N_q}_{M_q} = M_q! / \prod_{i=1}^{s} N^i_q!$, where $N^s_q \equiv N^V_q$. As a result, we have

$$\Xi = \prod_{q=1}^{t} Q_q(\{N^i_q\})\exp(\beta\mu^i_q) = \prod_{q=1}^{t} \frac{M_q!}{\prod_{i=1}^{s} N^i_q} \prod_{i=1}^{s-1} J^i_q \exp(\beta\mu^i_q), \tag{30.3}$$

Taking into account the fact that each type of site is filled with molecules of different sort, equation (30.3) yields the exact expression

$$\ln\Xi = \sum_{q=1}^{t}(1 + \sum_{i=1}^{s=1} \lambda^i_q)^{Mq}, \quad \lambda^i_q = \exp(\beta\mu^{i*}_q),$$
$$\mu^*_i = \mu_i + \beta^{-1}\ln(J^i_q). \tag{30.4}$$

The spreading pressure in the lattice gas model is given by

$$\pi = \beta^{-1}\ln\Xi / M = \sum_{q=1}^{1}(1 + \sum_{i=1}^{s-1}\lambda_q^i)^{Mq} / M. \qquad (30.5)$$

This quantity is a discrete analog of the ordinary pressure P in three-dimensional systems of volume V; for the rigid lattice in question, the change in volume is due to a change in the number of vacant sites of a fixed size.

The degree of filling of a site of type q by molecules of sort i, expressed by formula (30.2), follows from relation

$$\theta_q^i = \langle N_q^i \rangle / M_q = \partial(kT\ln\Xi) / \partial\mu_q^{i*}, \qquad (30.6)$$

where angular brackets <...> mean averaging over the ensemble, and the symbol <A> means the average value of A. Equation (30.2) is also obtained from the condition that determines the maximum contribution to the sum (30.3), $\partial(\ln\Xi)/\partial N_q^i|_{M,T,\{\mu\}} = 0$.

Equation (30.6) corresponds to the δ-shaped form of the distribution functions $P_q(\{N_q^i\}) = Q_q(\{N_q^i\})\exp\left(\sum_{i=1}^{s-1}\beta\mu_i\right)$ for sites of sort q, which 'cut out' from the total sum (30.3) maximal terms approximating the 'local' sums Ξ_q for sites q.

The equilibrium filling of the entire surface is obtained by weighing the sites of different types by means of the distribution functions F_q characterizing the probability of finding sites of the type q on the surface, $1 \le q \le t$, t is the number of types of different centres, their number is denoted by M_q, so that $F_q = M_q/M$ and $\sum_{q=1}^{t} F_q = 1$. Then the partial filling of the surface by molecules of sort i is written as

$$\theta_i = \Sigma_{q=1}^{t} F_q \theta_q^i. \qquad (30.7)$$

The equilibrium fluctuations of the local partial fillings η_{qp}^{ij} characterize the mean square deviations of the fillings by the i and j molecules of a pair of sites of the q and p sorts. They are defined as $\eta_{qp}^{ij} = \left\langle \left(N_q^i - \langle N_q^i \rangle\right)\left(N_p^j - \langle N_p^j \rangle\right)\right\rangle$.

For the adsorption of molecules of one sort ($s = 2$) on a non-uniform surface ($t > 1$), the formulas given correspond to $\eta_{qp} \equiv \eta_{qp}^{AA}$ [11]:

$$\eta_{qp} = M\Sigma_{q=1}^{t} F_q\theta_q(1-\theta_q) = \Sigma_{q=1}^{t} M_q\theta_q(1-\theta_q). \qquad (30.8)$$

which generalizes the well-known expression for the adsorption of one sort of molecules on a uniform surface (t = 1) $\eta \equiv \eta_{11} = M\theta(1-\theta)$ [11,16].

In the case of adsorption of a mixture of molecules (s > 2) on a uniform surface, the expressions for equilibrium fluctuations $\eta_{11}^{ij} \equiv \eta_{qp}^{ij}$ are constructed by analogy with the known formulas [2,3]. Equilibrium fluctuations η_{qp}^{ij} in the case of adsorption of a mixture of molecules on a non-uniform surface were not discussed prior to the work [14]. The structure of such expressions η_{qp}^{ij} is considered below in view of size effects.

Mathematical apparatus. The mathematical description of the discreteness requires the use of the appropriate apparatus [17,18], based on the use of the difference calculus instead of the traditional continuous calculus. This circumstance, apparently, was first pointed out in [9] – the molecule can not be divided into smaller parts. It was specified in [14, 15] that the ordinary difference (non-symmetrized) derivatives [9] should be replaced by symmetrized derivatives.

For the analysis of small systems, in connection with the increase in the role of fluctuations, it is necessary to take into account the difference between the distribution functions of the δ-shape and the effect of the boundedness of the surface or volume of the system under consideration. Both of these factors must influence the probabilities of the filling of the sites θ_q^i and the equilibrium fluctuations of these fillings.

Derivatives for small systems. The definition of the derivative of the function $F(x)$, where x is the argument of the continuous function F, is written in the differential calculus as [19]

$$\frac{dP(x)}{dx} = \lim_{h \to 0}\left[\frac{P(x+h)-P(x)}{h}\right], \tag{30.9}$$

where h is the increment of the continuous variable x. In numerical calculations, the value of the increment h remains finite, but much less than the region of the characteristic change of the function under consideration, so the original definition of the derivative is rewritten in the difference form $dP(x)/dx = \Delta P(x)/\Delta x = [P(x + h) - P(x)] / h$, provided that the used value of h corresponds to the equality $dP(x)/dx \approx \Delta P(x)/\Delta x$ with the specified accuracy.

LGMs always operate with a discrete set of molecules, and the difference increments are a natural way of counting in this approach. The magnitude of the argument increment for a discrete variable N can not be arbitrary. The minimum increment value ΔN is ± 1, since the molecule can not be broken up. By analogy with the expression (30.9) of the difference derivative $\Delta P(N)/\Delta N$, for any function $P(N)$ of the discrete argument N, the relation $\Delta P(N)/\Delta N = [P(N + \Delta N) - P(N)]/\Delta N$, where $\Delta N = \pm 1$.

The choice of the sign of $\pm h$ in the increment for the differential calculus does not matter. In mathematics courses [19] it is proved that the limit does not depend on the method of limiting transition inside the domain of definition of a continuous function. For small systems with a discrete change in the value of N, the fixation of the sign (± 1) affects the final result. Usually, a plus sign is conventionally taken as an increment. This choice was adopted in the works on calculus in finite differences [16] and in combinatorics [20,21]. Nevertheless, in practical work on numerical calculations in the case of differential calculus, it is shown [18] that such an asymmetric choice of the method of calculating the difference derivative loses exactly in comparison with the symmetric definition of the derivative when the rule is used: $dP(x)/dx = [P(x + h) - P(x-h)]/2h$.

Such symmetrization in the differential calculus is possible by virtue of the continuity of the argument x for any values of the increment h. When using the analog of the symmetric definition for the difference derivative, one should use the expression [14,15]

$$\Delta P(N)\Delta N = \left[P(N+1) - P(N-1)\right]/2. \tag{30.10}$$

To illustrate the differences in these methods for determining derivatives, we consider how the expressions for the chemical potential in the canonical ensemble with discrete values of the number of particles N and the number of sites of the system M on which these molecules are distributed are constructed ($s = 2$). We denote by $F_{N,M}$ the free Helmholtz energy: $F_{N,M} = E_{N,M} - TS_{N,M}$, where $E_{N,M}$ is the energy, and $S_{N,M} = k\ln\Omega_{N,M}$ is the entropy of the system. In accordance with formula (30.3) for an ideal system $E_{N,M} = -\beta^{-1}\ln J$, J is the internal statistical sum of the molecule, and $\Omega_{N,M} = M!/(N!\ N_V!)$, where the number of unoccupied sites $N_V = M - N$.

The chemical potential of a molecule for a given problem can be written in four ways: through the differential derivative (A) and the three difference derivatives: (B) the asymmetric method, $\Delta N = +1$, (B) the asymmetric method, $\Delta N = -1$, and (Γ) the symmetric method, $\Delta N = \pm 1$.

(A) Differential derivative: $\mu_{N,M} = \partial F_{N,M} / \partial N_{|M}$. This approach uses the expansion in the series ln*N!* by the Stirling formula

$$\ln N! = \\ = N(\ln N - 1) + \ln(2\pi N)/2 + (12N)^{-1} - (360N^2)^{-1} + o(N), \quad (30.11)$$

where $o(N)$ is the remainder of the series.

Restricting ourselves to the account in the Stirling formula of the first smallness correction, we have

$$F_{N,M} = -N\beta^{-1}\ln J - \beta^{-1}\{M\ln M - N\ln N - N_V\ln N_V + \ln[M/(2\pi NN_V)]/2\},$$

what gives

$$\mu_{N,M} = -\beta^{-1}\ln J - \beta^{-1}\times \\ \times\{\ln(N_V/N) - (1/N - 1/N_V)/2 - (1/N^2 - 1/N_V^2)/12 + ...\} \quad (30.12)$$

(B) The traditional asymmetric difference derivative with sign (+1) [9] is written as $\mu_{N,M} = F_{N+1,M} - F_{N,M}$. This gives

$$\mu_{N,M} = -(N+1)\beta^{-1}\ln J - \beta^{-1}\ln\Omega_{N+1,M} + \\ +N\beta^{-1}\ln J + \beta^{-1}\ln\Omega_{N,M} = -\ln J - \beta^{-1}\{\ln(N_V/N) - \\ 1/N + 1/(2N^2) - 1/(3N^3) + 1/(4N^4) + ...\} \quad (30.13)$$

(C) Asymmetric difference derivative with sign (−1):

$$\mu_{N,M} = F_{N,M} - F_{N-1,M}. = -N\beta^{-1}\ln J - \\ -\beta^{-1}\ln\Omega_{N,M} + (N-1)\beta^{-1}\ln J + \beta^{-1}\ln\Omega_{N-1,M} = -\ln J - \\ -\beta^{-1}\{\ln(N_V/N) + 1/N_V - 1/(2N_V{}^2) + \\ +1/(3N_V^3) - 1/(4N_V^4) + ...\} \quad (30.14)$$

(D) Symmetric difference expression for the chemical potential [14]:

$$\mu_{N,M} = [F_{N+1,M} - F_{N-1,M}]/2 = [-(N+1)\beta^{-1}\ln J - \beta^{-1}\ln\Omega_{N+1,M} + \\ +(N-1)\beta^{-1}\ln J + \beta^{-1}\ln\Omega_{N-1,M}]/2 = -\beta^{-1}\ln J - \beta^{-1}\{\ln(N_V/N) - \\ -(1/N - 1/N_V)/2 - (1/N^2 - 1/N_V^2)/4 + \\ +(1/N^3 - 1/N_V^3)/6 + ...\}. \quad (30.15)$$

A comparison of formulas (30.12)–(30.15) shows that the first two terms relating to macroscopic contributions in all variants of chemical potential determinations are the same, and the contributions from the size of the system depend on the method of determining the derivative. This difference begins with the third term relating to the first correction for the size of a system of type M^{-1} or N^{-1}. The asymmetric definitions of the difference derivatives (30.13) and (30.14) differ from the differential derivative (30.12), beginning with the first correction from the size of the system. Between themselves, both difference definitions (30.13) and (30.14) differ drastically. Thus, for small systems, the increment sign $\Delta N = \pm 1$ plays an important role. Simultaneously it is shown that the transition to a more accurate symmetric difference derivative leads to a different result (30.15): it differs from expressions (30.13) and (30.14) already in the third (i.e. first correction) term, which indicates the incorrectness of the asymmetric definitions. The differences (30.15) from the differential derivative (30.12) begin only with the fourth summand (ie in the second order of smallness), while the third terms in both expansions are the same. An increase in the accuracy of calculating the chemical potential is associated with an increase in the number of correction terms. The contributions from the subsequent correction terms of the expansion in powers of M^{-k} or N^{-k} in more precise expressions (30.15) have lower coefficients than in formula (30.12).

Thus, the way of constructing derivatives strongly affects the values of the chemical potential $\mu_{N,M}$. Similar constructions for the spreading pressure also lead to conclusions about the need to use only symmetric difference derivatives.

Higher derivatives. These symmetric definitions of difference derivatives should be used consistently and when calculating higher orders of derivatives [15]. In particular, the second difference derivative, which is necessary for finding the fluctuations, must be constructed as the first symmetric difference from the symmetrized first difference derivatives. This gives

$$\frac{\Delta^2 P(N)}{\Delta N^2} = \frac{\Delta}{\Delta N}\left(\frac{\Delta P(N)}{\Delta N}\right) = \frac{\Delta}{\Delta N}\left[\frac{1}{2}\big(P(N+1) - P(N-1)\big)\right], \qquad (30.16)$$

where increments $\Delta N = \pm 1$ refer to each of the arguments in the terms inside the bracket. The consistent application of rule (30.10) in formula (30.16) leads to expression

$$\frac{\Delta^2 P(N)}{\Delta N^2} = \frac{1}{4}\left[\left(P(N+2) - P(N)\right) - \left(P(N) - P(N-2)\right)\right]$$
$$= \frac{1}{4}\left[P(N+2) - 2P(N) + P(N-2)\right]. \tag{30.17}$$

Recall that in analogy with the recording of the first derivative, in the differential calculus the second derivative can be formally written in the asymmetric form $d^2P(x) / dx^2 = [P(x + 2h) - 2P(x + h) + P(x)]/h^2$ or in the symmetric form $d^2P(x)/dx^2 = [P(x + h) - 2P(x) + P(x-h)]/h^2$. It is obvious that the accuracy of the second (symmetrized) expression for the second derivative in point x is much higher than the accuracy of the first asymmetrized expression. For the second symmetric derivative constructed through the asymmetric first difference derivatives for $\Delta N = \pm 1$ we have

$$\frac{\Delta^2 P(N)}{\Delta N^2} = \frac{1}{2}\left[P(N+1) - 2P(N) + P(N-1)\right]. \tag{30.18}$$

The difference between the definitions (30.17) and (30.18) is in the different length of the interval on which the second derivative is constructed. The longer the interval length, the higher the accuracy of the expression $\Delta^2P(N)/\Delta N^2$. The formula (30.18) has an interval length equal to 2 between three values of N and $N\pm1$, on which the second derivative is determined, whereas in the completely symmetric discrete definition (30.17) the second derivative has an interval length between five values of N equal to 4.

In the general form, the difference derivative of the n-th order necessary for calculating the higher contributions of the fluctuations will be written in the form

$$\frac{\Delta^n P(N)}{\Delta N^n} = \frac{1}{2^n}\sum_{k=0}^{n}(-1)^{n-k} C_n^k P(N+n-2k), \tag{30.19}$$

where the length of the interval for calculating the derivatives is $2n$. In the particular case $n = 3$, formula (30.19) has the form

$$\frac{\Delta^3 P(N)}{\Delta N^3} = \frac{1}{8}\left[P(N+3) - 3P(N+1) + 3P(N-1) - P(N-3)\right]\cdot$$

31. An ideal system, one component [15]

Uniform surface. We begin our discussion of the thermodynamic properties of small systems with the simplest case of adsorption of molecules of one kind (N is their number) distributed over sites of a uniform lattice with size M size in the absence of lateral interactions between molecules ($s = 2$). In this example, we will consider the basic problems of the analysis of small systems. To obtain expressions for the mean-square fluctuations, it is necessary to find the second derivative in the vicinity of the maximum of the distribution function $\ln P(N, M)$. Further, knowing the fluctuations of the distribution function, one can obtain the characteristics of interest taking into account the contribution of the fluctuations and determine their effect on the calculated characteristics (and on the observed experimental data).

For the problem under discussion, formula (30.3) is rewritten as

$$\Xi = \sum_{N=1}^{M} P(N,M),\ P(N,M) = Q(N,M,T)\exp(\beta\mu N),$$
$$Q(N,M,T) = J^N M!/(N!N_V!), \tag{31.1}$$

where J is the internal statistical sum of the adsorbed particle, $N_V = M-N$, here μ is the chemical potential of the molecule, fixed by the state of the thermostat and let us introduce $\mu^* = \mu + \beta^{-1}\ln J$.

The traditional procedure for constructing macroscopic equations for the equilibrium distribution of molecules consists in replacing the sum over N in the expression (31.1) by the maximum term of the sum $\Xi^* = P(N^*)$, where $P(N^*) = Q(N^*, M, T)\exp(\beta\mu N^*)$, which approximates the value of the sum Ξ.

$$\ln P(N,M) = \ln\Omega(N,M) + \beta\mu^* N =$$
$$= \ln M! - \ln N! - \ln N_V! + \beta\mu^* N. \tag{31.2}$$

The desired expression for $P(N^*)$ is determined from the condition $\partial\ln P(N)/\partial N = 0$. The use of differential derivatives is justified for $M, N \to \infty$ and $N/M = \text{const}$, when the number of molecules is described within the framework of continuous calculus. The search for $P(N^*)$ in (31.2) is based on the use of the Stirling formula for a large number of molecules in the asymptotic form $\ln N! = N(\ln N-1)$.

For small systems, the calculation of the maximum term Ξ^* is based on the use of a more accurate expansion (30.11), in which $u(N) = \ln(2\pi N)/2 + (12N)^{-1} - (360N^2)^{-1} + o(N)$, $\Delta \ln P(N, M) / \Delta N = 0$. This gives the following equation for the most probable value of N^*, which is a function of temperature and a fixed value of the number of sites M and chemical potential μ:

$$\beta\mu^* = \ln\left[\frac{N}{N_V}\right] - \frac{1}{2}\ln\left[\frac{N(N_V+1)}{(N+1)N_V}\right]. \tag{31.3}$$

The first term corresponds to the well-known Langmuir equation for a macroscopic uniform surface. The second term takes into account the limited size of the surface. It more fully reflects the dimensional contribution to expression (30.15) for $\beta\mu_{N,\,M}$ (without expansion in a series). As the values of M and N increase in comparison with unity (M and $N \gg 1$), the second contribution decreases and, in the macroscopic size limit, it is zero.

The second derivative is calculated at constant values of M and μ^*

$$\frac{\Delta^2 \ln P(N,M)}{\Delta N^2} = \frac{1}{4}\ln\left[\frac{N(N-1)N_V(N_V-1)}{(N+1)(N+2)(N_V+1)(N_V+2)}\right]_{N^*}. \tag{31.4}$$

Under the sign of the logarithm, all the factors in the numerator are smaller than the corresponding factors in the denominator, so a negative value is on the right. Formally, for N, $N_V >> 2$, this expression (31.4.) becomes zero, which means a sharp narrowing of the distribution function – it goes into the δ-function.

For analytical estimates, in order to take into account the size factors, it is necessary to leave the corresponding contributions in powers of N^{-k} (similar to N_V^{-k}), where $k = 1, 2, 3$, etc., in the series expansion, as in (30.15). In particular, the first correction with terms $k = 1$ for the dimensional contribution is written as

$$\frac{\Delta^2 \ln P(N,M)}{\Delta N^2} = -\frac{M+2}{NN_V}\bigg|_{N^*}, \tag{31.5}$$

that $M \gg 2$ remains $(-M/NN_V)$ on the right-hand side of the formula (31.5), which coincides with the expression of [9] obtained in continuous calculus. This demonstrates the coincidence of the results in the first order in M^{-1} for discrete and continuous descriptions in the second order of expansion for $P(N)$ in the number of molecules, as in the first order of expansion for the first derivatives. Those. the allowance for the fluctuation contributions in the first order in M^{-1} preserves the coincidence of the two calculation methods.

In the expansion of the expressions (31.2)–(31.4) in a series, one must take into account the same accuracy of the expansion as in the logarithmic factors: $\ln(1+x) = \sum_{m=1}^{k}(-1)^{m+1}x^m/m$, where $-1 < x < 1$, with a given value k, and in the expression for $u(...)$ by the Stirling formula (30.11). After agreeing the two calculation methods on the basis of discrete and continuous calculus in small systems, one can use the usual integral relations and take into account the change in the form of the distribution function due to the contribution of Gaussian fluctuations, which is expressed as [8]

$$\ln P(N) = \ln\left[P(N^*,M)\right]-\eta(N-N^*)^2,$$
$$\eta(N^*,M) = -\frac{1}{2}\frac{\Delta^2 \ln P(N,M)}{\Delta N^2}\Big|_{N=N^*}. \quad (31.6)$$

In the first order of smallness of the size contributions, we obtain

$$\eta = M\theta(1-\theta)/(1-D),$$
$$D=\frac{1}{2M}\left(\frac{M+N^2+(M-N)^2}{(N+1)(M-N+1)}\right). \quad (31.7)$$

For a macroscopic uniform surface, (31.7) is transformed into $\eta = M\theta(1-\theta)$ [16,21].

It follows from (31.6) that

$$\Xi = \int_{-\infty}^{+\infty} P(N^*,M)\exp[-\beta_1(N-N^*)^2]d(N-N^*) =$$
$$= P(N^*,M)\pi^{1/2}\eta^{-1/2}, \quad (31.8)$$

or

$$\ln\Xi = \ln[P(N^*,M)] + 1/2\ln\pi - 1/2\ln[\eta(N^*,M)]. \quad (31.8a)$$

For the calculations of $\ln[P(N^*, M)]$, using formulas (30.11) and (31.6), we obtain

$$\ln \Xi = M\ln M - N\ln N - N_V \ln N_V + \beta\mu^* N + u(M) - u(N) - \\ -u(N_V) + 1/2\ln\pi - \frac{1}{4}\ln\left[\frac{N(N-1)N_V(N_V-1)}{(N+1)(N+2)(N_V+1)(N_V+2)}\right] \quad (31.9)$$

Spreading pressure. The last expression (31.9) can be used to calculate the so-called spreading pressure, which is traditionally applied in the LGM and serves as the equation of state $\pi = \beta^{-1}\ln \Xi/M$. It follows the equation for the pressure of the macrosystem with allowance for the fluctuation contributions and all the dimensional corrections

$$\beta\pi = \ln[M/N_V] - \ln\left[(N_V+1)N\eta(N^*,M)/(N+1)N_V\right] /2M + \\ + \left[1/2\ln\pi + u(M) - u(N) - u(N_V)\right]/M. \quad (31.10)$$

For small systems, substitution of formula (31.3) into the expression for $P(N, M)$ (31.2) give

$$\ln P(N,M)|_{N*} = \ln\Omega(N^*,M) + \beta\mu^* N^* = M\ln(M/N_V) + \\ + u(M) - u(N) - u(N_V) - \frac{N}{2}\ln\left[\frac{N(N_V+1)}{(N+1)N_V}\right].$$

where all the terms are calculated for $N = N^*$, which allows us to obtain an expression for the average spreading pressure $\beta P_a = \ln P(N, M)|_{N*}/M$

$$\beta\pi_a = \ln(M/N_V) + \left[u(M) - u(N) - u(N_V) - \frac{N}{2M}\ln\left[\frac{N(N_V+1)}{(N+1)N_V}\right]\right]. \quad (31.11)$$

In the macroscopic limit, both values of P and P_a go over into the well-known relationship between the number of adsorbed molecules and the spreading pressure [22,23]: $\pi = \pi_a = -kT\ln(1-N^*/M)$. We note that the formula (31.11), in contrast to (31.10), refers to the δ-shaped form of the distribution function $P(N, M)$ in a small system without taking into account the influence of fluctuations.

Degree of filling the surface. The average values of the number of adsorbed molecules are found from the known value of Ξ by the Gibbs formula [4,8]

$$N = \partial(kT\ln\Xi)/\partial\mu_{|T,M} = M\lambda^*/(1+\lambda^*), \qquad (31.12)$$

since the magnitude of the chemical potential is a continuous quantity, whence we obtain the well-known Langmuir equations (30.2).

Calculation of the average number of adsorbed particles with allowance for fluctuations is carried out according to an analogous formula via the constraints (31.3) and (31.9), reflecting the contributions of the dispersion of the distribution function Ξ

$$\begin{aligned} <N> &= \sum_{N=0}^{M} NP(N,M) / \sum_{N=0}^{M} P(N,M) = \partial\{kT\ln\Xi\}/ \\ /\partial\ln\lambda &= \partial\{\ln[P(N^*,M)] + 1/2\ln\pi - 1/2\ln[\eta(N^*,M)]\}/ \\ /\partial\ln\lambda &= N^* - \Delta N, \text{ where } \Delta V = \partial\ln\left[\eta\left(N^*,M\right)\right]/2\partial\ln\lambda. \end{aligned} \qquad (31.13)$$

These formulas readily allow us to answer the question of what molecular densities are most noticeable in approximation with terms of the first order of smallness M^{-1}. They give

$$\theta_M = \langle N\rangle / M = \theta_\infty + (M-2N)/(2M^2). \qquad (31.14)$$

From the last formula it follows that for $N = M/2$ we have $\theta_M = \theta_\infty$, i.e. when the surface is half filled, the dimensional corrections are completely absent for any of the largest areas. The maximum value of the size correction is realized for small ($\theta_M = \theta_\infty + (2M)^{-1}$) and large ($\theta_M = \theta_\infty - (2M)^{-1}$) fillings. However, taking into account the eigenvalue of the degree of filling of the surface (that is, in comparison with θ_∞), the contribution of the fluctuations is most noticeable for small $\theta \to 0$.

On small faces of microcrystals, the influence of fluctuations can be commensurable with the degree of filling θ. For example, for $M = 100$ and the degree of filling $\theta_\infty = 10^{-2}$ (that is, of the order of ~1%) we have that the filling of the surface can differ up to one and a half times ($\theta M = 3\theta_\infty/2$) or the value $\Delta\theta = \theta_M - \theta_\infty = 0.5\theta_\infty$ is 50% of the degree of filling of the macroscopic section. A similar influence of fluctuations ($\Delta\theta = 0.5\theta_\infty$) is realized with a simultaneous increase in the area of the section of the surface M and a decrease in the degree of its filling (θ_∞) ~ $1/M$. This factor can be important when considering experimental data for small microcrystals of sensors and catalysts. With increasing surface area, the value of the fluctuation contribution $\Delta\theta$ for a fixed value of θ_∞ decreases as $1/M$. (In the general case, for small values of M, it is necessary to use more precise expressions than formula (31.14).)

Numerical examples. Below are illustrations of the obtained equations for regions of different sizes [24]. The chemical potential is related to the number of adsorbed molecules by different adsorption isotherms in Fig. 31.1. These six curves are shown in Fig. 31.1 in the relative coordinates $\Delta\theta = (\theta-\theta_0)/\theta_0$ for $M = 100$. Here $\ln(aP) = \ln(N/N_V) + A_k$, $k = 1-6$, $a = J\beta/J_0$, where the index $k = 1-4$ corresponds, respectively, to the expressions for the corrections (30.12)–(30.15), and $k = 5$ is the first size correction into these expressions $A_5 = (1/N-1/N_V)/2$ corresponding to the first term and $k = 6$ is the exact expression (31.3) $A_6 = \ln\{[(N+1)N_V]/[N(N_V+1)]\}/2$. The values of θ_0 on the abscissa in the range from zero to unity refer to macroscopic values. Hill's recommendations on the use of non-symmetric difference derivatives $\Delta N = +1$ (30.13) and $\Delta N = -1$ (30.14) give the maximal differences from the exact solution (31.3). The calculation of the first correction turned out to be the most approximate to the exact solution, while taking into account the third corrections both in the discrete (30.15) and in the continuum description (A) has larger deviations than curve 5 for the variant ($k = 5$). Thus, a simple increase in the number of the terms in the series in M^{-k}, $k > 1$, does not automatically lead to an increase in the accuracy of the calculation.

The inset in Fig. 31.1 shows the curves for the variants ($k = 5$) and ($k = 6$) in the cases $M = 10^m$, where the exponents $m = 1$, 3 and 4 are indicated on the curves. The solid lines correspond to the exact contributions of A_6, and the dashed lines correspond to the first corrections from the size effects A_5. With decreasing M, the range in which the values of the dimensionless density θ reflecting allowance for the boundedness of the size of the surface area $1/M \leq \theta \leq (M-1)/M$ is sharply reduced.

To determine the effect of density fluctuations, one must know the variance $\eta = -D/2$ of the distribution function $P(N, M)$, which is expressed in terms of the second difference derivatives as $D = \Delta^2 P(N, M)/\Delta N^2{}_{|T,M}$. The following expressions for the exact value of D^E (31.4) were obtained in [15] and taking into account the contribution of only the first correction D^1. The formula for D^E corresponds to the exact expression A_6 in the isotherm equation (31.3). In the value of D^1, the effects of the bounded size of the faces of microcrystals are

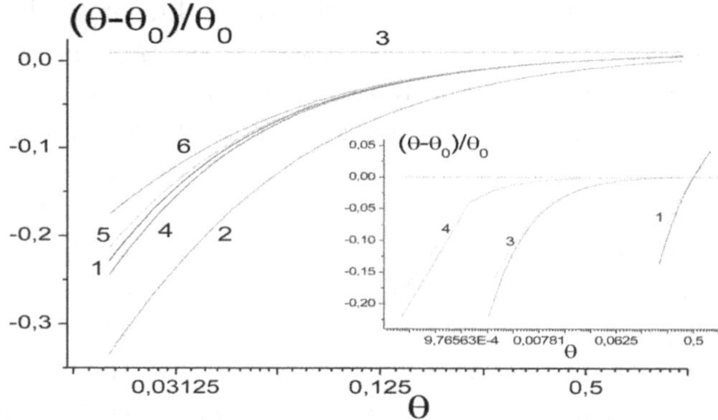

Fig. 31.1. Dependence $(\theta-\theta_0)/\theta_0$ on θ in logarithmic scale: (a) – for $M = 100$ using A_k, $k = 1, .., 6$, the values of k are shown on the curves. On the inset – similar curves are shown by solid lines for A_5, contour – for A_6 at $M = 10^m$, $m = 1, 3, 4$, the values of m are shown on the curves

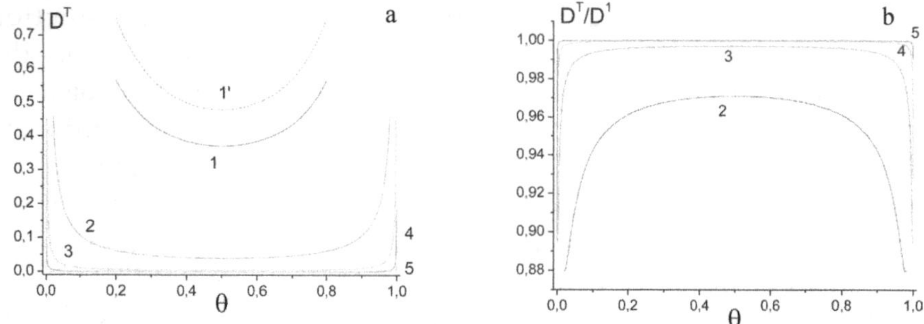

Fig. 31.2. The dependence of D^E (a) and the ratio D^E/D^1 (b) on θ for a uniform surface at $M = 10^m$ using A_6, the values of m are shown on the curves; curve 1′ in Fig. 31.2 *a* corresponds to D^1 using A_5 for $M = 10$.

taken into account in the first order in M^{-1} (calculation is performed using expression A_5).

Figure 31.2 *a* shows the concentration dependences of the coefficients D^E and D^1. With increasing M, the dispersion decreases sharply. Thus, for $M = 10^5$, practically remains zero throughout the entire density range, both for macroscopic dimensions. For $M = 10^4$, in the region of small and large fillings, there are differences from zero values for 2–3% density. Further, with a decrease in M to 10^3, the domain of difference of D^E from the zero value extends to 20%. At even smaller values of M, the value of D^E differs from the macroscopic size in the entire density range, starting at $M =$ 500–600. The curves for $M = 10$ show the maximum dispersion differences for small sections. The dashed curve 1′ refers to a similar dependence obtained with the first dimensional correction. Those. approximate solutions overestimate the variance. The D^E / D^1 ratios are shown in Fig. 31.2 *b* for $M = 10^2 - 10^5$.

The curves given for the adsorption of particles on a uniform surface indicate the importance of taking into account the fluctuation effects in the case of small faces of microcrystals.

Non-uniform surfaces. In real materials, the non-uniformity of the surface due to the presence of different faces and their edges is well known [13, 25, 26]. Therefore, for small particles, the contributions of different faces can not be neglected. The description of non-uniform ideal systems is reduced to the summation of the contributions of any characteristics calculated for individual faces. All formulas obtained above can easily be generalized by summation with weights F_q and replacement of the complete filling θ by local filling $\theta_q = N_q/M_q$, relating to the sites of type q, $1 \le q \le t$, t – the number of types of sites (here, they are not written). Complete adsorption isotherm written as $\theta(P) = \sum_{q=1}^{t} F_q q_q(P)$, wherein $F_q = M_q/M$ is the fraction of sites of type q on the whole surface, $\sum_{q=1}^{t} F_q = 1$.

Differences in the binding energies Q_q of the particle with surface portions of type q must be distinguished explicitly via local Henry constant $a_q = a_0 \exp(\beta Q_q)$, here a_0 is the pre-exponent having the back pressure dimension. Local fillings are related to the chemical potential as [14,15]

$$ln\left(a_q P\right) = ln\left(N_q / N_q^V\right) - \frac{1}{2}\ln\left[\frac{N_q(N_q^V+1)}{(N_q+1)N_q^V}\right] \tag{31.15}$$

where $N_q^V = M_q - N_q$, the second summand is a generalization of expression (31.3) with exact accounting of all size contributions (expressions A_6 for different faces).

The filling of different faces of the crystal takes place in accordance with the values of Henry's constants, so the effect of fluctuations for small and large fillings on different faces leads to a qualitatively new situation. Its meaning is illustrated in Fig. 31.3, which shows the number of adsorbed molecules on two small faces of the crystal ($\Delta Q = Q_1 - Q_2$). The abscissa is the degree of filling of the surface θ of the small crystal. The dotted curve refers to a macroscopic lattice. One can see the influence of the size factor on the local filling of different faces of a non-uniform surface. The results of the calculation show that 1) the differences in the degree

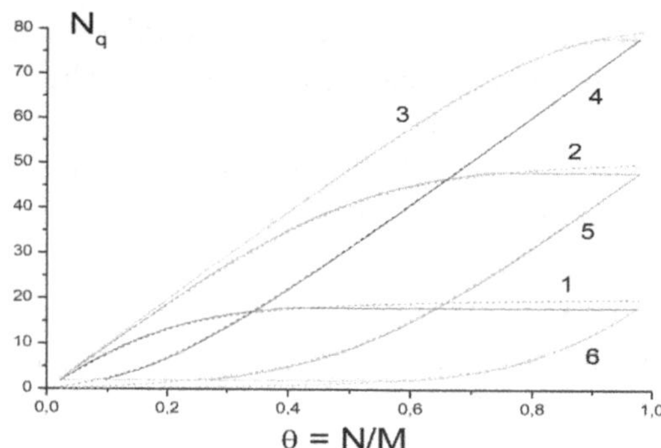

Fig. 31.3. Dependences of N_q, q = 1,2 on $\theta = N / M$ for t = 2 for the values of the molecular parameters $\beta\varepsilon$ = 1, Q_1 = 3ε, ΔQ = 3ε. Curves 1, 2, 3 correspond to N_1, curves 4, 5, 6 correspond to N_2 for F_1 = 0.2, 0.5, 0.8, respectively. The solid lines denote the curves for M = 100, the dotted lines the curves for $M = \infty$.

of filling of small and macroscopic systems are quite appreciable, and 2) the extent of the region θ in which the deviation data are observed is large enough, and one can speak about the possibility of their experimental detection.

The equation for the density fluctuations of adsorbed molecules on the non-uniform surface of small particles in the first order in N^{-1} will be written as

$$\eta = \sum_{q=1}^{t} M_q \eta_q,\ \eta_q = \theta_q(1-\theta_q)/(1-D_q),$$
$$D_q = \frac{1}{2M_q}\left(\frac{M_q + N_q^{\ 2} + (M_q - N_q)^2}{(N_q+1)(M_q - N_q + 1)}\right), \tag{31.16}$$

where η_q is the contribution of local density fluctuations in the section of sites of type q. With increasing M_q, the second term of the denominator D_q in η_q turns to zero, and the known expression [8] is obtained.

Analysis of this expression shows: 1) The maximum local dimensional density fluctuations exist for $\theta_q \rightarrow 0$ and $\theta_q \rightarrow 1$. (For $\theta_q \sim 1/2$, maximum density fluctuations not related to dimensional effects, including, for macroscopic systems.) 2) An important role in the value of η is played by the relations M_q/M_p. 3) The density fluctuations on a non-uniform surface oscillate – the maximum number of oscillations is equal to the number of types of centres

t, which makes it possible to estimate them from the experiment if there is a complete separation of the contributions of η_q from different parts of the surface.

Figure 31.4 shows concentration dependences of the density fluctuations of adsorbed molecules on a non-uniform surface consisting of two types of centres. The calculation is made for three surface compositions. The difference between the curves for small ones with $M = 100$ (solid lines) and macroscopic (dotted lines) of systems is observed at any densities. In the field of medium fillings the differences are small, but for small and large fillings they increase. For clarity, curves *5*, relating to local fluctuations corresponding to the sites of the first type of curve 1, are presented. The variation of the surface composition sharply affects the form of the concentration dependence of the mean-square fluctuation η. The presence of peaks in the value of η is associated with a transition from preferential filling of one site to another and from the ratio in the binding energies ΔQ. If the differences are small (curve 4), then both sections are filled simultaneously and there are no maxima. This fact was noted earlier for macrosystems in [11].

The accuracy of calculations. Let us discuss two questions on increasing the accuracy of calculating adsorption characteristics on non-uniform surfaces. The first question is related to the fact that the increase in fluctuations with decreasing size of a small system leads to the influence of fluctuations on the average values of the filling of the faces. Such a fact is absent in macroscopic systems, but for small systems it can be shown that the discrete nature of statistics

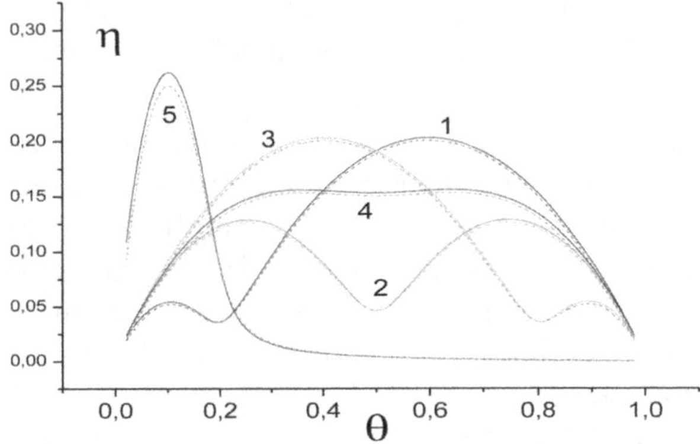

Fig. 31.4. Dependences of η on θ for $t = 2$ for the values of the molecular parameters $\beta\varepsilon = 1$, $Q_1 = 3$; curves 1, 2, 3 correspond to $\Delta Q = 6$ and $F_1 = 0.2$, 0.5, 0.8, respectively, curve 4 corresponds to $\Delta Q = 3$ and $F_1 = 0.5$. Curves 5 refer to local fluctuations η_1 corresponding to the sites of the first type of curve 1 for $F_1 = 0.2$ and $\Delta Q = 6$. Solid lines denote curves for $M = 100$, dotted lines for curves for $M = \infty$.

leads to violation of well-known expressions for the moments of Gaussian distribution functions [16], which approximate the discrete distribution functions in small systems [15]. The nature of such a violation is related to the asymptotic ($M \to \infty$) character of the Gaussian distribution functions and the complexity of determining the value of M from which this approximation becomes quite accurate.

The average number of adsorbed particles with an allowance for fluctuations is obtained from the well-known Gibbs formula [1,4]: $\langle N \rangle = \partial\{kT\ln\Xi\}/\partial\ln(P)$, which takes into account the variance of the distribution function $P(N, M)$. This Gibbs equation follows directly from the distribution functions and does not depend on the type of variables (discrete or continuous) for the number of molecules.

$$\begin{aligned}\langle N_q \rangle &= \partial\left\{\ln[P_q(N_q, M_q)] + 1/2\ln\pi - 1/2\ln(\eta_q)\right\} / \partial\ln(P) = \\ &= N_q - \Delta N_q,\end{aligned} \tag{31.17}$$

where $\langle N_q \rangle$ is the average number of particles on the surface for a given chemical potential μ, and $\Delta N_q = \partial\ln(\eta_q)/[2\partial\ln(P)]$ is the change in this number due to fluctuations. Figure 31.5 *a* shows the correction curves for local adsorption isotherms for two types of centres with allowance for the fluctuation density contributions related to the surface of the number of sites $M = 100$. Vertical and horizontal sections of the curves demonstrate the course of the changes in ΔN_q. For small densities, the corrections are negative, which leads to an increase in the degree of filling of each section of the surface, and for large degrees of filling, the fluctuation corrections reduce the overall filling. The value $q = 0$ corresponds to the mean corrections that refer to the whole surface: $\Delta N_0 = \sum_{q=1}^{t} F_q \Delta N_q$.

The analysis showed that with increasing surface size, the influence of fluctuations decreases rapidly. For values exceeding $M = 10^5$, we can speak of negligibly small deviations from the macroscopic behaviour of the system. In the particular case of a uniform surface and restricting to taking into account only the first size correction, we obtain the expressions for ΔN in the work [8].

The second question is related to the specifics of allowance for fluctuations on non-uniform surfaces. The structure of equations (31.9) and extension of equation (31.4) at any $t > 1$ shows [15,27] that the main problem in calculating the effects of fluctuations on non-uniform surfaces is the so-called limiting or 'forbidden' values of $N_q^b = 0, 1, M_q-1, M_q$ for which the right-hand sides of the mentioned expressions are not defined. Figure 31.1 shows how, with a decrease in M, the range of the value of the dimensionless density that corresponds to a given size of the surface area $1/M \leq \theta \leq (M-1)/M$ decreases. In Figs. 31.3 and 31.4 it is shown that when the local density approaches small and large fillings, the role of fluctuations always increases, and it is with these values of N_q that local bonds are absent. In the general case, an increase in the number of types of centres t inevitably leads to an increase in the total number of surface sites, which include the 'forbidden' numbers of $N_q^b = 0, 1, M_q-1, M_q$ molecules for large and small fillings of the faces of each type of centre. Their number is $N^b = 4t$, so with increasing t the total number of boundary values N^b increases.

The more non-uniform the surface with respect to bond energies, the more significant this factor is, since during the filling of one face the other faces will remain practically free or vice versa, almost completely filled. In both situations, the role of fluctuations increases. This situation is always realized when there are differences in local fillings related to the difference in the values $\beta\Delta Q = \beta (Q_1 - Q_2) > 3$, for example, for $t = 2$ (see Fig. 31.4).

The presence of 'forbidden' N_q^b values should be monitored in all calculations and this represents one of the problems in calculating adsorption on non-uniform surfaces. It is obvious that the direct summation of contributions in a large partition function for non-uniform surfaces [15] will take into account all configurations without exception, but this is a very laborious procedure. The transition to the use of information about maximum contributions and their variances significantly simplifies the calculation. But even in this case the problem for non-uniform surfaces remains complicated because of the large contribution of the 'forbidden' values of N_q^b. Therefore, it seems possible to introduce approximations of the boundary values of the occupation numbers on each of the faces in order to carry out the calculation in the entire density range with an allowance for fluctuations. We recall that, as a limiting value for small N (i.e., for an ideal gas or vacancies), we have $\langle \Delta N^2 \rangle = 1$ [1–3]. As the simplest method of approximating the density description in the region of 'forbidden' values of N_q^b, it is possible to define the limiting values of ΔN_q so that they are also equal to one. This will correspond to a doubled value of the current density in the region of limiting small and large fillings. Calculations of the fluctuation corrections ΔN_q with this approximation are shown in Fig. 31.5 *b*.

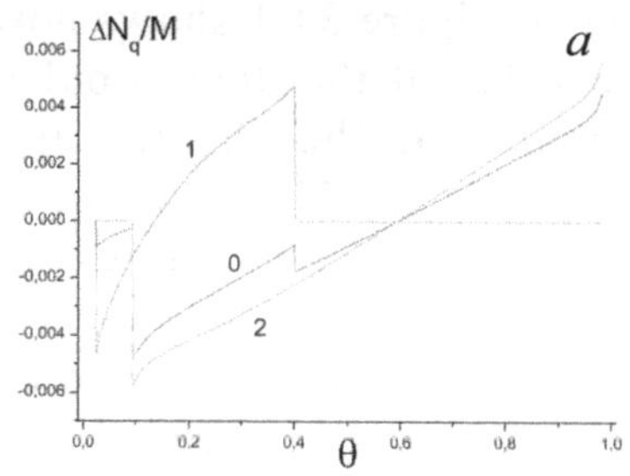

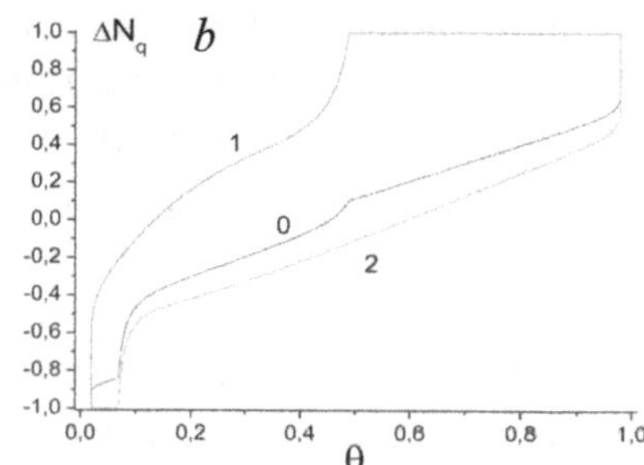

Fig. 31.5. Dependences of the initial values $\Delta N_q/M$ (a) and the corrected values ΔN_q (b), $q = 0, 1, 2$, on θ for $t = 2$ for the values of the molecular parameters $\beta\varepsilon = 1$, $Q_1 = 3\varepsilon$, $\Delta Q = 3\varepsilon$, $F_1 = 0.2$ on the surface $M = 100$. The values of q are shown on the curves.

Thus, it is obtained that the greatest relative influence of density fluctuations is manifested with small fillings of each face of the particle. The magnitude of the fluctuating contribution for large filling is the same as for small filling. As the total filling of the non-uniform surface increases, the root-mean-square (rms) value of density fluctuations have an oscillating character at any face dimensions, taking into account oscillations due to differences in the binding energies of molecules at different sites of the surface. An estimate is obtained – the surface size containing not less than 10^5 adsorption centres can be conditionally considered as the size larger than which the fluctuation corrections are small.

32. An ideal system, two components [15]

Uniform surface. Let us illustrate the specifics of constructing equations for the adsorption of mixtures on a uniform lattice ($t = 1$) containing two sorts of molecules with allowance for vacant sites, $s = 3$. Let N_i, $i = 1$ and 2 denote the number of particles of the first and second sort on the lattice of M sites. We denote the number of free lattice sites by $N_V = M - N_1 - N_2$ (the index V refers to the last of the number of occupied states of the sites s). The search for the maximum term in the large partition function Ξ is carried out by varying the numbers N_1 and N_2 for a fixed value of M and given chemical potentials μ_1, μ_2 and the temperature of the system T. $\ln P(N_1, N_2, M)$ will be considered:

$$\ln P(N_1, N_2, M) = \ln\Omega(N_1, N_2, M) + \beta\sum_{i=1}^{s-1} N_i\mu_i^* =$$
$$= \ln M! - \ln N_1! - \ln N_2! - \ln N_V! + \beta\sum_{i=1}^{s-1} N_i\mu_i^*,$$

and find the difference symmetric derivatives with respect to the numbers N_1 and N_2. When calculating the partial derivatives, only one variable changes, therefore, according to formula (30.10), we write

$$\frac{\Delta \ln P(N_1,N_2,M)}{\Delta N_1} =$$
$$= \frac{1}{2}\left[\ln P(N_1+1,\, N_2,M) - P(N_1-1,N_2,M)\right] =$$
$$= \beta\mu_1^* + \frac{1}{2}\ln\left[\frac{N_V(N_V+1)}{N_1(N_1+1)}\right] = D_1 + \delta P_1; \tag{32.1}$$
$$P_1 = \beta\mu_1^* + \ln(N_V / N_1),\ \delta P_1 = \frac{1}{2}\ln\left[\frac{N_1(N_V+1)}{N_V(N_1+1)}\right].$$

In the last equation, the macroscopic contribution P_1 and the correction for the limited volume of the system (size contribution) ΔP_1 are distinguished. For macrosystems with N_1, $N_V \to \infty$, the size contribution is zero. An analogous equation for the second partial difference symmetric derivative with respect to molecules of the second kind can be written in the form

$$\frac{\Delta \ln P(N_1,N_2,M)}{\Delta N_2} = \frac{1}{2}\left[\ln P(N_1,N_2+1,M) - P(N_1,N_2-1,M)\right] = P_2 + \delta P_2,$$

$$P_{(i)} = \beta\mu_i^* + \ln\left(N_V / N_i\right), \qquad \delta P_{(i)} = \frac{1}{2}\ln\left[\frac{N_i(N_V+1)}{N_V(N_i+1)}\right] \tag{32.2}$$

Equation (32.2) is written in a general form – it refers to any ideal system containing an arbitrary number of components (s–1). If the set of numbers of molecules of such a mixture is denoted by $\{N_i\}$, where $1 \le i \le s-1$, and the number of free sites $N_V = M - \sum_{i=1}^{s-1} N_i$, then we can express in general form the conditions on the maximum summand in the sum for Ξ: $\dfrac{\Delta \ln P(\{N_i\},M)}{\Delta N_i} = P_i + dP_i = 0$, whence follows

$$\beta\mu_i^* = \ln(N_i / N_V) - \frac{1}{2}\ln\left[\frac{N_i(N_V+1)}{N_V(N_i+1)}\right]. \tag{32.3}$$

The system of equations (32.3) describes the partial isotherms of the adsorption of a multicomponent mixture on a uniform surface that is limited in area. In the particular case of adsorption of a single-component substance, we pass to the equation (31.3). The solution of the system (32.3) determines the set of the most probable values of the numbers of the adsorbed molecules of the mixture $\{N_i^*\}$. This system is nonlinear with respect to the relationship between the given values of the chemical potentials of the mixture components $\{\mu_i\}$ and the numbers of adsorbed $\{N_i^*\}$ molecules. With an increase in the surface area, the value of ΔP_i decreases, and for macroscopic systems we have the well-known expressions $\beta\mu_i^* = \ln(N_i/N_V)$ (Langmuir partial isotherms).

The second partial derivatives are taken at constant values of the chemical potentials $\{\mu_i^*\}$ and the size of the system M

$$\frac{\Delta^2 \ln P(N_1,N_2,M)}{\Delta N_1^2}=\frac{\Delta}{\Delta N_1}\frac{\Delta \ln P(N_1,N_2,M)}{\Delta N_1}=$$
$$=\frac{1}{4}\ln\left[\frac{P(N_1+2,N_2,M)P(N_1-2,N_2,M)}{P(N_1,N_2,M)^2}\right]= \tag{32.4}$$
$$=\frac{1}{4}\ln\left[\frac{N_1(N_1-1)N_V(N_V-1)}{(N_1+1)(N_1+2)(N_V+1)(N_V+2)}\right].$$

This equation is identical in its form to equation (31.4) for the adsorption of one substance. The differences are in different ways of calculating N_V. For one substance, the value of N_V is uniquely determined by N_1 and M, whereas for a multicomponent mixture the value of N_V depends additionally on the values of N_k of the other components of the mixture.

Equation (32.4) shows that for any number of components of the mixture, the structure of the expression for the second derivative with respect to any component i is preserved.

It is convenient to represent equation (32.4) in the form of two contributions from the component i ($P_{2(i)}$) and the number of vacant sites ($P_{2(V)}$)

$$\frac{\Delta^2 \ln P(\{N_i\},M)}{\Delta N_i^2}=P_{2(i)}+P_{2(V)},$$
$$\text{where } P_{2(i)}=\frac{1}{4}\ln\left[\frac{N_i(N_i-1)}{(N_i+1)(N_i+2)}\right]. \tag{32.4a}$$

The formula for $P_{2(V)}$ is the formula (32.4a) for $P_{2(i)}$, in which the index i is replaced by the index V. This expression is fundamental for ideal systems, since all mixed derivatives are expressed through it, which are defined as

$$\frac{\Delta^2 \ln P(N_1,N_2,M)}{\Delta N_2 \Delta N_1} = \frac{\Delta}{\Delta N_2}\frac{\Delta \ln P(N_1,N_2,M)}{\Delta N_1} = \\ = \frac{1}{4}\ln\left[\frac{P(N_1+1,N_2+1,M)P(N_1-1,N_2-1,M)}{P(N_1-1,N_2+1,M)P(N_1+1,N_2-1,M)}\right] = \\ = \frac{1}{4}\ln\left[\frac{N_V(N_V-1)}{(N_V+1)(N_V+2)}\right] \equiv P_{2(V)} \tag{32.5}$$

Thus, for any number of mixture components, all the mixed second difference derivatives have the same form. This is due to the fact that the degrees of filling of any surface sites with different molecules of the mixture are related to each other by normalization conditions. Their 'engagement' is due to competition for filling free surface sites.

In the transition to macroscopic systems, all the second derivatives vanish because of the narrowing of the width of the distribution function $P(N_1, N_2, M)$ and its transition to the δ-shaped form. For bounded systems, it is necessary to use the normal distribution (s–1) of order. For the case $s = 3$, we will represent the distribution function in the form (since the number M here is fixed and to simplify the record we omit it)

$$P(N_1,N_2) = P(N_1^*,N_2^*)\exp[-\Sigma_{i=1}^{s-1}\Sigma_{j=1}^{s-1}\eta_{ij}(N_i-N_i^*)(N_j-N_j^*)],$$

where the parameters of the normal two-dimensional distribution η_{ij} are determined by the equations (32.4) and (32.5) with $\{N_i\}=\{N_i^*\}$

$$\eta_{ij} = -1/2\frac{\Delta^2 \ln P(N_1,N_2,M)}{\Delta N_j \Delta N_i}. \tag{32.6}$$

In this case, a large statistical sum is written as

$$\Xi^* = \int_{-\infty}^{+\infty}\int_{-\infty}^{+\infty} P(N_1^*,N_2^*)\exp\left[-\sum_{i,j=1}^{s-1}\beta_{ij}(N_i-N_i^*)(N_j-N_j^*)\right] = \\ = d(N_i-N_i^*)d(N_j-N_j^*) = P(N_1^*,N_2^*)\pi\det(\beta_{ij})^{-1}, \tag{32.7}$$

or the logarithm, the maximum term of the partition function, taking into account the fluctuations from which all the thermodynamic expressions are obtained, has the form

$$\ln \Xi^* = \ln\left[P(N_1^*, N_2^*)\right] + \ln \pi - \ln \det(\eta_{ij}). \qquad (32.71a)$$

The first term refers to the macroscopic characteristics of the system. All size properties of small systems are reflected through the elements of the matrix $\det(\eta_{ij})$.

The case of an ideal mixture of adsorbed molecules on a uniform surface corresponds formally to a well-developed theory of fluctuations in the bulk phase for small filling [10]. The differences associated with high densities in the LGM change the specific expressions for the equations for the maximum of the distribution function $(s-1)$ of order and their variance. But the general methodology is preserved, although the calculations become much more complicated, so we do not dwell on these questions.

As the simplest examples demonstrating the role of the size of a surface, we give analytical expressions in the first order of smallness of the contributions (for which the difference and continuous derivatives coincide) for 1) the determinant $\det(\eta_{ij})$, and 2) the corrections for the partial filling of the surface of a two-component mixture more accurate results should be obtained numerically.)

1) It can be proved that the expression for the determinant $\det(\eta_{ij})_{s-1}$ of dimension $(s-1)$, referring to the $(s-1)$ number of components of the mixture, is expressed as

$$\det(\eta_{ij})_{s-1} = M / \left(\prod_{i=1}^{s} N_i\right). \qquad (32.8)$$

This formula determines the denominator to calculate all the mean-square fluctuations, which are expressed in terms of the corresponding contributions of the inverse and attached matrices from the original matrix η_{ij} by the known technique [2,3,10].

2) The expressions for the fluctuation corrections for the partial fillings of a uniform surface by a two-component mixture are written as

$$\Delta\theta_i = (N_V - N_i) / 2M(N_V + N_i), \qquad (32.9)$$

where $\Delta\theta_i = \theta_i - \theta_i^\infty$ and $\theta_i = \langle N_i \rangle / M$, and θ_i^∞ is the partial filling of the surface by molecules of sort i for the macrosystem. Differences in the partial contributions of components are determined by the

Henry constants. The fluctuation correction for complete filling of the surface has the form $\Delta\theta = \Delta\theta_1 + \Delta\theta_2$. For $s = 2$, formula (32.9) becomes the formula (31.14).

Calculation of the mixture. Everywhere below, the calculations are performed at a temperature corresponding to $\beta\varepsilon = 1/2$, where ε is the interaction parameter between the molecules of the first sort. For an ideal system, the parameter $Q_1 = 4\varepsilon$ serves as a measure of the binding of the first sort molecule to the surface. It is assumed in the calculations that the binding energy of molecules of the first sort with the surface, which corresponds to a sufficiently strong coupling of the adsorbed particle to the surface. The binding energy of molecules of the second sort with the surface is assumed to be: $Q_2 = \gamma Q_1$, where $\gamma = 1.4$ [28].

The influence of the size effects on the adsorption of a binary gas mixture on a uniform surface has been investigated numerically by varying the value of M in the range from 10^5 to 10. The argument is given by the total filling of the surface θ and the molar fraction of the first component $x_1 = \theta_1/(\theta_1 + \theta_2)$. Using partial differential equations (32.3), the partial pressures of the P_i components in the vapour phase and the total pressure in the system $P = P_1 + P_2$ are calculated.

Figure 32.1 *a* shows the relative deviations at $(\theta_i - \theta_i^\infty)/\theta_i^\infty$ different sites from the value θ_i^∞ for the macroscopic system (1) with $M = \infty$. These deviations depend on both the M value and the mole fraction of the first component in the mixture. The calculations were carried out for three values $x_1 = 0.15$ (*1*), 0.5 (*2*) and 0.85 (*3*) at $M = 10^4$. Relative deviations sharply increase for small degrees of filling (for $P \rightarrow 0$). In order to demonstrate the dependence of the relative difference on the value of M, the inset data in an enlarged form shows the results of calculations for $x_1 = 0.15$ for $M = 10^4$ (*4*), 10^3 (*5*), and 10^2 (*6*) for small values of the pressure P.

Figure 32.1 *b* considers partial isotherms for $M = 10^4$. Their form corresponds to the well-known curves of Langmuir isotherms. The isotherms for other values of M slightly deviate from them. At $x_1 = 0.15$ and $x_1 = 0.85$, the partial isotherms 1 and 2 change in places. At $x_1 = 0.5$, both partial isotherms coincide with each other (therefore there is one curve 3).

Figure 32.2 shows that all the η_{ij} values are positive, and the diagonal elements exceed off-diagonal ones. Thus, the determinant is always positive.

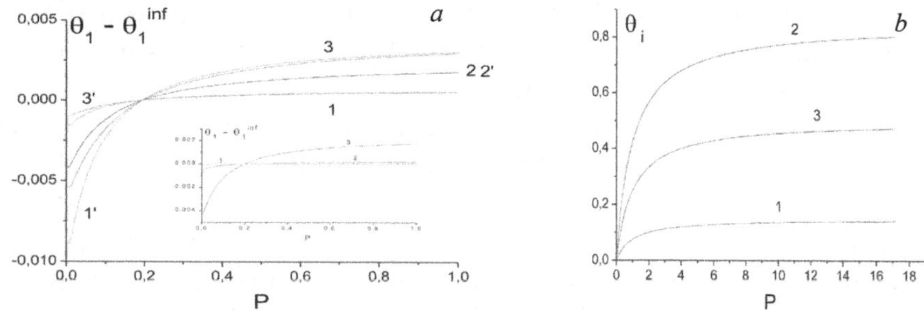

Fig. 32.1a. Deviations $\theta_i - \theta_i^\infty$ as a function of the pressure P at $M = 10^2$ with a change in the molar composition of the mixture $x_1 = 0.15$, 0.5 and 0.85 in the region of low pressures. The notation on the curves, respectively, is 1, 2, 3 for $i = 1$ and 1′, 2′, 3′ for $i = 2$. On the inset: the same deviations at $x_1 = 0.5$ for $M = 10^4$ (1), 10^3 (2) and 10^2 (3).

Fig. 32.1b (right). The dependencies of local densities θ_i on total pressure P. Symbols on the curves: 1 – θ_1 for $x_1 = 0.15$ and θ_2 for $x_1 = 0.85$, 2 – θ_2 for $x_1 = 0.15$ and θ_1 for $x_1 = 0.85$, 3 – $\theta_1 = \theta_2$ for $x_1 = 0.5$.

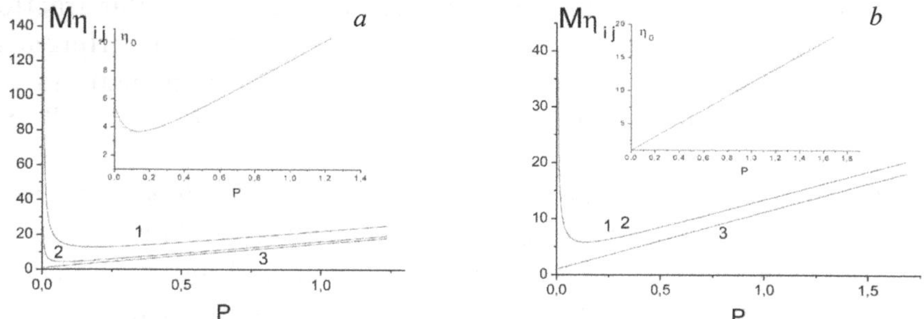

Fig. 32.2. Dependence of the values of the elements of the dispersion matrix η_{ij} on pressure P, at $x_1 = 0.15$ (a) and $x_1 = 0.5$ (b). The notation on the curves: 1 – η_{11}, 2 – η_{22}, 3 – $\eta_{12} = \eta_{21}$. On the insets: the number of conditionality η_0 of the matrix η_{ij}.

The size dependences of the fluctuations can be considered on the graphs connecting the values of the elements $M\eta_{ij}$ of the matrix of root-mean-square deviations with the total degree of filling of the surface θ, shown in Fig. 32.2 $x_1 = 0.15$ (a) and 0.5 (b). The curves are presented in the above form, so that the values of $M\eta_{ij}$ are commensurable. Thus, the magnitudes of the elements of the matrix decrease with increasing size of the section as M^{-1}.

The matrix η_{ij} is symmetric and positive definite. The number of conditionality of the matrix $\eta_0 = \max |\lambda_i| / \min |\lambda_i| \geq 1$, where λ_i are the eigenvalues of this matrix. In numerical methods, the number of conditionality of the matrix determines the sensitivity of the solution of the system of linear equations to the errors of the initial data.

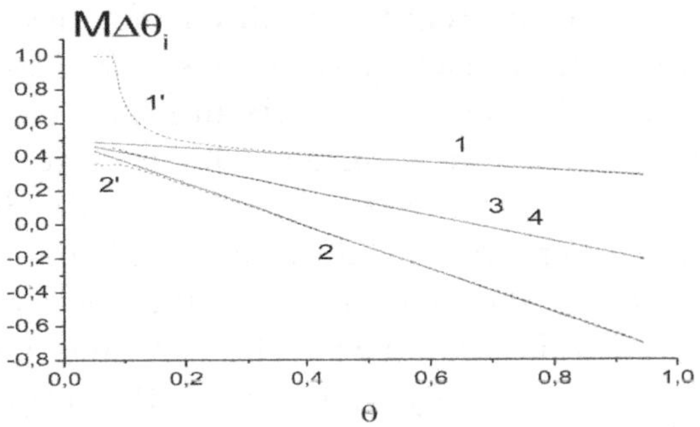

Fig. 32.3. Fluctuation corrections for local densities $\Delta\theta^i$ as a function of the degree of filling θ for $M = 10^4$ are continuous lines and $M = 10^2$ are dotted lines. The notation on the curves 1 – $\Delta\theta^1$, 2 – $\Delta\theta^2$ at $x_1 = 0.15$, 3 – $\Delta\theta^1$, 4 – Δθ2 at $x_1 = 0.5$ for $M = 10^4$. The same notation, but with a prime for $M = 10^2$.

The higher η_0, the more sensitive is the system to the errors of the original data (i.e. to the values of the elements of the matrix).

The effect of the density fluctuations $\Delta\theta_i$ on the values of partial degrees of filling is demonstrated by Fig. 32.3. The reduced values of the fluctuation corrections $M\Delta\theta_i$ are shown. These contributions are maximal in the region of small and large degrees of filling. The eigenvalues of the fluctuation corrections $\Delta\theta_i$ depend on the size of the region as M^{-1}.

The values of the corrections also depend on the molar fractions of the components of the mixture. The larger the differences in molar fractions, the more they differ from each other. By reducing the molar fraction of the first component in the mixture and the total amount of adsorbed material, one can reach N_1 values that correspond to the forbidden values of 0 and 1 in the formulas (32.4) and (32.5). Similarly, for a small fraction of the second component and for small fractions of free sites at large fillings $\theta \to 1$. In this case, the conditions for using the continual description are violated, even with symmetric difference derivatives, and the limiting values of the corresponding filling degrees correspond to a maximum value of $\Delta N_i = 1$ [24], which corresponds to the fixation of all molar fractions upon reaching one of N_i, $i = 1, 2$, and V, values $\Delta N_i = 1$. This factor is important for small systems and its role increases during the transition to non-uniform surfaces.

Non-uniform surfaces. This general case of ideal systems retains the properties indicated above for the adsorption of one substance on the non-uniform surfaces and for the adsorption of mixtures on a uniform surface. Because of the independence of the adsorption process at the sites of different types, all characteristics are summed up by contributions from each type of sites q, and by virtue of the 'meshing' of the molecules of the mixture with each other when competing for free sites of a surface of the same type, we have that the preceding formulas must be rewritten with additional lower indices of sites of type q. The description of non-uniform ideal systems is reduced to the summation of the contributions of any characteristics on individual faces. The formulas obtained above can be easily generalized by summing with weights F_q for the contributions of different types of centres q ($M_q = MF_q$) and replacing the partial fillings θ_i by the local partial fillings $\theta_q^i = N_q^i/M_q$ related to sites of type q, $1 \le q \le t$, t is the number of types of sites.

The generalization of formulas (32.3) for local partial isotherms to the case of non-uniform surfaces is written as

$$a_q^i P_i = \frac{\theta_q^i}{\theta_q^V}\left[\frac{\theta_q^V(\theta_q^i + 1/M_q)}{\theta_q^i(\theta_q^V + 1/M_q)}\right]^{1/2}, \tag{32.10}$$

The elements of the dispersion matrix η_{qp}^{ij} in which the mean values θ_q^i are found from the solution of the system of equations (32.10), are determined in the form

$$\eta_{qp}^{ij} = -1/2\frac{\Delta^2 \ln P(N_q^i, N_p^j, M)}{\Delta N_p^j \Delta N_q^i} = \begin{cases} D_q(i) + D_q(V), & i = j, q = p \\ D_q(V), & i \ne j, q = p \\ 0, q \ne p, \text{any } i, u, j \end{cases} \tag{32.11}$$

where $D_q(i) = \frac{1}{4}\ln\left[\frac{\theta_q^i(\theta_q^i - 1/M_q)}{(\theta_q^i + 1/M_q)(\theta_q^i + 2/M_q)}\right]$, including $= V$.

As for a uniform surface, the size effects appear in small areas of individual faces due to their limited size and due to fluctuations of these local partial degrees of density filling. In general, the degree of filling of type q sites by molecules of type i can be represented in the form $\theta_q^i(\text{fl}) = \theta_q^i + \Delta\theta_q^i$, where the degree of filling θ_q^i is found from the solution of the system of equations (32.10), and the fluctuation corrections of the local partial fillings $\Delta\theta_q^i$ are found from the following expression

$$\Delta\theta_q^i = -\frac{1}{2M_q}\frac{\partial \ln \det}{\partial \ln(p_i)} = -\frac{1}{2M_q}\sum_{q=1,k=1}^{t,(s-1)}\frac{\partial \ln \det}{\partial \theta_q^k}\frac{\partial \theta_q^k}{\partial \ln(a_q^i p_i)} \quad (32.12)$$

This formula is a generalization to the non-uniform surfaces of the fluctuation correction for a pure substance, in which the double sum explicitly takes into account the number of independent components of the mixture $1 \leq i \leq (s-1)$ located on sites of the type q, $1 \leq q \leq t$.

The noted peculiarities of the influence of the boundedness of segments of uniform faces are mainly preserved for a non-uniform surface. To illustrate this fact, we give, for example, Fig. 32.4, as a generalization of the calculations in Fig. 32.3, for local partial fluctuation corrections to the degree of filling of $\Delta\theta_q^i$ on a non-uniform surface consisting of two types of centres: $t = 2$, $F_1 = F_2 = 0.5$, $Q_2^i = Q_1^i/2$.

The local partial and average partial isotherms of adsorption are shown in Fig. 32.5. These isotherms have a simple Langmuirian form of saturation curves. By varying the proportion of sites of different types and the binding energies, a wide spectrum of the isothermal binary mixture can be obtained (curves 3 and 6 refer to the average partial isotherms).

A characteristic feature of the curves in Fig. 32.4 is the presence of large sections with a fixed limiting value of the deviation of the fluctuation increment of the local partial density. This feature is enhanced with a decrease in the mole fraction of any of the components of the mixture. This tendency is preserved with increasing number of types of different centres t.

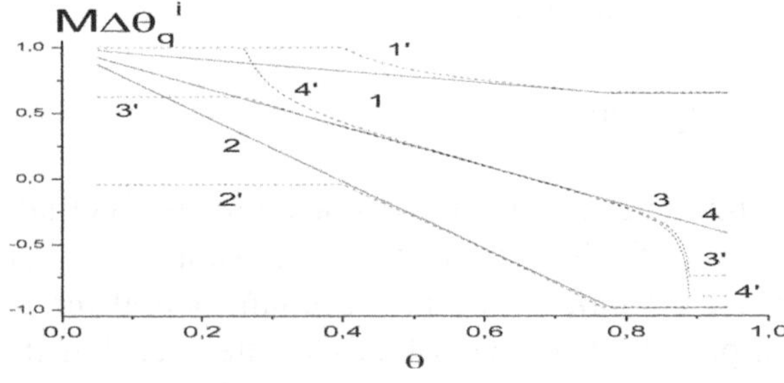

Fig. 32.4. Fluctuation corrections for local densities $\Delta\theta_q^i$ as a function of the degree of filling θ for $M = 10^4$ are continuous lines and $M = 10^2$ are dotted lines. The notation on the curves 1 – $\Delta\theta_q^1$, 2 – $\Delta\theta_q^2$ at $x_1 = 0.15$, 3 – $\Delta\theta_q^1$, 4 – $\Delta\theta_q^2$ at $x_1 = 0.5$ for $M = 10^4$. The same notation, but with the prime for $M = 10^2$.

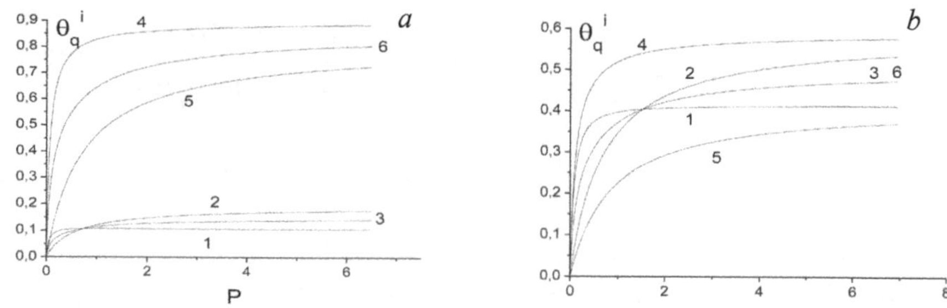

Fig. 32.5. Dependences of the local densities θ^i_q on the total pressure P at $x_1 = 0.15$ (a) and $x_1 = 0.5$ (b). The notation on the curves: 1 – θ^1_1, 2 – θ^1_2, 3 – θ^1, 4 – 1 – θ^2_1, 5– θ^2_2, 6 – θ^2.

Equations (32.11) show that the so-called limiting or 'forbidden' values $N^b_q = 0, 1, M_q - 1, M_q$ play an important role in calculating the effects of fluctuations on non-uniform surfaces, for which the right-hand sides of expressions (32.11) are not defined . An increase in the number of types of centres t inevitably leads to an increase in the total number of surface sites, which include the 'forbidden' numbers of molecules $N^b_q = 0, 1, M_q{-}1, M_q$, for large and small fillings of the faces of each type of centre. Their number is $N^b = 4t$, so with increasing t the total number of boundary values N^b increases.

The more non-uniform the surface with respect to bond energies, the more significant this factor is, since during the filling of one face the other faces will remain practically free or vice versa, almost completely filled. In both situations, the role of fluctuations increases. In this paper, as the simplest method for approximating the density description in the region of 'forbidden' values of N^b_q, the limit values of ΔN^i_q are used so that they are equal to one – the limiting value for the minimum value of $N^i_q = 1$ (that is, for an ideal gas of particles or vacancies).

33. Non-ideal systems

Mean-field approximation. Let us consider the fluctuation effects in the simplest case of adsorption of one substance ($s = 2$) on different faces of a single crystal, taking into account lateral interactions in the mean-field approximation. This leads to the fact that the parameters of the lateral interaction ε_{qq} can depend on the type of the face q, $1 \le q \le t$, t is the number of site types [28], but each face is considered isolated. The symbol M is used to denote the total number of surface sites consisting of sections of size M_q, $\sum^t_{q=1} M_q = M$. In this version of the theory, the Hamiltonian (30.1) can be rewritten in the form [13]

$$H = \sum_{f=1}^{M} \sum_{i=1}^{2} \left(v_q^i \gamma_f^i \eta_f^q - \frac{1}{2} \sum_{g,j} \varepsilon_{qq} \gamma_f^i \gamma_g^j \eta_f^q \eta_g^q \right). \tag{33.1}$$

where the index g refers to z_{qq} the number of nearest neighbours of the site f on the face q.

We will consider the distribution of molecules in the grand canonical ensemble (μ, M, T), where the symbol μ is the chemical potential of the molecules in the bulk phase. Each site can be occupied or free. We denote by N_q the number of adsorbed molecules at sites of type q. The number of free sites of type q is denoted by $N_q^V = M_q - N_q$.

The expression for the local functions Q_q (N_q, M_q, T) for faces of type q is written as

$$Q_q\left(N_q, M_q, T\right) = M_q! / \left(N_q! N_q^V!\right) J_q^N \exp(z_{qq} \beta \varepsilon_{qq} N_q^2 / 2M_q)). \tag{33.2}$$

This implies the change of equation (31.2) to the following expression

$$\begin{aligned} \ln P_q(N_q) &= \Sigma_{q=1}^t \{\ln M_q! - \ln N_q! - \\ &-\ln N_q^V! + \beta \mu_q^{i*} N_q + z_{qq} \beta \varepsilon_{qq} N_q^2 / (2M_q)\}. \end{aligned} \tag{33.3}$$

Omitting the intermediate calculations, we write out the equation for the local isotherm, taking into account both lateral interactions and size effects [15]

$$\beta \mu_q^* = \ln\left[\frac{N_q}{N_q^V}\right] - \frac{1}{2} \ln\left[\frac{N(N_V + 1)}{(N+1)N_V}\right] - z_{qq} \beta \varepsilon_{qq} N_q / M_q. \tag{33.4}$$

The second derivative is defined as

$$\frac{\Delta^2 \ln P_q(N_q, M_q)}{\Delta N_q^2} = \frac{1}{4} \ln\left[\frac{N_q(N_q - 1) N_q^V (N_q^V - 1)}{(N_q + 1)(N_q + 2)(N_q^V + 1)(N_q^V + 2)}\right] + z_{qq} \beta \varepsilon_{qq}. \tag{33.5}$$

It follows from Eqs. (33.4) and (33.5) that in the absence of correlation effects between adsorbed molecules, the contributions constructed for the size effects retain their form. However, the presence in these equations of the contribution from lateral interactions can substantially change the solutions of the equations obtained in comparison with the solution of the equations of Section 31.

As an example, let us indicate how the equation (31.14) changes by the fluctuation correction $\Delta\theta_q = \theta_q - \theta_q^\infty$ in the first order in M_q^{-1}:

$$\Delta\theta_q = \frac{M_q^2(M_q - 2N_q)}{2[M_q^{\ 2} - \beta\varepsilon_{qq}z_{qq}N_q(M_q - N_q)]^2} = \frac{(1-2\theta_q)}{2M_q[1-\beta\varepsilon_{qq}z_{qq}\theta_q(1-\theta_q)]^2}. \tag{33.6}$$

In the second equation, the size component is explicitly distinguished. As the size of the face increases, the differences in the fillings decrease. The second equation also allows us to rewrite this correction by selecting the equation $\beta\varepsilon_{qq}z_{qq}\theta_q(1-\theta_q) = 1$ for the spinodal curve in the denominator, which separates the 'metastable' region and the region of thermodynamic instability of the stratifying molecules. The magnitude of the correction on a section of type q depends on the nature of the intermolecular interaction.

If $\varepsilon_{qq} < 0$, which corresponds to the case of ordered chemisorbed molecules, then with the contribution of the lateral interaction $|\beta\varepsilon_{qq}z_{qq}|$ the value of the denominator increases with respect to the denominator for an ideal system and the effect of lateral interaction coincides with the effect of increasing the size of the face. If $\varepsilon_{qq} = 0$, then we return to equation (31.14).

In the case of attraction between $\varepsilon_{qq} > 0$ molecules, the denominator decreases with increasing $\beta\varepsilon_{qq}z_{qq}$, so the influence of the lateral interaction and the increase in the size of the face are directed in opposite directions. As the current value of the local density θ_q approaches the given face q to the spinodal curve, the behaviour of the correction depends on the ratio of the quantities M_q and the square bracket. It can be seen that the decrease in the denominator of the density occurs very rapidly (according to a quadratic law), therefore, for any fixed value of M_q there exists a density such that the correction (33.6) increases sharply. Obviously, this cooperative behaviour of the system depends on the size of the region according to equation (33.4). A more precise description of the phase behaviour of molecules for small systems requires the use of more accurate approximations than the approximation of the molecular field.

Quasi-chemical approximation. The formulation of the problem remains, but now we need to take into account the correlation effects [29]. The mean value of the number of pairs of sites of different types qp is denoted by M_{qp}. The connection between the number of pairs of sites of different types has the form $\Sigma'^{t}_{p=1} M_{qp} + 2M_{qq} = z_q M_q$ where z_q is the number of neighbours of a site of type q, the sign of the prime of the sum means the absence of a term with $p = q$, $1 \leq p \leq t$. If we introduce the numbers z_{qp} of sites of type p near the site of type q that characterize the local structure of the non-uniform surface, then their relation to the numbers of pairs of M_{qp} sites of type qp is given as $z_{qp} = (1+\Delta_{qp})M_{qp}/M_q$ or $M_{qp} = z_{qp}M_q/(1+\Delta_{qp})$, where Δ_{qp} is the Kronecker symbol. The total balance of pairs of sites is written in the form of two sums $\Sigma^{t}_{q=1} M_{qq} + \Sigma^{t^*}_{qp=1} M_{qp}$.

As above, J_q is the partition function of a particle in a site of type q; $1 \leq q \leq t$, J is the partition function of the molecule in the gas phase, $\mu = \beta^{-1}\ln(\beta P/J)$, $\beta = (kT)^{-1}$, P is its pressure; ε_q is the binding energy of a particle with a site of type q on the surface of the adsorbent. The parameters of the lateral interaction ε_{qp} depend on the type of the pair qp, $1 \leq q, p \leq t$, on which there are two adjacent adsorbed molecules.

The expression for the partition function of the non-uniform system in the QCA, taking into account interactions only between nearest neighbors, is written as $Q = Q_1 Q_2$, where

$$Q_1 = \prod_{q=1}^{t} \sum_{N_q, N_{qq}=0}^{M_q, M_{qq}} Q_{qq}, \quad Q_2 = \prod_{qp=1}^{t^*} \sum_{N_{qp}=0}^{M_{qp}} Q_{qp}, \tag{33.7}$$

The first factor Q_1 refers to the distribution of adsorbed molecules at centres of types q: the sum over particles A is taken over all possible values of N_q from zero to the full filling of all sites of M_q of a given type q, and their pairs ij = AA, AV, VA and VV on pairs of sites of one type qq. The second factor Q_2 refers to the cross pairs of sites of different types qp ($q \neq p$), on which the same particles ij = AA, AV, VA and VV are found. The number of such pairs of sites for $q \neq p$ is renumbered (the order of the indices does not play the role of $M_{qp} = M_{pq}$) and their total number is denoted by t^*. The summation is carried out over the numbers of pairs N_{qp} of neighbouring particles AA from zero to M_{qp}.

These two types of factors in (33.7) consist of the following contributions

$$Q_{qq} = C_{M_q}^{N_q} J_q^{N_q} \left(C_{M_q}^{N_q}\right)^{-z_{qq}} \frac{M_{qq}!}{N_{qq}^{AA}!N_{qq}^{VV}![(N_{qq}^{AV}/2)!]^2} \exp(\beta\varepsilon_{qq} N_{qq}^{AA}), \qquad (33.8)$$

$$Q_{qp} = \left(C_{M_q}^{N_q}\right)^{-z_{qp}} \left(C_{M_p}^{N_p}\right)^{-z_{pq}} \frac{M_{qp}!}{N_{qp}^{AA}!N_{qp}^{AV}!N_{qp}^{VA}!N_{qp}^{VV}!} \exp(\beta\varepsilon_{qp} N_{qp}^{AA}) \qquad (33.9)$$

where $N_q^V = M_q - N_q$ is the number of free sites of type q, and $C_{M_q}^{N_q} = \dfrac{M_q!}{N_q!N_q^V!}$ is the number of combinations of M_q sites of type q by the number of adsorbed N_q particles at these sites.

Correcting factors $\left(C_{M_q}^{N_q}\right)^{-z_{qq}}$ and $\left(C_{M_q}^{N_q}\right)^{-z_{qp}} \left(C_{M_p}^{N_p}\right)^{-z_{pq}}$ are necessary for the refinement of the entropy factor, since the number of independent pairs is overestimated in the QCA [3, 13]. In the case of $\beta\varepsilon_{qp} \to 0$, they lead to an exact solution corresponding to a chaotic distribution of molecules over sites of different types $\tilde{N}_{qp}^{ij} = M_{qp} N_q^i N_p^j / (M_q M_p) = z_{qp} N_q^i N_p^j / [(1+\Delta_{qp}) M_p]$, where i, j = A, V [13, 30].

The balance of the number of pairs N_{qp}^{ij} of different types entering into expressions (2) and (3) for statistical weights in the quasi-chemical approximation is expressed as

$$N_{qq}^{AV} = N_{qq}^{VA} = z_{qq} N_q^A - 2N_{qq}^{AA}, \quad N_{qq}^{VV} = z_{qq} M_q / 2 - z_{qq} N_q^A + N_{qq}^{AA} \qquad (33.10)$$

for pairs of sites consisting of sites of the same type qq, and

$$\begin{aligned} N_{qp}^{AV} &= z_{qp} N_q^A - N_{qp}^{AA}, N_{qp}^{VA} = z_{pq} N_p^A - N_{qp}^{AA}, \\ N_{qp}^{VV} &= z_{pq} M_p - z_{pq} N_p^A - z_{qp} N_q^A + N_{qp}^{AA} \end{aligned} \qquad (33.11)$$

for pairs of sites consisting of sites of different types ($q \neq p$). The sum of all pairs of particles of different sorts satisfies the normalization condition $N_{qp}^{AA} + N_{qp}^{AV} + N_{qp}^{VA} + N_{qp}^{VV} = z_{qp} M_q = z_{pq} M_p = M_{qp} = M_{pq}$.

To analyze the fluctuational contributions, we consider the probability of the system $P(\{N_q, N_{qp}\}) = Q\exp[\beta\mu N]$ to be in concrete states in the grand canonical ensemble, where $N = \sum_{q=1}^{t} N_q$:

$$\ln P(\{N_q,N_{qp}\}) = \Sigma_{q=1}^{t} \ln P_q(\{N_q,N_{qq}\}) + \\ +\Sigma_{qp=1}^{t^*} \ln P_{qp}(\{N_q,N_p,N_{qp}\}), \tag{33.12}$$

where two types of summands correspond to the contributions of Q_1 and Q_2.

Let us find the minimum conditions for ln$P$ ($\{N_q, Nq_p\}$), depending on the independent variables $X(= N_q, N_{qp}, N_{qp})$ and for its second derivatives characterizing the dispersion matrix of the distribution function $P(\{N_q, N_{qp}\})$. For small particles it is necessary to use symmetric difference derivatives instead of the usual differential derivatives [14,15]. The extremum condition $\Delta \ln P(\{N_q,N_{qq}\})/\Delta N_q = 0$ condition gives the equation for the local adsorption isotherm, which connects the chemical potential of the system ($\mu_q = \mu + \beta^{-1}\ln(J_q)$) specified by the thermostat, with the number of molecules on each face N_q (or the degree of local filling $\theta_q = N_q/M_q$)

$$\beta\mu_q = \frac{1}{2}\times \\ \times\left[\left(1-\sum_p z_{qp}\right)\ln\frac{N_q^A(N_q^A+1)}{N_q^V(N_q^V+1)} + \ln\frac{(N_{qq}^{VV}-z_{qq})![(N_{qq}^{AV}+z_{qq})/2!]^2}{(N_{qq}^{VV}+z_{qq})![(N_{qq}^{AV}-z_{qq})/2!]^2} + \right. \\ \left. +\sum_{p\neq q}{}'\ln\frac{(N_{qp}^{VV}-z_{qp})![(N_{qp}^{AV}+z_{qp})!]}{(N_{qp}^{VV}+z_{qp})![(N_{qp}^{AV}-z_{qp})!]}\right]. \tag{33.13}$$

The extremum conditions for each type of N_{qq}^{AA} $1 \le q \le t$, and N_{qq}^{AA}, $1 \le (qp) \le t^*$ $(\Delta \ln P(\{N_q,N_{qp}\}) / \Delta N_{qp}^{AA} = 0)$ pairs are followed by equations on the relationship between pair functions:

$$\beta\varepsilon_{qq} = \frac{1}{2}\ln\frac{N_{qq}^{AA}(N_{qq}^{AA}+1)N_{qq}^{VV}(N_{qq}^{VV}+1)}{[N_{qq}^{AV}/2(N_{qq}^{AV}/2+1)]^2}, \\ \beta\varepsilon_{qp} = \frac{1}{2}\ln\frac{N_{qp}^{AA}(N_{qp}^{AA}+1)N_{qp}^{VV}(N_{qp}^{VV}+1)}{N_{qp}^{AV}(N_{qp}^{AV}+1)N_{qp}^{VA}(N_{qp}^{VA}+1)}. \tag{33.14}$$

These equations transform into known macroscopic expressions for $N_{qq}^{AA} \gg 1$. Equations (33.13) and (33.14) define a system of equations taking into account the limited size of the faces of microcrystals and the lateral interaction of molecules.

The description of the adsorption of molecules in small regions of a non-uniform surface with an allowance for lateral interactions and taking into account quadratic equilibrium fluctuations is constructed by analogy with the known formulas [2,3]. We construct formulas for the elements of the dispersion matrix $\eta_{km} = -1/2\dfrac{\Delta^2 \ln P\left(\{N_q, N_{qq}\}\right)}{\Delta X \Delta Y}$ where the symbols X and Y refer to all independent variables N_q, N_{qq} and N_{qp}, (the index k is determined by a list of θ_q, θ_{qq}, θ_{qp}) whose mean values are found from solutions of the system of equations (33.13) and (33.14).

Knowing η_{km} one can obtain the corrections ΔX_k (i.e. $\Delta\theta_q$, $\Delta\theta_{qq}$ and $\Delta\theta_{qp}$) to the degrees of filling of sites of different types and their pair probabilities due to fluctuations as

$$\Delta X_k = -\frac{1}{2}\frac{\partial \ln \det}{\partial \ln(\lambda_k)} = -\frac{1}{2}\sum_{b=1}^{T_D}\frac{\partial \ln \det}{\partial X_b}\frac{\partial X_b}{\partial \ln(\lambda_k)} \tag{33.15}$$

where det is the determinant of the dispersion matrix made up of the elements η_{km}, its dimension $T_D = 2t + t^*$ is equal to the number of independent variables N_q, N_{qq} and N_{qp}, and express the degree of filling of sites with an allowance for the fluctuations X_k (fl) = $X_k + \Delta X_k$. In the formula (33.15) we use $\lambda_k = \exp(\beta\mu)$ for N_q^A; $\lambda_k = \exp(\beta\mu_{qq})$ for N_{qq}^{AA}; $\lambda_k = \exp(\beta\mu_{qp})$ for N_{qp}^{AA}. To calculate the derivatives, we use the system $\partial X_b/\partial \ln \lambda_k$ of equations (33.13) and (33.14).The derivatives $\partial \ln \det/\partial X_b$ are calculated numerically.

A small drop. The theory was applied to calculate the influence of fluctuations on the characteristics of drops, including the value of the minimum radius R_0 of the drop formed [31]. Taking into account the reality of the quantities $z_{qp}(R)$ (see Section 24), the gamma functions [32] were used instead of the factorials in the calculations using formulas (33.12)–(33.15). The calculation was carried out for both an equilibrium and a metastable drop.

It is found that in the vicinity of the appearance of thermodynamically stable drops for different temperatures, while the volume of individual monolayers and their fluctuations are limited, the size of the drop is larger than when using macroscopic averages for local densities and mean pairs of particles [33]. The nature of this change in the dimensions of R_0 is reflected in Table 33.1 (the value of T_{crit} refers to the critical temperature in a macroscopic volume). The data of the table demonstrate a rather strong influence of the limited size of the system on the values of the drop radii corresponding to the condition that the surface tension σ of the drop is zero, i.e. when

Table 33.1. The minimum dimensions of the liquid phase of the drop (R_0) in the metastable vapour at different temperatures $\tau = T / T_{crit}$ [31]

τ	0.60	0.66	0.73	0.79	0.86
R_0	8	5	5	6	9
R_0(fluct)	16	15	13	12	13

the volume of spherical monolayers is limited, the drop remains thermodynamically unstable for a larger size.

The analysis showed that dimensional effects (smallness of the system and its fluctuations) appear at values of the radii of drops R of the order up to 40λ. This value is in full agreement with the lower limit of drop sizes, for which it is incorrect to apply the equations of thermodynamics at radii smaller than $R_t \sim 41\lambda$ [12] (see next Section).

34. The lower limit of the applicability of thermodynamics [12]

The mathematical apparatus used in thermodynamic equations to describe the curvature of any local area of the interface [19, 34] is an apparatus of differential geometry that operates second-order continuum derivatives. In all the equations of thermodynamics and the mechanics of continuous media for volume phases and interfaces, it is assumed that there are no fluctuations. A comparison of the continual and discrete descriptions should lead to conditions under which the contributions of fluctuations disappear. For a more rigorous estimate of R_t, we compare the thermodynamic properties of a certain surface region from the point of view of (A) thermodynamics and (B) molecular theory.

(A) From the point of view of the theory of a continuous medium, we consider the section representing an elementary minimal area, by means of which the 'total surface' is covered (by Borel's lemma). Recall that, according to this lemma [19], if the complete closed interval [*a*, *b*] is covered by an *infinite* system of open intervals (that is, without inclusion of limit points of a given interval), then it is always possible to extract a *finite* subsystem of open intervals from it, which also covers the entire interval [*a*, *b*]. This means that for a continuum of interior points of the finite interval under consideration one can construct a covering consisting of a finite number of interior intervals. This formulation of Borel's lemma is written out for simplicity in the case of one dimension. Its generalization to a two-dimensional surface is verbatim, replacing the term 'interval' by the term 'region', and has the same meaning.

(B) From the point of view of a discrete medium, the same matching region is the minimum region on which the thermodynamic function, its first and second derivatives must be determined (the second derivative is also needed to describe the curvature of the surface). According to the approach to discrete systems described above, such a region is a local domain containing in one dimension a minimal number of sites $L = 4\lambda$. For a cylindrical surface, we have one dimension and two dimensions for a spherical surface.

Comparison of the two methods of describing the minimum portion of the surface (A) and (B) gives the answer to the desired radius of the drop.

Cylindrical surface. In any cross section of a cylindrical drop, we have a circle that is approximated by a correctly inscribed polygon containing N_s – the number of sides, with the base length $L = 4\lambda$ (Fig. 34.1 *a*). We denote by the symbol ξ the accuracy of the coincidence of the length of the polygon and the circumference of the circle, then it is easy to see that $R_t/\lambda = N_sL$, where $N_s = \pi/(6\xi)^{1/2}$ is determined by the accuracy ξ of the description of the length of circumference by the broken line. From the physical point of view, the accuracy ξ determines the number of particles on the dividing surface, more precisely, the line in the cross section of the cylinder.

The procedure for writing a polygon into a circle quickly converges. When $N_s = 4$, we have the accuracy $\xi \sim 10\%$, and for $N_s = 6$ – the accuracy is $\xi \sim 4.5\%$, which corresponds to $R_t = 16$ and 24λ, for which $\eta \sim 1.8\%$ and 1.2%. Obviously, these values refer to fairly crude approximations. The accuracy of calculating the concentration profiles of drops $\{\theta_q\}$ is, as a rule, no lower than $\sim 0.1\%$, therefore, increasing the accuracy of the number of particles on the dividing line, we have for $N_s = 12$ (accuracy $\xi \sim 1\%$, $R_t = 48\lambda$ and $\eta = 0.6\%$) and $N_s = 24$ (accuracy $\xi \sim 0.5\%$, $R_t = 96\lambda$ and $\eta = 0.3\%$). We note that in the latter case the rms fluctuation η is even greater than the accuracy of calculation of the concentration profile.

Thus, with an increase in the accuracy of matching the number of particles on the dividing line, the size of the drop corresponding to the lower limit of the applicability of thermodynamics increases quite rapidly.

Spherical surface. For a spherical drop, the second dimension must be taken into account. In this case, the area $S_0 \sim L^2$ corresponds to an elementary area. We approximate the surface of a spherical drop

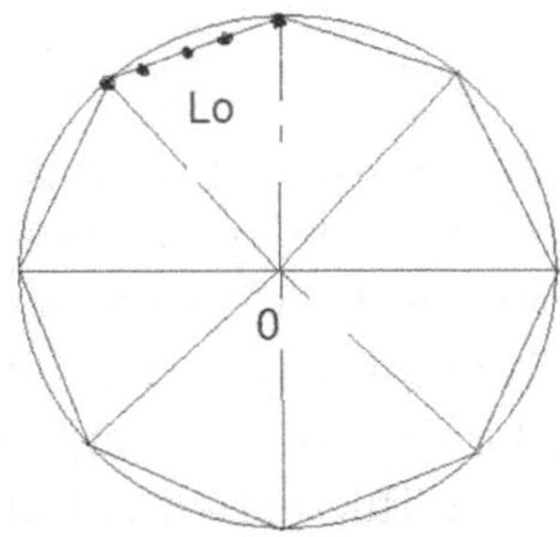

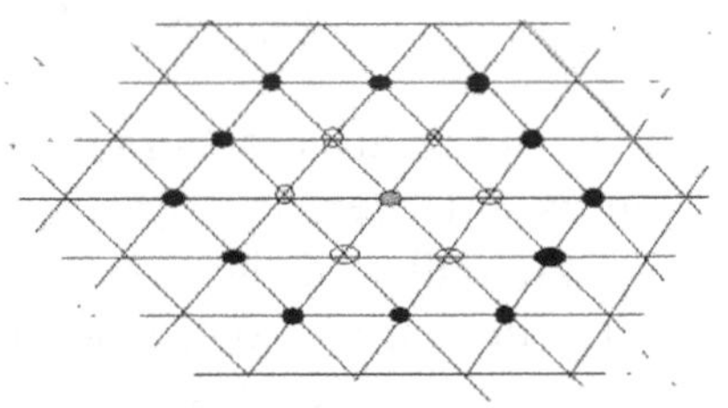

Fig. 34.1b. Elementary platform for the surface of the sphere. The gray circle is the centre, the light circles are the first neighbours, the dark circles are the second neighbours on the lattice z = 6.

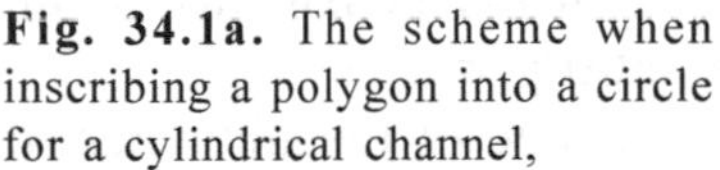

Fig. 34.1a. The scheme when inscribing a polygon into a circle for a cylindrical channel,

of radius R by the set of such elementary areas (as 'mathematical points – analogues of coverings in Borel's lemma). The most dense two-dimensional packing of molecules corresponds to a structure of $z = 6$ with a communication length between the cells λ. This structure has an anisotropic area formed by cells in the redistribution of two coordination spheres around the central cell. The maximum diameter of an elementary area with such a structure is equal to a segment of length L (Fig. 34.1 b). The minimum size of the site is $2\times3^{1/2}L$. This leads to a value $S_0 = 6\times3^{1/2}\lambda^2$.

Each flat elementary area can serve as the basis for a cone, which is drawn from the centre of the sphere. To simplify the estimates, we assume that the elementary area has the form of a circle, then the surface of the spherical part of the cone resting on this area is a figure of rotation and can be easily determined [19]. We denote it by S_c. The total number of cones on the surface of a sphere of radius R is equal to the number of molecules $N_c = 4\pi R^2/S_c$, coinciding with the number of sites approximating the sphere. You can enter the effective radius of the circle, equal in area of the considered area, defining it as $R_{ef} = (S_0/\pi)^{1/2}$. Then the quantity $S_c = 2\pi R^2\ [1-\{1-(R_{ef}/R)^2\}^{1/2}]$. Expanding the expression for S_c under the radical to the second term of smallness, we obtain the relationship between (S_0-S_c) and the ratio (R_{ef}/R).

Taking into account the proportionality of the number of molecules and the area of the flat elementary area, we require that the difference in the areas of all flat areas and all the spherical parts of inscribed cones satisfy condition

$$(S_0 - S_c)\ /\ S_0 < \xi_s, \tag{34.1}$$

where ξ_s is the relative accuracy of the surface area of a drop of a given radius R_t (where $\xi_s \sim \xi^2$). This condition is easily achieved by increasing the radius of the sphere. The condition for the smallness of the density fluctuations at the boundary of the minimum sizes of the applicability of thermodynamics is verified indirectly in comparison with the value of η – the relative mean square fluctuation [2] of the number of molecules N_c located on the surface of a sphere of radius R_t. The quantity η must certainly be less than the possibility of an accurate experimental determination of quantity σ.

As a result, the expression connecting the size of the diameter of the sphere R_t and the accuracy of the description of the surface area of the sphere ξ_s: $R_t/\lambda = [6 \times 3^{1/2}/(2\pi\xi_s)]^{1/2} = 1.29/\xi_s^{1/2}$ follows from the condition (34.1) measured in units of cell length λ.

This expression corresponds to the following sets of values: $\xi_s \sim 0.1\%$, $R_t = 41\lambda$ and $\eta \sim 0.7\%$; $\xi_s \sim 0.05\%$, $R_t = 82\lambda$ and $\eta = 0.34\%$; $\xi_s \sim 0.01\%$, $R_t = 129\lambda$ and $\eta \sim 0.2\%$.

Taking into account that the accuracy of calculations of the system of equations in the molecular theory [35–37] is not lower than 10^{-3} or 0.1%, we get $R_{t1}/\lambda = 41$, which, apparently, satisfies the existing experimental methods [22,23,38,39]. In the case of a significant increase in the experimental possibilities for a more accurate determination of the surface of tension, it is necessary to reduce the criterion value to $\xi = 10^{-4}$, which leads to an increase in the radius to $R_{t2}/\lambda = 129$ and corresponds to a value of $\eta \sim 0.2\%$. .

The obtained estimate is not related to the details of the molecular distribution and does not depend on either the choice of the surface of tension or the temperature. Near the critical point, large spontaneous fluctuations of the molecules are realized, and the probability of the existence of an isolated spherical drop in this situation is negligibly small. Nevertheless, this estimate correlates well with the results of numerical calculations [40]: up to temperatures $\tau = 0.99$, the width of the transition layer, both for a plane boundary and for drops, does not reach this value. Therefore, it should be assumed that for $R_{t1} < 41\lambda$ the thermodynamic description is not justified. For argon atoms, this corresponds to $R_{t1} \sim 16$ nm. Assuming that at $\xi_s \sim 0.05\%$, both the discrete nature of the matter and the contributions of the fluctuations can be neglected, this gives grounds for using the thermodynamic description for $R_{t2} > 80–100\lambda$ (or $R_{t2} \sim 40$ nm for argon atoms). Thus, the minimum size of the drop radius R_t, from which the thermodynamic description can be used, is 16 to 40 nm. These estimates are in good agreement with the initial postulates of thermodynamics about the need for the presence of sufficiently large amounts of matter in the system.

From surface to volume. Traditionally, a qualitative discussion of the role of fluctuations uses the estimate of the relative mean-square fluctuation for the Poisson distribution $\eta_P = \langle N^2 \rangle^{1/2}/N = N^{-1/2}$ [1–4]. The main condition in this distribution is the absence of correlations between molecules. The same estimate corresponds to an ideal gas of molecules. It is constructed for a macroscopically small subsystem in which any number of molecules can exist, provided that it is small in comparison with the total number of molecules in the complete system from which this subsystem is allocated.

Above in Table 29.1 and in the analysis of ξ and ξ_s, when M was varied, η was used, which is the analogue of the numbers η_P, which was calculated for the number of molecules N in the system in the form of a small compact body (rather than isolated molecules). Let us discuss the correspondence between the values of η and η_P for small systems. To do this, let us consider how the magnitude of the mean-square fluctuation η of a uniform system containing M centres changes in the problem of equilibrium filling of a monolayer on a dividing surface. These calculations refer to a small system of non-interacting molecules [24]. (Similar results are obtained by the mean-field approximation, which takes molecular interactions into account, but in a crude manner [15].)

Figure 31.2 *a* shows the concentration dependences of the function D directly related to the value of η as $D = -2\eta$ [15,24], which depends on the size of the region M and on the number of molecules N in the given region ($N_V = M–N$). The relation with the quantities η_P follows from the equality $N = \theta M$. The maximum values of D refer to the region of small and large degrees of filling θ. In the region of average fillings, the value of η is much lower. Fluctuations for a small number of molecules in small systems behave in a similar way, as in macroscopic systems, increase rapidly with decreasing N. As the density increases, for small systems, the same increase in fluctuations is observed as for a small number of molecules. In this case, $N \sim M$, which allows us to use the quantity η instead of η_P for the usual Poisson distribution as in Table 29.1. This fact is related to the presence of vacancies, whereas in real liquids there should be a fluctuation of the discharge regions identical to the fluctuations in the number of low-density molecules–phase inversions correspond to processes of bubble formation in a liquid. (For macrosystems, fluctuations in the region of large densities are usually not discussed.)

For $M = 10^4$, in the region of small and large fillings, there are differences from zero values for 2–3% density. Further, with a decrease in M to 10^3, the difference region D extends from the zero value to 20%. At even smaller values of M, the value of D differs from the macroscopic size in the entire density range, starting at $M = 500–600$. The curves for $M = 10$ show the maximum dispersion differences for small sections. The dotted curve 1′ in Fig. 31.2 refers to a similar relationship obtained with the first dimensional correction $D_1 = [-M/(NN_V)]$. Thus, approximate estimates overestimate the variance.

In order that the density fluctuations can be completely ignored, it is necessary to decrease D in the entire range of θ. With increasing M, the dispersion decreases sharply. So, for $M = 10^5$, the dispersion remains close to zero in the entire density range, then this size can be considered analogous to the value of R_{t1}. For a dividing drop surface, the given M corresponds to a radius $R/\lambda \sim 46$. This value is close to the earlier estimate $R/\lambda = 41$ for $\xi_s = 0.1\%$, which corresponds to $\eta \sim 0.7\%$ (note that $\eta \sim 1\%$ corresponds to $R/\lambda = 28$, see Table 29.1).

The above calculation confirms the universality of the estimates obtained. They are based on the discrete nature of the substance, are not related to the details of the molecular distribution, and do not depend on either the choice of the surface of tension or the temperature. However, a special consideration is required near the critical point.

Internal regions of phases. To go to macrophases, it is necessary to increase the drop radius in order to neglect the contribution of σA to the Gibbs potential (Section 29). Such a transition is consistent with the condition, for example, that the contribution of σA is 0.3 or 0.1%. According to Table 29.1 this refers to the number of molecules of the system $M_{\text{mac}} = 4\times10^7-10^9$, which exceeds the number of molecules in the surface regions of drops with radii R_{t1} and R_{t2} from 4×10^2 to 5×10^4 times. It was shown above that the use of the simplest estimate for determining the density fluctuations η in the dense phase (instead of the number of molecules in the gas phase η_P) makes it possible to control the magnitude of the fluctuations depending on the size of the system. For macroscopic volume phases, one can correlate the continual and discrete descriptions of a small region of a three-dimensional lattice around the selected site, and repeating the procedure used above for surface properties by word-by-word, obtain estimates for the quantities R^v_{t1} and R^v_{t2} for the volume phase, which in their sense are completely analogous to the values of R_{t1} and R_{t2}, but without taking into account the state of the surface. Restricting ourselves to the continuum constraints $R^v_{t1} = (3R^2_{t2})^{1/3}$ (similarly for R^v_{t2}) we have the radii of the inner regions $R^v_{t1} = 17\lambda$ and $R^v_{t2} = 29\lambda$.

For macrophases, the question of the size of the internal local region is related to the number of molecules in the region $dV = V_m(R^v_t)$, which is implied in all methods of a continuous medium:

1) In the theory of the formation of new phases (nucleation) and the theory of chemical processes in condensed phases, expressions are constructed for the probability of realizing these processes in a fluctuation manner. It is important to distinguish between the concept of fluctuations as the realization of a multiparticle event consisting of a multitude of phase molecules (for example, the appearance of a phase nucleus, or a collective reorganization of a medium in the process of electron transfer), and as a factor requiring corrections to allow for the limited size of the region in which the elementary process proceeds, describe these processes.

In the first case, it is sufficient to know the probability of multiparticle configurations, whereas in the second case it is additionally required to take into account the size effects of the fluctuations with respect to the indicated multiparticle configurations.

2) For problems of non-equilibrium thermodynamics, it is necessary to have an estimate of the local region in which the concept of local equilibrium is defined. All non-equilibrium flows are constructed in the form of an expansion of the chemical potential gradient, which is the basic thermodynamic characteristic, and it is necessary to determine the conditions for its correct calculation. At present, the concept of a local area does not have a specific connection to the size V_m, which does not allow linking molecular models to the real conditions of most processes.

3) Intermediate systems comprising 10^5 to 10^9 molecules constitute a wide class of transition systems with a characteristic linear size from $10^2\lambda$ to $10^3\lambda$ that can have spatial non-uniformity in each direction, and the efficiency of modelling such non-uniform systems largely depends on the possibility of using continual or discrete models.

The constructed estimates give answers to the questions posed.

1. The estimates of the radii R^v_{t1} and R^v_{t2} refer to the isolated region within the phase. To analyze the properties of local volumes with a radius less than R^v_{t1}, fluctuations must be taken into account. If the radius of the region is larger than R^v_{t2}, then when analyzing the properties of isolated local volumes, one can neglect the allowance for fluctuations. When there is a large number of such local areas within the system, this is not an isolated small system. Local areas are not distinguished in any way, therefore contributions from them are all equivalent. As a result, in a large ensemble of identical regions with a radius less than R^v_{t1}, small-scale fluctuations can be ignored – it suffices to confine oneself to the average values characterizing the most probable distribution.

2. If we consider local internal regions near macrophase interfaces (for example, in membranes), then the boundaries themselves must in any case be separated from the properties of the internal volume. The contribution of fluctuations depends on the set of local regions in the near-surface regions, i.e. of the cross-sectional area. If the surface is macroscopic, then density fluctuations are also not taken into account – averaging over the cross section removes the effects of fluctuations. However, for small cross-sectional areas, when the areas are commensurable or exceed R^v_{t1} or R^v_{t2} by at least an order of magnitude, fluctuations effects are necessary for $R < R^v_{t1}$ and they can be neglected for $R > R^v_{t2}$.

3. It follows that for a drop of radius $R_{t1} \leq 41$, the effect of density fluctuations on the surface can not be neglected, although fluctuations can be ignored at the centre of the drop with the characteristic size of the sphere $R^v_{t2} = 29$. But in order to obtain the thermodynamic characteristics of a drop, it is necessary to take into account its total volume, and for regions with $R > R^v_{t2}$ fluctuations must already be taken into account because these regions include a surface. In the more general case, even the proximity of the boundaries to the local region (without its inclusion) requires the inclusion of fluctuations.

Thus, in macrophases, the allowance for fluctuations depends on the degree of uniformity of the internal volume of the system: in uniform regions the role of fluctuations is negligible, and in non-uniform macro-regions, the necessary condition for neglecting fluctuations is that the size of the non-uniform region is large, i.e. the total volume exceeds M_{mac}. Otherwise, an independent analysis of the role of fluctuations is required.

We note that the non-uniformity of the system is a natural state of the phase in the case of deformation of solids and in porous bodies, as well as with the combined influence of pores and deformations during the transport of molecules through solid membranes. The estimation of R^v_{t2} actually determines the concept of V_m for the volume phase introduced in the mechanics of continuous media. Accordingly, this situation is generalized to the situation of chemical reactions and nucleation processes, when the role of fluctuations can be neglected, if the total volume of the system is macroscopic, and not for non-equilibrium states: for $R > R^v_{t2}$, local fluctuations can be neglected, and for $R < R^v_{t1}$ – it is impossible. The obtained sizes indicate that for characteristic linear sizes of the interior regions up to $10^2\lambda$, a continual description for transport processes can not be used. While for internal regions of the order of $10^3\lambda$ and more, a continual description of such processes will be justified.

The restriction on the minimum number of macrophase molecules $M_{mac} = 4\times10^7$–10^9 is important for the molecular dynamics and Monte Carlo methods, in which periodic boundary conditions are used, with respect to the 'central' design cell in the modeling of the volume phase. The size of the cells must be larger than R^v_{t2} in order to completely neglect the fluctuations. For smaller cells, theoretical methods for small systems are, in fact, a tool for monitoring numerical methods. Note that, traditionally, in modelling by Monte Carlo and molecular dynamics methods, the bulk properties of small systems are mainly considered, since these methods are technically difficult to perform an analysis of surface properties due to a strong fluctuation.

Dilute gases. Figure 31.2 shows that the dependence $\eta(\theta)$ with decreasing θ in the region of small θ and with increasing θ in the region of large θ has a similar form. In order to proceed to the traditional treatment of a rarefied gas, we recalculate the number of particles M by the volume occupied by the rarefied gas, i.e. is $\sim10^3$ times greater than the volume of the dense phase. Then the substance in the drop with the radius R_{t1} corresponds to the number of 2.1×10^4 molecules, and the drop of R_{t2} corresponds to 10^5 molecules, which corresponds to $(2.1-10)\times10^7$ cells. The latter value corresponds to the lower value of the estimate $M_{mac} = 4\times10^7$.

In the gas phase, the main dimensions are the mean distance between molecules ρ and the mean free path ℓ. The first value is used in equilibrium characteristics, and the second is used in kinetic characteristics. The connection with the degree of filling of the volume θ of these quantities is expressed as $\rho = \lambda/\theta^{1/3}$ and $\ell = \lambda/(2^{1/2}\theta)$ [41,42].

Hence for the rarefied gas range $\theta = 10^{-4}–10^{-3}$ it follows that the value of ρ varies from 10 to 22λ. Thus, the region with a radius R^v_{t1} is commensurable with the volume of the gas in which the molecules are at an average distance ρ, but this region is smaller than R^v_{t2}. If we compare the total number of molecules in a cube with a side equal to the mean free path, then $N = \theta L^3 = \theta/(2^{1/2}\theta)^3 = (2^{1/2}\theta^2)^{-1}$, or $N = 7.1\times10^5–7.1\times10^7$. The indicated sizes go beyond the permissible values of R^v_{t2}, and this amount corresponds to the meaning of the lower macrophase size.

The meaning of this comparison follows from the traditional condition in the theory of a rarefied gas that the relaxation of the gas to equilibrium occurs at characteristic times of the order of the time of a single collision of molecules. It should be recalled that the very concept of time and mean free path is statistical, and they refer to macroscopic regions in the size and in the number of molecules. The fact that within the volume of a cube with a size equal to the mean free path is $N = 7.1\times10^7$ at $\theta = 10^{-4}$, is in complete correspondence with the kinetic theory of gases.

35. Micro-non-uniform systems

In addition to the fluctuation estimates that affect the characteristics of the system, an important role is played by energy non-uniformities that change the relationship between the number of particles and the values of the thermodynamic functions. In this section, two such examples are considered: the presence of edge atoms between the faces of small microcrystals that affect the values of the surface tension of the entire system and the limited size of the volume of the cavity in which the included or adsorbed phase can be located.

Estimation of the conditions for the application of Wulff's theorem. The equilibrium form of a single crystal, which is in contact with its saturated vapour or melt, is determined by Wulff's theorem, which is written as σ_i/h_i = const [43–49]. Thus, the equilibrium shape of a single crystal is characterized by the fact that its faces are removed from a certain point (the Wulff point) by distances proportional to the surface tension of the faces. This point is unique for a single crystal. The theorem expresses the concrete equilibrium condition of the crystal under isothermal conditions for the constant volume of the system (T = const and V = const). An important basis of this theorem is the assumption of the existence of the most equilibrium concept of 'surface tension' of the faces of an anisotropic solid σ_i, where different faces i have different surface tensions. In Wulff's theorem, only the contributions of individual faces are considered, and the contributions of the edge faces are neglected (they are taken into account in Ref. [49] – they change the position of the Wulff point). In our analysis, the question is posed differently: for what size of a single crystal of a given form can we assume that the contributions from the edges can be neglected practically (the contribution of a finite number of vertices is omitted). That is, we consider the size effect in an non-uniform system in which the edges, characterized by linear tension, change the surface tension of the entire system.

This question is considered on the example of the simplest forms of single crystals: tetrahedral, cubic and octahedral [50]. The cubic form consists of faces (100) and 12 edges, and the remaining two consist of faces (111) and have 6 and 8 edges, respectively, which greatly simplifies the analysis [51]. Any interface is a non-uniform region. To describe the distribution of molecules in the lattice gas model (LGM), structural functions of cell distributions in the transition macroscopic region are constructed, which are characterized by a set of nearest neighbor numbers z_{qp}, denoting the number of neighbouring sites of the layer p around the sites of the layer q; $\sum_{p=q=1}^{q+1} z_{qp} = z$. Thus, for a cubic structure with $z = 6$, the numbers z_{qp} are equal to 4 for $p = q$ and 1 for $p = q+1$, $\gamma = 1$; and for the FCC structure with $z = 12$, the numbers z_{qp} are equal to 6 for $p = q$ and 3 for $p = q\pm1, \gamma = 1/\sqrt{2}$.

Surface tension characterizes the excess free energy of the transition region. The surfaces of small microcrystals are non-uniform systems: on their surfaces there are different faces and edges between the faces. We represent $\sigma_{av}S_{av} = \Sigma_i\sigma_iS_i + \Sigma_m\xi_mL_m$, where $S_{av} = \Sigma_iS_i + \Sigma_mL_m$, S is the total surface of the microcrystal, σ_{av} is the average value of the surface tension, ξ_m is the linear tension of the edge of type m. For our simplest structures there are only single types of faces, and the signs of the sums are removed. The average value of the surface tension σ_{av} will be written as $\sigma_{av} = \sigma S\,(1+ \xi L/\sigma S)/S_{av}$. The condition for Wulff's theorem is satisfied if the second term ($\xi L/\sigma S \ll 1$) can be neglected – this will be the energy estimate for the condition of applicability of Wulff's theorem. The simplest geometric estimate of the fulfillment of this inequality can be the ratio $L/S \ll 1$, since the quantities S and ξ are of the same order. To do this, we need to consider how the length of the edge and the area change with the growth of the crystal.

The unit of length should be the lattice parameter λ, related to the potential parameter as $\lambda = 2^{1/6}\sigma_{AA}$, which characterizes the average distance between atoms. We shall measure the crystal size in the numbers of the N atoms forming the single crystal under consideration. In this case, length measurements using the lattice parameter λ and the number of N atoms give the smallest values of the number of atoms located on the edges (N_1) and on the faces (N_2) above the indicated figures: $L/S = N_1/N_2$.

In the field of Fig. 35.1, the N_1/N_2 ratio characterizes the geometric estimate of the contribution of all the edge atoms to all surface atoms of the faces of the lattices under consideration. It can be seen that the value of the L/S ratio can be neglected only in the case of N exceeding $N \sim 150000 \sim (53)^3$ atoms in the crystal – the ratio should be compared with the accuracy of the experimental measurements, and at least should be less than 1%.

For the energy estimate, we take into account that between the atoms of the edge and the atoms of the face there exists a transition region κ as between the surface and the volume. When passing to small crystals by analogy with drops [52, 53], we assume that the width of the transition region decreases approximately by a factor of two. Thus, κ is a quantity of the order of two monolayers. Then the region of edge atoms increases from two sides $(1 + 2\kappa)$ times. At the same change in the assignment of sites, the number of atoms on the faces decreases. In this case, the ratio N_1/N_2 increases approximately 5.1 times. At the same time, the ratio of the energy contributions ξ/σ per site from the free energy of the sites on the edge and on the faces

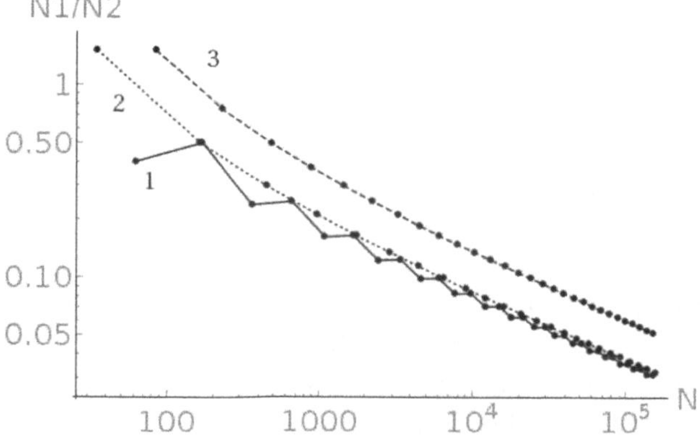

Fig. 35.1. The ratio of the number of edge atoms N_1 to the number of surface atoms N_2 on the faces, depending on the number of atoms of the entire crystal, N (for geometric estimation) [51].

decreases primarily due to a change in the number of bonds for the edge atoms with respect to the average number of sites of the face (which is approximately 1/3 to 1/2). Then the total energy change $\xi L/\sigma S$ can be estimated as a value of the order of 2. The energy estimate gives an overestimate for a given level of deviation accuracy by a factor of two in comparison with the geometric estimate, which is N approximately 3×10^5 or $(67)^3$.

The obtained estimates are in agreement with the earlier obtained results on the analysis of density fluctuations on the drop surface, which made it possible to obtain the lower bound for the applicability of the thermodynamic relationships [12]: for the range of the region radius from 41λ to 90λ (which corresponds to the number of particles from 3.8×10^5 to 29×10^5, accuracy of the description of surface tension is about 0.1%). For a bulk phase, fluctuations in a sphere with a radius of 29λ can be neglected, which is $N \sim 10^5$, and the calculation of van der Waals type clusters has shown [54] a transition to a stable FCC lattice structure at $N \sim 2.5\times10^5$ for $T = 0$ K without allowance for vibrational contributions, which can somewhat reduce this value of N, if we take into account the oscillations.

The estimates obtained for the number of atoms in single crystals are approximately several times higher than estimates based on the obtained value of R_{t2}, which indicates the qualitative agreement of different estimates. It follows from such estimates that as the Miller index increases, the areas of faces decrease, and the contributions of their edge atoms increase. This aspect is ignored in the continual consideration of Wulff's rule [46–48]. This aspect is very important in the transition to more complex forms of surfaces. Thus, in the rhombic dodecahedron, the contribution of the edges decreases noticeably only with an increase in the total number of atoms by approximately an order of magnitude, compared with the numbers obtained above for the simplest geometries considered: tetrahedral, cubic, and octahedral.

A drop in a spherical pore [53]. The limited volume factor plays a role in: 1) adsorption of molecules on small open faces of microcrystals, 2) adsorption of molecules in bound and isolated pores of polydisperse material, and 3) spinodal decomposition of the fluid at short times for large supersaturations of the volume phase, when each local region operates on the average the same. A general approach to the phase state of matter in limited volumes was formed in [55]. Below we consider an example of the effect of volume in the simplest case of a 'drop in a spherical pore' on the phase state of a substance and its surface tension within a pore in the absence of influence of pore walls. In this system there is a single dimensional parameter – the radius of the system.

The parameters of the system are variables: V – volume, N – number of particles and T – temperature. Under isothermal conditions, a decrease in the volume of the system is associated with a change in the quantity V. In this case, the quantity N, or the ratio N/V, can remain constant. In the first case, as the volume decreases, the density of the system changes, and in the second case the number of molecules decreases. With a decrease in the volume of the system, the contribution of the vapour–liquid transition region begins to manifest itself, which in the macroscopic limit is much smaller than the contributions from each of the phases, so the magnitude of the volume of the vapour–liquid transition region should affect the characteristics of the phase state. This, in turn, should influence the surface tension values σ.

The description of equilibrium drops in a macrovolume of a vapour is discussed in detail in Section 24. Let us denote the volume of the system by R_{sys}. If the volumes of the transition region and the coexisting phases are commensurate, we introduce the weights f_q that fix the volume of a region of type q: $1 \le q \le \kappa$. Here, in addition, there are two values of q related to the phase of the liquid $q = 1$ and the phase of the vapour $q = \kappa$, in comparison with the system (24.8). Now we can determine the average density of the system $\theta_{av} = \sum_{q=1}^{\kappa} f_q\theta_q$, where f_q is the fraction of sites in the phase region or the spherical monolayer q; by the definition $f = N_q/N$, N_q is the number of sites in the region q of radius R_q, $N = \sum_{q=1}^{k} N_q$. Quantity θ_{av} is retained when the volume V of a system is simultaneously varied and the amount of substance N is changed, or changes when the volume V is changed and the amount of substance N is maintained.

We confine ourselves to an analysis of the first case. As stated above, fixing the size of the drop R means fixing the size of the region with a constant fluid density. The value of the transition region (κ–2) refers to the region with variable density $V(\kappa-2) = \sum_{q=2}^{\kappa-1} f_q$ and the remaining part of the system $V_\kappa = V(R_{sys})) - V(\kappa-2)$ is the vapour volume. Thus, the volume of the vapour not only depends on the volume of the drop, but also on the volume of the transition region, which in turn depends on the temperature.

Equations (24.8) can be written for the fluid phase sites $q = 1$ and the vapour $q = \kappa$, and the equality condition of pressures must also be satisfied $\pi_{q=1} = \pi_{q=\kappa}$. These two equations complement the system (24.8) when the complete system is closed, if the average value of the density θ_{av} is fixed. Setting this value is necessary to eliminate uncertainty by varying the width of the transition region κ. This allows us to use the functional connection between the densities of the non-uniform system, generated by the condition of constancy of the chemical potential in the complete limited system, in order to trace the effect of a change in the volume of the entire system on its characteristics without violating the biphase condition. This formulation eliminates the need to have a complete view of the isotherm and determine the position of the Maxwell secant, and provides analysis of any system radii from macroscopic (on the order of 1 cm) to several nanometers.

To calculate the surface tension, we used the definition (24.15). As a criterion for the position of the reference line, we select the condition that there is no adsorption of molecules inside the transition layer, then the fixation of this position corresponds to the equimolecular separating line ρ_e. Calculations were made for the simplest argon liquid [56]. We consider a drop of radius R in a bounded system of radius R_{sys} at the reduced temperatures $\tau = T/T_{melt} = 0.68$ and 0.89. Molecules interact within the first coordination sphere on the FCC lattice ($z = 12$). The calculation of pairs of sites of different sorts *zqp* on a curved lattice is performed with a lattice parameter $\alpha = 1$. For concreteness, we confine ourselves to states of the system with a matter density $\theta_{av} = \theta^* \pm \delta$, where $\theta^* = 0.5$, $\delta = 0.01$. Dependences of the properties of the system on its size were studied.

Figure 35.2 shows the mean density of the system θ_{av} (curve 1) for the average density of the transition region (curve 2) and the average phase density (curve 3) for varying R_{sys}.

In the course of the solution, the average density of the system (curve 1), according to the thermodynamic condition, remains constant up to the second significant number. At the same time, as the size of the system decreases, the average density over phases (curve 3) increases, and the average density along the transition region (curve 2) decreases. The average density in the transition region decreases with a decrease in the size of the system, since molecular reallocations occur in such a way that the reference surface ρ_e is displaced closer to the drop (see below).

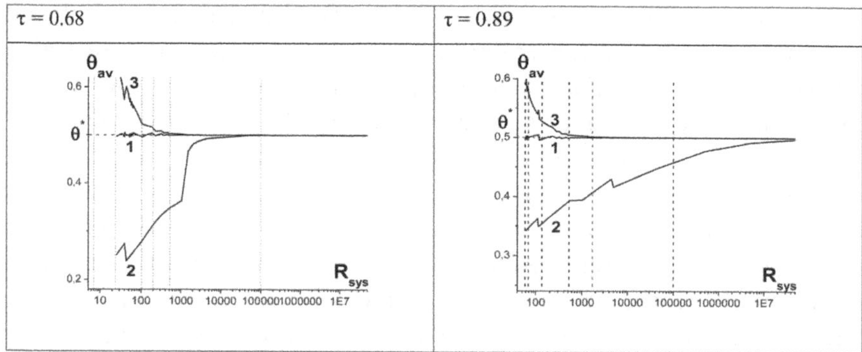

Fig. 32.2. Dependence of the densities averaged over the entire system (1), in phases (3) and in the transition region (2) on the radius of the system R_{sys}.

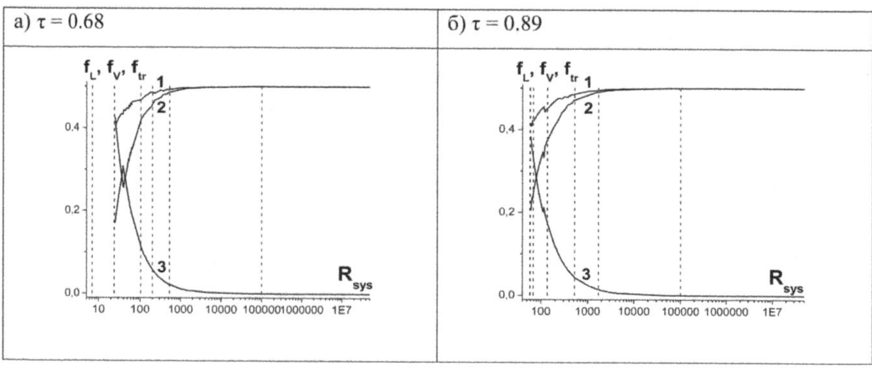

Fig. 35.3. Dependence of the fractions of the sites occupied by the liquid f_L or the vapour f_V phase or the transition region f_{tr} on the radius of the system R_{sys}.

The average density in the phases increases, primarily due to the increasing excess of the fraction of sites occupied by the drop over the fraction of vapour sites with a decrease in the size of the system, as shown in Fig. 35.3, and to a lesser extent due to the fact that the vapour density increases faster than the drop density decreases (see below). Figure 35.3 characterizes the fractions of the drop sites (curve 1), the fractions of the vapour sites (curve 2) and the transition region (curve 3).

With a decrease in the size of the system, the fraction of sites occupied by the drop and vapour decreases, and the fraction of the transition region increases. The vertical lines in Figs. 35.2 and 35.3 indicate the values at which, with a decrease in the size of the system, the deviation from the set value of the average density first becomes greater than 1) 0.0001 (R^1), 2) 0.001 (R^2), 3) 0.005 (R^3),

4) 0.01 (R^4), and also the radii below which there is no solution for the discrepancy for the mean density with an error less than 0.01 (R^*) and below which the width of the transition region (R^{**}) is artificially lowered, because it occupies almost the entire volume.

These values are given in Table 35.1, which also gives the corresponding radii of the drop R, the values of the width of the transition region κ, and the normalized surface tension $\sigma(R_{sys})/\sigma_{bulk}$ for all the indicated radii except for R^* and R^{**}, since for them there is no solution with a given accuracy with respect to the supplied thermodynamic condition.

Analysis showed that with a decrease in the size of the system: 1) the density of the liquid decreases, and the faster the higher the temperature; 2) the vapour density increases and the faster the lower the temperature; 3) the width of the transition region and the position of the reference surface are reduced, and at higher temperatures it is somewhat faster, but at low temperatures it shows greater discreteness; 4) the surface tension decreases and the faster, the higher the temperature; 5) the chemical potential and internal pressure increase, and with the higher speed, the lower the temperature.

Figure 35.4 gives the dependences of the normalized surface tension of the drop in the bounded (curves 1 and 2) and unrestricted systems (curves *3* and 4) on its radius R at τ = 0.68 (curves 1, 3) and 0.89 (curves 2, 4) .

Table 35.1.

$\tau = 0.68$						
	R^1	R^2	R^3	R^4	R^*	R^{**}
R_{sys}	10^5	523	200	103	24	7
R	8×10^4	412	162	65	–	–
κ	10	8	8	8	–	–
$\sigma(R_{sys})/\sigma_{bulk}$	1.00	0.86	0.80	0.74	–	–
$\tau = 0.89$						
	R^1	R^2	R^3	R^4	R^*	R^{**}
R_{sys}	105	523	137	69	59	58
R	8×10^4	411	105	51	–	–
κ	19	14	14	13	–	–
$\sigma(R_{sys})/\sigma_{bulk}$(unlim)	0.98	0,81	0,67	0,58	–	–

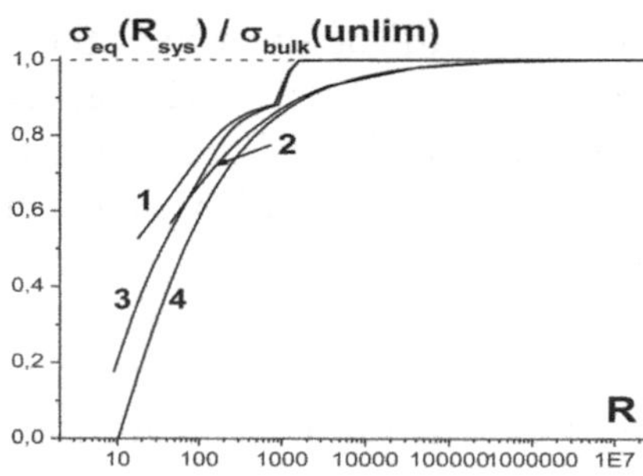

Fig. 35.4. Dependence of the normalized surface tension on the radius of the drop.

According to Fig. 35.4, the surface tension of a drop in a bounded system (curves 1 and 2) is greater than in an unbounded system (curves 3 and 4), which is apparently associated with high pressures in a limited system. The curves 1 and 2 break due to loss of solutions, and curve 3 due to the roughness of the parameter $\alpha = 1$ in the region of small dimensions. Curve 4 shows that the surface tension turns to zero – the two-phase state of the phase breaks up, but the value of this two-phase existence limit should be higher for a more correct parameter value than $\alpha = 1$.

Thus, a direct molecular calculation of the characteristics of equilibrium drops shows that when the size of the system decreases (at condition $\theta_{av} = 0.5$), the critical temperature decreases and the surface tension decrease and the internal pressure and the chemical potential increase.

The phase rule and micro-non-uniform systems. In concluding this Section, we will point to an example of the influence of the micro-non-uniformity of the system on the phase states of matter. In [57], the correctness of using the macroscopic rule of Gibbs phases in micro-non-uniform systems, in particular, in polydisperse adsorption systems, was considered. Adsorption systems are characterized by a strong short-range surface potential. The joint influence of the surface potential and intermolecular interaction forms subregions of the general system in which phase transitions of the first sort are realized. The surface potential of the walls tends to stratify the adsorbate molecules along two-dimensional subdomains of the general space in order to form phases of approximately the same density in these two-dimensional subregions with the same potential. This trend is analogous to the behaviour of bulk phases with respect to the phase rule in external fields.

The diagrams of stratification in micro-non-uniform systems are characterized by a multiplicity of the vapour–liquid phase transitions of the first sort [34,58]. Accordingly, each curve of stratification has its own critical point. Such systems include adsorption, both on open non-uniform surfaces and inside porous bodies.

It is shown that the type and number of phase diagrams of adsorption systems are determined by the surface potential, which depends on the composition and structure of the non-uniform surfaces of the adsorbent. The number and parameters of the critical points depend on the composition and structure of the non-uniform systems. The maximum possible number of critical points is due to the presence of macroscopic regions of different types. In the case of a locally ordered structure of regions consisting of adsorption centres of different types, one critical point with modified values of critical parameters is realized, as in uniform systems. If the influence of the surface potential predominates over the intermolecular interaction, then there are no critical points in the system.

Those. in micro-non-uniform systems many phase states can be realized, similar to the set of phase states created by extended external fields (gravitational, electric, magnetic) in macroscopic volumes. At the same time, the presence of interfaces between phases in non-uniform systems and their curvature does not introduce any additional influence as that of the 'external' fields on the character of stratification.

References

1. Gibbs J.W. Thermodynamics. Statistical mechanics. Moscow: Nauka, 1982.
2. Landau L.D., Livshits E.M., Theoretical physics. V. 5. Statistical physics. Moscow: Nauka, 1964.
3. Hill T.L., Statistical Mechanics. Principles and Selected Applications. – Moscow: Izd. Inostr. lit., 1960. – 486 p. [N.Y.: McGraw–Hill Book Comp. Inc., 1956].
4. Gibbs J.W.. Elementary principles in statistical mechanics, developed with especial references to the rational foundations, NewYork, 1902.
5. Fowler R.H. Statistical Mechanics. Cambridge: Cambridge Univer. Press, 1936.
6. Klimontovich Yu.L., Statistical physics. Moscow: Nauka, 1982.
7. Heer C.V., Statistical mechanics, kinetic theory and stochastic processes. Moscow: Mir, 1976. [Academic Press New York, London, 1972].
8. Hill T.L., J. Chem. Phys. 1962. V. 36. P. 3182.
9. Hill T.L., Thermodynamics of Small Systems. Part 1. New York Amsterdam: W. A. Benjamin, Inc., Publ., 1963.
10. Hill T.L. Thermodynamics of Small Systems. Part 2. New York Amsterdam: W. A. Benjamin, Inc., Publ., 1964.
11. Oh B.K., Kim S.K. J. Chem. Phys. 1977. V. 67. P. 3427.

12. Tovbin Yu.K., Zh. fiz. khimii. 2012. V. 86. No. 9. P. 1461. [Russ. J. Phys. Chem. A 86, 1356 (2012)]
13. Tovbin Yu.K., Theory of physico-chemical processes on the gas-solid interface, Moscow: Nauka, 1990. [CRC, Boca Raton, Florida, 1991].
14. Tovbin Yu.K., Zh. fiz. khimii. 2010, Vol. 84. No. 11. P. 2182. [Russ. J. Phys. Chem.A. 86, 1356 (2012)].
15. Tovbin Yu.K., Khim. fizika. 2010, V. 29. No. 12. P. 74.[Russ. J. Phys. Chem. B. V. 4. № 6. P. 1033 (2010)].
16. Feller W., Introduction to Probability Theory and its Applications. Volume 1. Moscow: Mir, 1984.
17. Gelfond A.O., Calculus of finite differences. Moscow: Nauka, 1967.
18. Godunov S.K., Ryaben'kii V.S.. Difference schemes (introduction to theory). Moscow: Nauka, 1973.
19. Fikhtengol'ts G.M. Course of differential and integral calculus. Moscow: Fizmatgiz, 1963. V. 1. and V. 3.
20. Kofman A., Introduction to applied combinatorics. Moscow: Nauka, 1975.
21. Kolchin V.F., Sevastyanov B. A., Chistyakov V. P., Random Allocations (Nauka, Moscow, 1978; Winstonand Sons, Washington, DC, 1978).
22. Adamson A. W., The Physical Chemistry of Surfaces,Mir, Moscow, 1979. [Wiley, New York, 1976].
23. Summ B.D., Goryunov Yu.V., Physico-chemical basis of wetting and spreading. Moscow: Khimiya, 1976.
24. Tovbin Yu.K., Rabinovich A.B., Zh. fiz. khimii. 2010. V. 84. No. 12. P. 2366. [Rus. J. Phys. Chem. A, 2010, Vol. 84, No. 12, P. 2166].
25. Roberts M.W., McKee C.S., Chemistry of the metal-gas interface. Oxford: Clarendon Press, 1978.
26. Somorjai, G.A., Chemistry in two-dimension surface. N.Y., Ithaca: Cornell Univ. Press L., 1981.
27. Tovbin Yu.K., Rabinovich A.B., Zh. fiz. khimii. 2011. V. 85. No. 11. P. 2118. [Rus. J. Phys. Chem. A, 2011, V. 85, No. 11, P. 1977].
28. Tovbin Yu.K., DAN SSSR. 1977. V. 235. P. 641.
29. Tovbin Yu.K., Zh. fiz. khimii. 2012. V. 86. No. 7. P. 1301. [Rus. J. Phys. Chem. A, 2012, V. 86, No. 7, P. 1180] .
30. Tovbin Yu.K., Fizikokhimiya poverkhnosti i zashchita materialov. 2011. V. 47. No. 2. P. 115. [Protection of Metals and Physical Chemistry of Surfaces, 2011, V. 47. No. 2. P. 141].
31. Tovbin Yu.K., Rabinovich A.B., Zh. fiz. khimii. 2013. V. 87. No. 2. P. 337. [Russ. J. Phys. Chem. A, 2013, Vol. 87, No. 2, pp. 329].
32. Lebedev N.N., Special functions and their applications. Moscow: Fizmatlit, 1963.
33. Tovbin Yu.K., Rabinovich A.B. Zh. fiz. khimii. 2011. V. 85. No. 8. P. 1514. [Russ. J. Phys. Chem. A, 2011, Vol. 85, No. 8. P.1398].
34. Sychev V.V., Differential equations of thermodynamics. Moscow: Vysshaya shkola, 1991.
35. Tovbin Yu.K., Zh. fiz. khimii. 2010. V. 84. No. 2. P. 231. [Rus. J. Phys. Chem. A, 2010, V. 84. № 2. P. 180].
36. Tovbin Yu.K., Ibid. 2010. V. 84. No. 4. P. 797. [Rus. J. Phys. Chem. A, 2010, V. 84. № 4. C. 705].
37. Tovbin Yu.K., Ibid. 2010. V. 84. No. 10. P. 1882. [Rus. J. Phys. Chem. A, 2010, V. 84. № 10. C. 1717].

38. Jaycock M. Parfitt J., Chemistry of interfaces.– Moscow: Mir, 1984. – 270 p. [Wiley, New York, 1981].
39. Zimon A.D. Fluid adhesion and wetting. Moscow: Khimiya, 1974.
40. Tovbin Yu.K., Rabinovich A.B., Izv. AN. Ser. khim. 2009. No. 11. P. 2127. [Russ. Chem. Bull. 58, 2193 (2009)].
41. Tovbin Yu.K., Teoret. osnovy khim. tekhnologii. 2005. V. 39. No. 5. P. 523. [Theor. Found. Chem. Engin. 2005. V. 39. No 5. P. 493].
42. Tovbin Yu.K., The Molecular Theory of Adsorption in Porous Solids, Moscow, Fizmatlit, 2012. [CRC Press, Taylor & Francis Group, 2017].
43. Bazarov I.P., Thermodynamics. Moscow: Vysshaya shkola,1991.
44. Semenchenko K. V., Selected Chapters of Theoretical Physics. Moscow: Prosveshchenie, 1966.
45. Hilton H., Zentralblatt fur Mineralogie,Goelogie und Palaeontologie, 1901. S. 753.
46. Liebmann H., Z. Kristallogr., 1914. B. 53. S. 171.
47. Laue M., Z. Kristallogr., 1943. B. 105. S. 124.
48. Chernov A.A., Givargizov E.I., Bagdasarov X.C., et al., Advanced crystallography. V. 3. Crystal formation. Moscow: Nauka, 1980.
49. Rusanov A.I., Phase equilibria and surface phenomena. Leningrad: Khimiya, 1967.
50. Sirotin, Yu.I., Shaskol'skaya M.P., Fundamentals of crystal physics. Moscow: Nauka, 1979.
51. Titov S.V., Zaitseva E. S., Tovbin Yu. K., Zh. fiz. khimii. 2017. V. 91. No. 12. P. 2155. [Rus. J. Phys. Chem. A, 2017, Vol. 91, No. 12, P. 2481].
52. Tovbin Yu.K., Rabinovich A.B., Izv. AN. Ser. khim. 2010. No. 4. P. 663. [Rus. Chem. Bull. 2010. T. 59. No. 4. P. 677]
53. Tovbin Yu.K., Zaitseva E.S., Fiziko-khimiya poverkhnosti i zashchita materialov. 2017. V. 53. No. 5. P. 451. [Protection of Metals and Physical Chemistry of Surfaces, 2017, Vol. 53, No. 5, P. 765].
54. Doye J.P.K., Calvo F., J. Chem. Phys. 2002. V. 116. P. 8307.
55. Tovbin Yu.K., Zh. fiz. khimii. 2018. V. 92. No. 1. P. 36. [Russ. J. Phys. Chem. A 92, 29 (2018)].
56. Hirschfelder J. O., Curtiss Ch. F., Bird R. B., Molecular theory of gases and liquids. – Moscow: Inostr. Lit., 1961. – 929 p. [Wiley, New York, 1954].
57. Tovbin Yu.K., Zh. fiz. khimii. 2013. V. 87. No. 6. P. 928. [Rus. J. Phys. Chem. A, 2013, V. 87, No. 6, P. 906].
58. Tovbin Yu.K., Zh. fiz. khimii. 2009. V. 83. No. 10. P. 1829. [Rus. J. Phys. Chem. A. 2009. V. 83. No. 10. P. 1647.

5

Non-equilibrium processes

This chapter discusses issues related to the non-equilibrium state of the system. Chapter 1 presents the main provisions of the thermodynamics of non-equilibrium processes, which are equally related to different aggregate states and phases. The attraction of concepts about relaxation times in the non-equilibrium thermodynamics turned out to be necessary for understanding the connection between the Kelvin equation and the equilibrium state of small drops of a liquid in a vapour (Section 7). Moreover, the fundamentals of non-equilibrium thermodynamics should be analyzed to understand the fundamentals of thermodynamics associated with the very concept of equilibrium and an implicit postulate regarding time constraints in the notion of Gibbs 'passive forces'.

This chapter also discusses two other issues that go beyond traditional non-equilibrium thermodynamics: under what conditions does the local equilibrium condition used in all thermodynamic constructions is broken and how to describe in a self-consistent way the rate of stages in three aggregate states (at any densities, including transient region of the interface), so that the molecular description gives a complete interpretation of the assumptions of non-equilibrium thermodynamics in all aggregate states.

We begin with a discussion of the characteristic relaxation times for mass, impulse, and energy transfer for the gas phase and for mass transfer in solid phases to show the range of relaxation times in three aggregate systems.

36. Relaxation times

The processes of relaxation characterize the processes of the transition of the system to the equilibrium state and are described by kinetic equations. The transfer of mass, impulse, and energy is described by the equations of continuum mechanics, which for vapour–liquid systems are traditionally called hydrodynamic equations [1,2], and in solids – elasticity equations [2,3]. They often give a formal answer that a mathematically correct solution of these equations corresponds to an infinite time of full equalization of all characteristics. However, any changes are considered with a certain degree of accuracy and, naturally, there are no infinite relaxation times. These discussed processes in the gas have their own specific ranges of values, and their estimates for the characteristic length L of the order of 1 cm are considered below [4–6].

Transfer properties. Impulse transfer is responsible for the process of establishing equilibrium pressure. The basic physical mechanism of pressure equalization is the wave of gas density generated by the initial perturbation $\partial^2P/\partial t^2 = c^2\partial^2P/\partial x^2$, i.e., the equation describing this process is the wave equation of the hyperbolic type, here for simplicity it is one-dimensional along the x axis, and c is the velocity of the sound wave. Its solution in an infinite medium has a solution of the type of propagating waves $P(x,t) = f_1(x + ct) + f_2(x-ct)$. Therefore, the scale of the pressure relaxation time is the time during which this wave passes the path L (here from one wall of the vessel to the other wall), i.e. $\tau_p \sim L/c$. In air, the velocity of sound is $c \approx 300$ m/s or 3×10^4 cm/s, and the time for establishing the pressure is of the order of $\tau_p \sim 3\times10^{-5}$ s.

Equalization of the concentrations of gas mixture components is a process of diffusion type, described in the simplest case by an equation of parabolic type (as above, we look at one-dimensional motion) $\partial n/\partial t = D\partial^2 n/\partial x^2$. In an unbounded medium, this equation has a simple solution of the type of a spreading Gaussian distribution $n(t,x) = \exp\{-x^2/4Dt\}/(4\pi Dt)^{1/2}$ (if the perturbation at the initial time $t = 0$ had the form $n(0,x) = \delta(x)$). Therefore, the average square of the size of the cloud of particles creating a perturbation of a uniform system is equal to $\overline{x^2} = \int x^2 n(t,x)dx = 2Dt$. If the system is bounded by walls, then the size of the degradation region of increased density $\left(\overline{x^2}\right)^{1/2}$ reaches the value L after the time $\tau_n \sim L^2/2D$ elapses. In the three-dimensional case $\left(\overline{r^2}\right)^{1/2} = \left(\overline{x^2}\right)^{1/2} + \left(\overline{y^2}\right)^{1/2} + \left(\overline{z^2}\right)^{1/2} = 3\left(\overline{x^2}\right)^{1/2}$ we shall have a consequence $\tau_n \sim L^2/6D$. According to tabular data, the diffusion coefficients D for gases such as O_2, N_2, CO_2, etc. are $D = \sim 0.14–0.2$ cm²/s, therefore at $L \sim 1$ cm, we obtain that $\tau_n \sim 1/6 \cdot 0.2 \approx 1$ s.

The equalization of temperature T is also described by an equation of the parabolic type, but instead of the coefficient D there is a coefficient of thermal diffusivity K, which is equal to the coefficient of thermal conductivity divided by the product of the gas density by its specific heat. For an air-type gas, this quantity is of the order of $K \sim 0.2$ cm²/s, so that $\tau_T \sim L^2/6K \approx 1$ s (for air, the coincidence of the order of this quantity with coefficient D is purely random).

These simple estimates of the relaxation times demonstrate: 1) the time estimate τ is produced by methods that go beyond quasi-static thermodynamics (as always, the criteria for any approximation should be determined within the framework of a more general consideration); 2) the relaxation processes of various thermodynamic characteristics are different in the mechanism of relaxation itself, therefore, the times τ can differ significantly in magnitude since they depend in different ways on macroscopic parameters (for example, $\tau_p \sim L$, $\tau_n \sim L^2$, etc.); 3) only the 'gas' version of the system is considered above, the relaxation processes in which are determined by the macroscopic equations of motion of a continuous medium. In the study of relaxation processes involving the consideration of external fields (electromagnetic, surface, etc.), the estimation requires a transition to a microscopic level, since it is necessary to involve certain parameters already at the atomic–molecular level (time of molecule rotation in the field, absorption time and emission of a photon, etc.). An analogous situation arises also in the consideration of relaxation processes in the local regions of thermodynamic systems associated with taking into account the interaction of the particles with one another during the transition to condensed phases. In particular, for a solid body, the ratios of the relaxation times (Section 12) are transformed into strongly pronounced inequalities, the differences in which between impulse and mass relaxation times reach up to 10–15 orders of magnitude.

Diffusion of vacancies. Let us consider the relaxation time t^* of the concentration equalization process in solid spherical samples of radius R_s due to the transfer of atoms by the vacancy mechanism in the course of the process, which is as close to equilibrium as the temperature decreases from the melting point to the current value of T. Let us denote by t_0 the time necessary for establishment in the crystal equilibrium distribution of vacancies $C_0(T_t)$ from the moment of its formation at $T = T_t$. We denote by δT the step in temperature with its decrease. The cooling process starts from the surface $R = R_s$ (on which a new concentration of vacancies

$C_0(T_t-\delta T)$ is established) and extends deep into the sphere. The thermal equilibrium is established much faster than the concentration equilibrium, so the process proceeds at a fixed temperature. We denote the concentration of vacancies in the centre of the sphere $R = 0$ at the initial instant of time (equal to the equilibrium vacancy concentration at temperature T_t) by C_1. During the establishment of the vacancy equilibrium after a certain time t at the centre of the sphere, the concentration should become sufficiently close to the value of $C_0(T_t-\delta T)$ in order to consider the entire sample having one vacancy concentration equal to $C_0(T_t-\delta T)$ corresponding to the temperature $T_t-\delta T$. Further, the process of cooling the sample proceeds in the same stepwise manner.

For a spherical sample, this formulation of the problem corresponds to a one-dimensional problem for the radial diffusion of vacancies with the boundary condition $C_0(T)$) for $R = R_s$, which is described by equation

$$\frac{\partial C}{\partial t} = D\left(\frac{\partial^2 C}{\partial R^2} + \frac{2}{R}\frac{\partial C}{\partial R}\right), \tag{36.1}$$

where D is the diffusion coefficient, C is the concentration of vacancies at time t at a point with radius R, $0 \le R \le R_s$. If at the initial instant of time $t = 0$ the concentration of vacancies on a spherical surface is $C = C_0$, and the concentration $C = C_1$ for any R, then according to [7,8], the solution at any time t will have the form:

$$\frac{C - C_1}{C_0 - C_1} = 1 + \frac{2R_s}{\pi R}\sum_{n=1}^{\infty}\frac{(-1)^n}{n}\sin\left(\frac{n\pi R}{R_s}\right)\exp\left(-\frac{Dn^2\pi^2 t}{R_s^2}\right).$$

If we denote by C_1^* the concentration of vacancies in the centre of the sphere at any time, then it is defined as the limit at $R \to 0$:

$$\frac{C_1^* - C_1}{C_0 - C_1} = 1 + 2\sum_{n=1}^{\infty}\frac{(-1)^n}{n}\exp\left(-\frac{Dn^2\pi^2 t}{R_s^2}\right).$$

For large times (in this case, all the terms of the sum multiplied by exponents with exponents proportional to $n^2 t$, are much less than the first term corresponding to $n = 1$) close to the time of exhaustion of non-equilibrium vacancies at a given temperature T, we give a finite expression for the time necessary for the difference between the quantities C_0 and C_1^* was sufficiently small – i.e. the whole sample was in a new equilibrium state, in the following form

$$D\pi^2 t = R_s^2 U, \quad U = -\ln\left((C_0 - C_1^*)/[2(C_0 - C_1)]\right). \tag{36.2}$$

Under the assumption of the differential nature of cooling $C_1 - C_0 \sim 10^{-2} C_1$ and the considerable degree of exhaustion of the density of non-equilibrium vacancies at the centre of the sphere, where δ_c is a given small value $C_1^* - C_0 = \delta_c C_0$, this solution relates the time t to the temperature dependence of the diffusion coefficient. It follows from formula (36.2) that with decreasing R_s by an order of magnitude the value of the time t decreases by two orders of magnitude, and simultaneously, taking into account the dependence $D = D_0 \exp(-\beta E)$ [8–13], which decreases the temperature for which can be obtained the equilibrium state of vacancies in the entire sample for a given time interval.

Diffusion of the label. In the complete equilibrium of the system, the flow of matter is characterized by a self-diffusion coefficient, which really represents a flux of 'labelled' isotopes. The described process can be used to analyze the conditional diffusion process of a label if it is assumed that the diffusion coefficient refers to a label and the equation describes its transport from the surface to the centre at an equilibrium state of matter density and vacancies, respectively. (The process is modelled in the fact that the time of changing the vacancy density with a change in temperature does not enter into it). In this case, Eq. (36.1) is related to the analysis of the thermal motion of atoms as a function of temperature. The diffusion coefficient is written, as before, in the form $D^* = D_0 \exp(-\beta E^*)$ [8-13], where $\beta = (RT)^{-1}$, R is the gas constant, T is the temperature, E^* is the activation energy of self-diffusion, D_0^* is the pre-exponent of this coefficient. As the temperature decreases, the value of D^* decreases. The process of label mixing is associated with the concentration of vacancies, which also decreases with decreasing temperature throughout the body and as the temperature decreases the vacancy density θ_V decreases as $\theta_V = 1 - \theta_A = [1 + K_a \exp(\Delta H_c / RT)]^{-1} \sim K_c \exp(-\Delta H_c / RT)$ where $K_a \approx K_c^{-1}$, $K_c = J_A^0 / J_A$ is the pre-exponent associated with the partition functions of the atom in the solid J_A and in the vapour J_A^0, ΔH_c is the energy of removal of the atom into vacuum (during sublimation) θ_A is the number of sites of the crystal lattice occupied by the particle A, $\Delta H_c = z\varepsilon_{AA}$, z is the number of the nearest neighbours, and ε_{AA} is the energy of their interaction. When T is reduced to zero, the equilibrium defect crystal must be free of vacancies: $\theta_V \to 0$. This conditional statement of the problem allows us to use Eq. (36.2) for a model-free analysis of the relaxation times using experimental data on self-diffusion of the label [8–13]. Diffusion of the label makes the process of thermal motion 'observable'. With a decrease in the share of vacancies thermal motion is inhibited. All other methods for calculating the process of diffusion and relaxation times require the mandatory introduction of model parameters and the provision of a self-consistent description of the transport process with a description of the phase state of the element. The latter condition is realized quite rarely [10, 13].

The question of the possibility of realizing thermal motion and achieving a conventional value for mixing the label for a 'reasonable' experiment time at different temperatures is a natural question. The times of establishing a uniform label distribution in the crystal, starting from the time of its solidification at the temperature of the triple point T_t (or the melting point) were estimated in [14–16], with a step-by-step decrease in temperature to a certain temperature for a number of elements. The maximum time during which an equilibrium state can be reached in a real experiment is defined as the process time from T_t to the lowest achievable temperature in the experiment. We denote this temperature by T^* and the corresponding time by $t^* = t(T^*)$. The time t^* answers the question of the possibility of realizing the equilibrium state of the label for the 'reasonable' time of the experiment. The time to establish the uniformity of the label within the entire sample depends on its size. The reading starts from a radius of 1 cm. The smaller the R_s, the lower temperatures can cool the sample, observing the thermal motion (1 year corresponds to the order of 3.15×10^7 s). Below we will characterize the temperature through the ratio $\delta = T/T_t$.

Figure 36.1 shows the dimensionless time curves for reaching the label equilibrium for ten single crystals as a function of the dimensionless temperature at $R_s = 1$ cm [15]. The calculation was carried out at $\delta T = 1$ K and $\delta_c = 10^{-4}$. The time to equilibrium at the melting point is denoted by t_0. In the calculations it was estimated by the formulas (36.2) as the time of establishment of the equilibrium concentration of the label with decreasing temperature from T_t to $T_t - \delta T$. With decreasing temperature the curves increase almost exponentially.

The values of the parameters E and D_0, found by the authors of [8–20] from measurements in the indicated temperature ranges, are given in Table 36.1. These data were used for finding t_0 and

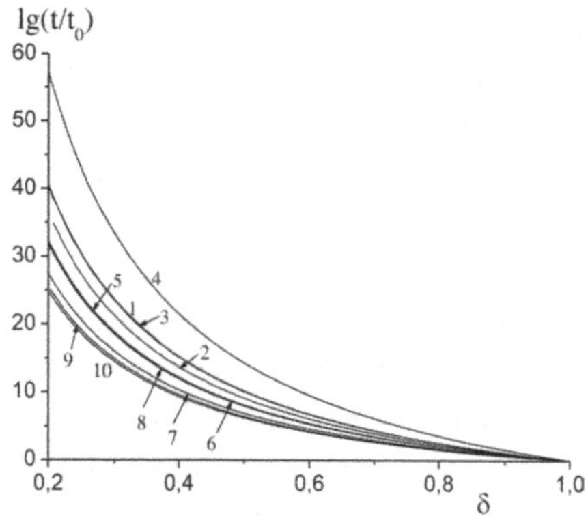

Fig. 36.1. The generalized dependence of the reduced time for the establishment of vacancy equilibrium t/t_0 in a spherical sample of radius R_s = 1 cm on the reduced melting point $\delta = T/T_t$ for single crystals: 1 – Ar, 2 – Kr, 3 – Xe, 4 – Si, 5 – Ag, 6 – Al, 7– Au, 8 – Cu, 9– Na, 10 – K.

Table 36.1. Literature data on measurements of the diffusion coefficient in temperature ranges (the upper limit is maximally close to melting point T_t) and the parameters of diffusion coefficient D_0 and E for a number of crystals

Crystal	T_t, K	$T_{min} - T_{max}$, K	D_0, cm²/s	E, kcal/mol	Reference
Ar	83.8	72–83	4.0	3.9	8,18,19,20
Kr	116	90–115	5.0	4.8	8,18,19,20
Xe	133	130–Tt	9.7	7.4	8,18,19,20
Na	370.7	274–367	0.242	104.5	8,17,21
K	336.1	273–333	0.033	9.4	8,17,22
Al	931.5	729–916	1.71	33.9	8,17,23
Ag	1235	914–1228	0.67	45.9	8,17,24
Au	1337	1031–1333	0.084	41.6	8,17,25
Cu	1356	1163–1334	0.78	48.9	8,17,26
Si	1699	1498–1673	0.8	109.8	8

t^*. The values of t_0 for samples of radius 1 cm are given in Table 36.2, as well as the times of emergence of the equilibrium state t_0 for samples with sizes from 1 cm to 1 nm (the first number in each cell) t^*.The second number is the time to reach equilibrium when the temperature is lowered from the minimum temperature in the experiment under T_{max} consideration T_{min}. Values t_0 (with the exception of Na and K) are commensurate with the duration of one year ~3.5×10^7 s or exceed it. Thus, even at the melting point, in order to reach the equilibrium state of the label, times that exceed the time of ordinary laboratory measurements are needed. Moreover, this refers to the minimum temperature of the experiment – the time t^* exceeds t_0 by approximately two or three orders of magnitude. Therefore, for any further lowering of the temperature, the curves in Fig. 36.1 (the decimal logarithms on the ordinate axis!) correspond to sufficiently large times $t > t^*$, which can not always be realized in the laboratory.

Table 36.2. The calculated characteristics of the process of establishing the label equilibrium t_0, t^* and δ^* for samples of crystals of different sizes R_s by the vacancy mechanism

Crystal	Para-meter	R_s							
		1 cm	1 mm	0.1 mm	10 μm	1 μm	100 nm	10 nm	1 nm
	t_0, s	$2.59\cdot10^{9}$	$2.59\cdot10^{7}$	$2.59\cdot10^{5}$	$2.59\cdot10^{3}$	$2.59\cdot10^{1}$	$2.59\cdot10^{-1}$	$2.59\cdot10^{-3}$	$2.59\cdot10^{-5}$
Ar	t^*, s	$1.32\cdot10^{11}$	$1.32\cdot10^{9}$	$1.32\cdot10^{7}$	$1.32\cdot10^{5}$	$1.32\cdot10^{3}$	$1.32\cdot10^{1}$	$1.32\cdot10^{-1}$	$1.32\cdot10^{-3}$
	δ^*	0.86	0.73	0.64	0.57	0.52	0.46	0.43	0.39
	t_0, s	$1.85\cdot10^{8}$	$1.85\cdot10^{6}$	$1.85\cdot10^{4}$	$1.85\cdot10^{2}$	$1.85\cdot10^{0}$	$1.85\cdot10^{-2}$	$1.85\cdot10^{-4}$	$1.85\cdot10^{-6}$
Kr	t^*, s	$8.27\cdot10^{10}$	$8.27\cdot10^{8}$	$8.27\cdot10^{6}$	$8.27\cdot10^{4}$	$8.27\cdot10^{2}$	$8.27\cdot10^{0}$	$8.27\cdot10^{-2}$	$8.27\cdot10^{-4}$
	δ^*	0.78	0.66	0.58	0.52	0.46	0.42	0.40	0.35
	t_0, s	$8.51\cdot10^{8}$	$8.51\cdot10^{6}$	$8.51\cdot10^{4}$	$8.51\cdot10^{2}$	$8.51\cdot10^{0}$	$8.51\cdot10^{-2}$	$8.51\cdot10^{-4}$	$8.51\cdot10^{-6}$
Xe	t^*, s	$2.19\cdot10^{11}$	$2.19\cdot10^{9}$	$2.19\cdot10^{7}$	$2.19\cdot10^{5}$	$2.19\cdot10^{3}$	$2.19\cdot10^{1}$	$2.19\cdot10^{-1}$	$2.19\cdot10^{-3}$
	δ^*	0.81	0.70	0.62	0.55	0.52	0.45	0.41	0.38
	t_0, s	$4.81\cdot10^{6}$	$4.81\cdot10^{4}$	$4.81\cdot10^{2}$	$4.81\cdot10^{0}$	$4.81\cdot10^{-2}$	$4.81\cdot10^{-4}$	$4.81\cdot10^{-6}$	$4.81\cdot10^{-8}$
Na	t^*, s	$7.51\cdot10^{8}$	$7.51\cdot10^{6}$	$7.51\cdot10^{4}$	$7.51\cdot10^{2}$	$7.51\cdot10^{0}$	$7.51\cdot10^{-2}$	$7.51\cdot10^{-4}$	$7.51\cdot10^{-6}$
	δ^*	0.76	0.60	0.50	0.43	0.38	0.34	0.30	0.28
	t_0, s	$5.56\cdot10^{6}$	$5.56\cdot10^{4}$	$5.56\cdot10^{2}$	$5.56\cdot10^{0}$	$5.56\cdot10^{-2}$	$5.56\cdot10^{-4}$	$5.56\cdot10^{-6}$	$5.56\cdot10^{-8}$
K	t^*, s	$1.83\cdot10^{8}$	$1.83\cdot10^{6}$	$1.83\cdot10^{4}$	$1.83\cdot10^{2}$	$1.83\cdot10^{0}$	$1.83\cdot10^{-2}$	$1.83\cdot10^{-4}$	$1.83\cdot10^{-6}$
	δ^*	0.81	0.65	0.54	0.46	0.40	0.36	0.32	0.29
	t_0, s	$4.04\cdot10^{7}$	$4.04\cdot10^{5}$	$4.04\cdot10^{3}$	$4.04\cdot10^{1}$	$4.04\cdot10^{-1}$	$4.04\cdot10^{-3}$	$4.04\cdot10^{-5}$	$4.04\cdot10^{-7}$
Al	t^*, s	$7.85\cdot10^{9}$	$7.85\cdot10^{7}$	$7.85\cdot10^{5}$	$7.85\cdot10^{3}$	$7.85\cdot10^{1}$	$7.85\cdot10^{-1}$	$7.85\cdot10^{-3}$	$7.\ 85\cdot10^{-5}$
	δ^*	0.78	0.65	0.56	0.49	0.44	0.40	0.36	0.33
	t_0, s	$9.65\cdot10^{7}$	$9.65\cdot10^{5}$	$9.65\cdot10^{3}$	$9.65\cdot10^{1}$	$9.65\cdot10^{-1}$	$9.65\cdot10^{-3}$	$9.65\cdot10^{-5}$	$9.65\cdot10^{-7}$
Ag	t^*, s	$6.94\cdot10^{10}$	$6.94\cdot10^{8}$	$6.94\cdot10^{6}$	$6.94\cdot10^{4}$	$6.94\cdot10^{2}$	$6.94\cdot10^{0}$	$6.94\cdot10^{-2}$	$6.94\cdot10^{-4}$
	δ^*	0.74	0.63	0.54	0.48	0.43	0.39	0.36	0.32
	t_0, s	$4.95\cdot10^{7}$	$4.95\cdot10^{5}$	$4.95\cdot10^{3}$	$4.95\cdot10^{1}$	$4.95\cdot10^{-1}$	$4.95\cdot10^{-3}$	$4.95\cdot10^{-5}$	$4.95\cdot10^{-7}$
Au	t^*, s	$5.89\cdot10^{9}$	$5.89\cdot10^{7}$	$5.89\cdot10^{5}$	$5.89\cdot10^{3}$	$5.89\cdot10^{1}$	$5.89\cdot10^{-1}$	$5.89\cdot10^{-3}$	$5.89\cdot10^{-5}$
	δ^*	0.77	0.63	0.53	0.46	0.41	0.36	0.33	0.30
	t_0, s	$6.42\cdot10^{7}$	$6.42\cdot10^{5}$	$6.42\cdot10^{3}$	$6.42\cdot10^{1}$	$6.42\cdot10^{-1}$	$6.42\cdot10^{-3}$	$6.42\cdot10^{-5}$	$6.42\cdot10^{-7}$
Cu	t^*, s	$1.31\cdot10^{9}$	$1.31\cdot10^{7}$	$1.31\cdot10^{5}$	$1.31\cdot10^{3}$	$1.31\cdot10^{1}$	$1.31\cdot10^{-1}$	$1.31\cdot10^{-3}$	$1.31\cdot10^{-5}$
	δ^*	0.86	0.71	0.60	0.52	0.46	0.41	0.38	0.34
	t_0, s	$6.16\cdot10^{14}$	$6.16\cdot10^{12}$	$6.16\cdot10^{10}$	$6.16\cdot10^{8}$	$6.16\cdot10^{6}$	$6.16\cdot10^{4}$	$6.16\cdot10^{2}$	$6.16\cdot10^{0}$
Si	t^*, s	$3.65\cdot10^{16}$	$3.65\cdot10^{14}$	$3.65\cdot10^{12}$	$3.65\cdot10^{10}$	$3.65\cdot10^{8}$	$3.65\cdot10^{6}$	$3.65\cdot10^{4}$	$3.65\cdot10^{2}$
	δ^*	0.89	0.79	0.71	0.65	0.60	0.55	0.51	0.48

Size effects. This approach allows you to follow the size effects. The decrease in the size of the sample sharply reduces the time for the establishment of vacancy equilibrium. It is possible to speak about the practically rapid establishment of equilibrium only for samples with a radius of order and less than 1 μm (with the exception of Si).

The third number in each cell of the second column of Table 36.2 – the value $\delta^* = T^*/T_t$. It is obtained by simulating the cooling of a sample of radius R_s (for given values of the parameters C_0, C_1, δT, δ_c) for the same time for which the minimum temperature was reached in the actual experiment at R_s = 1 cm (see Table 36.1). The lower R_s, the lower temperatures to which it was possible to cool the sample during the actual experiment (real experiments were conducted with a poorly defined dispersion composition). This means that these values δ^* characterize the maximum reduction in temperature that can be reached in the experiment, with the achievement of a label equilibrium. At lower temperatures, under the experimental conditions, it would be impossible to achieve equilibrium. All examples indicate the existence of a temperature range from zero to T^* K, in which complete equilibrium of the sample is not achieved.

It should be noted that with the reduction of R_s to 10 nm for small systems it is no longer possible to apply the laws of classical thermodynamics [or the continuous medium model] [27] and from the continual description of migration it is necessary to go over to discrete models for describing the equilibrium state of the crystal and the migration of atoms. The value R_s = 1 nm is given for a methodical purpose – it is devoid of physical meaning, not only with respect to the diffusion equation, but also with respect to the spherical shape of the sample. Finally, as the size of the microcrystals decreases, it should be borne in mind that the contribution of surface diffusion along the grain boundaries increases sharply, and the material becomes finely dispersed, i.e. its properties differ from the bulk properties of single crystals. Even for R_s = 10 nm, the temperature range from which the equilibrium state is achievable in principle starts from δ^* = 0.3–0.4 (we are not talking about quantum crystals).

Mass transfer coefficient of the substance. The relationship between the mass transfer coefficient and the label coefficient is usually established on the basis of the postulates of non-equilibrium thermodynamics, which leads to the expression for the binary mixture connecting the two factors as [8,10–13] $D_i = D_i^* g_i$, i.e. for each component, the mass transfer coefficient D_i is expressed as the product of the self-diffusion coefficient D_i^* and the thermodynamic factor g_i. We note that this expression can formally be applied to a pure substance – the diffusion of vacancies also takes place in it.

When the label is transferred, there is no time factor related to the process of equalization of the vacancies themselves over the sample – their diffusion accelerates the transfer of atoms. Therefore, an estimate using the experimental values of the self-diffusion coefficient D^* gives an overestimated range of relaxation times. In order to eliminate it, we must pass to the account of the influence of the thermodynamic factor g. Then equation (36.1) will correspond to the real mass transfer of the density or vacancies of the pure element. Here the question arises as to the method of creating a concentration gradient. The postulate of non-equilibrium thermodynamics [28-30] on the presence of local equilibrium, also fixes the condition for the equilibrium distribution of vacancies between local regions ($g = g_e$). In one-component systems, any non-equilibrium distribution of atoms, uniquely due to the material balance, is associated with vacancy disequilibrium $\theta_A+\theta_V= 1$ (at any site there must be a vacancy or atom). Therefore, the hypothesis of the equilibrium distribution of vacancies in the presence of a density gradient does not correspond to a pure substance.

The kinetic approach [31–33] to the transport of matter shows that the local equilibrium can be realized in the presence of an equilibrium relationship between the pair function θ_{AA} and the concentration θ within the local region, but under the condition that there is no equilibrium vacancy distribution between neighboring regions. The density gradient θ leads to a change in the values of the pair function θ_{AA} in different concentration regions due to gradθ. As a result, different types of concentration factors g_e and g_k are obtained depending on the accuracy of the transfer dynamics description. From the kinetic point of view, it is more rigorous to take into account the influence of the density gradient on the change in the pair function in neighboring regions through the thermodynamic factor g_k. In this case, two ways of calculating g_k can be formulated: a direct analysis of equations (20.4) and (20.5), taking into account the effect of the substance flux on the change in the connection between the functions θ_{AA} and θ [31], or if the equilibrium coupling between the pair function θ_{AA} and concentration θ [32,33]. Below we restrict ourselves to the second method of calculating g_k: the first method is very cumbersome and it is difficult to extend it to relaxation to equilibrium in non-uniform

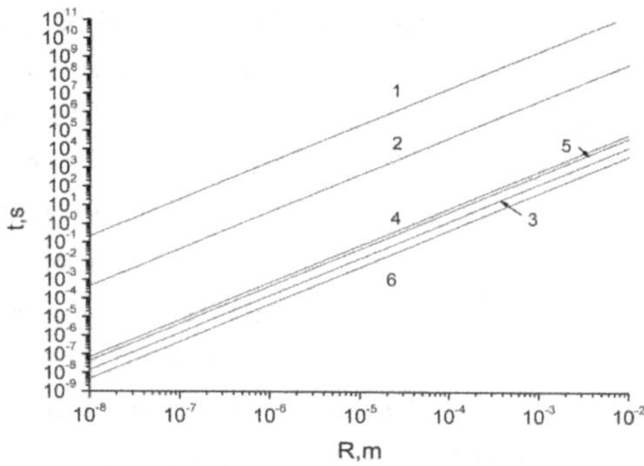

Fig. 36.2. Dependence of the time of establishment of equilibrium on the radius *R, m*, for samples of Ni and Cu in the process of mass transfer when using different values of g_i in calculations: 1 – g_e(Ni), 2 – g_e(Cu), 3 – g_k(Ni), 4 – g_kCu), 5 – g_k(Ni), 6 – g_v(Cu) at T = 1273K.

systems. For strongly non-equilibrium processes, both methods for calculating g_k are equally violated and direct use of the equations for the kinetics of paired distribution functions is required (20.5).

Figure 36.2 compares the following options for the calculation of the thermodynamic factor for copper and nickel: 1) the traditional assumption about the equilibrium distribution of components in substitutional alloys g_e, when the fraction of vacancies in the values of the activity coefficients of the alloy components can be neglected; 2) the kinetic version of the theory under the condition of local equilibrium between pair and unary distribution functions in the non-equilibrium flow g_k, and 3) the model expression for the local vacancy density g_V, in which it is assumed that the vacancy formation energy is approximately equal to half the activation energy of the label transfer [8,10]. The calculation is made for T = 1273 K, which corresponds to vacancy concentrations $\theta_V^{\text{Ni}} = 6.8 \cdot 10^{-8}$ and $\theta_V^{\text{Ni}} = 1.57 \cdot 10^{-4}$. This gives three types of g_i for nickel when calculating the mass transfer coefficient: $g_e = 0.99998$, $D_e = 0.31 \cdot 10^{-15}$ m²/s, $g_k = 0.147 \cdot 10^8$, $D_k = 0.457 \cdot 10^{-8}$ m²/s, $g_V = 0.47 \cdot 10^7$, $D_V = 0.146 \cdot 10^{-8}$ m²/s, and, respectively, for copper: $g_e = 0.979$, $D_e = 0.14 \cdot 10^{-12}$ m²/s, $g_k = 0.624 \cdot 10^4$, $D_k = 0.9 \cdot 10^{-9}$ m²/s, $g_V = 0.92 \cdot 10^5$, $D_V = 0.133 \cdot 10^{-7}$ m²/s. Figure 36.2 shows that taking into account the thermodynamic factor sharply reduces the value of the mass transfer coefficient. Here this difference reaches up to seven orders.

Despite such a sharp acceleration of the transfer process due to taking into account mobility of vacancies compared with the mobility of the label at an equilibrium number of vacancies, the overall effect of lowering the temperature can lead to a wide range of values δ^* exceeding zero temperatures. This issue has not been studied practically today. Here, various situations are possible, connected with the need to jointly take into account the effect of temperature on the equilibrium vacancy density and the diffusion coefficient. The vacancies have different manifestations themselves in diffusion processes for pure components and alloys. It is no accident that the experimental data in [13] were approximated by the empirical dependences of the mass transfer coefficients on temperature and density. In many ways, this depends on the mismatch between the activation energies and the heats of sublimation.

The result obtained on the impossibility of achieving zero temperatures in real conditions does not contradict the Nernst heat theorem. As T tends to zero θ_V also tends to zero, and in the low-temperature limit the equilibrium defect crystal must be freed of vacancies. This expression is obtained from the analysis of statistical sums, in which the number of configurations realized for all possible arrangements of atoms and vacancies along lattice sites is counted [8–13]. Each of the configurations is realized by permuting atoms and vacancies, but the real trajectory of their displacements is not concretized, just as the time necessary for such a rearrangement is not concretized. Since there are no restrictions on the way of enumeration of all configurations, the final result agrees with the Nernst theorem. In the actual process of lowering the temperature in a single crystal, during its relaxation, there is a single diffusion mechanism for the redistribution of atoms and vacancies. It can be shown [33] that the limiting result of such a process at infinite times leads to the same expression for the vacancy density, as well as the direct calculation of all configurations. However, the relaxation times become so large (see Figure 36.1 and the data in Table 36.2) that during the experiment for about ten years such states at low temperatures are not realized.

37. Motions in three aggregate states

The molecular theory based on the LGM allows one to describe, from a single point of view, the characteristics of matter in three aggregate states. This property of the molecular theory applies to both equilibrium and non-equilibrium processes. This brings it to the level of generality of thermodynamics, which does not differentiate its methods between any phases and aggregate states. The unified point of view not only unifies the structure of the kinetic equations of physical and chemical processes within each of the phases, but also allows us to consider from the same positions all processes at the interfaces of different phases.

A characteristic feature of the use of kinetic equations is the different scale of the characteristic times of the transport stages and chemical reactions depending on the properties of the particles, as well as the strong influence of the environment on the rates of these elementary stages in the condensed phases. In order not to encumber the presentation, as in Sections 19 and 20, we divide the elementary stages of transport and chemical reactions in vapour-liquid systems and in solids (including their boundaries). In solid bodies, the migration of atoms occurs by the vacancy or interstitial mechanisms, whereas the vapour–liquid systems are interpreted as a diffusion of molecules in diffuse regions in dense phases, as a free passage in rarefied gas phases.

As a basis for the development of the theory, the concept of the thermal velocity of molecules is used, which is used in all aggregate states, beginning with a solid body in which there are vacancies (low concentrations), up to the gas where vacancies constitute the bulk of the volume. For the intermediate liquid phase, the concepts of vacancy regions are introduced, through which the molecule can move at a thermal velocity. The physical prerequisite for the concept of vacancies in a liquid is a reduced density in comparison with a solid. This rarefaction has a fluctuation character on the average in the volume of the system, and its magnitude can be comparable with the volume of the molecule, which is sufficient for the shift of the molecule. This point of view is suitable for any density and molecules of any size, as well as for their rotational motion.

The concept of the thermal velocity of molecules is a traditional 'gas' point of view in the kinetic theory. For dense phases, this concept is preserved with decreasing length of the displacement of the molecule, which allows us to speak about the concreteness of each trajectory (its extent and direction). The calculation of the thermal velocity itself uses a probabilistic description, which is natural from the kinetic theory of the fluid. The thermal velocity is introduced in terms of the hopping probabilities $W_a = W_{fg}^{AV}(\chi)$ by the considered distance χ. The meaning of the hopping probability $W_{fg}^{AV}(\chi)$ is the average of the continuous process of motion of the molecule A (located at the initial instant t_0 at site f) into the free cell g, which the particle reaches in time τ. The average is taken along all its trajectories, taking into account lateral interaction with all neighbors and collision with them. The length of the jump by the vacancy mechanism depends on the density: $\chi \gg 1$ is characteristic for vapour, $\chi \sim 1$ for liquid and solid. The calculation of the probability of hopping W_a is traditional in the LGM model discussed in Section 20 (see Section 56).

We recall briefly the main provisions of 'microscopic' hydrodynamics, which in principle solved the main problem of constructing balance fluxes of impulse, energy, and mass in two (vapour–liquid) from the three discussed aggregate states [34]. The connection of the solid phase is associated with a greater localization of molecules than in the liquid. This question was considered in [34] in the discussion of flows near solids, and in [32] when considering the processes of dissolution in solids or their internal rearrangement. The generality of the construction procedure $W_{fg}^{AV}(\chi)$ is most obvious when $\chi = 1$. The exponential form of writing expressions for W_a (20.2) is a consequence of the thermalization of the molecule with a thermostat. Solid wall atoms and neighbouring molecules act in the role of a thermostat for any molecule. The thermostat is considered immobile in the absence of a flow of molecules and mobile if a convective flow of molecules is realized in the system (the walls of the pores are always considered immovable). In the initial positions of the LGM it is always assumed that the thermostat is stationary, and only the processes of redistribution of molecules inside the thermostat are realized in it.

Using the example of a surface monolayer, one can explain the difference between these two concepts if one considers the lattice structure of surface adsorption centres with variable height of the activation barrier for hopping from one centre to another [34]. When the barrier is high, the hopping of each particle is determined by the temperature and the probability that the molecule will gain enough energy to overcome the barrier. This energy comes from the substrate, which is a given thermostat. All other molecules of the adsorbed monolayer are located in a similar situation. Their mutual influence is manifested through lateral interactions that change the height of the barrier, but the molecules themselves move individually all the time, staying in different centres. After the molecule jumps, its excess energy is given back to the thermostat (this is the case of the diffusion mechanism of any flow).

We shall decrease the height of the hopping barrier. In the limit, it can be reduced to such an extent that it formally determines only the position of the molecule. In this case, the state of the molecule is mainly determined by its lateral interactions with neighbors, and their mutual arrangement plays a major role in the state of the adsorbed film. The thermostat for an arbitrarily chosen molecule is the substrate (the potential of the substrate remains common to all molecules) and surrounding molecules. The molecule receives energy for movement both from the substrate and from neighboring molecules, just as the molecule discharges excess energy to the substrate and to the neighbours. In this case, any organization of the flow will be accompanied by a general redistribution of molecules, in which the isolated molecule and its neighbours participate equally in collective motion (therefore the thermostat is considered mobile).

Averaging the thermal velocities of the molecules in the cell volume (i.e., in the direction of the forward and backward directions along the axes $\alpha = x,y,z$), which is the production of the resulting flux, allows us to introduce a new micro-level characteristic $\boldsymbol{u}_f$, which is an analog macroscopic flow velocity. The average value of microhydrodynamic velocity is defined as

$$\mathbf{u}_f \equiv \mathbf{u}_A(f) = \int \mathbf{v}_f^A \theta_f^A \left(\mathbf{v}_f^A\right) d\mathbf{v}_f^A / \theta_f^A. \qquad (37.1)$$

The components $u_{f\alpha}$ describe the average velocity of molecules in cell f in the direction α at a characteristic length scale equal to the diameter of the molecule. This makes it possible to take into account the smallest flows (average displacements of molecules) in the convective flow of matter, which appears in the continuity equation. Accordingly, for its description it is necessary to introduce all the traditional transport coefficients.

If we formally consider the forward and backward jumps of molecules in the fixed thermostat in the form of an algebraic sum

$$I_f = \lambda \sum\nolimits_{g \in z(f)} \left[U_{fg}^{VA} - U_{fg}^{AV} \right], \qquad (37.2)$$

then they form diffusion [31, 35, 36] rather than convective flows with respect to the thermostat. Therefore, by setting a pressure gradient (or molecular densities) in a stationary thermostat, the diffusion coefficient can be determined from an analysis of such a non-equilibrium flow.

The kinetic theory defines for a single-component substance the only type of flow of moving molecules – a convective flow described by the continuity equation. To consider the convective flow, it is necessary to match the movement of the thermostat with the thermal motion of the molecules. This is achieved by using an equation for the distribution of pairs, describing their relative position in the coordinate space (as in the theory of a liquid) and depending on the state of the thermostat. The presence of a convective flow at a microscopic level changes the process of redistribution of molecules in comparison with a stationary thermostat, therefore at the microlevel there are two channels for the transport of molecules: micro-convective (described by microhydrodynamic velocity $\boldsymbol{u}_f$) and thermal.

To determine the relationships of these flows, we should consider the local stationary state in the course of the evolution of the pair function. Then one can find a correction to the equilibrium pair distribution function, through which all real distributions of the molecules are recalculated. Considering the shifts of neighbouring regions of matter, an analysis of the non-equilibrium impulse flux can be used to find the shear viscosity (similarly, the bulk viscosity at deformation/compression of the volume under consideration). This allows us to calculate the convective flow rate using a Navier–Stokes-type equation in which the viscosity coefficients are calculated within the framework of the molecular theory based on the LGM, but taking into account the motion of the thermostat. A similar principle is used to construct microscopic equations for the energy transfer process.

In solid bodies, atoms/molecules do not have degrees of freedom translationally and the kinetic equations discussed below pass into the equations of Section 20. However, this does not exclude the separation of types of atomic motions in solids into diffusion jumps at the microlevel (37.2) and macroscopic motions of planes (37.1) that are well known in the Darken interpretation of the Kirkendall effect [10, 13, 31] or in the displacements of the pore boundaries (the Frenkel effect) [31, 37].

In general, this approach leads to a system of difference equations for ordinary dynamic variables in hydrodynamics (densities, three components of the vector $\mathbf{u}_f$ and temperature) defined at lattice sites. Taking into account the possibility of a detailed description of the molecular distribution of molecules over a wide range of time from picosecond to microseconds without the loss of all molecular information and calculation of microhydrodynamic information on the indicated dynamic variables and transport coefficients, this method is alternative with respect to the molecular dynamics method.

Below we will consider the questions of the way out for the condition of local equilibrium, traditionally used in hydrodynamic equations, including in microscopic hydrodynamics, and self-consistent description of processes in different modes and different aggregate states for any degrees of deviations from equilibrium.

38. Equations of conservation of molecular properties

Transition to transport equations. The kinetic equations for unary and paired distribution functions are given in Section 19. These equations are constructed by closing the Bogolyubov chain [38, 39], which takes into account collisions (Boltzmann terms), intermolecular interactions (Vlasov contributions), and exchange properties (mass, impulse, energy) between different cells due to the thermal motion of molecules in space [34]. Exchange flows are composed of direct movements of molecules into neighbouring free cells and from the transfer of properties (impulse and energy) through collisions with their neighbours. Such a system of equations represents the most detailed description of molecular systems, and with its help, when the scale is larger or rough, one can proceed to transport equations on any space-time scales depending on the problem under consideration. These equations can be applied to three aggregate states. As a result, a complete system of transport equations will describe all time intervals (from picoseconds at the microlevel to seconds at the macrolevel), spatial scales and the concentration range from gas to liquid and solid. With their help, one can justify the equations of non-equilibrium thermodynamics from Section 9, which are also used for any aggregate states, and in addition obtain expressions for the dissipative coefficients.

The transport equations are obtained from the system of kinetic equations [34,40,41] by the unary (19.2) and pair (19.3) DFs in the complete phase space $x_f = (r_f, v_f)$, multiplying these equations on the left and right by the remaining properties (mass, impulse and energy): $S_f^{(r)}(S_f^{(r)} = m_f(r=1), m_f v_{fj}, j=1,2,3(r=2-4); mv_f^2/2(r=5))$ and $S_{f\psi}^{(r,n)} = S_f^{(r)} S_g^{(n)}(S_g^{(n)} = m_{g,} m_g v_{gi}, mv_g^2/2)$ and subsequent averaging over the thermal velocities of the molecules v_{fj} and v_{gi}. To illustrate the idea, the evolution of only one component is considered. The resulting system is the equation of conservation of the mean unary and pair moments at local velocities, describing the transfer of different properties. This is the standard way of describing the flows for unary DFs [42–45]. To analyze the applicability of the local equilibrium condition, we need the means for paired DFs.

The averages for any local values of A_f and A_{fg} with respect to local velocities are defined as

$$\begin{aligned}\langle A_f\theta_f\rangle &= \int A_f\theta_f(\mathbf{r}_f,\mathbf{v}_f)d\mathbf{v}_f= \\ &=\theta_f(\mathbf{r}_f)\int A_f\Psi(\mathbf{v}_f)d\mathbf{v}_f=\theta_f(\mathbf{r}_f)\langle A_f\rangle.\end{aligned} \tag{38.1}$$

$$\begin{aligned}\langle A_{fg}\theta_{fg}\rangle &= \iint A_{fg}\theta_{fg}(\mathbf{r}_f,\mathbf{r}_g,\mathbf{v}_f,\mathbf{v}_g)d\mathbf{v}_f d\mathbf{v}_g = \\ &=\theta_{fg}(\mathbf{r}_f,\mathbf{r}_g)\iint A_{fg}\Psi(\mathbf{v}_f,\mathbf{v}_g)d\mathbf{v}_f d\mathbf{v}_g=\theta_{fg}(\mathbf{r}_f,\mathbf{r}_g)\langle A_{fg}\rangle,\end{aligned} \tag{38.2}$$

or $\langle A_f\rangle = \int A_f\Psi(\mathbf{v}_f)d\mathbf{v}_f$ and $\langle A_{fg}\rangle = \iint A_{fg}\Psi(\mathbf{v}_f,\mathbf{v}_g)d\mathbf{v}_f d\mathbf{v}_g$ – the mean is taken over the velocity part of the unary and paired DFs with weight functions $\psi(\mathbf{v}_f)$ and $\psi(\mathbf{v}_f,\mathbf{v}_g)$.

The unary distribution function in (38.1) can be represented as $\theta_f(\mathbf{r}_f,\mathbf{v}_f)=\theta_f(\mathbf{r}_f)f_f^0\xi_f(\mathbf{v}_f)$ where the function $\xi_f(\mathbf{v}_f)$ characterizes the non-equilibrium correction to the Maxwellian distribution function. Dynamic variables $\mathbf{x}_f$ are variables of the phase space for each particle of the system. When averaging over the coordinates inside the cells and the velocities $\mathbf{x}_f$, equations with a spatial resolution of the particle positions in the cells of the order of their molecular size are obtained [34, 40]. Such a 'coarsening' of the coordinates often proves to be sufficient for a uniform volume phase, but for strong external fields, for example, in the field of the surface potential, one should keep the dependence on the coordinates $\theta_f(\mathbf{r}_f)$.

For paired DFs, the presence of correlated thermal velocities

$$\theta_{fg}(\mathbf{r}_f,\mathbf{r}_g,\mathbf{v}_f,\mathbf{v}_g)\equiv\theta_{fg}(\mathbf{r}_f,\mathbf{r}_g)\Psi(\mathbf{v}_f,\mathbf{v}_g)=\theta_{fg}(\mathbf{r}_f,\mathbf{r}_g)f_f^0f_g^0\xi_{fg}(\mathbf{v}_f,\mathbf{v}_g)$$

is taken into account in (38.2), where $\Psi(\mathbf{v}_f,\mathbf{v}_g)=f_f^0f_g^0\xi_{fg}(\mathbf{v}_f,\mathbf{v}_g)$ is the dependence of the pair DF on the thermal velocities, f_g^0 is the Maxwellian equilibrium distribution function with respect to velocities at the site g, $\xi_{fg}(\mathbf{v}_f,\mathbf{v}_g)$ is the correlation function of the velocity distribution of the molecules at the sites f and g with respect to the Maxwellian function velocities.

We will take into account that the averaging over spatial and velocity variables is carried out independently. In order to simplify the DF recoding, we additionally include in the averaging symbol ⟨...⟩ the procedure averaging over the coordinates $\mathbf{r}_f^i$ and $\mathbf{r}_g^j$ inside the cells f and g, then

$$\left\langle A_f \theta_f^i \right\rangle = \theta_f^i \left\langle A_f \right\rangle, \ \left\langle A_{fg} \theta_{fg}^{ij} \left(\mathbf{r}_f^i, \mathbf{r}_g^j \right) \right\rangle = \\ = \left\langle \theta_{fg}^{ij} \left(\mathbf{r}_f^i, \mathbf{r}_g^j \right) \right\rangle \left\langle A_{fg} \right\rangle = \theta_{fg}^{ij} \left\langle A_{fg} \right\rangle. \tag{38.3}$$

where θ_f^i and θ_{fg}^{ij} are the probabilities of finding a particle i at the site f and two particles i, j at the sites f and g (the usual lattice pair DF [32–34]).

We outline the general idea, omitting intermediate transformations. The transition to the transport equations is related to the separation of contributions to the kinetic equations (19.2) and (19.3) into contributions of different nature [40,41]: Boltzmann (associated with particle collision), Vlasov (unconnected with particle collision) and 'migration' corresponding to the escape of particles from cells through free adjacent cells. Thus, for unary DFs (19.2), it is written as

$$\left(\frac{\partial}{\partial t} + \mathbf{v}_f \frac{\partial}{\partial \mathbf{r}_f} \right) \theta_{(1)} \left(\mathbf{x}_f \right) = I_f = I_{f\zeta} + I_{fg} \left(M \right) - I_{fg} (M), \\ I_{f\zeta} = I_{f\zeta} \left(B \right) + I_{f\zeta} (V) \tag{38.4}$$

where on the left side are the usual terms in the absence of external fields for the evolution of the density of molecules in the Boltzmann equation. The right-hand side of equation (38.4) is called the collision integral I_f, which is divided according to the indicated contributions: Boltzmann ($I_{f\zeta}(B)$) and Vlasov ($I_{f\zeta}(V)$) and $I_{fg}(M)$ describing the migration flow. The Vlasov term describes the interaction of a particle at a site f with all its neighbors $h \in z_f$, but $h \neq \zeta$ [40], with particles that do not collide. That is, the directions of particle velocities at the sites f and ζ are oriented towards each other, providing collision, and the particle velocity directions at the sites f and h are oriented in different directions, except for collisions in the considered time interval. The magnitude of this time interval is determined by the time variation in the left-hand side of the kinetic equation in the derivative $\partial/\partial t$. By its structure, the terms in the term $I_{f\zeta}(V)$ are completely similar to the collision integral $I_{f\zeta}(B)$. The term $I_{fg}(M)$ describes the thermal displacements of particles between cells: from the occupied cell f the particle goes to the free cell g. Sites $g \in z_f^*$ are located around site f, excluding sites ζ occupied by particles A, with which there is a collision (B) and excluding sites h that are near to f and with them there is Vlasov interaction (V). The transfer of properties in the exchange term through particle collisions is taken into account by the molecules at the site ζ with which the particle collides at the site f (i.e., through the analog of the Boltzmann contribution at high densities). The term $I_{gf}(M)$ describes the process associated with the thermal motion of the particles in the opposite direction. These two last terms in (38.4) will be called exchange contributions.

From the kinetic equation (19.3), contributions for the collision integral I_{fg}, which is on the right in the kinetic equation for the pair DFs [41], are similarly distinguished. This collision integral I_{fg}, consists of the following contributions:

$$I_{fg} = I_{(f)g} + I_{(g)f} + I_{(f\psi)g}(M) - \\ -I_{(\psi f)g}(M) + I_{f(g\psi)}(M) - I_{f(\psi g)}(M), \tag{38.5}$$

where $I_{(f)g} = I_{(f\zeta)g}(B) + I_{(f\zeta)g}(V)$. Here, the $I_{(f\zeta)g}(B)$ term describes the Boltzmann collisions of the particle f from the pair fg with particles at the sites ζ that do not coincide with the site g. The particle at site g is near and can influence by its potential the collision of particles f and ζ. The same effect through the interaction potential is exerted by particles that are near at sites h that do not coincide with the sites g and ζ. This situation is described by Vlasov's contribution $I_{(f\zeta)g}(V)$. The contribution $I_{(g)f}$ for another particle at site g from the given pair fg has a similar meaning. The remaining terms (38.5) describe the exchange terms. The term reflects $I_{(f\psi)g}(M)$ two exchange mechanisms by moving particles from site f to site ψ or passing a property from site f to site ψ, as described above for the unary function. The parentheses indicate participants in the process – the particle at site g is nearby and can affect the course of the transfer of properties, but there is no direct collision with it.

The term $I_{f(g\psi)}(M)$ has the same meaning as the term $I_{(f\psi)g}(M)$. The brackets indicate the transfer mechanisms of the properties transfer between the sites g and ψ, and the particle f is close by. Finally, the term $I_{(f\psi)g}(M)$ describes the process inverse to the process given by the term $I_{f(g\psi)}(M)$.

After separating the contributions and multiplying the kinetic equations by local properties, we obtain the corresponding transport equations. The separation of contributions by different types is necessary for constructing expressions for dissipative coefficients, which have different forms on different spatial scales.

Complete system of transport equations. If we confine ourselves to the traditional assumption for hydrodynamics [1,36,46] that the transferred properties $S_f^{(r)}$ and $S_{fg}^{(r,n)}$ themselves remain practically constant on the characteristic scale of the change in the coordinates: $\partial S_f^{(r)} / \partial r_{fj} = 0$ (This eliminates the strong non-uniformity of the density of the system inside the region *f*), similarly $\partial S_{fg}^{(r,n)} / \partial r_{fj} = 0$ and $\partial S_{fg}^{(r,n)} / \partial r_{gj} = 0$, then the following system of equations of conservation of unary and pair properties is constructed

$$\frac{\partial \langle \theta_f S_f^{(r)} \rangle}{\partial t} + J_f^{(r)} = I_f^{(r)}, \quad J_f^{(r)} =$$
$$= \sum_{j=1}^{3} \frac{\partial \langle \theta_f v_{fj} S_f^{(r)} \rangle}{\partial r_{fj}} - \theta_f \left\langle \frac{F(f)_j}{m} \frac{\partial S_f^{(r)}}{\partial \mathrm{v}_{fj}} \right\rangle, \tag{38.6}$$
$$I_f^{(r)} = \left\langle \left[\sum_{g \in z_f} \left(S_g^{(r)} U_{fg}^{VA} - S_f^{(r)} U_{fg}^{AV} \right) + \sum_{g \in z_f} \left(S_g^{(r)} U_{fg}^{AA^*} - S_f^{(r)} U_{fg}^{A^*A} \right) \right] \right\rangle$$

$$\frac{\partial \langle \theta_{fg} S_{fg}^{(r,n)} \rangle}{\partial t} + J_{fg}^{(r,n)} = I_{fg}^{(r,n)},$$
$$J_{fg}^{(r,n)} = \sum_{j=1}^{3} \left[\frac{\partial \langle \theta_{fg} v_{fj} S_{fg}^{(r,n)} \rangle}{\partial r_{fj}} + \frac{\partial \langle \theta_{fg} v_{gj} S_{fg}^{(r,n)} \rangle}{\partial r_{gj}} - \theta_{fg} \left\langle \frac{F(f)_j}{m} \frac{\partial S_{fg}^{(r,n)}}{\partial \mathrm{v}_{fj}} \right\rangle - \theta_{fg} \left\langle \frac{F(g)_j}{m} \frac{\partial S_{fg}^{(r,n)}}{\partial \mathrm{v}_{fj}} \right\rangle \right],$$
$$I_{fg}^{(r,n)} = \Big\langle \sum_{h \in z_f^*} \left(S_{hg}^{(r,n)} U_{(hf)g}^{(AV)A} - S_{fg}^{(r,n)} U_{(hf)g}^{(VA)A} \right) + \sum_{h \in z_f^*} \left(S_{hg}^{(r,n)} U_{(hf)g}^{(A^*A)A} - S_{fg}^{(r,n)} U_{(hf)g}^{(AA^*)A} \right) + \tag{38.7}$$
$$+ \sum_{h \in z_g^*} \left(S_{fh}^{(r,n)} U_{f(gh)}^{A(VA)} - S_{fg}^{(r,n)} U_{f(gh)}^{A(AV)} \right) + \sum_{h \in z_g^*} \left(S_{fh}^{(r,n)} U_{f(gh)}^{A(AA^*)} - S_{fg}^{(r,n)} U_{f(gh)}^{A(A^*A)} \right) \Big\rangle$$

where t is time, the sum over j is taken along the axes x, y, z (the index j = 1−3) – it refers to the components j of contributions from the cell variables f or g, $F(f)_j$ is the component j of the external force in the cell f; instead of the traditional mass density ρ ($\rho = m\theta/v_0$), a numerical density θ [40,41] is used. The left-hand sides of $J_f^{(r)}$ and $J_{fg}^{(r,n)}$ equations determine the hydrodynamic flows. Equations (38.6) and (38.7) differ from the expressions in [40,41] by the presence of contributions to the left of the external forces, and also by the explicit form of the collision integrals, expressed in terms of the thermal velocities of migration of the molecules to the free neighboring cells U_{fg}^{AV} (see Section 56) and the collision rate in case of occupation of these cells U_{fg}^{A*A} [34]:

$$U_{fg}^{ij} = K_{fg}^{ij}\theta_{fg}^{ij}\exp(-\beta\varepsilon_{fg}^{ij})\Lambda_{fg}^{Ij},\Lambda_{fg}^{ij}\Lambda\left(S_f^iS_g^j\right)^{z-1}, \tag{38.8}$$

where the expressions for the non-ideality functions S_f^i are defined on the basis of TARR (theory of absolute reaction rates) in Sections 53 and 55. To calculate the terms in Eq. (20.5), the expressions $U_{hfg}^{(ij)A} = U_{hf}^{ij}\Psi_{fg}^{jA}$, $\Psi_{fg}^{jA} = t_{hg}^{jA}\exp\left(\beta\delta\varepsilon_{fg}^{jA}\right)/S_{fg}^{j}$ are used in QCA. The structure of the equations is discussed in more detail in Appendix 2.

The dimension of the complete system of transport equations b (38.6) and (38.7): $b = b_1 + b_2$, consists of the number of equations for the transfer of unary properties $S_f^{(r)}$, $b_1 = 5$, and the number of equations for the transfer of paired properties $S_{fg}^{(r,n)}$, $b_2 = 15$. The remaining properties of pairs (write these variables without permuting the indices f and g) are the following functions: 5 functions of the type $\left\langle m_f S_g^{(n)}\right\rangle$ (mass – property $S_g^{(n)}$); 6 functions of type $\langle m_f v_{fi} m_g v_{gk}\rangle$ (impulse – impulse); 3 functions of the type $\langle m_f v_{fi}/m_g v_g^2/2\rangle$ (impulse – energy); and one function $\langle m_f v_f^2/2m_g v_g^2/2\rangle$ (energy–energy). We denote by y_j the dynamic variable for unary and pairwise moments, $j \in b$. To indicate that the dynamic variable y_j refers to a particular group of these variables, we use the notation $j \in b_1$ or $j \in b_2$.

39. The hierarchy of Bogolyubov times

Practically in all theoretical approaches in the kinetics of physicochemical processes in various phases based on the equations of chemical kinetics [47–50], hydrodynamics [1, 2, 35, 36], non-equilibrium thermodynamics [28–30], and kinetic theory [42–35], for any differences in the non-equilibrium with respect to concentrations, the assumption of local equilibrium for paired DFs and the smallness of deviations from it are possible when proceeding to the description of processes at the macroscopic level. In this case, the number of parameters of the state of the system in dynamics coincides with their number in equilibrium.

From the molecular point of view, the local execution of the combined equation for the First and Second Laws of thermodynamics is due to the unique dependence of the pair distribution function of the molecules θ_2 on the density (or on the unary DF) at a fixed temperature at any time t. Usually such a dependence is expressed in the form of algebraic or integral equations connecting the First and Second Laws DFs without the time argument $\theta_2(\theta)$. This leads to the fact that at a fixed temperature and numerical density θ of the particles the pair DF is time-independent and $d\theta_2/dt = 0$.

An analysis of the condition for the realization of this condition was considered in [51] for a one-component system, which allowed us to introduce a criterion fixing the absence of the condition of local equilibrium; when the critical value of this criterion is reached, the description of thc transport process within the framework of classical non-equilibrium thermodynamics becomes incorrect.

The organization of the flow of pairs between the regions *f* and *g*. Consider the flow of molecules between regions with the coordinates of the planes *f* and *g* (Fig. 39.1). Suppose that the linear size of the region in which the local equilibrium is realized is L, so that $dV \sim L^3$. In Fig. 39.1 *a* this area is selected as a rectangle; therefore, for the coordinates *f* and *g*: $d\theta_2(f)/dt = 0$ and $d\theta_2(g)/dt = 0$, and a flow of molecules (concentrations) and their pairs occurs between these regions with local equilibrium. If the distance between the coordinates *f* and *g* is less than the size L, then there is a common region for local equilibrium of the pairs with respect to the local density. In this case, there is no transfer of pairs of molecules between the planes *f* and *g*, and, consequently, of the density transfer: the local equilibrium region is incorrectly chosen. If the distance between the coordinates *f* and *g* exceeds the size L, then formally the flow of pairs is realized only between two regions of local equilibrium.

A more accurate way of describing the flow of pairs between the coordinates *f* and *g* is obtained if it is functionally related to the value of the density gradient, without specifying the size of the local

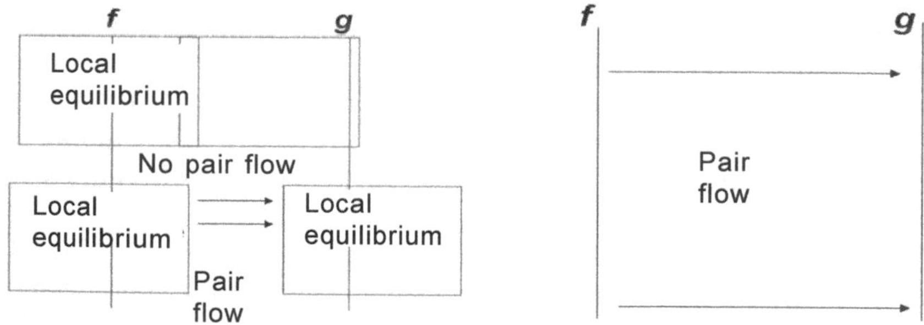

Fig. 39.1. Formal representation of the flows of molecules and their pairs between two regions of local equilibrium with coordinates *f* and *g*.
(right) Flows of molecules and their pairs between the two coordinates *f* and *g* in the case of a functional connection of the flow of pairs to the molecular density gradient.

equilibrium region L (Fig. 39.1 *b*). This possibility gives an idea of the hierarchy of characteristic relaxation times between the first and second DFs in the local equilibrium condition, proposed by N.N. Bogolyubov [38]: the time dependence of the pair DF is expressed in terms of the time dependences of unary DFs

$$\theta_2(x_f, x_g, t) = \theta_2(\theta(x_f, t),\ \theta(x_g, t)), \tag{39.1}$$

here $x_f = (r_f, v_f)$ and we consider the total phase space of the coordinates r_f and velocities v_f in the neighborhood of the coordinate f; In this case, local quasi-equilibrium of the pairs

$$d\theta_2(\theta(x_f, t),\ \theta(x_g, t)) / \partial t = A = 0, \tag{39.2}$$

where A is some function that must be constructed from the kinetic equation for the pair DF with the condition of a small deviation of the pair DF θ_2 from its equilibrium θ_2^e value for a given density θ. Therefore, $\theta_2(t) = \theta_2^e + \delta\theta_2(\theta(t))$, where $\delta\theta_2(\theta(t))$ is the correction to the value of the equilibrium paired DF due to the influence of the flow of molecules changing the density as a function of time θ(t). Or, the main difference between the notions of a stationary (quasi-equilibrium) and equilibrium is the appearance of $\delta\theta_{(2)}(\theta(t))$ a correction depending on the properties of the flow, with their general property – the independence of the pair DF on time $d\theta_2/dt = 0$.

40. The criterion for local equilibrium

Consider the reduction of the constructed system of equations due to the transition to the local equilibrium condition from the positions of the hierarchy of times of N.N. Bogolyubov [38]. According to the expression (39.2) for a pair DF in the total phase space, this transforms the kinetic system of equations (38.7) for paired moments in the equations that are not explicitly dependent on time.

Local equilibrium for a complete system of equations. This condition means $d\left\langle S_{fg}^{(r,n)}\theta_{fg}\right\rangle / dt = 0$, moreover, as, then it allows to $\left\langle S_{fg}^{(r,n)}\theta_{fg}\right\rangle = \left\langle S_f^{(r)}\theta_f S_g^{(n)}\theta_g\right\rangle$ say that the time dependence of the moment related to site f is only on average and $\left\langle S_f^{(r)}\theta_f\right\rangle$ to site g on average $\left\langle S_g^{(n)}\theta_g\right\rangle$ (but this does not mean the loss of spatial correlations between molecules at the sites f and g). Then for the summands $J_{fg}^{(r,n)}$ and $I_{fg}^{(r,n)}$ in the system (38.7) we have

$$d\left\langle S_{fg}^{(r,n)}\theta_{fg}\right\rangle / dt = A_{fg}^{(r,n)} = J_{fg}^{(r,n)} - I_{fg}^{(r,n)} = 0 \qquad (40.1)$$

where the functions $A_{fg}^{(r,n)}$ must be constructed at the following positions, which determine small perturbations of dynamic variables in the local equilibrium region.

When describing the flows of unary properties in the volume of the system, the usual gradient expressions are used from Eq. (38.6):, where $y_j(1) = y_j(0) + \text{grad}(y_j(0))\lambda/2$ where λ is the cell size, the symbols 0 and 1 indicate the numbering of the sites in a certain direction. Such expansions determine the main contributions of the flows for $j \in b_1$ (through $\text{grad}\left\langle \theta_g S_g^{(n)}\right\rangle$), taking into account the dependence of the thermal velocities (38.8) on unary and paired DFs.

When describing the flows of paired properties, the usual gradient expressions are also used: $y_j(-1,0) = y_j(0,1) - \text{grad}(y_j(0,1))\lambda/2$ and $y_j(1,2) = y_j(0,1) + \text{grad}(y_j(0,1))\lambda/2$, where $j \in b_2$. As explained above, the paired dynamic variables $y_j = \left\langle S_{fg}^{(r,n)}\theta_{fg}\right\rangle$, $j \in b_2$, can be represented at local equilibrium in the form $y_j = y_j^e + \delta y_j$, where y_j^e is the value of the given variable in the state of strict equilibrium (i.e., in the absence of flow), and the correction δy_j to equilibrium values y_j^e is due to the influence of the flow and we confine ourselves to the linear approximation $\text{grad}(y_j) = \text{grad}(y_j^e)$; the perturbations of the values of the variables due to the flows are small. The smallness of these quantities coincides in order with the smallness of the gradients of the properties that cause this flow.

Using the superposition approximation for third-order hydrodynamic moments $J_{fg}^{(r,n)}$ in the left-hand sides and substituting the gradient expansions of all the properties $J_{fg}^{(r,n)}$ on the right-hand sides for each of the equations of the system (38.7) allows us to construct expressions $A_{fg}^{(r,n)} = 0$. Such expansions, also taking into account the dependence of the thermal velocities (38.8) on the unary and paired DFs, isolate additional contributions from the flows (through $\delta\left\langle S_{fg}^{(r,n)}\theta_{fg}\right\rangle$) from them. Whence, dropping the cumbersome technical details of taking into account the specifics of the transfer of properties along two mechanisms (through displacements and collisions) and solving the algebraic system of equations $A_{fg}^{(r,n)} = 0$, we write out the main result – the structure of the connection between the gradients of dynamic variables $\left\langle \theta_f S_f^{(n)}\right\rangle$ and non-equilibrium corrections to the moments of paired DFs

$$\delta\left\langle S_{fg}^{(r,n)}\theta_{fg}\right\rangle = \sum_{n=1}^{b_1} B_{fg}^{(r,n)} \operatorname{grad}\left\langle \theta_g S_g^{(n)}\right\rangle. \quad (40.2)$$

The coefficients $B_{fg}^{(r,n)}$of this expansion represent additional contributions to the dissipative coefficients of the transfer of properties. In the particular case $r = n$ they are analogs of the well know diffusion coefficients at $r = n = 1$ [31,52] (by analogy with diffusion in multicomponent systems – although here we are looking at a one-component system). For $r \neq n$, such coefficients are called crosses: they characterize the flows of the property r under the influence of the gradient of the property n.

In [38], the concept of a hierarchy of characteristic relaxation times between the first and second DFs under the condition of local equilibrium was introduced for a gas with a Maxwellian velocity distribution function, i.e. without taking into account the correlations between the molecular velocities. This same assumption has always been used for any laminar flow in dense phases [45-45]. The presence of correlations of the hydrodynamic velocities appeared in the Karman–Howarth equation at the mesolevel [36,46], whereas here they arise at the microlevel [40]. In [31], the concept of a hierarchy of characteristic times under the condition of local equilibrium is formulated for the pair moment of masses with respect to the processes of mutual diffusion of the components of the alloys.

The obtained structure of expressions (40.2) for the pair moments ($j \in b_2$) shows that the correlation values of the velocity properties will be of the order of the magnitude of the gradient causing this flow. In fact, it is through such small fluctuations that nucleation and the development of macroscopic correlations are realized, both in size and intensity. The very existence of large correlations is possible only in the case of a local equilibrium. The introduced criterion for the non-equilibrium of the partial paired moments separates the regions of thermodynamic parameters related to local equilibrium. If there were no velocity correlations for the equilibrium state, then in the presence of a flux these correlations are non-zero. Their role depends strongly on the density and on the hierarchy of the characteristic relaxation times (see below).

For dense phases, equilibrium density correlations play an important role in the thermodynamic characteristics of the system and, with respect to them, density fluctuations can play a decisive role in the course of diffusion processes. The proposed approach generalizes the equations [34] used to study the dynamics of low-intensive flows in strongly non-uniform porous systems. It is of fundamental importance that the dimension of the reduced system remains the same as in ordinary hydrodynamics or non-equilibrium thermodynamics equal to $b_1 = 5$. We note that the cross dissipative coefficients for the transfer of unary properties appear from the kinetic equation to paired DFs even in the absence of accounting for collisions with neighbouring particles [53] (that is, $U_{hfg}^{(A^*A)A} = U_{hfg}^{(VA)A} = 0$).

Local disequilibrium. In intense fluxes $d < S_{fg}^{(r,n)}\theta_{fg} > /dt \neq 0$, the correlations can be much larger than in accordance with formula (40.2), and therefore it is necessary to have a criterion indicating the inability to use the reduced equations in place of the complete system of equations (38.6), (38.7), i.e. go to the full system with dimension b. If we can single out unary DFs and the explicit (relaxation) dependence of the pair DF on the time as independent arguments, then instead of (39.1) we have:

$$\theta_{(2)}(x_f, x_g, t) = \theta_{(2)}\left(x_f, x_g, t, \theta_{(1)}(x_f, t),\ \theta_{(1)}(x_g, t)\right) \tag{40.3}$$

Here the choice of independent variables in the full-pair DF corresponds to the structure of the time hierarchy between the distribution functions of different dimensions: the independent variables for the time dependences are the non-equilibrium functions $\theta_f(x_f, t)$, $\theta_g(x_g, t)$ and the paired DFs.

The evolution of a pair DF in the total phase space describes how

$$\frac{d\theta_{(2)}(t,x_f,x_g)}{dt} = \frac{\partial\theta_{(2)}(t,x_f,x_g)}{\partial t} + \frac{\partial\theta_{(2)}(t,x_f,x_g)}{\partial\theta_{(1)}(x_f,t)}\frac{d\theta_{(1)}(x_f,t)}{dt} + \\ + \frac{\partial\theta_{(2)}(t,x_f,x_g)}{\partial\theta_{(1)}(x_g,t)}\frac{d\theta_{(1)}(x_g,t)}{dt}. \tag{40.4}$$

Hence it follows that the condition of non-equilibrium of the states of the system that go beyond the condition of local equilibrium is the inequality

$$\partial\theta_{(2)}(x_f, x_g, t)/\partial t \neq 0, \tag{40.5}$$

connected with the explicit time dependence of the pair DF.

Denoting the three terms on the right-hand side of (40.4), respectively $d\theta_{(2)}(t,x_f,x_g)/dt = d_{fg} + D_{fg} + D_{gf}$, we introduce the minimum value $D_{\min} = \min\left(|D_{fg}|, |D_{gf}|\right)$ use it to write the expression for the ratio of the relaxation and synchronous variation of the pair DF: $\omega = \left|d_{fg}\right| / D_{\min}$. This ratio characterizes the relaxation contribution of the direct variation of the second DF with time in comparison with the change due to a synchronous change in the pair DF from the unary DF, which varies with time (as in eq, (39.1)). It is natural to take it for the degree of disequilibrium. The value $\omega = 0$ corresponds to the case of local equilibrium.

Similarly, when passing to other moments in equations (38.7), we can introduce the same time arguments for the moments of the pair DF

$$\left\langle S_{fg}^{(r,n)}\theta_{(2)}(x_f,x_g,t)\right\rangle = \left\langle S_{fg}^{(r,n)}\theta_{(2)}(x_f,x_g,t,\theta_{(1)}(x_f,t),\ \theta_{(1)}(x_g,t))\right\rangle \quad (40.6)$$

Then the derivative for the pair moment is written as follows

$$\frac{d\left\langle S_{fg}^{(r,n)}\theta_{fg}\right\rangle}{dt} = \frac{\partial\left\langle S_{fg}^{(r,n)}\theta_{fg}\right\rangle}{\partial t} + \frac{\partial\left\langle S_{fg}^{(r,n)}\theta_{fg}\right\rangle}{\partial\left\langle S_{f}^{(r)}\theta_{f}\right\rangle}\frac{d\left\langle S_{f}^{(r)}\theta_{f}\right\rangle}{dt} + \\ + \frac{\partial\left\langle S_{fg}^{(r,n)}\theta_{fg}\right\rangle}{\partial\left\langle S_{g}^{(n)}\theta_{g}\right\rangle}\frac{d\left\langle S_{g}^{(n)}\theta_{g}\right\rangle}{dt}. \quad (40.7)$$

Introducing in (40.7) $d\left\langle S_{fg}^{(r,n)}\theta_{fg}\right\rangle / dt = d_{fg}^{(r,n)} + D_{fg}^{(r,n)} + D_{gf}^{(n,r)}$, then with their help one can obtain a system of relations responsible for $\left\langle S_{fg}^{(r,n)}\theta_{fg}\right\rangle$ the direct relaxation effect on the change in the time of the moment in comparison with its synchronous variation $\omega_{fg}^{(r,n)} = \left|d_{fg}^{(r,n)}\right| / D_{fg,\min}^{(r,n)}$, where $D_{fg,\min}^{(r,n)} = \min(\left|D_{fg}^{(r,n)}\right|, \left|D_{gf}^{(n,r)}\right|)$.

Such partial criteria, which fix the absence of the condition of partial local equilibrium, when certain critical values are reached, indicate an incorrect description of the transport process in the framework of classical non-equilibrium thermodynamics.

More complicated cases of constructing analogues of criteria for the non-equilibrium state of the system in the case of a multicomponent mixture were considered in [54]. This allows us to consider processes with a strong deviation from local equilibrium and to discuss the concept of ‘passive forces’ introduced by Gibbs [55].

41. Strongly non-equilibrium states and the structure of transport equations

The formulated structure of the complete system of equations (38.6) and (38.7) allows, as a solution, to obtain, as a function of time, the average values of the five dynamic variables $\langle S_f \rangle$ and their 15 pair combinations $\langle S_{fg} \rangle$ (see Appendix 2). These equations represent a microscopically modified hydrodynamic theory of mass, impulse, and energy transfer due to allowance for fluctuations arising when the flow velocity increases, and the kinetic theory of mean square fluctuations of dynamic variables describing the processes of mass, impulse, and energy transfer. The theory relies on closed expressions for the decoupling of multiparticle probabilities through unary and paired DFs and does not have a small thermodynamic parameter (in contrast to the approaches discussed in Ref. [56]), and can therefore be used for various strongly non-equilibrium processes. Introducing in the usual way [57], deviations of one dynamic variable $\Delta S_f = S_f - \langle S_f \rangle$, and their standard deviations as $\Delta S_{fg} = \langle S_f - \langle S_f \rangle\rangle \, (S_g - \langle S_g \rangle) = \langle S_{fg} \rangle - \langle S_f \rangle \langle S_{fg} \rangle$, as a result of the solution of the constructed system of equations, the time dependences of the root-mean-square fluctuations of the dynamic variables S_f will be obtained. In contrast to the correlation functions that can be introduced for one or two different times [1,36,46], all the new 15 dynamic variables $\langle S_{fg} \rangle$ refer to one time point that corresponds to the time scale of the evolution of the variables $\langle S_f \rangle$.

The resulting system of transport equations for unary properties (38.6) is supplemented by equations for the pair properties (38.6). The general property of the resulting new transport equations for pair properties is that for each unknown function new unknown functions appear with increasing dimension in hydrodynamic velocities, or for both sites of the pair due to the displacement of this property by the hydrodynamic flow, or for one of the sites. In the constructed system of equations, this effect is due to the non-equilibrium nature of the pair DF.

This increase in the dimension of the hydrodynamic variables is a known property of the coupled equations at the hydrodynamic level [46,58], which is completely analogous in its nature to the linkage for the DF at the microscopic level [38] (but known prior to the establishment of an analogous fact for molecular DFs [38]). In both cases, the problem arises of decoupling the higher DF from the lower ones, because otherwise the dimension of the system increases indefinitely. To close it, we need a single approximation of all the equations constructed to preserve the description of the evolution of the transfer of all properties by a self-consistent method (see Appendix 2).

Apparently, for the first time the attraction of additional invariants S_{fg} (except for the five known S_f) in the kinetic theory of ideal gases was proposed in [59, 60]. In this way, the kinetic equations for the unary and paired DFs were constructed. However, it was not possible to really advance in this direction, because the basis was the modification of the unary non-equilibrium distribution function (in the manner of the Maxwellian function with additional invariants in the exponent). Moreover, this way had no prospects for dense systems taking into account intermolecular interactions. In the approach [40,41], qualitatively different from [59, 60], there is not only the transition to dense systems, but also to the concept of correlations, both between thermal velocities and between microhydrodynamic velocities, which are implicitly dependent on each other. Recall, once again, that a strictly equilibrium distribution is realized only in the absence of flows. The traditional assumption of the independence of thermal velocities from local hydrodynamic velocities is associated with the consideration of sufficiently large sizes of regions, as discussed in Section 34. In this case, an almost equilibrium distribution of pairs is established inside the region R_v.

The appearance of correlated hydrodynamic variables can be associated with a violation of the local equilibrium condition when mechanical, energy and/or chemical equilibrium is disrupted. In terms of physical meaning, the condition of local equilibrium uniquely relates the probabilities of unary and paired distribution functions (DF) at a constant temperature. Any violation of local equilibrium should lead to a disruption of the equilibrium coupling between pair and unary DFs and for the description of fluxes it is necessary to use kinetic equations for both unary and pairwise DFs.

The constructed system of transport equations for strongly non-equilibrium processes can be treated as at the initial instant of time under mechanical, temperature and concentration perturbations with respect to the initial state, and when they arise during the evolution of the process. For a single-component substance, the local equilibrium is primarily disturbed by an increase in the linear flow velocity.

Such closure procedures should be solved for each specific statement of the problem. For these purposes, the constructed system of equations represents a rigorous basis for a correct analysis of the nature of disengagement. In concrete formulations, this system of equations is drastically reduced by the allocation of space-time scales, but it remains common for any matter densities. In the general case (with the exception of rarefied gases), the following characteristic scales of the relaxation times for impulses, temperature and density [57] can be considered fulfilled: $\tau_{\text{impulse}} \ll \tau_{\text{temperature}} \ll \tau_{\text{density}}$.

In the presence of local equilibrium, not only the equations (38.6) go over into the traditional system of equations of hydrodynamics, but under these conditions all effects of the correlation of the hydrodynamic variables in equations (38.7) disappear and all the introduced means for the pair functions vanish. This makes it possible to use the conditions for realizing local equilibrium as a necessary and sufficient criterion for neglecting the inclusion of correlation effects between the considered combinations of local mean characteristics, which justifies the existence of a local generalized equation (8.3) for the First and Second laws of thermodynamics [30].

Simplifications of the dynamic part of the system (38.6) and (38.7) can also be realized if the flows are not very intense, but are not in equilibrium with a strong non-uniformity of the system, for example, for the wall regions of solids – they have a strong gradient of velocities and densities (although the flow velocity along the surface is not necessarily zero) [34]. In this case, the first transfer equation in (38.7) becomes the equation of [34,61,62] $\dfrac{\partial\langle<\theta_{fg}>\rangle}{\partial t}+\sum_{j=1}^{3}\left\{\dfrac{\Delta<\theta_{fg}v_{fj}>}{\Delta r_{fj}}+\dfrac{\Delta<\theta_{fg}v_{gj}>}{\Delta r_{gj}}\right\}=I_{fg}^{(1,1)}$ describing the dynamics of low-intensity flows in porous systems.

Analogous simplifications can also be realized in subsequent equations of the system for paired DFs, if the fluxes are not very intense, but non-equilibrium due to the strong non-uniformity of the system. Conversely, in the situation with non-equilibrium flows of sufficient intensity, an analogue of the Karman–Howarth equation is obtained from equations (38.6) (see Appendix 2), which is well known in the theory of turbulence [36]. The equations of Appendix 2 (A2.9) and (A2.10) agree with the results of fluctuation hydrodynamics [64] in the sense that the fluctuations of the velocities and the heat flux do not depend on each other and can be considered independently.

Kinetic equations of similar structure are used in problems of plastic deformation in solids [65–67]. In these papers, we propose the analogy of turbulence in solids and liquids. In solid bodies deformation is regarded as a non-equilibrium transition in an ensemble of defects such as microscopic shifts. This allowed us to consider a number of structural–scaling transitions in the description of thermodynamics and kinetic effects in materials in the bulk submicro- (nano-) crystalline state, including the processes of plastic deformation of solids.

The question of the kinetic equations for the evolution of solids, apparently, is one of the most complex, since in comparison with gas and liquid systems, deformation states play an additional important role in them, which determine spatio-temporal correlations in the local mechanical and density characteristics [65–72]. Taking into account the rapid thermal relaxation in solids and the establishment of the total temperature, the system of equations written out is sharply reduced when describing 'frozen' or metastable states. In this case, the so-called flicker noise is often observed, which is associated with the rearrangement of the solid matrix over long times [73, 74]. The physical nature of this phenomenon can be different [75–77], and requires analysis in specific processes. The constructed equations make it possible to pass from phenomenological constructions [74] to atomic–molecular models of such processes.

In a general case, the equations for constructed for the pair correlations between the masses, velocities, and energies of molecules describe a broad class of cooperative processes at intermediate (supramolecular) levels in three aggregate states. The possibilities of the new system of equations are oriented to the study of physicochemical processes in turbulent flows and in frozen matrices of solids, in which the state of the system can differ very much from the equilibrium states.

42. Relaxation times and passive forces

Different state parameters differ by their relaxation times. In real conditions, different relationships between them are possible [57,64]: from $\tau_{\text{imp}} \le \tau_{\text{ener}} \le \tau_{\text{mass}}$ for rarefied gases up to

$$\tau_{\text{imp}} \ll \tau_{\text{ener}} \ll \tau_{\text{mass}}, \tag{42.1}$$

which allows partial reduction of the complete system of equations (38.6), (38.7) for any degree of deviation from equilibrium, since part of the fastest dynamic variables will correspond to the stationarity condition $d\left\langle S_{fg}^{(r,n)}\theta_{fg}\right\rangle / dt = 0$ with zero time derivative.

The constructed equations allow us to discuss the concept of so-called 'passive forces' well known in thermodynamics, introduced by Gibbs [55]. In the complete system of equations (38.6) and (38.7) all forces present in real systems are explicitly taken into account: intermolecular interactions and external potentials. The kinetic theory of the atomic–molecular level does not work with fictitious forces. As the strongly non-equilibrium system approaches the equilibrium state, new forces can not appear in it. In the equilibrium state, dynamic equilibrium is realized, and all reversible processes have equal in magnitude but opposite directed velocities of elementary stages, which allows the system to maintain its unchanged state in time. Therefore, equations (38.6) and (38.7) exclude the appearance of any unknown forces in thermodynamic equilibrium, and the 'appearance' of passive ones can be due only to the lack of information and/or the inability (unwillingness) to establish the cause of the observed unchanged states of the system under study. By the time the 'passive forces' [55] were introduced (~1875), a number of factors that were established later were unknown. Gibbs pointed to the existence of two classes of situations for which he introduced 'passive forces'. These are 1) chemical reactions under conditions that do not allow the realization of chemical transformations, and 2) mechanical effects on solids, for example, in the example of sliding friction.

In the first case, many situations are known where the process is impeded by high activation barriers, which can not be overcome at low temperatures or in the absence of a catalyst. The example with the process of oxidation of hydrogen, chosen by Gibbs, is additionally significantly complicated by the multistage chain mechanism of the process [47,78]. However, today this is the usual 'working' non-equilibrium situation, connected with a large difference in the internal states of reagents, described by the modern theory of chemical transformations [47,49,50,78–80]. Naturally, at the time intervals (42.1), when the role of the chemical reaction is negligible, the state of the reagents can be described by thermodynamic constraints (without a time argument).

When transferring the mass, it is necessary to distinguish between methods for changing concentrations – reactions or diffusion. In the kinetic regime, the diffusion relaxation times are shorter than the relaxation times of the chemical transformations. In the diffusion regime, on the contrary, the relaxation times of the chemical transformations are less than the diffusion relaxation times. All this requires detailing the relaxation times of τ_{mas} both by process type and by each of the components i of the mixture, $1 \leq i \leq s$, where s is the number of components in the mixture: $\tau^{dif}_{den}(i)$ and $\tau^{reac}_{den}(i)$. Differences between $\tau^{dif}_{den}(i)$ and $\tau^{reac}_{den}(i)$ for each component are associated with changes in its energy in the course of redistributions in space or in internal states. The former usually requires less energy change due to lateral interactions, which are usually less than the energy of internal bonds. In the second case, the changes are associated with internal rearrangements and with disconnection.

In equilibrium thermodynamics, the quantity $\tau^{reac}_{den}(i)_{\max}$ refers to the times for the maximum value of the relaxation time related to the relaxation time of the component i, having the maximum value: $\tau \gg \tau^{reac}_{mas}(i)_{\max}$. If $\tau < \tau^{reac}_{mas}(i)_{\max}$, then this refers to the situation for which Gibbs introduced passive forces of the first kind associated with inhibition of chemical transformations. The range of relaxation times extends to $\tau < \tau^{dif}_{mas}(i)_{\min} << \tau^{reac}_{mas}(i)_{\max}$, where $\tau^{dif}_{mas}(i)_{\min}$ is the minimum relaxation time of the diffusion redistribution from all components of the mixture. This set of characteristic times allows us to give a strict interpretation of the separation of components into mobile and immovable and 'decipher' all versions of Gibbs' 'passive forces' [22].

When differentiating the relaxation times of masses in solids, we take into account that $\tau^{dif}_{mas}(i)$ can decay into two different stages: the relaxation of the density of component i with time $\tau^{dif}_{mas}(i)$ for mass transfer and the relaxation of the pair DF for components ij with time $\tau^{dif}_{mas}(ij)$. They are related to each other by the conditions of the hierarchy of time as $\tau^{dif}_{mas}(ij) < \tau^{dif}_{mas}(i)$ (the relaxation of the impulse proceeds much faster than the relaxation of temperature, and the latter is much faster than the relaxation processes of mass transfer). This specificity of solids leads to an increase in the number of dynamic variables, since paired DFs do not enter into the number of ordinary thermodynamic variables, therefore, in real conditions, it is necessary to prove experimentally the existence of a mass transfer equilibrium and equalization of the chemical potential throughout the system. Only then can the conditions for the application of postulates and equations of classical equilibrium thermodynamics be satisfied. In reality, especially for dense phases, such a proof is very rare, and in an overwhelming number of situations the equations of thermodynamics are applied postulatively.

In the second case, we are dealing with a non-equilibrium in the displacements of molecules in space, or with their deformation states in a solid, described by Eqs. (38.6) and (38.7). The question of the generality of methods for describing a rigid body and hydrodynamics within the framework of the theory of a continuous medium is discussed in [2]. The absence of the continuity equation in Lamé's equations for displacements in solids is due to the condition of invariance of the density, which contradicts all mechanochemical processes discovered later, in which the density changes due to diffusion and reactions. For the relative displacement of two solid macrobodies of the general system, discussed by Gibbs, it is necessary to spend efforts on overcoming frictional forces. Tribology is actively engaged in this process today, and the same molecular models [81] are used for it as for other catalysis, the physical chemistry of the surface and the chemical physics of solids and mechanochemistry [70,82,83]. The situation depends on the time ranges of mechanical influences: whether diffusion mixing of atoms of contacting bodies begins or not. We should note (see numerous references in [82,84,85]) the work on the development of a theory for the diffusion–viscous flow of solids, the evolution of structural imperfections due to diffusion processes in polycrystals, the formation of discontinuities/ruptures, as well as the formation of porous bodies and the dynamics of pores. In these papers, the theory included the continuity equation for a rigid body, as in hydrodynamics, and replaced the coefficient of sliding friction by the coefficient of viscosity [3] (which, as Gibbs noted, is not related to the concept of passive forces.) Therefore, today the level of solid state theory processes is significantly different from the ideas of the Gibbs times. For our time, the concept of 'passive force' is not only archaic, but also incorrect – without an explicit indication of the driving force, it is impossible to build kinetic equations.

From the point of view of the kinetic theory (38.6) and (38.7), all the problems of mechanochemistry satisfy the condition (42.1) [2,3,81–85]. To apply thermodynamic concepts to the description of solid-phase processes, a preliminary analysis is needed on the possibility of using thermodynamics (or on the absence of diffusion inhibitions). Estimates of section 36 of the characteristic relaxation times for the density of solid elements in spherical samples of radius from 1 cm to 1 nm have shown that even the simplest vacancy density equalization mechanism can not always be realized for real time of experiments [14–16].

From the differences in the characteristic relaxation times of the impulse, energy, and mass of the particles (42.1), it follows that reversible deformation changes in finite times can be realized only for fixed compositions and structures of solids. If, during prolonged mechanical disturbances of solids, changes occur in the distribution of atoms, then during the removal of mechanical disturbances, the process of relaxation of the spatial distribution of atoms of a solid, due to a change in the short-range order described by equations (38.7) for paired moments, is realized along other trajectories of the variation of dynamic system parameters. Therefore, always, if the equilibrium state was not reached in the forward direction, the movement in the forward direction under load, and the motion in the reverse directions without load occur along different trajectories. This is the natural hysteresis of any dynamic variables in non-equilibrium processes. Only the achievement at the final point of a strictly equilibrium state allows the process not to depend on the previous trajectory of the process. Refusal to bind to a strictly equilibrium state as a reference [86] means an arbitrariness in the choice of the reference state, and a return to it is practically impossible, because the state of a rigid body is described by a much larger number of variables in the system (38.7) than in a strictly equilibrium state (in fact, the entire list of deformation links of the entire body must be taken into account). Only in a state of strict equilibrium is the thermodynamic state of a solid body described by the usual number of thermodynamic variables. A formal sign of the disequilibrium of a mechanical system is the presence of non-diagonal components of the stress tensor – only the diagonal elements of the stress tensor must be in equilibrium [87].

In the case of variations in the pairwise DF (or paired moments) beyond the domain of definition of parameters satisfying the local equilibrium condition specified in Section 40, the process can never be reversible. For its description it is necessary to know the dynamics of a pair DF, which does not include the number of independent thermodynamic parameters of macroscopic systems, and without it the equations of continuum mechanics become non-closed.

Thus, the concept of ‘passive forces’ does not agree with the atomic-molecular theory that operates with real forces due to intermolecular interactions and external potentials. Instead, it is necessary to operate with the concepts of sets of relaxation times of elementary stages describing the dynamics of the process under consideration.

Formally, by analogy with situations for which Gibbs introduced passive forces, the same forces can be associated with metastable states in vapour–liquid systems, as well as for many situations with deformation effects of active phases in solid-phase systems (such as films, membranes, catalysts, nanocomposites, etc. of a wide variety of materials, including polymeric ones).

43. Non-equilibrium thermodynamic functions

We confine ourselves to a discussion of the structure of equations (20.4) and (20.5) for solid-phase systems supplemented by equations from system (38.6) and (38.7) describing their deformation in the course of physicochemical processes (these equations are analogous to the elasticity theory equations [2,3], but contain mechanical modules, depending on the evolution of the concentrations of components). Knowing the solution of these kinetic equations with respect to θ^i_f and $\theta^{in}_{fg}(r)$, it is possible to calculate all the thermodynamic functions depending on them as arguments, including non-equilibrium thermodynamic potentials at each instant of time.

We discuss this question using the example of the Helmholtz energy $F = E-TS$, where E is the energy and S is the entropy of the system. The energy of the system is expressed in terms of these non-equilibrium functions (for simplicity we look at a uniform system) as

$$E = N\left\{\sum_{i=1}^{s-1}\left[\theta_i\beta^{-1}\ln(a_i) + \frac{1}{2}\sum_{r=1}^{R} z(r)\sum_{j=1}^{s-1}\varepsilon_{ij}(r)\theta_{ij}(r)\right]\right\}. \qquad (43.1)$$

According to the definition of entropy in the classical case, the quantity S is determined by the expression ln $\Delta\Gamma$, where $\Delta\Gamma$ is called the statistical weight of the macroscopic state of the subsystem. The quantity S is a dimensionless quantity. Because the number of Δ states can not be less than unity, then the entropy can not be negative. The entropy of the whole system as a whole can also be written as the average value of the logarithm of the distribution function $S = -\sum_{\{\gamma_f^i\}} P(\{\gamma^i_f\}, t)\ln P(\{\gamma^i_f\}, t) + S_0$, where the list of all configurations $\{\gamma^i_f\}$ is equivalent to the phase space, and the constant S_0 is chosen so that in the absence of correlations it corresponds to the distribution $P(\{\gamma^i_f\},t)$ expressed by the product of the distributions of individual particles [64,73]. This definition refers to an arbitrary (equilibrium and non-equilibrium) state of the system. In the QCA this leads to the following formula:

$$S = Nk\sum_{i=A}^{s}\left\{\theta_i \ln(\theta_i) + \tfrac{1}{2}\sum_{r=1}^{R} z(r)\sum_{j=A}^{s}\left[\theta_{ij}(r)\ln\theta_{ij}(r) - \theta_i\theta_j \ln(\theta_i\theta_j)\right]\right\}. \quad (43.2)$$

Thus, expressions (43.1) and (43.2) give a notation for the non-equilibrium Helmholtz energy at any time, including the equilibrium state of the system. The difference between the non-equilibrium state and the equilibrium state lies in the method for calculating the functions θ_f^i and $\theta_{fg}^{in}(r)$, as a function of time, and in the limiting case for large times their distribution is described by the equations (18.4) and (18.1). It is this that determines the fundamental importance that the kinetic equations and equilibrium distribution equations are self-consistent (see Chapter 6).

Local pressure. For Gibbs energy we need to add contributions for local mechanical work – the local equation of state. This question is considered in detail in [88]. For particles with translational motion of components, the expression for local pressure will be written as

$$P_{kn}(f) = n_f kT_f - \frac{1}{v_f^0}\left\langle \sum_{h\in z(f)}\sum_{i,j=1}^{s-1}\theta_{f_k h}^{ij} r_{f_n h}^{ij}\frac{\partial\varepsilon_{fh}^{ij}(r)}{\partial r_{f_k h}^{ij}}\right\rangle \quad (43.3)$$

The calculation of expression (43.3) requires knowledge of the time dependences of unary and paired DFs. Differences in the components of the pressure tensor $P_{kn}(f)$ are determined by the nature of the distribution of neighbouring particles in space and in time, which in non-equilibrium is non-uniform along different directions. They depend on the current quantities $\{\theta_{fh}^{ij}\}$, which obey the transport equations (38.6) and (38.7). Their solution connects all non-equilibrium unary and paired DFs, which makes it possible to calculate the $P_{kn}(f)$ values of local pressure as a function of the non-equilibrium states of the system in the region f. In the course of solving the equations of the system (38.6) and (38.7), it is necessary to take into account the linkage of all local quantities for analogous local values in neighbouring sites and regions, which leads to the necessity of formulating the boundary conditions for the component concentrations, the components of the pressure tensor and the energy flux vector on the entire surface, which limits the system.

In contrast to the mechanics of a continuous medium, symmetrization is absent on the hydrodynamic scale of dimensions (see [88] and Appendix 2, for which the symmetry of the components of pressure tensors $P_{kn}(f) = P(f)\,\delta_{kn}$ [1–3]) occurs in the bulk phase and for micro-non-uniform systems in non-equilibrium states there is no symmetrization $P_{kn}(f) \neq P_{nk}(f)$. The lack of symmetrization is also manifested in the construction of dissipative coefficients in locally non-uniform porous systems even under the condition of local equilibrium [34, 61, 62].

In the case under discussion, expression (43.3) can vary depending on the properties of the system. The simplest example of a non-uniform system is the interface of coexisting phases. Let us denote by q the number of the monolayer within the transition region, $p = q \pm 1$ – the numbers of neighbouring layers, $z_{qp}(R_d)$ – the number of neighbours of the central site q with neighbouring layers p, $\sum_{p=q-1}^{q+1} z_{qp}(R_d) = z$ the definition of numbers $z_{qp}(R_d)$ is given in Section 24 ($R_d \equiv R$) [90]. For $R_d \to \infty$ we obtain an expression for the pressures inside a plane interface. The indices kn are related to the macroscopic symmetry of the drop boundary with the numbers of the layers q and p (here they are left for clarity).

$$P_{kn}(q\,|\,R_d) = n_q kT_q - \frac{1}{\mathrm{v}_f^0}\left\langle \sum_{p=q+1}^{q-1} z_{qp}(R_d) \sum_{i,j=1}^{s-1} \theta_{q_k p}^{ij} r_{q_n p}^{ij} \frac{\partial \varepsilon_{qp}^{ij}(r)}{\partial r_{q_k p}^{ij}} \right\rangle \qquad (43.4)$$

If in this expression we go from the local (non-symmetrized) DF to the symmetrized DF with respect to the local volume on the second size scale, as is used in the formulas [90, 91] with a continual description of the forces, then in the second term the coefficient 1/2 should appear. The difference between the symmetrized and asymmetrized characteristics can be easily explained by the example of the pair contribution to the total energy of the system: if the total system is considered, each bond fg enters it twice (so that it is not duplicated by the coefficient 1/2) if the linkage fg counts from the site f or g, then you do not need to enter a factor of 1/2.

Taking into account the coefficient 1/2, formula (43.4) coincides with the well-known Irving–Kirkwood expression [92], obtained by direct calculation of the forces between molecules of the mixture located on different sides of the selected plane of the fluid, or by the virial theorem [91,79,80]. For planar and spherically curved vapour –liquid interface, expression (43.4) can be rewritten by isolating the normal and tangential components of the pressure tensor ($\alpha = N,\ T$) for each layer q of the transition region

$$P_q^\alpha = n_q kT_q - \frac{1}{2\mathrm{v}_f^0}\left\langle \sum\nolimits_{p=q+1}^{q-1} z_{qp}(R_d)\cos^2(qp,\alpha)\sum_{i,j=1}^{s-1}\theta_{qp}^{ij} r_{qp}^{ij}\frac{\partial\varepsilon_{qp}^{ij}(r)}{\partial r_{qp}^{ij}}\right\rangle \quad (43.5)$$

the symbol (qp, α) denotes the angle between the direction of α and the direction of communication between the sites in the layers q and p (for simplicity, only the nearest neighbors are indicated).

If the system is completely uniform, then Eq. (43.3) constructed through asymmetrized DFs with respect to the distinguished site f goes into the equation for the bulk pressure of a multicomponent mixture, analogous to equation (18.5)

$$P = nkT - \frac{1}{6\mathrm{v}_0}\left\langle z\sum_{i,j=1}^{s-1}\theta_{ij} r_{ij}\frac{\partial\varepsilon_{ij}(r)}{\partial r_{ij}}\right\rangle. \quad (43.6)$$

Here, the indices of the coordinates of the particles in the isotropic volume are omitted, and the coefficients 1/2 and 1/3 are introduced in connection with the symmetrization of the pair DFs in the bulk phase and in the transition to an arbitrary orientation of the pair in space (or 1/2 in the two-dimensional system). Equation (43.6) is usually derived in two equivalent ways: from the virial theorem [95] and from the differentiation with respect to the volume of the partition function of the system Z ($F = -kTln\ Z$ is the free energy) [38], which corresponds to the thermodynamic definition of pressure as $\partial F/\partial V|_{T,N} = -P$ [1–3,55].

Equations (43.3) are obtained under the assumption that all neighbors of the central component i are mobile components and the terms from external $F_j^i(k)$ forces created by the fixed components of the system were not taken into account. If these forces are related to surface forces, equations (43.3) are modified by the fact that part of the surrounding volume is occupied by particles of fixed components that do not change the state of occupation of those sites that they fill (denoted by the symbol d). Then the pressure inside the mobile components that are in the field of the wall is written as

$$P_{kn}(f)=n_f kT_f-\frac{1}{v_f^0}\left\langle \sum_{h\in z(f)-w}\sum_{i,j=1}^{s-1}\theta_{f_kh}^{ij}r_{fh_n}^{ij}\frac{\partial\varepsilon_{fh}^{ij}(r)}{\partial r_{f_kh}^{ij}}\right\rangle-$$
$$-\frac{1}{v_f^0}\left\langle \sum_{d\in w}\sum_{i,j=1}^{s-1}\theta_{f_kw}^{id}r_{fw_n}^{id}\frac{\partial\varepsilon_{fw}^{id}(r)}{\partial r_{f_kw}^{id}}\right\rangle \tag{43.7}$$

Here there are two contributions from the intermolecular interaction of mobile components and from their interaction with the wall (the symbol w). In the last term, the atoms of the wall d belong to the near-surface region of the solid ($d \in w$), whereas the area occupied by the wall is subtracted from the total set of sites $z(f)$ surrounding the mobile particle($h \in z(f)-w$). In this way, it is possible to describe different types of non-uniform surfaces [32, 34], repeating word-for-word the conditions for the size of regions and the time scales described after formula (43.3).

In a state of non-equilibrium, the thermodynamic potentials lose their priority properties inherent in the equilibrium state, when their minima determine the most probable values of the distribution functions of all dimensions. Non-equilibrium thermodynamic potentials play the role of the characteristics of the total energy in the absence of equilibrium. In this case, they do not include the kinetic energy of the centre of mass of the whole system and its potential energy of position as a whole (this definition coincides with the usual definition of the total energy of the system in equilibrium). The kinetic approach lacks the priority of entropy. In the absence of the kinetic theory, the direction of the process was determined through the values of S. In the presence of kinetic theory, the course and direction of the process in full volume is determined from the solution of the equations by the current values of the dynamic parameters. *Entropy is an accompanying information* on the course of the dynamic process at any times and spatial scales. (Although its local production retains its meaning in any non-equilibrium state).

In particular, in order to operate with the value of entropy and / or its production, it is necessary to know the local temperature T, which is determined from the solution of the complete system of equations, including the transfer of energy (or temperature). The determination of the non-equilibrium thermodynamic potentials includes the contributions due to the product $T_{non\text{-}eq}S_{non\text{-}eq}$, provided that the quantity $T_{non\text{-}eq}$ is known at every point of the system in the non-equilibrium state from the solution of the kinetic equations for the energy/temperature transfer. The specificity of solid-phase processes due to large differences in the characteristic times of the processes of establishing thermal equilibrium (by relaxation of vibrational motions) and concentration changes (due to diffusion redistribution of components) is that in $T_{non\text{-}eq} \cong T$, where T is the average temperature of the system on the scale of laboratory samples (the case where $T_{non\text{-}eq} \neq T$ refers to intense perturbations and requires separate consideration). But the value of $S_{non\text{-}eq}$ remains unbalanced due to the 'frozen' diffusion process even in long-term observation.

44. Non-equilibrium surface tension

The formulas for the relationship between surface tension and unary and paired DFs are indicated in Sections 23 and 24. The knowledge of non-equilibrium thermodynamic functions and local components of the pressure tensor indicated in Section 43 allows one to calculate the non-equilibrium surface tension for any type of phase interface. The function $F(t)$ (43.1) and (43.2) is a non-equilibrium analog of the equilibrium free energy for a given time t [88, 90]. The same analogs are introduced for other Gibbs thermodynamic potentials G, entropy S, internal energy $U = E_{lat} + E_{vib} + E_{tr}$ ($E_{lat} = \sum_q E_q^{pot}$, and the last two contributions denote the internal energy of the vibrational and translational motions).

For a solid–vapour or liquid interface, the mobile subsystem is always in an equilibrium distribution. It is 'tuned' to a given non-equilibrium state of a solid (including an adsorbent). Formulas for the function $F(t)$, $G(t)$, $S(t)$, $U(t)$ can be expressed in the same way in terms of the local concentrations $\theta_q^i(t)$ and the pair functions $\theta_{qp}^{ij}(t)$, regardless of whether they are equilibrium or non-equilibrium [90]. The functions $\theta_q^i(t)$ and $\theta_{qp}^{ij}(t)$ must be determined by the kinetic equations of the processes that formed the boundary of the phases to a given instant of time. Therefore, the corresponding non-equilibrium analogs of surface tension can be calculated through them as an excess of the surface free energy $\sigma(t)$, which depends on time. These characteristics, being a function of time, pass to the limiting values of the equilibrium surface tension over long times (for this, it is important to fulfill the self-consistency condition – Chapter 6).

The concept of dynamic surface tension for a vapour–liquid σ_{dyn} system was introduced much later than the creation of thermodynamics for the particular case of creating a freshly formed system (without relaxation of the surface of a liquid [91,97,98]).

As indicated above, the concentration in the interphase layer of the solution in thermodynamic equilibrium differs from the concentration in the bulk phase. However, immediately after the formation of a fresh surface, it can have almost the same concentration as inside the bulk phase. The surface tension of such a surface is called the dynamic surface tension σ_{dyn} [97]. Its calculation is simple enough, although the calculation of the surface tension of the system at subsequent intermediate stages is impossible in the framework of traditional equilibrium concepts. For this, non-equilibrium approaches were used [98,99], since the mentioned interphase region is not in thermodynamic equilibrium, and the evolution of the dynamic surface tension should be calculated using the equations of thermodynamics of irreversible processes.

The concept of non-equilibrium surface tension is much broader than the particular case mentioned above, and it can refer to any kind of the non-equilibrium state of the system. Its magnitude varies depending on the evolution of the state of the system. The phase boundary attracts constant attention [100–114], but the overwhelming number of papers is based on the macroscopic equations used to find the effective parameters of time dependences for dynamic surface tension (see for example [114]). This is an important question for practice, but it leaves aside microscopic ideas about the essence of the relaxation process. The attraction of molecular modelling methods [115] meets the usual problems of their application for the interface boundaries, when the density difference at the boundary to two or three orders does not allow a reliable study of the given process. Therefore, below we briefly discuss the possibility of involving the molecular theory.

The non-equilibrium thermodynamic functions introduced above for bulk phases can be transferred to surface characteristics by the same principle. The surface tension is determined through the excess free energy of a real and a hypothetical system extended to the same separation surface (its principle must be consistent with the calculation of the equilibrium distribution of components (formulas of the Sections 21–24 and Appendix 1). The dynamic problem allows finding all current distributions and obtaining a non-equilibrium surface tension as functions of time.

Gibbs [55] introduced mechanical (γ) and thermodynamic (σ) determinations of the surface tension for solids. He defined γ and σ, respectively, as the work of forming a unit of surface area by mechanical action with an unchanged composition and surface structure (for example, when cutting a body), or in the process of dissolution–precipitation, in which no mechanical stress occurs. In fact, Gibbs allowed the existence of two ways of forming the same value of the new area of the formed surface δA in different ways, or it is a question of the effect of the evolution of the system on its identical final state (with a finite value of δA). This topic was least developed in thermodynamics, and its refinements are still being continued. It should be noted that work was performed both on thermodynamic refinements [116–119] and on modelling at the microscopic level [120–124]. The original meaning of the term 'surface tension' was formulated in the mechanics of continuous media, as a purely mechanical stress on the boundary of the body in question.

The difference between the value of γ and the dynamic surface tension σ_{dyn} introduced later is due to the absence of the effect of external loads in the σ_{dyn} value, whereas this aspect is not specified in relation to the value of γ. For a liquid, this state is instantaneous, whereas for a solid body, due to long relaxation times, its state can be 'frozen' for a long time. Both characteristics introduced by Gibbs, γ and σ, relate to the relaxation times of solid state states at large times. If we are talking only about purely mechanical states, then relaxation takes place quickly. But the possibility of changing the composition and structure, by analogy with the relaxation of σ_{dyn}, is also not discussed. All this leads to different interpretations of Gibbs' concepts. For a long time there was a problem even with the sign of the value of γ at the solid–gas interface. Only recently [119], the quantity γ was assigned to positive quantities, like σ, which should be positive under the stability conditions.

At that time, there was no other alternative way of changing the properties of a solid and its surface due to the diffusion of atoms, which was discovered much later [8–13,82]. This channel for changing the state of a solid body is the main channel in many situations and it is absolutely necessary for it to introduce relaxation times for the limiting subsystems. As in the discussion of the concept of passive forces, it is easier to formulate the questions that a microscopic theory should answer than to analyze its correspondence with existing thermodynamic interpretations.

First of all, this refers to the accounting for the defectiveness of a solid body. Most often in calculations of surface tension [120], taking into account the temperature influence of vibrational motions, we are talking about ideal structures in which defects are absent (including vacancies). The technique for calculating the vibrational states of highly defect structures remains to this day complex [16,125], although the theoretical problems of its solution are known [126–131]. Non-equilibrium analogs of thermodynamic potentials, reflecting the effect of material defectiveness, are more stringent than Gibbs' proposal to use the excess surface tension $U^s(t)$ as a definition of γ, and they not only reflect the essence of the non-equilibrium states of solids, but also give a rigorous way of calculating these characteristics through kinetic equations in the course of their temporal evolution.

Accounting for external mechanical loading complicates the situation, since an additional factor appears that affects the distribution of components both inside the solid body and on its surface, and, consequently, affects the distribution of the mobile phase with respect to the mechanically disturbed adsorbent. The introduction of non-equilibrium analogs of thermodynamic potentials requires a refinement of the very concept of the process of creating a new surface under the action of mechanical loads: is it about the value of $\sigma(t)$ relating to the created surface after removal of the applied load or in the course of the external load itself. Further, there should be a concretization whether the surface is created from the volume of the solid phase (for example, by cleavage), or it is created by applying a load to an already existing surface. In the second case, if, after removing the load, the surface does not relax to the initial value of the surface area, then we are dealing with an analog of plastic deformation, and it is necessary to indicate what changes occurred in the solid body to describe them. For this, during the mechanical load, it is necessary to detail the type and method of creating the contact interaction. Different variants are possible here, ranging from direct perturbations of neighbouring solids (or neighbouring phases) to indirect perturbations at the 'far' ends of the crystal, when the deformation interaction is transmitted along the crystal lattice to the entire volume and to the surface. Each non-equilibrium process is characterized by its kinetic scheme and dynamics specificity, which must reflect microscopic models – otherwise the process under the influence of mechanical disturbances is not determined [81].

These questions are naturally important for the analysis of ensembles of small bodies, which are described in the framework of two-level models (see Appendix 2) [54,88,90]. A microscopic theory based on two-level models is only formulated, although for macroscopic systems the problems of contact interactions are well developed [132–137].

In addition, the problem of the size of the created surface is also important and has a simple meaning only in the case of ideal (planar, spherical, etc.) geometries. Any more or less realistic mechanical disturbance causes rough surfaces of various kinds to arise, and for them the estimation of the surface area is an independent problem. This detailing should clarify the microscopic interpretation of real solid-phase objects, depending on the method of their formation [138]. The transition to a strictly statistical description of the non-equilibrium states of solids makes it possible to relate thermodynamic constructions to other measurements: structural, kinetic, mechanochemical, etc.

45. Relaxation of the interface

The equilibrium conditions of the two-phase system are considered in Section 6. We discuss them in more detail, taking into account the contribution of the phase interface, and in addition take into account the process of relaxation of the intense parameters. Let the equilibrium state be established in the isolated system ($d_eS/dt = 0$), and the progress of the process in the vicinity of the equilibrium point is considered $dS/dt = d_iS/dt>0$, and let the relaxation stage belong to the state of the interface (within each phase, a fast alignment of the properties is assumed). The main attention is paid to the very process of establishing a local equilibrium, in order to analyse the relaxation of the intensive parameters from it. (Such problems are not considered in traditional non-equilibrium thermodynamics [30], using the conditions of local equilibrium).

Taking into account the properties of non-equilibrium functions, which are analogs of equilibrium potentials (Section 43), we write, by analogy with Section 6, the expression for the time evolution of entropy. All functions of this expression (extensive and intensive) are functions that depend on the time argument. For example, $U(t)$ is a function of time in terms of the time dependence of unary and paired DFs, and in the long-time $\lim U(t)_{|t\to\infty} = U$, where U denotes the equilibrium value of the internal energy discussed in Section 6. Then, near the full-equilibrium neighbourhood, the expression for entropy evolution will be written as

$$\delta S(t) = \delta S_\alpha(t) + \delta S_\beta(t) + \delta S_\beta(t), \tag{45.1}$$

where the first two terms refer to the two phases α and β, and the term $\delta S_b(t)$ refers to the surface contribution (Section 21). Although all functions are functions of time, the conditions for fixing the internal energy, volume and number of particles do not change the form of expressions related to the choice of independent variables. As above, the virtual changes of the system parameters are expressed as $\delta U_\beta = -\delta U_\alpha$, $\delta V_\beta = -\delta V_\alpha$, $\delta N_\beta = -\delta N_\alpha$. Fixing the internal energy, volume and number of particles inside the complete system, and choosing as the independent variables the extensive characteristics of phase α, we have an expression for the time dependence of entropy

$$\delta S(t) = \left(\frac{1}{T_\alpha(t)} - \frac{1}{T_\beta(t)}\right)\delta U_\alpha(t) + \left(\frac{P_\alpha(t)}{T_\alpha(t)} - \frac{P_\beta(t)}{T_\beta(t)}\right)\times$$
$$\times \delta V_\alpha(t) - \left(\frac{\mu_\alpha(t)}{T_\alpha(t)} - \frac{\mu_\beta(t)}{T_\beta(t)}\right)\delta N_\alpha(t) + \frac{\sigma(t)}{T_b(t)}\delta A(t) \tag{45.2}$$

where $\sigma(t)$ is the non-equilibrium surface tension, $A(t)$ is the interface area, and $T_b(t)$ is the effective temperature of the transition region of the interface. This expression as a function of time t tends to its limiting state, which we denote by $\lim \delta S(t)_{|t\to\infty} = \delta S = 0$.

Our task is to consider the properties of the $\delta S(t)$ function as a function of the conditions on the ratio of the relaxation times for the pressure and the chemical potential at the limiting transition $t \to \infty$. The analysis using thermodynamic hypotheses in Section 7 showed that formally here it is possible to obtain two different limiting values for the intensive variables.

To simplify the analysis of the expression $\delta S(t)$, we consider the isothermal process of relaxation of the system. This allows us to write down $T_\alpha = T_\beta = T_b = \text{const} = T$, and formula (45.2) is rewritten as

$$\delta S(t) =$$
$$= \left[\left(P_\alpha(t) - P_\beta(t)\right)\delta V_\alpha(t) - \left(\mu_\alpha(t) - \mu_\beta(t)\right)\delta N_\alpha(t) + \sigma(t)\delta A(t)\right]/T \tag{45.3}$$

We simplify the notation by introducing the notation for differences between pressures and chemical potentials in two phases $\Delta P(t) = P_\alpha(t) - P_\beta(t)$ and $\Delta\mu(t) = \mu_\alpha(t) - \mu_\beta(t)$, then the time derivative for entropy evolution will be written in the form (45.4)

$$\frac{d\delta S(t)}{dt}=\frac{1}{T}\left[\frac{d\Delta P(t)}{dt}\delta V_{\alpha}(t)+\Delta P(t)\frac{d\delta V_{\alpha}(t)}{dt} - \frac{d\Delta\mu(t)}{dt}\delta N_{\alpha}(t) - \right.$$
$$\left.-\Delta\mu(t)\frac{d\delta N_{\alpha}(t)}{dt}+\frac{d\sigma(t)}{dt}\delta A(t)+\sigma(t)\frac{d\delta A(t)}{dt}\right] \quad (45.4)$$

To analyze the effect of the temporary relaxation of partial equilibria on the state of complete equilibrium of the system, it is necessary to separate the temporal evolution of extensive and intensive parameters: $\frac{d\delta S(t)}{dt}=\frac{d\delta S_{\text{int}}(t)}{dt}+\frac{d\delta S_{ext}(t)}{dt}$. Temporal changes of all extensive variables $d\delta S_{ext}(t)/dt$, i.e. $d\delta V_{\alpha}(t)/dt$, $d\delta N_{\alpha}(t)/dt$ and $d\delta A(t)/dt$ provided that their current values remain within specified limits within the system under consideration, they do not play any role in the evolution of the intensive parameters. These contributions are analogs of the external influence on the entropy fluxes in the isothermal process during the establishment of complete equilibrium with respect to particular (mechanical and chemical) equilibria with respect to the intensive parameters. They need to be subtracted from expression (45.4) in order to analyze only the time dependences of the intensive parameters, which in the limit $t \to \infty$ determine the state of complete two-phase equilibrium

$$\frac{d\delta S_{\text{int}}(t)}{dt}=\frac{d\delta S(t)}{dt}-\frac{d\delta S_{ext}(t)}{dt}=\frac{d\delta S(t)}{dt}-$$
$$-\left[\Delta P(t)\frac{d\delta V_{\alpha}(t)}{dt}-\Delta\mu(t)\frac{d\delta N_{\alpha}(t)}{dt}+\sigma(t)\frac{d\delta A(t)}{dt}\right]/T, \quad (45.5)$$

$$T\frac{d\delta S_{int}(t)}{dt}=\frac{d\Delta P(t)}{dt}\delta V_{\alpha}(t)-\frac{d\Delta\mu(t)}{dt}\delta \mathrm{N}_{\alpha}(t)+\frac{d\sigma(t)}{dt}\delta A(t). \quad (45.6)$$

We rewrite expression (45.6), taking into account that, by the definition of extensive variables $\delta V_{\alpha}(t)$, $\delta N_{\alpha}(t)$, and $\delta A(t)$ between them there are proportionality coefficients with respect to the contributions from these quantities to thermodynamic functions that are not related to the concrete state of the system: $V = K_{NV}N$, $K_{NV} = dN/dV = \text{const}_{NV}$ and $V = K_{AV}A$, $K_{AV} = dA/dV$, where $K_{AV} \ll 1$. Then

$$\frac{T}{\delta V_\alpha(t)}\frac{d\delta S_{int}(t)}{dt}=\left[\frac{d\Delta P(t)}{dt}-K_{NV}\frac{d\Delta\mu(t)}{dt}+K_{AV}\frac{d\sigma(t)}{dt}\right] \quad (45.7)$$

This form of recording reduced all the relationships to the relations between the relaxation times of the pressure τ_p, the chemical potential τ_μ, and the non-equilibrium surface tension τ_σ, respectively, for the three terms (45.7). By mechanical determination of σ the relaxation time of surface tension is related to the pressure relaxation time [88]. All quantities in (45.7) are described by different kinetic partial differential equations (hyperbolic for impulse transfer (pressure) and parabolic for mass transfer (chemical potential)). For qualitative analysis, let us recall the example of the simplest monomolecular chemical reaction of the transformation of the component $A \rightarrow$ *products*, the relationship between the relaxation time and the rate of change of the corresponding parameters.

The aspiration of concentration $\theta(t)$ to its equilibrium value θ_e at $t \rightarrow \infty$, can be characterized by an exponential dependence of the type $\theta(t) = \theta_e + C\exp(-t/\tau)$, where C is a constant related to the magnitude of the parameter $\theta_0 = \theta(t = 0)$ at the initial time $t = 0$ ($C = \theta_0 - \theta_e$), τ is the relaxation time (which in its physical sense corresponds to the average lifetime of the original reagent A). Within the framework of the mass action law, the evolution of the parameter θ is described by the kinetic equation of the first order $\frac{d\theta}{dt} = K_A(\theta_0 - \theta)$, where K_A is the rate constant of the reaction of component A. The solution of this equation has the form $\theta_0 - \theta = \theta_0\exp(-K_A t) = \theta_0\exp(-t/\tau)$, or $\tau = 1/K_A$. The larger the magnitude of K_A and the faster the parameter $\theta(t)$ changes with time, the shorter is its relaxation time. This principle is also preserved for other variables described by equations of a more complex type (see Section 36).

Let us consider the relationships between the relaxation times of pressure τ_P associated with the time dependence $d\Delta P(t)/dt$ and the chemical potential τ_μ (related to $d\Delta\mu(t)/dt$) with respect to the relaxation time of the surface contribution τ_σ (related to $d\sigma(t)/dt$), and discuss the limiting cases of establishing the equilibrium of a non-uniform system.

1. The case of coexisting equilibrium contacting phases. In Section 28 it is noted that any phase boundary does not affect the internal properties of the phases. The boundary properties can not be parameters of the state of the system, since they necessarily functionally depend on the parameters of the state of the bulk

coexisting phases. Therefore, the relaxation time of the surface tension τ_σ will be less than the relaxation time of the pressure τ_P in the coexisting phases ($\tau_P \gg \tau_\sigma$). Therefore, the third term in (45.7) can be neglected, and the experimental data (42.1) [57] from the ratio between the relaxation times of the pressure (impulse) and the chemical potential (mass) leads to the inequality $\tau_P < \tau_\mu$ (with the exception of the rarefied gas) or $\tau_P \ll \tau_\mu$ in dense phases. It follows that the time of the relaxation process of the system is completely determined by the mass transfer relaxation time and the evolution of the pressure is written as $\Delta P(t) = \Delta P(\Delta\mu(t))$ – in the form of a functional dependence on the evolution of the chemical potential. By construction, in the limit $\Delta\mu(t \to \infty) = 0$, which leads to the limit $\Delta P(t \to \infty) = 0$ for any sizes of coexisting phases, or $P_\alpha = P_\beta$. This is the case of the equilibrium drop described in Section 24 (phase inversion in Section 24 refers to vapour bubbles in the liquid phase).

2. The case of a foreign film separating two phases. If the foreign film excludes the exchange of molecules between phases that are in the internal equilibrium states, then the film relaxation time is much longer than the relaxation time of the mobile phases. This is a typical case of a film of dense material (latex, rubber, etc.) [117], which limits the mobile phase of vapour or liquid. Moreover, the relaxation time of the film itself, determined by its internal properties, is in no way connected with the state of matter in both phases. Here σ_m is the mechanical surface tension of the material. In the absence of an exchange of a substance there is no chemical equilibrium, i.e. $d\Delta\mu(t)/dt = 0$, and only mechanical equilibrium is considered when the right-hand side of equation (45.7) is written as

$$\frac{T}{\delta V_\alpha(t)} \frac{d\delta S_{int}(t)}{dt} = \frac{d}{dt}\left(\Delta P(t) + K_{AV}\sigma(t)\right) \tag{45.8}$$

The process of establishing equilibrium is determined by the relationship between the relaxation times of the pressure and the surface tension of the film. As $t \to \infty$, we have $\Delta P(t \to \infty) = 0$ as above, while $\sigma(t \to \infty) = \sigma_m$, which obviously does not coincide with σ_e discussed in Chapter 3.

In the limit, another state of the system is obtained, which differs from the strictly equilibrium state for a plane vapour–liquid boundary $P_\alpha = P_\beta$ by an amount $K_{AV}\sigma_m = \sigma_m dA/dV = 2\sigma_m/R$, i.e. $P_\alpha = P_\beta + 2\sigma_m/R$ (the Laplace equation). To analyze the size dependences of $\sigma_m(R)$, there must be a molecular model for taking into account the mechanical characteristics in the actual film between phases, determined by the properties of the film material.

3. If the foreign film does not prevent the transfer of molecules, but the relaxation time of the exchange of molecules (chemical potential τ_μ) is less than the relaxation time of pressure τ_P, then we consider again the case $\tau_P \gg \tau_\mu$ in which equation (45.8) is again obtained. The relation $\tau_P \gg \tau_\mu$ contradicts the experimental data (42.1) [57], and this case is considered formally. But it corresponds to the Kelvin equation in the accepted sequence of establishing first a mechanical and then a chemical equilibrium. The functional relationship $\Delta\mu(t) = \Delta\mu\ (\Delta P(t))$ is realized, when the chemical potential is adjusted to the value of the pressure established by the moment. As a result, again $P_\alpha = P_\beta + 2\sigma_m/R$, where the value σ_m can not be determined experimentally due to $\tau_P \gg \tau_\mu$.

This situation corresponds to the so-called case of metastable drops (see Appendix 1 and Chapter 7) when, in practice, the experimentally measured value of the surface vapour–liquid tension for a plane boundary is substituted for the unknown value of σ_m. In practice, this substitutes the properties of a foreign film by an amount σ for the interface of a real vapour–liquid system.

46. Influence of fluctuations on the rate of stages

The effect of equilibrium fluctuations on microcrystalline particles on the adsorption isotherms was shown in Chapter 4. Here, their effect on the adsorption–desorption rate and surface reaction rate on small microcrystalline particles is discussed. As above, for simplicity, the main attention is paid to the absence of lateral interactions between adsorbed molecules. Equations of the equilibrium distribution of molecules and their mean square displacements on a uniform surface are described by the equations of Section 31 and 32 (for a binary mixture). Equilibrium fluctuations affect the rates of elementary stages if fast surface diffusion takes place, for a fixed number of adsorbed particles their distribution can be regarded as practically equilibrium.

In the absence of interaction between adsorbed particles on a uniform surface, the rates of elementary stages are described within the framework of the law of mass action as

$$U_m = K_m W_m,\ W_m = \theta_i\,(m=1) \text{ and } \theta_i\theta_j\ (m = 2) \tag{46.1}$$

where m is the molecular order of the reaction, K_m is the rate constant, W_m is the concentration component of the rate of the stage, in the particular case of ideal reaction systems.

The knowledge of the average number of adsorbed molecules $N_i = M\theta_i$ makes it possible to calculate the rates of elementary reactions. According to the theory of fluctuations [57], in the expressions for the fluctuations of the rates of elementary reactions W_m is the concentration component of the reaction rate, depending on the number s of variables, taking into account the definition of $\Delta\theta_i = (N_i - \langle N_i\rangle)/M$, will be written as [139,140]

$$\Psi(U_m) = \sum_{i,k=1}^{s} \frac{\partial W_m}{\partial\langle N_i\rangle} \frac{\partial W_m}{\partial\langle N_k\rangle} \langle \Delta N_i \Delta N_k \rangle, \qquad (46.2)$$

where the calculation of the mean-square means $\langle \Delta N_i \Delta N_k\rangle$ is carried out taking into account the limited size of the system M.

Uniform surface. In [139] the value of the temperature corresponding to $\beta\varepsilon = 1/2$ is assumed, where ε is the interaction parameter between molecules of the first sort. For an ideal system, the parameter ε serves as a measure of the binding of the molecule of the first sort to the surface. It is assumed in the calculations that the binding energy of molecules of the first sort with the surface is $Q_1 = 4\varepsilon$, which corresponds to a sufficiently strong binding of the adsorbed particle to the surface. The binding energy of molecules of the second sort with the surface is assumed to be $Q_2 = \gamma Q_1$, where $\gamma = 1.4$.

Figure 46.1 shows the rates of monomolecular desorption processes of the first and second components U_1 and U_2 (*a*) and the fluctuations of the desorption rates Ψ_1 and Ψ_2 (*b*) ($m = 1$). Since the densities θ_i are entered linearly in the expressions (46.1), the role of the rate constants reduces to the coefficients before the concentration dependence W_m (for simplicity, their values are put to unity). The change in M has little effect on the behaviour of the rates, and therefore their behaviour is shown for small fillings, where these changes are more noticeable. At the same time, the dependence on x_1 is very significant. With increasing M, the range of values of θ in which the rate fluctuations are determined increases. In general, varying the size of the region M weakly shifts the curves relative to each other at a given scale.

These changes are somewhat more noticeable on fluctuations. The results are given for $x_1 = 0.15$ and $x_1 = 0.5$. For ease of comparison,

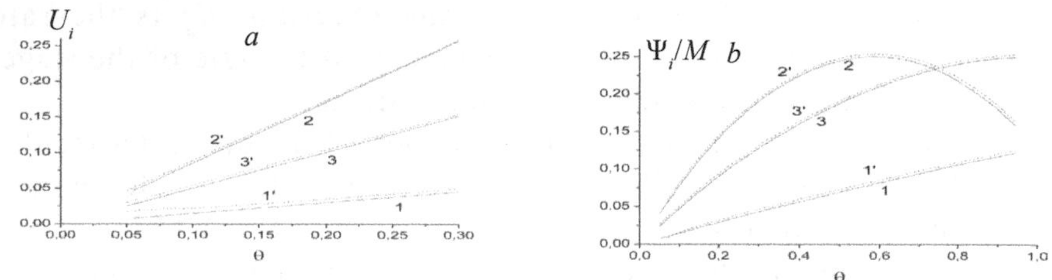

Fig. 46.1. The rates of the desorption processes U_i (a) and their fluctuations Ψ_i (b) $i = 1, 2$ ($m = 1$) as a function of density θ for $M = 10^4$ (solid lines) and $M = 10^2$ with density fluctuations (dashed lines). The notation on the curves: 1, 2 corresponds to $i = 1, 2$ for $x_1 = 0.15$, 3 – corresponds to $x_1 = 0.5$. The same notation, but with a prime for $M = 10^2$.

the desorption rate fluctuations are given in scale Ψ_i/M. If the molar fractions of the components are the same ($x_1 = 0.5$), then both the rates and the fluctuations are monotonous. But if the rates U_1 and U_2 are monotonic at any ratio of components, then for the molar fraction $x_1 = 0.15$ for component 1 (with a smaller fraction), the rate fluctuation curve has a monotonous form, whereas for component 2 (with a larger fraction) it passes through an extremum. For $M = 10^4$, the contributions of the density fluctuations to the values of the rate fluctuations are weakly noticeable at a given scale, whereas for $M = 10^2$ they are quite noticeable.

The behaviour of the relative deviations for elementary reaction rates calculated with allowance for only the size contribution and with an additional allowance for the density fluctuations from the size contribution is more evident than in the case of $M = \infty$. For monomolecular reactions, this comparison is shown in Fig. 46.2. The inset in Fig. 46.2 shows their absolute deviations. These changes agree with the course of the curves in Fig. 32.3, when the deviations change sign during the transition from small values of pressure to large.

Non-uniform surface. The description of non-uniform ideal systems is reduced to the summation of the contributions of any characteristics on individual faces. The formulas obtained above can be easily generalized by summing with weights F_q for the contributions of different types of centres q ($M_q = MF_q$) and replacing the partial fillings θ_i by the local partial fillings $\Delta\theta^i_q = N^i_q/M_q$ related to sites of type q, $1 \leq q \leq t$, t is the number of types of sites.

Figure 46.3 shows the corresponding desorption rates and their fluctuations on the same non-uniform surface (here $t = 2$, $F_1 =$

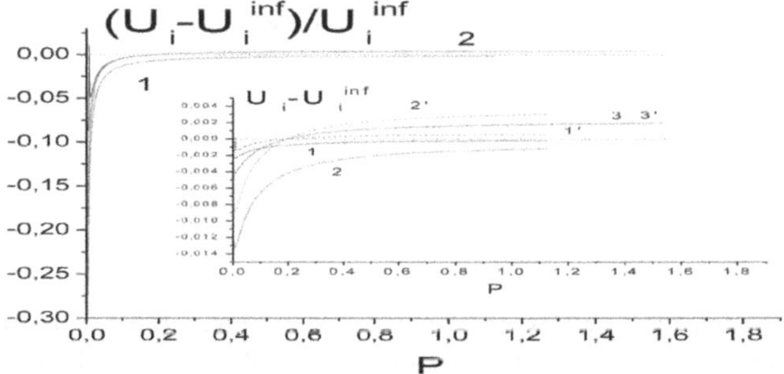

Fig. 46.2. The relative deviations of the reaction rates U_i at $M = 10^2$, the notation on the curves: 1 – corresponds to $x_1 = 0.15$, 2 – $x_1 = 0.5$. On the inset – their absolute deviations, without taking into account the contributions of $\Delta\theta^i_q$ – solid curves, taking into account these contributions – dotted line; the notation on the curves: 1, 2 corresponds to $i = 1, 2$ for $x_1 = 0.15$, 3 – corresponds to $x_1 = 0.5$. The same notation, but with a prime, for curves with allowance for the contributions of $\Delta\theta^i_q$.

$F_2 = 0.5$, $Q^i_2 = Q^i_1/2$). Figure 46.3 is similar to Fig. 46.1, however, the differences between the curves related to different regions and to the account of the effects of fluctuations are more clearly shown (Fig. 46.1 is given on a smaller scale). This is due to the fact that the surfaces with the same total area M are compared, which additionally differ in the size of the portions M_q having different binding energies.

Thus, an increase in the degree of non-uniformity of the surface leads to an increase in the role of the factor of boundedness of its individual regions, and an increase in the contributions of local fluctuations. The obtained results indicate the need to take into account fore the nanosized crystalline particles the size limitations of individual faces and density fluctuations of adsorbed particles, which affect both the equilibrium adsorption isotherms and the surface reaction rates.

Reactions of the Langmuir–Hinshelwood type. Fluctuations for the Langmuir–Hinshelwood type reaction $Z_qA + Z_pB \rightarrow$ *products in the gas phase* flowing on a non-uniform surface depend not only the surface composition, but also the structure of the non-uniform surface [140]. For a Langmuir–Hinshelwood type reaction, the rate of the surface reaction between the components of the first and second sort on a non-uniform surface will be written as [140]

$$U_{AB} = \sum_{q,p=1}^{t} F_{qp} K_{qp}^{AB} \theta_q^A \theta_p^B, \qquad \sum_{p=1}^{t} F_{qp} = F_q \text{ and } \sum_{q=1}^{t} F_q = 1, \tag{46.3}$$

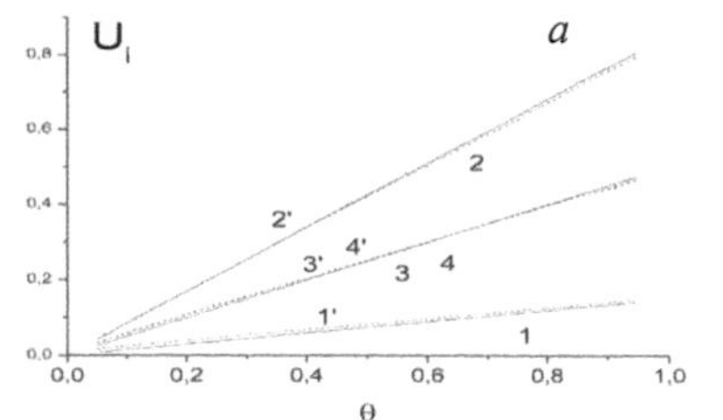

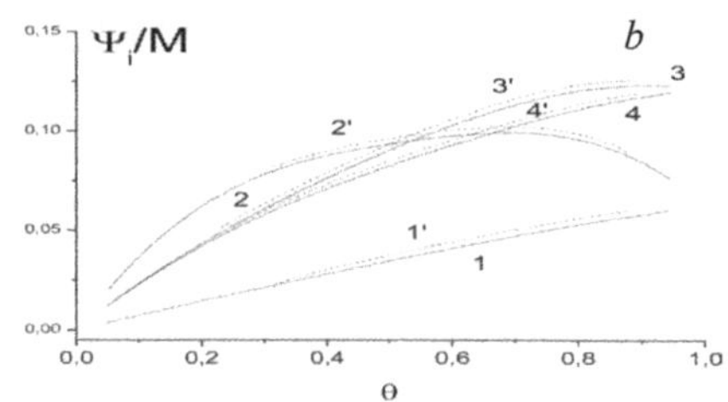

Fig. 46.3. The rates of the processes U_i (a) and their fluctuations Ψ_i (b) i = 1,2, as a function of the density θ for $M = 10^4$ – solid line and $M = 10^2$ – dotted line . The notation on the curves: 1, 2 corresponds to i = 1, 2 for x_l = 0.15, 3, 4 corresponds to i = 1, 2 for x_1 = 0.5. The same notation, but with a prime, for $M = 10^2$.

where K_{qp}^{AB} is the rate constant of the reaction between the first and second components located at the sites of the q-th and the p-th type in the case of ideal reaction systems, F_{qp} is the probability of finding a pair of centres of the q and p type on the surface, which are related to each other and the surface composition by the normalization relations. The equations (46.3) give normalized expressions for the rate, which do not reflect the real size of the system. In order to take into account the number of adsorption centres, it is necessary to take into account the connection of F_{qp} with the number of pairs of sites of different types $F_{qp} = (1+\Delta_{qp})M_{qp}/z_q M$, where Δ_{qp} is the Kronecker symbol and z_q is the number of the nearest neighbours for a site of type q. For simplicity of analysis, we will assume that there are three types of sites on the surface (t = 3), all sites have the same number of neighbours $z_q = z$, and only particles that are at sites of the first and second types (sites of the third type play the role of the carrier of active centres) enter the reaction, then in the expression (46.3) the summation over the types of the site extends to the pairs 11, 12, 21 and 22. Let the model reaction $A + B$ on the non-uniform surface be characterized by a specific matrix of local rate constants: $K_{11}^{AB}:K_{12}^{AB} = K_{21}^{AB}:K_{22}^{AB} = 1:5:1$.

To demonstrate the role of the surface structure in bounded systems, the following surface structures are examined:

1) $F_1 = F_2 = 1/2$, $F_3 = 0$; $F_{11} = F_{22} = 1/2$. The remaining $F_{qp} = 0$.

2) $F_1 = F_2 = 1/2$, $F_3 = 0$; $F_{11} = F_{22} = F_{12} = F_{21} = 1/4$. The remaining $F_{qp} = 0$. This lattice refers to the random lattice ($F_{qp} = F_q{*}F_p$).

3) $F_1 = F_2 = 1/2$, $F_3 = 0$; $F_{12} = F_{21} = ½$; The remaining $F_{qp} = 0$. The 'chessboard' type lattice

4) $F_1 = F_2 = F_3 = 1/3$; all $F_{qp} = 1/9$. The random lattice.

Figure 46.4 shows the concentration dependence of the Langmuir-Hinshelwood reaction rate on the pressure of the first component p_1, at a fixed value of the pressure of the second component, $p_2 = 1$. The field in Fig. 46.4 *a* shows the *A*+*B* reaction rates on a non-uniform surface of macroscopic size U_{AB} (∞). With an increase in the degree of filling of the first component the reaction rate passes through a maximum, and with the predominant filling of the surface with the first component, the rate begins to decrease. This is the typical course of the reaction rate of the Langmuir–Hinshelwood type [25]. The variant of the considered surface of the catalyst is marked with the corresponding number.

The highest values of the rate are observed in the case of the maximum number of pairs of different types 1 and 2, which is realized on the chessboard type structure. With a chaotic arrangement of active centres (curve 2) of the same surface composition, the rate is noticeably lower since the probability of finding active pairs of sites is smaller. With a decrease in the surface composition of the active centres with the same random arrangement of the active centres, the reaction rate decreases even more (curve 4).

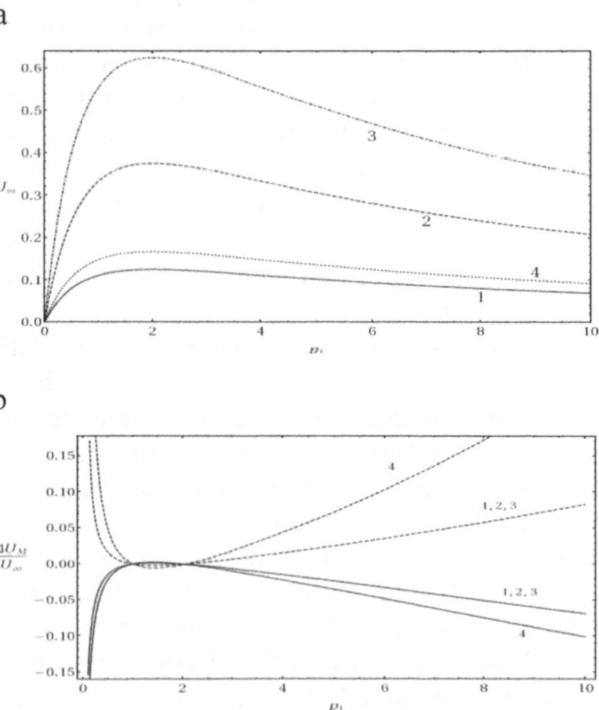

c

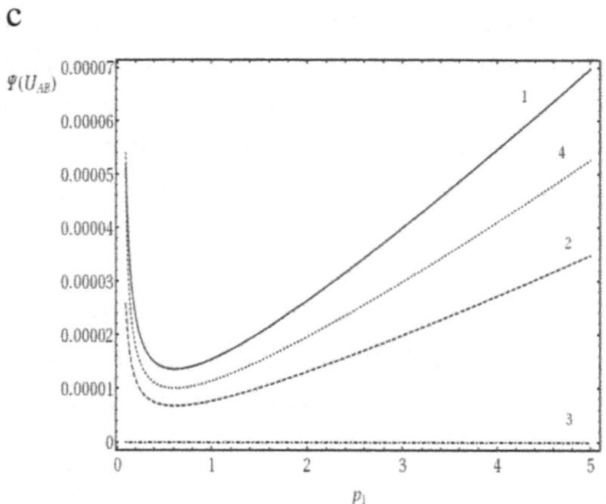

Fig. 46.4. (*a*) The dependence of the reaction rate U_{AB} (∞) in an unrestricted system on the pressure of the first component. (*b*) Dependence of the relative change in the reaction rate in the final system as compared to the infinite one $[U_{AB}(M)-U_{AB}(\infty)]/U_{AB}(\infty)$ without taking into account the fluctuations (solid lines) and $[U_{AB}(\mathrm{fl})-U_{AB}(\infty)]/U_{AB}(\infty)$, taking into account the fluctuations (dashed lines), on the pressure of the first component. (*c*) Dependence of the reaction rate fluctuation $\Psi(U_{12})$ on the pressure of the first component.

Curve 1 corresponds to the case of a 'spotted' arrangement of centres of the same type, in which there are no most active centres of type 1 and type 2 close to each other. We note the similarity of the values of the rates on the curves 1 and 4, although they refer to different compositions and structures of the surface of the catalyst. Relative changes in the surface reaction rates under the same conditions as for macroscopic systems (field in Fig. 46.4 *a*), in the case of bounded systems $[U_{AB}(M)-U_{AB}(\infty)]/U_{AB}(\infty)$ and with additional consideration of the effect of density fluctuations of reagents $[U_{AB}(\mathrm{fl})-U_{AB}(\infty)]/U_{AB}(\infty)$ are shown in Fig. 46.4 *b*. Here, the symbols $U_{AB}(M)$ and $U_{AB}(\mathrm{fl})$ denote the rates obtained in calculations in which local partial fillings respectively relate to bounded areas of the surface and to an additional account of the effects of fluctuations. In field 3*b*, the solid lines refer to the limited system $M = 100$ without fluctuations, and the dashed lines to the case in which the fluctuations are taken into account.

With a decrease in the size of the system the reaction rate decreases in those filling regions in which the partial fillings of the reagents are reduced – the region of small and large surface occupancies. Conversely, an additional consideration of the effects of fluctuations in a bounded system increases the reaction rate in full accordance with those changes in the partial fillings discussed in Section 32. Note that for given simplified model parameters, in the case of an ideal reaction system at $F_3 = 0$ the values of the relative deviations of the reaction rate for the first three types of surfaces coincide.

Figure 46.4 *c* shows the fluctuations of the $A+B$ reaction rate itself, which according to the formulas [139, 141] can be represented in the form

$$\Psi(U_{AB}) = \sum_{q,p=1}^{t} F_{qp} K_{qp}^{AB} \Psi(\theta_q^A \theta_p^B) = \sum_{q=1}^{t} F_{qq} K_{qq}^{AB} \Psi(\theta_q^A \theta_q^B). \qquad (46.4)$$

In the second equality, it is taken into account that for ideal reaction systems there is no mutual influence of reagents located at different sites. The expression for the fluctuation of the function is $\Psi(\theta_q^A \theta_q^B)$ constructed according to the usual rules for calculating fluctuations in individual areas of the surface [57, 142]: $\Psi\left(U_{AB}\right) = \sum_{q=1}^{t} K_{qq} F_{qq} \left(\theta_q, \eta_{qq} \theta_q\right) / (M^2 F_q^2)$ where the matrix product of the vector $\theta_q = \left(\theta_q^B, \theta_q^A\right)$ and the fluctuation matrix (the average of the matrix on the vector θ_q) are given in the parentheses.

The values of the rate fluctuations calculated in the normalized form (through partial degrees of filling) for a bounded system $M = 100$ with an allowance for density fluctuations of partial $\theta_q^i(M)$ are shown in Fig. 46.4 *c*. The numbers of the curves correspond to the four types of surfaces considered. In the case of a completely ordered arrangement of centres of different types, there are no pairs of the same type, so curve 3 has a zero value. As above, the displacements of the degrees of surface filling in the region of small and large fillings increase the fluctuations of all quantities – in this case the fluctuations of the rate of the reaction itself.

We note that the transition to a greater number of types of surfaces increases the degree of surface non-uniformity, which leads to an increase in the role of the factor of boundedness of its individual regions and an increase in the contributions of local fluctuations.

47. Fluctuations of rates in small non-ideal reaction systems

In the general case, the molecular–statistical theory for arbitrary non-equilibrium states with an allowance for size fluctuations requires detailed analysis for different characteristic relaxation times of the realized stages α of the general process. Below we shall limit ourselves to an outline of the basis of this approach. It has been pointed out above that the very concept of local equilibrium refers to the relationship between local concentrations and paired DFs (Sections 39 and 40). When transferring the kinetic equations to small bodies, one should take into account the relationships of local volumes, the value of M, as indicated above for equilibrium fluctuations, and the number of molecules N_q^i within a given volume.

In the LGM the function $P(\{\gamma_f^i\},\tau)$ is the concentration part of the total non-equilibrium distribution function with respect to coordinates and impulses, which remains after averaging over the impulses (Sections 20 and 43). The form of this function turns out to be universal for any aggregate state of the system; therefore, there exists an explicit expression (43.2) for the non-equilibrium entropy in QCA (for simplicity, the formula (43.2) is written out for a uniform region). This expression allows us to construct a one-to-one correspondence between formula (43.2) and statistical sum Q (16.8) (or a specific form in the QCA (33.7)), which reflects the statistical weight of any local equilibrium configurations. The universality of the expression for entropy means that one can directly use this kind of connection between S (and the text before (43.2)) and the statistical weight in Q for non-equilibrium states. This allows us to generalize the concept of a partition function to non-equilibrium states if we confine ourselves to a small scale of the time interval (this is the value of dt in the left-hand sides of equations (20.4), (20.5)) and restrict ourselves to those changes in the system that can occur in time dt. Then, instead of calculating the total partition function Q, one can select that part of the reaction subsystem from the general non-equilibrium system that participates in a particular equilibrium for a given instant of time. Such time-local equilibria fully correspond to the idea of the activated complex of the stage, which is in equilibrium with the reaction subsystem, in the theory of absolute reaction rates (TARR) [32–34]. This interpretation uniquely relates the stage-by-stage description of the dynamics of the spatial redistributions in the LGM described by equations (20.4) and (20.5), and the change in the non-equilibrium function $P(\{\gamma_f^i\},\tau)$.

The reformulation of the equilibrium type changes the expression for the function $\ln P(\{N_q^i, N_{qp}^{ij}\})$ that was introduced for ideal systems in the sections 30–32, by simply replacing the chemical potential in them with the original reactants at $P(\{N_q^i, N_{qp}^{ij}(r)\}) \equiv Q\exp(\beta\sum_{i=1}^{s-1}\mu_q^{i*}(\alpha)N_q^i)$ where $\mu_q^{i*}(\alpha)$ is the chemical potential of the activated complex: it has its internal degrees of freedom $J_q^{i*}(\alpha)$ and the energy of lateral interaction with surrounding neighbours ε_{ij}^* [32]. In this formulation, the search for difference derivatives is carried out at fixed ratios N_q^i/M_q and $N_{qp}^{ij}(r)/M_{qp}(r)$, since the appearance of the activated complex occurs in the elementary stage when one reagent is converted into another.

As a result, repeating verbatim the procedure for deriving expressions for elementary process rates in the TARR framework for non-ideal reaction systems [32], we obtain [143], taking into account the procedure of the sections 30, 32 for the mixture and 33, that

$$\beta\mu_q^{i*} = \frac{1}{2}\left[\left(1 - \sum_{r=1}^{R}\sum_{p=1}^{t(r)} z_{qp}(r)\ln\frac{N_q^i(N_q^i+1)}{M_q(M_q+1)}\right) + \sum_{r=1}^{R}\ln\frac{(M_{qq}(r)-z_{qq}(r))![(N_{qq}^{iV}(r)+z_{qq}(r))/2!]^2}{(M_{qq}(r)+z_{qq}(r))![(N_{qq}^{iV}(r)-z_{qq}(r))/2!]^2} + \right.$$
$$\left. + \sum_{r=1}^{R}\sum_{p=1,p\neq q}^{t^*(r)}{}'\ln\frac{(M_{qp}(r)-z_{qp}(r))![(N_{qp}^{iV}(r)+z_{qp}(r))!]}{(M_{qp}(r)+z_{qp}(r))![(N_{qp}^{iV}(r)-z_{qp}(r))!]}\right] \qquad (47.1)$$

where μ_q^{i*} is the chemical potential of the activated complex of the single-site stage $i \to$ *product*. Note that in square brackets the first term is normalized to the number of sites of type q; while the second and third terms are normalized to the corresponding numbers of pairs of sites qq and qp. This expression is in complete agreement with expression (33.13) for the equilibrium distribution of molecules in volume-limited systems: equating the elementary process rates in the forward and backward directions, we obtain expression (33.13). For example, for the adsorption stage $\beta\mu_q^i = \beta(\mu_q^{i*} - \mu_q^{V*})$. Substitution of the formula (47.1) in equation (20.1) leads to the appearance of the terms $U_f^i(\alpha)$ in (20.4), (20.5). Similarly, the appearance of local rates of two-site stages is considered.

The generalization of the equations for the pair functions (33.14) leads to the following relationships written out in general form for all pairs in terms of a coefficient $\Delta_{qp}^{ij} = 1 - \Delta_{qp}(1-\Delta_{ij})/2$ that depends on two Kronecker symbols for the types of sites and sorts of particles in the pair ($\Delta_{qp}^{ij} = 1/2$ for $q = p$ and $i \neq j$; and $\Delta_{qp}^{ij} = 1$ for other cases),

$$\beta\delta\varepsilon_{qp}^{ij}(r) = \frac{1}{2}\ln\frac{\Delta_{qp}^{ij}N_{qp}^{ij}(r)(\Delta_{qp}^{ij}N_{qp}^{ij}(r)+1)}{\Delta_{qp}^{iV}N_{qp}^{iV}(r)(\Delta_{qp}^{iV}N_{qp}^{iV}(r)+1)} \qquad (47.2)$$

As above, the symbol $\delta\varepsilon_{qp}^{ij}(r) = \varepsilon_{qp}^{*ij}(r) - \varepsilon_{qp}^{ij}(r)$ reflects the difference in energy between $\varepsilon_{qp}^{*ij}(r)$ the activated complex and $\varepsilon_{qp}^{ij}(r)$ the ground states of the central molecule i interacting with its neighbours.

Equation (47.2) connects all pairs (*ij*) with pair (*iV*) through the energy parameter.

For dynamic variables in small volumes, the same $X_k(\text{fl}) = X_k + \Delta X_k$ bonds should be performed between the average values of the dynamic variables describing the most probable state of the molecules and the fluctuation corrections (33.15) reflecting the boundedness of the system that were in the equilibrium state, i.e. in dynamics it is a question of considering the connection $\partial X_k(\text{fl})/\partial t = \partial X_k/\partial t + \partial \Delta X_k/\partial t$ throughout the process in the form where the kinetic equations (20.4) and (20.5) are used to calculate the most probable states of the molecules X_k.

The calculation of the fluctuation corrections requires a consideration of the dynamic analogue of expression (33.15), reflecting only those changes in the system states during the time *dt* which are compatible with the stages in the equations (20.4) and (20.5). In this case, it is required to calculate the time evolution of the dynamic corrections ΔX_k by means of functional derivatives with respect to the variables ΔX_k on the basis of the expressions (47.1) and (47.2)

$$\partial \Delta X_k/\partial t = -\frac{1}{2}\sum_{b=1}^{T_D}\frac{\partial \partial \ln \det *}{\partial t \partial X_b}\frac{\partial X_b}{\partial \ln(\lambda_k *)} \tag{47.3}$$

where the det* = det*($\{X_k\}$) determinant is constructed from the second derivatives with respect to the functions (47.1), (47.2) (in place of (33.13) and (33.14) in (33.15)); Here the time derivative of $\partial \ln \det^* = \partial X_b$ at each step is taken from the right-hand sides of the system of kinetic equations (20.4) and (20.5); and the values λ_k^* have the following meaning: $\lambda_k^* = \exp(\beta\mu_q^{i*})$ for N_q^i; $\lambda_k^* = \exp(\beta\delta\varepsilon_{qq}^{ij}(r))$ for $N_{qp}^{ij}(r)$; $\lambda_k^* = \exp(\beta\delta\varepsilon_{qp}^{ij}(r))$ for $N_{qp}^{ij}(r)$.

From the well-known fact that the entropy in a non-equilibrium state is less than the entropy in the equilibrium state, it follows naturally that fluctuations in the equilibrium state exceed fluctuations in the non-equilibrium states *X*, so that the above expressions for equilibrium fluctuations can serve as upper estimates for the size kinetic fluctuations. All size estimates on the range of application of thermodynamics and on the minimum phase size obtained from equilibrium distributions remain in force. In this case, the flow of any property is related to the size of the cross-section area S_{ar}, through which the flow passes, and the length of the region *L* along the flow determines the gradient of the property (density) under consideration. The smaller the length L, the smaller the volume $M = S_{ar}L$, in which the fluctuations are of interest.

Lattice and vibrational contributions of subsystems in dense phases differ sharply with their characteristic relaxation times (by more than 10 orders of magnitude). The lattice subsystem is non-equilibrium at low temperatures, and the vibrational contributions are practically always in equilibrium. Taking into account these regularities, it is possible to construct the size dependences for unary and paired DFs, and through them to calculate the thermophysical and thermodynamic characteristics of local solid bodies (mechanical modules, heat capacities, thermal conductivity and friction coefficients) and average for the entire small body. Under intense perturbations, one can also go on to describe the transfer of impulse under deformations and fractures [144].

The availability of kinetic equations opens the possibility of introducing dynamic exit criteria for local and complete equilibria with the goal of a transition from the thermophysical characteristics to thermodynamic ones. The equations (20.4) and (20.5) can be easily concretized when considering particular problems of dynamics and the transition to fluctuating equilibrium characteristics. We list the most general formulations of the problem: the relaxation of internal uniform drop regions to the equilibrium state; the allowance for fluctuations in the process of relaxation and in the equilibrium state; relaxation in microcrystals for a rigid lattice, taking into account the connection with the gas phase at the solid–vapour or liquid interface; adsorption on solid non-re-arrangeable surfaces of solids; adsorption on solid-tunable surfaces of solids, absorption in interstitials and other issues.

References

1. Landau L.D., Lifshitz E.M. Theoretical physics.VI. Hydrodynamics. Moscow, Nauka, 1986.
2. Sedov L.I., Continuum mechanics. Moscow, Nauka, 1970. V. 1.
3. Landau L.D., Lifshitz E.M. Theoretical physics. VII. Theory of Elasticity. Moscow, Nauka, 1987.
4. Reif F., Statistical physics. Moscow, Fizmatlit, 1977. [McGraw-Hill Book Comp. 1964]
5. Gurov K.P., Phenomenological thermodynamics of irreversible processes (Physical basis). Moscow, Fizmatlit, 1978.
6. Kvasnikov I.A., Thermodynamics and statistical physics. V. 1: Theory of equilibrium systems: Thermodynamics. Moscow, Editorial URSS, 2002.

7. Crack J., The Mathematics of Diffuion. Oxford: Oxford Univer. Press. 1975.
8. Mehrer H., Diffusion in solids. Dolgoprudny. Ed. House. Intellekt, 2011. [Springer – Verlag Berlin Heidelber, 2007].
9. Frenkel J., Introduction to the theory of metals. Moscow and Leningrad, GITTL, 1950.
10. Bokshtein B,S,, Bokshtein S,Z,, Zhukhovitsky A,A,m Thermodynamics and kinetics of diffusion with solids. Moscow, Metallurgiya, 1974.
11. Kitel Ch., Introduction to Solid State Physics. Moscow, Nauka, 1978.
12. Girifalco, L.A., Statistical Mechanics of Solids. Moscow, Mir, 1975. [Oxford, New York: Oxford Univ. Press, 1973].
13. Borovskii, I.B., Gurov, K.P., Machukova, I.D., and Ugaste, Yu. E., Processes of Inter-Diffusion in Alloys, Moscow: Nauka, 1973.
14. Tovbin Yu.K., Komarov V.N., Rus. Chem. Bull., 2013, V. 62, No. 12 P. 2620.
15. Tovbin Yu. K., Komarov V.N., Fiz. Tverd. Tela. 2014. V. 56. No. 2. P. 337. [Phys. Solid State 56, 341 (2014)].
16. Tovbin Yu.K., Titov S.V., Komarov V.N., Fiz. Tverd. Tela, 2015. Vol. 57. No. 2. P. 342. [Phys.Solid State 57, 360 (2015)].
17. Neumann G., Tuijn C., Self-Diffusion and Impurity Diffusion in Pure Metals: Handbook of Experimental Data. Pergamon Materials Series 14, 2009.
18. Tishchenko N.P.. Physica Status Solidi (a). 1982. 73, 279 (1982).
19. Berne A., Boato G., De Paz M., Nuovo Cimento 24, 1179 (1962).
20. 20. Berne A., Boato G., De Paz M., Nuovo Cimento 46, 182 (1966).
21. Nachtrieb N.A., E. Catalano, J.A. Weil. J. Chem. Phys. 20, 1185 (1952).
22. Mundy J.N., Barr L.W., F.A. Smith., Phil. Mag. 15, 411 (1967).
23. Lundy, T.S., Murdock, J.F., J. Appl. Phys. 33, 1671 (1962).
24. Rothman N.L. et al., Phys. Stat. Sol. 39, 635 (1970).
25. Herzig Ch., Eckseler H., Bussmann W., Cardis D., J. J. Nucl. Mater. 69/70 61 (1978).
26. Rothman S.J., Peterson N. L., Physica status solidi (b) 1969 V. 35. No. 1. P. 305.
27. Tovbin Yu.K., Zh. fiz. khimii. 2012. V. 87. No. 9. P. 1461. [Russ. J. Phys. Chem. A 86, 1356 (2012)].
28. Prigogine I., Introduction to the thermodynamics of irreversible processes. Izhevsk: 'Regular and chaotic dynamics', 2001.
29. de Groot S., Mazur P., Non-equilibrium thermodynamics. Moscow, Mir. 1964.
30. Haase R., Thermodynamics of irreversible processes. Moscow, Mir, 1967. [Dr. Dietrich Steinkopff, Darmstadt, 1963].
31. Gurov K.P., Kartashkin B.A., Ugaste Yu.E., Mutual diffusion in multiphase metal systems. Moscow, Nauka, 1981.
32. Tovbin Yu.K., The theory of physical and chemical processes at the gas-solid interface. Moscow, Nauka, 1990. 1990. [CRC, Boca Raton, Florida, 1991].
33. Tovbin Yu.K., Progress in Surface Science. 1990. V. 34, No. 1-4. P. 1-236.
34. Tovbin Yu.K., The Molecular Theory of Adsorption in Porous Solids, Moscow, Fizmatlit, 2012. [CRC Press, Taylor & Francis Group, 2017].
35. Mason E., Malinauskas A., Transport in porous media: a model of a dusty gas. Moscow, Mir, 1986.
36. Bird R. B., Stewart W., Lightfoot E. N., Transport Phenomena. – Moscow: Khimiya, 1974. – 687 p. [John Wiley and Sons, New York, 1965].
37. Geguzin Ya.E. Diffusion zone. Moscow, Science. 1979, 344 p.
38. Bogolyubov N.N. Problems of dynamic theory in statistical physics. Moscow, Gostekhizdat, 1946.
39. Gurov K.P. Foundations of the kinetic theory. Moscow, Nauka, 1967. 460 p.

40. Tovbin Yu.K., Zh. fiz. chemistry. 2014. T. 88. No. 2. P. 261. [Rus. J. Phys. Chem. A, 2014, Vol. 88, No. 2, pp. 213].
41. Tovbin Yu.K., Teoret. Fundamentals of chemical. technol. 2013. T. 47. No. 6. P. 672.
42. Chapman S., Kauling T. Mathematical theory of non-uniform gases. Moscow, Izd-vo inostr. lit., 1960. 510 p.
43. Huang, K., Statistical Mechanics. Moscow, Mir, 1966. 520 p.
44. Ferziger, J., Kaper, G., Mathematical theory of transport processes in gases. Moscow, Mir. 1976. p. [North-Holland, Amsterdam,1972]
45. Hirschfelder J. O., Curtiss Ch. F., Bird R. B., Molecular theory of gases and liquids. – Moscow: Inostr. Lit., 1961. – 929 p. [Wiley, New York, 1954].
46. Monin A.S., Yaglom A.M., Statistical hydrodynamics. Moscow, Nauka, 1965. Part 1; 1967. Part 2.
47. Eremin E.N., Fundamentals of chemical kinetics. Moscow, Vysshaya shkola, 1976. 374 p.
48. Glasston S., Laidler K.J., Eyring H., Theory of absolute reaction rates. Moscow, IL, 1948 [Princeton Univ. Press, New York, London, 1941].
49. Nikitin E.E., Theory of elementary atomic-molecular processes in gases. Moscow, Khimiya, 1970.
50. Eyring H., Lin S.G., Lin S.M., Fundamentals of chemical kinetics. Moscow, Mir, 1983.
51. Tovbin Yu.K., Zh. fiz. khimii. 2015. V. 89. No. 9. P. 1347. [Rus. J. Phys. Chem. A, 2015, Vol. 89, No. 9, P. 1507].
52. Manning J. Kinetics of the diffusion of atoms in crystals. Moscow, Mir, 1971.
53. Lebed' I.V., Khim. fizika. 1996. V. 15. No. 6. P. 64.
54. 54. Tovbin Yu.K., Zh. fiz. khimii. 2017. 91. No. 3. P. 381. [J. Phys. Chem. A 91, 403 (2017)]
55. Gibbs J.W., Thermodynamics. Statistical mechanics. Moscow, Nauka, 1982.
56. Zubarev D.N., Non-equilibrium statistical thermodynamics. Moscow, Nauka, 1971.
57. Landau L.D., Livshits E.M., Theoretical physics. V. 5. Statistical physics. Moscow, Nauka, 1964.
58. Keller L., Fridman A.D., Proc. 1st Intern. Congr. Appl. Mech., Delft., 1924.
59. Zhigulev V.N., Tr. TsAGI. 1969. V. 1135. P. 45.
60. Zhigulev V.N., Teor. Matem. Fiz., 1971. V. 7. No. 1. P. 101.
61. Tovbin Yu.K., Khim. Fizika. 2002. V. 21. No. 1. P.83.
62. Tovbin Yu.K., Zh. fiz.khimii. 2002. V. 76. No. 1. P.76. [Russ. J. Phys. Chem. 2002. V. 76. № 1. P. 64].
63. Tovbin Yu.K., Khim. Fizika.. 2004. V. 23. №12, C. 82.
64. Landau, L.D., Lifshitz E.M. Theoretical physics. IX. Statistical physics. Part 2. Moscow, Nauka, 1966.
65. Naimark O.B., Pis'ma Zh. Eksper. Teor. Fiz. 1998. V. 67. No. 9. P. 714.
66. Naimark O.B., Pis'ma Zh. Eksper. Teor. Fiz. 1997. V. 23. No. 13. P. 81.
67. Naimark O.B., et al., Fiz. mezomekhanika. 2009. V. 12. No. 4. P. 47.
68. Panin V.E., Egorushkin V.E., Ibid. 2008. V. 11. No. 2. P. 9.
69. Butyagin P.Yu., Usp. khimii. 1984. V. 53. No. 11. P. 1769.
70. Butyagin P.Yu., Chemical physics of solids. Moscow, Publishing House of Moscow State University, 2006.
71. Tomashev N.D., Chernova G.P., Theory of corrosion and corrosion-resistant structural alloys. Moscow, Metallurgiya, 1986.
72. Zhilyaev A.P., Pshenichnyuk A.I., Superplasticity and grain boundaries in ultrafine-grained materials. Moscow, Fizmatlit, 2008.

73. Klimontovich Yu.L., Turbulent motion and the structure of chaos. Moscow, Nauka, 1990.
74. Timashev S.F., Flicker-noise spectroscopy. Information in chaotic signals. Moscow, Fizmatlit, 2007.
75. Kogan Sh.M., Usp. fiz. nauk. 1985. V. 145. No. 2. P. 285.
76. Weissman M.B., Rev. Mod. Phys. 1988. V. 60. P. 537.
77. Zarkhin L.S., et al., Usp. khimii. 1989. V. 58. P. 644.
78. Semenov N.N., Chain reactions. Leningrad. ONTI, 1934.
79. Zel'dovich Ya.B., et al., Mathematical theory of combustion and explosion. Moscow, Nauka, 1980.
80. Frank-Kamenetsky D.A., Diffusion and heat transfer in chemical kinetics. Moscow, Nauka, 1987.
81. Dedkov G.V., Usp. fiz. nauk, 2000. V. 176. No. 6. P. 585.
82. Lifshitz I.M., Selected works. Physics of Real Crystals and Disordered Systems. Moscow, Nauka, 1986.
83. Suzdalev I.P., Nanotechnology: physical chemistry of nanoclusters, nanostructures and nanomaterials. KomKniga, Moscow, 2006.
84. Cheremskoi P.G., Slezov V.V., Betekhtin V.I., Pores in a solid body. Moscow, Energoatomizdat, 1990.
85. Zhilyaev A.P., Pshenichnyuk A.I., Superplasticity and grain boundaries in ultrafine-grained materials. Moscow, Fizmatlit. 2008.
86. Rusanov A.I., Thermodynamic basis of mechanochemistry. St. Petersburg, Nauka, 2000.
87. Sirotin Yu.I., Shaskol'skaya M.P., Fundamentals of crystallophysics. Moscow, Nauka, 1979.
88. Tovbin Yu.K., Zh. fiz. khimii. 2017. V. 91. No. 8. P. 1243. [Rus. J. Phys. Chem. A, 91, No. 8, P. 1357 (2017)].
89. Tovbin Yu.K., Zh. fiz. khimii. 2010. P. 84. No. 2. P. 231. [Rus. J. Phys. Chem. A, 84, No. 8, P. 1788 (2017)].
90. Tovbin Yu.K., Zh. fiz. khimii. 2014. T. 88. No. 11. P. 1788. [Rus. J. Phys. Chem. A, 88, No. 11, P. 1965 (2014)].
91. Ono S., Kondo S., Molecular theory of surface tension. Moscow, IL, 1963. [Handbuch der Physik, Vol X (Springer) 1960].
92. Rowlinson J., Widom B., Molecular theory of capillarity. Moscow, Mir, 1986. [Oxford: Clarendon Press, 1982].
93. Irving J.H., Kirkwood J.G., J. Chem. Phys. 1950. V. 18. P. 1950.
94. Schofield P., Chem. Phys. Lett., 1966. V. 62. P. 413.
95. Grant M., Desia R.C., Molec. Phys. 1981. V. 41. P. 1035.
96. Hirschfelder J., Curtis C., Byrd R. Molecular theory of gases and liquids. Moscow, IL, 1961.
97. Rice O.K., J. Phys. Chem. 1927. V. 31. P.207.
98. Prigogine I., Defay R., J. Chim. Phys. 1949. V. 46. P. 367.
99. Defay R., J. Chim. Phys. 1954. V. 51. P. 299.
100. Summ B.D., Fundamentals of Colloid Chemistry. 2007. Moscow, Akademiya.
101. Shikhmurzaev Y.D., Capillary Flows with Forming Interfaces. 2007, Taylor & Francis.
102. Shikhmurzaev Y.D., J. Fluid Mech. 1997. V. 334. P. 211.
103. Blake T.D., J. Colloid Interface Sci. 2006. V. 299. P. 1.
104. Dussan E.B., J. Fluid. Mech. 1976, V. 77, 665.
105. Rasmessen D.H.J., J. Chem. Phys. 1986. V. 85. P. 2272.

106. Hua X.Y., Rosen M.J., Journal of Colloid and Interface Science. 1988. V. 124. No.2. P. 652.
107. Dynamics of surfactant self-assemblies: micelles, microemulsions, vesicles, and lyotropic phases. ed. R. Zana, Taylor & Francis, 2005.
108. Ross S., Morrison I.D., Colloidal systems and interfaces. Wiley-Interscience publication, New York, Toronto, 1988.
109. Adamczyk Z., Journal of Colloid and Interface Science. 2000. V. 229. No. 2. P. 477.
110. de Gennes P.G., Rev. Mod. Phys. 1985, 57, 827–863.
111. Blake T.D., Shikhmurzaev Y.D., J. Colloid Interface Sci. 2002. V. 253. P. 196–202.
112. Khaidarov G.G., et al., Vestnik Sankt-Peterb. Univ., Series 4. 2011. Issue 1. P. 3–8.
113. Khaidarov G.G., et al., Vestnik Sankt-Peterb. Univ. Series 4. 2011. Issue. 1. P. 24–2.
114. Filippov L.K., Filippova N.L., Journal of Colloid and Interface Science. 1997. V. 187. P. 352.
115. Lukyanov A.V. Likhtman A.E., J. Chem. Phys. 2013, 138, 034,712.
116. Eriksson J.C., Surface Sci. 1969. V. 14. P. 221.
117. Adamson A. W., The Physical Chemistry of Surfaces,Mir, Moscow, 1979. [Wiley, New York, 1976].
118. Rusanov A.I., Surface Science Reports. 1996 Vol. 23, Nos. 6–8.
119. Rusanov A.I., Surface Science Reports. 2005. Vol. 58. P. 111.
120. Benson G.G., Yun K.S.S., The solid-gas interface. Ed. E. Alison Flood, Marcel Dekker, Inc., New York, 1967.
121. Dunning W., Interphase boundary gas-solid. Ed. E. Flad. Publishing house Mir, Moscow 1970. P. 230.
122. Sparnaay M.J., Surface Sci. Reports. 1984. V. 4. N 3/4. P. 103–269.
123. Cammarata R.C., Progress in Surface Sci. 1994. V. 46. No. 1. P. 1.
124. Ibach H., Surface Science Reports. 1997. V. 29. P. 193.
125. 125. Tovbin Yu.K., Zh. Fiz. Khimii. 2014. V. 88. No. 7–8. P. 1266. [Rus. J. Phys. Chem. A, 88, No. 7-8, P. 1438 (2014)].
126. Born M., Huang K., Dynamic theory of crystal structures. Moscow, IL, 1958.
127. Leibfried G., Microscopic theory of the mechanical and thermal properties of crystals. Moscow and Leningrad, GIFML, 1963. [Handbuch der Physik, Band 7, Springer, Berlin, 1955, Vol. 2].
128. Maradudin A.A., Theoretical and Experimental Aspects of the Effects of Point Defects and Disorder on the Vibrations of Crystals (Academic Press, New York, London, 1966; Mir, Moscow, 1968).
129. Kosevich A,M,, Fundamentals of the Mechanics of the Crystal Lattice. Moscow, Nauka, 1972.
130. Khachaturyan A.G., The theory of phase transitions and the structure of solids. Moscow, Nauka, 1974.
131. Katznelson A.A., Olemskoy A.I., Microscopic theory of non-uniform structures. Moscow, Izd-vo MGU, 1987.
132. Coussy O., Poromechanics. John Wiley & Sons, Ltd. Chichester, 2004.
133. Fischer-Cripps A.C., Introduction to contact mechanics. 2ed., Springer Science + Business Media, LLC. 2007.
134. Popov V.L., Mechanics of contact interaction and physics of friction. Moscow Fizmatlit, 2013.
135. Aizikovich S.M., et al., Contact problems of the theory of elasticity for non-uniform media. Moscow, Fizmatlit, 2006.
136. Aleksandrov V.M., Chebakov M.I., Introduction to the mechanics of contact interactions. – Rostov-on-Don: Publishing House of the Central Research and Develop-

ment Centre, 2007.

137. Johnson K.L., Contact mechanics. Cambridge Univer. Press, Cambridge London, New York, New Rochelle, 1985.
138. Suzdalev I.P., Nanotechnology: physical chemistry of nanoclusters, nanostructures and nanomaterials. KomKniga, Moscow. 2006.
139. Tovbin Yu.K., Rabinovich A.B., Zh. fiz. khimii. 2011. V. 84. No. 12. P. 2366. [Rus. J. Phys. Chem. A, 84, No. 12, P. 2166 (2011)].
140. Tovbin Yu.K., Titov S.V., Zh. fiz. khimii. 2013. V. 87. No. 1. P. 77. [Rus. J. Phys. Chem. A, 87, No. 1, pp. 93 (2013)].
141. Tovbin Yu.K., Zh. fiz. khimii. 2010. V. 84. No. 11. P. 2182. [Rus. J. Phys. Chem. A, 84, No. 11, P.1993 (2013)].
142. Hill T.L., Statistical Mechanics, Principles and Selected Applications, NewYork, McGraw-Hill Book Comp., Inc., 1956.
143. Tovbin Yu.K., Zh. fiz. khimii. 2015. V. 89. No. 3. P. 551. [Rus. J. Phys. Chem. A, 89, No. 3, P. 547 (2017)].
144. Ionov V.N., Selivanov V.V., Dynamics of destruction of a deformed body. Moscow,Mashinostroenie, 1987.

6

Elementary stages of the evolution of the system

Fundamentally important for constructing probabilities $W_\alpha(\{I\}\rightarrow\{II\})$ is the use of models describing the elementary process within the framework of the theory of absolute reaction rates (TARR) or the theory of the transition state [1-3]. For the first time, TARR was used for the W_α functions in Eq. (20.1) for non-ideal systems in adsorption studies [4–8] and label transfer in a liquid [9]. Later, this theory was extended to a wide variety of processes [10–17]. The velocities of all elementary motions will be described in the framework of this theory.

In the kinetic theory, for any phases, the key is the problem of self-consistency of the expressions for the rates of elementary reactions (stages) and the equilibrium state of the reaction system. The essence of this statement is that, equating the expressions for the reaction rates of any of the stages occurring in the forward and backward directions, equations describing the equilibrium distribution of the molecules of the given system must be obtained. Within the framework of the LGM (lattice gas model), it is possible to find the conditions under which these self-consistency conditions for the description of reaction rates and system equilibrium are fulfilled [10, 14–17]. This question is discussed successively for elementary processes occurring on one (section 50) and two (section 52) sites of the lattice system, taking into account the interaction in quasi-chemical approximation (QCA) for nearest neighbours, and then (section 54) for any interaction radius between the neighbours R.

For an entire non-uniform lattice structure, the self-consistency conditions will be satisfied if they are satisfied for each site and for each pair of sites. Therefore, when we consider the self-consistency of the expressions for the rates of stages, we will operate on specific individual sites, i.e. before using the averaging procedure with different distribution functions for the composition and structure of the non-uniform structure [14–17]. Section 49 shows how the rates for single-site processes are obtained – these are the terms $U_f^i(\alpha)$ in equation (20.4). These terms are not closed – in their right-hand parts there are unknown correlators of a higher (second) dimension. To close Eq. (20.4), it is necessary to construct new kinetic equations for unknown correlators (20.5) or to express them through correlators of lower dimensionality – the latter can be done if the pair functions belong to the equilibrium distribution (Section 16). This principle is common to a system of interacting particles.

48. The rate of elementary stages

The basic idea of TARR is the relationship between the concentration θ^* (the number of activated complexes (AC) per unit volume) with the rate of the elementary process $U = \theta^* \nu$, where ν is the frequency of crossing the activation barrier (s^{-1}). For translational motion [1–3], $\nu = u_t/b$, where u_t is the average velocity of motion of an AC of mass m^* equal to $u_t = (kT/2\pi m^*)^{1/2}$, b is the length of the activation barrier.

The concentration of AC at the top of the barrier can be written as follows: $\theta^* = \theta F_t$, where θ is the concentration of AC after the replacement of one vibrational degree of freedom of movement by translational motion, $F_t = (2\pi m^* kT)^{1/2} b/h$; h is the Planck constant. Hence we obtain that the translational velocity $U = \theta kT/h$. Or, introducing in the usual manner the specific rate of the elementary process $K_i = U/\theta_i$ for the motion of the activated complex formed from particle i, we have that $K_i = \theta\, kT/(h\theta_i)$.

The ratio $\theta/\theta_i = F^*/F_i$ is expressed [1–3] in terms of sums over the states AC (F^*) and molecules in the ground state (F_i), which leads to the standard form for the rate constant of the elementary motion $K_i = kTF^*/(F_i h)$ for the gas phase.

As a result, the type of motion is concretized only in the expressions for the sums over the states of the AC and the molecule in the ground state. For a rarefied phase $F^*/F_i = F_t$. For non-ideal reaction systems this ratio varies, and the influence of neighbouring molecules is additionally required.

The reaction rate in the theory of the transition state is defined as [1–3]:

$$U_i = \chi w \theta_i^* / b, \tag{48.1}$$

where χ is the transmission coefficient, w is the average velocity of passage of the barrier of length b, $w = (2\pi M^*\beta)^{-1/2}$, M^* is the mass of the complex, U_i is the rate referenced to one site of the structure. The problem reduces to the calculation of the concentration of AC θ_i^* for the monomolecular process and θ_{ij}^* for the bimolecular process.

For ideal reaction systems (in the absence of the influence of lateral interactions) the rates of elementary stages, mono- and bimolecular reactions are described within the framework of the law of mass action

$$U_i = K_i\theta_i\,, \qquad U_{ij} = K_{ij}\theta_i\theta_j\,. \tag{48.2}$$

where K_i and K_{ij} are the rate constants of elementary processes (stages) that characterize the specific rates of elementary processes:

$$K_i = K_{ij}^0 \exp\left(-E_i / k_B T\right) = \kappa \frac{kT}{h}\frac{F_i^*}{F_i}\exp\left(-E_i / kT\right), \tag{48.3}$$

$$K_{ij} = K_{ij}{}^0 exp\left(-E_{ij} / k_B T\right) = \kappa \frac{kT}{h}\frac{F_{ij}^*}{F_i F_j}\exp\left(-E_{ij} / kT\right),$$

where K_i and K_{ij}^0 are the pre-exponentials of the rate constants, E_i and E_{ij} are the activation energies of the reaction $i \rightarrow$ *product* and $i + j \rightarrow$ *products*; F_i and F_j are the statistical sums (sums over the internal states) of the original molecules, F_i^* and F_{ij}^* are the statistical sums of AC calculated for all degrees of freedom except for the 'reaction path'.

It is assumed here that the number of sites of the structure does not change during the reaction, and the concentration of particles can be characterized as the 'degree of filling' of the surface θ_i. In the derivation of the equations (48.1)–(48.3), it is assumed that the equilibrium distribution of molecules in the reaction system is realized, and that the chemical transformation stage is limited. It also assumes: 1) the absence of diffusion transport at the macroscopic level (uniform distribution over the macrovolume), (2) the absence of the influence of external fields, (3) the absence of a diffusion-controlled regime of the reaction at the molecular level, (4) the absence of influence of intermolecular interactions, and (5) the fraction of particles reacting per unit time is so small that it does not distort the equilibrium distribution of molecules on the surface.

However, the use of thermodynamic approaches in problems of kinetics for non-ideal systems (Section 10), which expresses the use of the concepts of activity coefficients for reagents and activated complexes, leads to contradictions (Section 64). Below we discuss the expressions for the rates of elementary stages in molecular theory based on the LGM (lattice gas model) [14–17] for non-ideal systems, which is applicable to all aggregate states of substances and interfaces, and will ensure their self-consistency with a description of the equilibrium state of the reaction system in the QCA.

In the molecular theory of non-ideal reaction systems, it is necessary to consider successively the entire spectrum of configurations of neighbouring molecules that can influence the course of the reaction for the reagents at the central sites under consideration and weigh the probability of realizing the elementary stage for each of the neighbours' configurations on the surface. Lateral contributions affect the probability of formation of ACs through a change in the activation energy, so the number of neighbours and the way they are located are important for the rate of the process.

49. One-site processes

The multiparticle (cooperative) nature of elementary processes is clearly manifested in the calculation of their rates. The reason for this is illustrated by the examples of mono- ($iZ \rightarrow product$) and bimolecular ($iZ+jZ \rightarrow products$) reactions that are realized in a two-component system (s = 2, i = particle A and vacancy V, here Z is the symbol of the reagent-containing site). The generalization of the expressions (48.2) and (48.3) in the case when the interaction of neighbouring particles is taken into account leads to a change in the concentration dependences of the rates of elementary processes.

In the absence of interaction between particles, the rate of the monomolecular reaction ($iZ \rightarrow product$) is expressed in terms of the equations (48.2) and (48.3), where θ_i is the concentration of component i (its mole fraction) which characterizes the probability that any lattice site is occupied by a particle of sort i. Let us discuss the process by the example of adsorption. If $i = A$, then U_A denotes the desorption rate; if $i = V$, then U_V is the adsorption rate. In this case, K_A is the desorption rate constant and K_V is the adsorption rate constant. To unify the form of the recording U_i, we will include the partial pressure of the molecules of the gas phase P_A in the effective adsorption rate constant K_V.

Consider a uniform lattice each site of which has z neighbours. For flat lattices, the sites correspond to adsorption centres, for which $z = 3$, 4, or 6; in the bulk lattices the sites correspond to the absorption centres and for them $z = 4$, 6, 8 or 12. In the simplest case, the influence of the interaction of neighbouring particles manifests itself primarily in the fact that, depending on the local composition of the neighbours around the central particle i, the height of the activation barrier of the elementary one-site reaction varies. The number of the nearest neighbours around the central particle i will be denoted by z. The arrangement of neighbouring particles A around the reagent i is given by the symbols n and σ, where n is the number of particles A in the first coordination sphere z around particles i; $\sigma(n)$ is the index that determines the way the particles are arranged for a fixed n (Fig. 16.1).

Denote by $E_i(n\sigma)$ the interaction energy of the particle i with its neighbours in the ground state, and by $E_i^*(n\sigma)$ the interaction energy of the activated reaction complex with its neighbours in the transition state. The difference $E_i^*(n\sigma) - E_i(n\sigma)$ is the change in the height of the activation barrier for a given local configuration $n\sigma$. At a fixed position of all neighbouring particles, the rate of the elementary process at one central site is written as follows:

$$U_i(n\sigma) = K_i\theta_i(n\sigma)\,F_i(n\sigma), \quad F_i(n\sigma) = \exp\left\{\beta[E_i^*(n\sigma) - E_i(n\sigma)]\right\}, \quad (49.1)$$

where $\theta_i(n\sigma)$ is the probability of existence of the configuration of neighbours $n\sigma$ around the reagent i. This quantity is the probability of a complex event consisting in the simultaneous arrangement of n particles A and $(z-n)$ vacancies in the configuration $n\sigma$ around the central particle i. To calculate the function $\theta_{ij}(n\sigma)$, it must be expressed in terms of reagent concentrations θ_i. The method of calculating $\theta_{ij}(n\sigma)$ through the concentration of reagents depends on the intermolecular interaction accounting approximation used (see Section 16).

The expression for the average velocity of the considered reaction U_i on a uniform surface (lattice) and pertaining to the unit surface is obtained by averaging (49.1) over all ways of arrangement of the particles. If we denote by $\alpha_i(n\sigma)$ the statistical weight of the configuration $n\sigma$ and assume that the rate constant K_i depends on the influence of neighbours only through the value of the activation energy determined by the function $F_i(n\sigma)$, then we get

$$U_i = \Sigma_n \Sigma_\sigma \alpha_i(n\sigma) U_i(n\sigma) = K_i \Sigma_n \Sigma_\sigma \alpha_i(n\sigma)\theta_i(n\sigma) F_i(n\sigma), \quad (49.2)$$

We assume that all the energy characteristics of the system can be expressed through pair interactions of the nearest neighbours [14]. We denote by ε_{ij} the interaction parameter of particles i and j in the ground state. Obviously, $\varepsilon_{ij} = \varepsilon_{ji}$. If one of the particles is a vacancy, then $\varepsilon_{iV} = 0$. The formation of the activated complex is associated with the transition of the particle from the ground state to the activated one to which the change in the position of the reagent relative to the positions of its neighbours corresponds. In this case the interaction energy of the particles changes. We denote by ε^*_{ij} the interaction parameter of the particle i, which is in the activated state with the neighbouring particle j in the ground state. For the activated complex $\varepsilon^*_{ij} \neq \varepsilon^*_{ji}$, since these parameters refer to different reagents and their neighbouring particles.

Taking into account only pair interactions, we write out the expression for the energy of the ground state for a two-dimensional lattice $z = 4$ (analogous expressions are satisfied for a two-dimensional lattice with $z = 3$ and three-dimensional lattices with $z = 4, 6, 8$):

$$E_i(n\sigma) = n\varepsilon_{iA} + (z - n)\varepsilon_{iV}, \quad (49.3)$$

i.e. $E_i(n\sigma)$ does not depend on the location of neighbouring particles. Both terms correspond to the interaction of the particle i with its neighbours (particles A and vacancies), $0 \leq n \leq z$.

For the activated complex, the analogue of expression (49.3) will be written as

$$E^*_{ij}(n\sigma) = n\varepsilon^*_{iA} + (z-n)\varepsilon^*_{iV},$$

This leads to the expression for the function $F_{ij}(n\sigma)$:

$$F_i(n\sigma) = \exp\left\{\beta\left[n\delta\varepsilon_{iA} + (z-n)\delta\varepsilon_{iV}\right]\right\}, \; \delta\varepsilon_{ij} = \varepsilon^*_{ij} - \varepsilon_{ij}. \quad (49.4)$$

Formula (49.4) characterizes a linear change in the energy in the exponent of the function $F_i(n\sigma)$ with a change in the number of neighbouring particles. Then in this approximation

$$U_i = K_i \Sigma_n \Sigma_s \alpha_i(n\sigma)\theta_i(n\sigma) \exp\left\{\beta\left[n\delta\varepsilon_{iA} + (z-n)\delta\varepsilon_{iV}\right]\right\}, \quad (49.5)$$

The calculation of the single-site reaction rate was reduced to determining the weight contributions $\alpha_i(n\sigma)$ and calculating the many-particle functions $\theta_i(n\sigma)$. In the general case of a non-equilibrium particle distribution, it is necessary to solve the corresponding kinetic equations. In the case of an equilibrium distribution of reagents, which is realized for the limiting stage under consideration in the kinetic regime (with a large activation energy of the reaction), these many-particle functions can be calculated in concrete equilibrium approximations for the interaction of particles. In order to take into account, as closely as possible, the real properties of the cooperative behaviour of interacting particles, it is necessary to take into account the correlation effects in their spatial distribution. This requirement is achieved if in calculating the functions $\theta_i(n\sigma)$ we use the pair distribution functions θ_{ij} characterizing the probability of finding particles i and j at neighbouring sites (the so-called quasi-chemical approximation (QCA)). As a result, the probabilities of multiparticle particle configurations $\theta_i(n\sigma)$ are approximated through the local probabilities θ_i for detecting particles i (unary distribution functions or the probability of detection of particles of a particular sort) and through the pair distribution functions. We express the equilibrium function of a particular distribution $\theta_i(n\sigma)$ in the following form:

$$\theta_i(n\sigma) = \theta_i t_{iA}^n t_{iV}^{z-n}, \tag{49.6}$$

where $t_{ij} = \theta_{ij}/\theta_i$ is the conditional probability of finding the particle j near the particle i; $\theta_{iA} + \theta_{iV} = \theta_i$ or $t_{iA} + t_{iV} = 1$. For the case under discussion, the transition to pair probabilities θ_{ij} means that the weights of configurations with different σ for n = const are equiprobable. As a consequence, the statistical weight of the distributions of particles A at neighbouring sites relative to the central particle i will be written as the number of combinations of z sites with n particles A: $\Sigma_\sigma \alpha_i(n\sigma) = C_z^n$. Then the formula (49.5) can be rewritten as

$$U_i = K_i\theta_i \sum_{n=0}^{z} C_z^n \Phi_{iA}^n \Phi_{iV}^{z-n} = K_i\theta_i S_i^{\,z},$$
$$S_i = \Phi_{iA} + \Phi_{iV},\ \Phi_{ij} = t_{ij}\exp(\beta\delta\varepsilon_{ij}). \tag{49.7}$$

In the function S_i the summation is carried out over all sorts of neighbouring particles. The S_i functions are the factors of the non-ideality function of the reaction system $\Lambda_i^* = S_i^z$, which reflect the influence of each neighbouring particle on the height of the activation barrier of the reaction. In the absence of interaction, the factors $S_i = 1$ and formula (49.7) goes over into the reaction rate for an ideal reaction system (Section 10).

50. Self-consistency of the rates of single-site stages with an equilibrium distribution of molecules

We confine ourselves to the case of fast mobility of reagents in a mixture consisting of s components, when they are equilibrated over the sites of a uniform surface [10]. The stages of adsorption, desorption, or chemical transformations involving adsorbed particles can be limiting. Let the elementary stage of the chemical transformation $A \leftrightarrow B$ be realized, which corresponds to the equality of rates in the forward and backward directions $U_A(\alpha) = U_B(\alpha)$ – the symbol α corresponds to a certain stage number in the general multi-stage process. In conditions of complete equilibrium all stages must be in dynamic equilibrium. With the example of expressions in QCA, we show that the expressions obtained for the reaction rates are self-consistent with a description of the equilibrium state of the reaction systems. That is, equating the rates of direct and reverse directions of each stage of the general process, we obtain the equations of the equilibrium distribution of particles.

The structure of the equations for the reaction rates $U_A(\alpha) = K_A(\alpha)V_A(\alpha)$, where $K_A(\alpha)$ is the reaction rate constant, and $V_A(\alpha)$ is the concentration component of the reaction rate. For an ideal reaction system $V_A(\alpha) = \theta_A$ and for a non-ideal reaction system it is described by the formulas (49.7).

The ratio of the rate constants determined by formulas (48.3) gives the equilibrium constant K_e and excludes the partition function of the activated complex which appears in the pre-exponents of the rate constants of the direct and reverse reactions. But still there is the characteristic of the activated complex $\varepsilon^*_{ij} \equiv \varepsilon_{ij}(\alpha)$, which is included in the non-ideality function $S_i(\alpha)$

$$K_1 = \frac{K_A(\alpha)}{K_B(\alpha)} = \frac{V_B(\alpha)}{V_A(\alpha)} = \frac{\theta_B[S_B(\alpha)]^z}{\theta_A[S_A(\alpha)]^z}. \tag{50.1}$$

To prove the self-consistency of the expressions, it is necessary that the ratio of the non-ideality functions $S_B(\alpha)/S_A(\alpha)$ does not depend on the parameters of the interaction of the activated complex with the neighbours ε^*_{ij}. To this end, we consider the equations in QCA [18]

$$\theta_A = \theta_B \exp\left[\beta(v_B - v_A)\right] S_B(A)^z,$$
$$S_B(A) = \sum_{j=1}^{V} t_{Bj} \exp\left[\beta(\varepsilon_{Aj} - \varepsilon_{Bj})\right] \quad (50.2)$$

$$\theta_{AC} = \theta_{BC} \exp\left[\beta(v_B - v_A + \varepsilon_{AC} - \varepsilon_{BC})\right] S_B(A)^{z-1}. \quad (50.3)$$

here $S_B(A)$ in the lower index is the sort of the considered central particle, and the symbol in brackets is the 'reference' sort of the mixture – often this symbol refers to the vacancy, and $t_{ij} = \theta_{ij}/\theta_i$. Equation (50.2) is the equation (18.4), and equation (50.3) is a different form of the equation (18.1) [14,18].

Eliminating the exponent $\exp[\beta(v_i - v_s)]$ from the equations (50.2) and (50.3), we obtain a connection between the pair and unary DFs

$$t_{AC} = t_{BC} \exp\left[\beta(\varepsilon_{AC} - \varepsilon_{BC})\right] / S_B(A). \quad (50.4)$$

This expression shows that the denominator does not depend on the type of the neighbouring molecule C, and the normalization condition is satisfied

$$\sum_{j=1}^{s} t_{Aj} = \sum_{j=1}^{s} t_{Bj} \exp\left[\beta(\varepsilon_{Aj} - \varepsilon_{Bj})\right] / S_B(A) = 1. \quad (50.5)$$

Using (50.3), we prove the following identity

$$\sum_{j=1}^{s} t_{nj} \exp\left[\beta(\varepsilon_{pj} - \varepsilon_{nj})\right] = \sum_{j=1}^{s} t_{ij} \exp\left[\beta(\varepsilon_{pj} - \varepsilon_{ij})\right] / \sum_{j=1}^{s} t_{ij} \exp\left[\beta(\varepsilon_{nj} - \varepsilon_{ij})\right], \quad (50.6)$$

which allows one to move from one sort of central molecule to another. To prove this identity, it suffices to put $A = n$, $B = i$, $C = j$ in (50.4) and substitute it in the left-hand side of equation (50.6). This identity will be abbreviated as $S_n(p) = S_i(p)/S_i(n)$. (In the particular case, as a consequence of (50.6), we get that $S_n(p) = S_p(n)^{-1}$ or $S_n(p)S_p(n) = 1$).

In formula (50.1), from the equality of the rates of elementary reactions, equating the rates of the direct and inverse one-site stages $ZA \leftrightarrow ZB$, the ratio

$$S_B(\alpha) / S_A(\alpha) = \sum_{j=1}^{s} t_{Bj} \exp[\beta(\varepsilon^*_{Bj} - \varepsilon_{Bj})] / \sum_{j=1}^{s} t_{Aj} \exp[\beta(\varepsilon^*_{Aj} - \varepsilon_{Aj})].$$

It involves one activated complex between reaction products A and B, so its properties are the same, or $\varepsilon^*_{Aj} = \varepsilon^*_{Bj} \equiv \varepsilon^*_{(AB)j}$. This allows us to obtain from the identity (50.6) that

$$\left(\frac{\sum_{j=1}^{s} t_{Bj} \exp[\beta(\varepsilon^*_{Bj} - \varepsilon_{Bj})]}{\sum_{j=1}^{s} t_{Aj} \exp[\beta(\varepsilon^*_{Aj} - \varepsilon_{Aj})]}\right)^z \equiv \left(\frac{S_B(AB^*)}{S_A(AB^*)}\right)^z = \tag{50.7}$$

$$= \left(\sum_{j=1}^{s} t_{Bj} \exp[\beta(\varepsilon_{Aj} - \varepsilon_{Bj})]\right)^z = \left(S_B(A)\right)^z$$

which is exactly equal to the concentration component of the isotherm (50.2). This shows that regardless of the path of the process through a chemical reaction or through the exchange of molecules with a thermostat, the equilibrium state is the same.

In the chaotic approximation, the analog of equation (50.6) has the form

$$\sum_{j=1}^{s} \theta_j \exp\left[\beta(\varepsilon_{pj} - \varepsilon_{nj})\right] = \sum_{j=1}^{s} \theta_j \exp\left[\beta(\varepsilon_{pj} - \varepsilon_{ij})\right] / \sum_{j=1}^{s} \theta_j \exp\left[\beta(\varepsilon_{nj} - \varepsilon_{ij})\right] \tag{50.8}$$

and for the mean field approximation, which is obtained from the random approximation by the limiting transition $\beta\varepsilon_{ij}\to 0$, but without the limiting values of such a transition, the analog of equation (50.6) will be expressed as

$$\exp\left[\sum_{j=1}^{s} \beta(\varepsilon_{pj} - \varepsilon_{nj})\theta_j\right] = \exp\left[\sum_{j=1}^{s} \beta(\varepsilon_{pj} - \varepsilon_{ij})\theta_j\right] / \exp\left[\sum_{j=1}^{s} \beta(\varepsilon_{nj} - \varepsilon_{ij})\theta_j\right]. \tag{50.9}$$

These expressions also lead to the conclusion of self-consistency for single-site reactions due to the identity of the interactions of the activated complex with neighbours in the forward and reverse directions of the reaction, or $\varepsilon^*_{Aj} = \varepsilon^*_{Bj} \equiv \varepsilon^*_{(AB)j}$.

51. Two-site processes

Similarly, we can consider the rate of the bimolecular reaction $iZ+jZ$, which is expressed in terms of the equations (48.2) and (48.3), where θ_i and θ_j are the concentrations of the components i and j, which characterize the probabilities that any lattice site is occupied by a particle of sort i and j. If $i = j = $ A, then U_{AA} is the desorption rate; at $i = A$, $j = V$ (or vice versa) U_{AV} is the migration or hopping rate of particle A; if $i = j = $ v, then U_{VV} is the adsorption rate. In this case, K_{AA} is the desorption rate constant, K_{AV} is the migration rate constant and K_{vv} is the rate constant of adsorption. As above, in order to unify the recording form, the values of U_{ij} will include the partial pressure of molecules of the gas phase P_{A2} in the effective adsorption rate constant.

The influence of the interaction of neighbouring particles is manifested primarily in the fact that, depending on the local composition of neighbours around the central pair ij, the height of the activation barrier of the elementary reaction between reagents i and j varies. The number of the nearest neighbours around the central pair ij will be denoted by z^*. We will characterize the arrangement of the neighbouring particles A around the reagents ij by the symbols n and σ, where n is the number of particles A in the first coordination sphere z^* around the pair of particles ij (respectively, $(z^*–n)$ of sites is occupied by vacancies); $\sigma(n)$ is the index that determines the way the particles are arranged for a fixed n (Fig. 51.1).

We denote by $E_{ij}(n\sigma)$ the interaction energy of the reactants ij with all its neighbours in the ground state, and by $E^*_{ij}(n\sigma)$ the interaction energy of the activated reaction complex $i + j$ with its neighbours in the transition state. The difference $E^*_{ij}(n\sigma) - E_{ij}(n\sigma)$ is the change in the height of the activation barrier for a given local configuration $n\sigma$. For a fixed position of all neighbouring particles, the rate of the elementary process at two central sites will be written as follows:

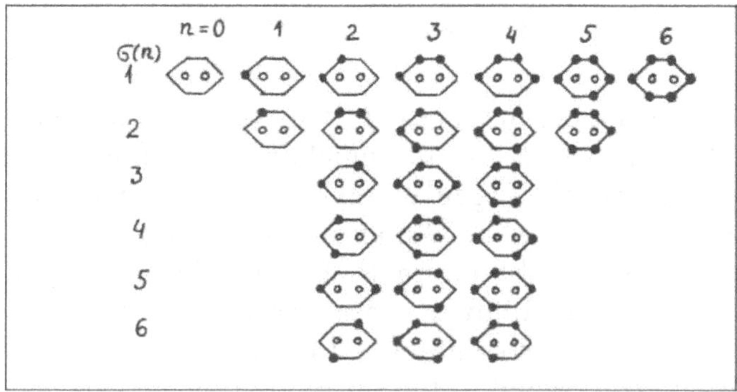

Fig. 51.1. The configurations of particles A in the first coordination sphere around two central particles on a square lattice; light circles — central particles, black circles — adsorbed particles A.

$$U_{ij}(n\sigma) = K_{ij}\theta_{ij}(n\sigma)\, F_{ij}(n\sigma),$$
$$F_{ij}(n\sigma) = \exp\left\{\beta\left[E_{ij}^{*}(n\sigma) - E_{ij}(n\sigma)\right]\right\}, \qquad (51.1)$$

where $\theta_{ij}(n\sigma)$ is the probability of the neighbours configuration $(n\sigma)$ around the reagents ij. This quantity is the probability of a complex event consisting in the simultaneous arrangement of n particles A and $(z^{*}-n)$ vacancies in the configuration $n\sigma$ around the central pair of particles ij. To calculate the function $\theta_{ij}(n\sigma)$, it must be expressed in terms of reagent concentrations θ_i and θ_j. The method of calculating $\theta_{ij}(n\sigma)$ through the concentration of reagents depends on the approximation of the intermolecular interactions used.

The expression for the average velocity of the considered reaction U_{ij} on a uniform surface (lattice) and pertaining to the unit surface is obtained by averaging (51.1) over all the arrangements of the particles. If we denote by $\alpha_{ij}(n\sigma)$ the statistical weight of the configuration $n\sigma$, then we obtain

$$U_{ij} = \Sigma_n \Sigma_\sigma \alpha_{ij}(n\sigma) U_{ij}(n\sigma) = K_{ij} \Sigma_n \Sigma_\sigma \alpha_{ij}(n\sigma)\theta_{ij}(n\sigma)\, F_{ij}(n\sigma), \qquad (51.2)$$

For a planar lattice $z = 4$, the complete set of configurations is shown in Fig. 51.1. Here it is assumed that the rate constant K_{ij} depends on the influence of neighbours only through the value of the activation energy determined by the function $F_{ij}(n\sigma)$.

As above, taking into account only pair interactions, we write out the expression for the energy of the ground state for a two-dimensional lattice $z = 4$:

$$E_{ij}(n\sigma) = \varepsilon_{ij} + n_1\varepsilon_{iA} + \left(z-1-n_1\right)\varepsilon_{iV} + n_2\varepsilon_{jA} + \left(z-1-n_2\right)\varepsilon_{jV}, \qquad (51.3)$$

i.e. $E_{ij}(n\sigma)$ does not depend on the method of location of neighbouring particles. Here n_1 and n_2 are the numbers of neighbouring particles A around the initial reagents i and j, which correspond to the configuration number of neighbouring particles σ, $n_1 + n_2 = n$. The first term ε_{ij} reflects the interaction of the initial reagents in the initial state. The two following terms correspond to the interaction of the particle i with its neighbours (particles A and vacancies). The last two terms correspond to the interaction of the particle j with its neighbours. The value $(z-1)$ appears due to the condition that for each reagent i or j one of the z adjacent sites is occupied by another reagent of the given bimolecular reaction j or i; and the remaining

$(z-1)$ neighbouring sites are occupied by neighbouring particles, $0 \le n_{1(2)}$ or n_1, $n_2 \le z-1$.

For the activated complex, the analogue of (51.3) is written as

$$E^*_{ij}(n\sigma) = n_1\varepsilon^*_{iA} + (z-1-n_1)\varepsilon^*_{iV} + n_2\varepsilon^*_{jA} + (z-1-n_2)\varepsilon^*_{jV}, \quad (51.4)$$

where there is no interaction between the 'reagents' i and j, since for the activated complex the interaction between the parts of the complex enters its internal energy. The other terms have the same structure. This leads to the expression for the function $F_{ij}(n\sigma)$:

$$F_{ij}(n\sigma) = \exp\{\beta[-\varepsilon_{ij} + n_1\delta\varepsilon_{iA} + (z-1-n_1)\delta\varepsilon_{iV} + n_2\delta\varepsilon_{jA} + (z-1-n_2)\delta\varepsilon_{jV}]\},$$

where $\delta\varepsilon_{ij} = \varepsilon^*_{ij} - \varepsilon_{ij}$, and in accordance with the configuration number σ, the connection between $n_1 + n_2 = n$ is fulfilled. Then in this approximation

$$U_{ij} = K_{ij}\exp(-\beta\varepsilon_{ij})\Sigma_n\Sigma_\sigma\alpha_{ij}(n\sigma)\theta_{ij}(n\sigma)\exp\{\beta[n_1\delta\varepsilon_{iA} + (z-1-n_1)\delta\varepsilon_{iV} + n_2\delta\varepsilon_{jA} + (z-1-n_2)\delta\varepsilon_{jV}]\}, \quad (51.5)$$

In the quasi-chemical approximation of the interaction between particles [10, 14, 15], the equilibrium distribution function $\theta_{ij}(n\sigma)$ is expressed as

$$\theta_{ij}(n\sigma) = \theta_{ij}t^{n_1}_{iA}t^{n_2}_{jA}t^{z-1-n_1}_{iV}t^{z-1-n_2}_{jV}, \quad (51.6)$$

where $t_{ij} = \theta_{ij}/\theta_i$ is the conditional probability described after formula (49.6). As above, the weights of configurations with different values of σ for n = const are equiprobable (for example, $\theta_{ij}(31) = \theta_{ij}(34)$ in Fig. 51.1). As a consequence, for the statistical weights of the particle distributions of A at neighbouring sites relative to both central reagents i and j, we can write $\Sigma_\sigma\alpha_{ij}(n\sigma) = C^{n_1}_{z-1}C^{n_2}_{z-1}$, where each of the factors is the number of combinations of $z-1$ sites in n_1 (or n_2) by the particles A.

Taking into account the above, the final formula can be written as

$$U_{ij} = K_{ij}\exp(-\beta\varepsilon_{ij})\theta_{ij}\sum_{n_1=0}^{z-1} C_{z-1}^{n_1}\Phi_{iA}^{n_1}\Phi_{iV}^{z-1-n_1}\times$$
$$\times\sum_{n_2=0}^{z-1} C_{z-1}^{n_{21}}\Phi_{jA}^{n_2}\Phi_{iV}^{z-1-n_2} = K_{ij}\exp(-\beta\varepsilon_{ij})\theta_{ij}\left(S_iS_j\right)^{z-1}. \quad (51.7)$$

The influence of intermolecular interactions is manifested through a change in the probability of encounter of reagents described by the function θ_{ij} ($\theta_{ij} \neq \theta_i\theta_j$), and through the non-ideality functions of the reaction system.

In the case of a two-dimensional lattice $z = 6$, expression (51.3) must be rewritten, since there are two sites simultaneously located at a distance of the first neighbours from both reagents. We denote the number of such neighbouring particles by n_3, and their energy contribution through $E_{ij}(n_3\sigma)$. Taking these contributions into account, formula (51.3) can be rewritten in the form

$$\begin{aligned} E_{ij}(n\sigma) &= \varepsilon_{ij} + n_1\varepsilon_{iA} + \left(z-3-n_1\right)\varepsilon_{iV} + \\ &+ n_2\varepsilon_{jA} + \left(z-3-n_2\right)\varepsilon_{jV} + E_{ij}(n_3\sigma), \\ E_{ij}(n_3\sigma) &= n_3(\varepsilon_{iA}+\varepsilon_{jA}) + \left(2-n_3\right)(\varepsilon_{iV}+\varepsilon_{jV}), \end{aligned} \quad (51.8)$$

where $0 \leq n_{1(2)} \leq z-3$, $0 \leq n_3 \leq 2$, and $n_1 + n_2 + n_3 = n$. Similarly, for a three-dimensional lattice $z = 12$.

For an activated complex, the analogue of expression (51.8) is written as

$$E^*_{ij}(n\sigma) = n_1\varepsilon^*_{iA} + (z-3-n_1)\varepsilon^*_{iV} + n_2\varepsilon^*_{jA} + (z-3-n_2)\varepsilon^*_{jV} + E^*_{ij}(n_3\sigma),$$
$$E^*_{ij}(n\sigma) = n_3(\varepsilon^*_{iA} + \varepsilon^*_{jA}) + (2-n_3)(\varepsilon^*_{iV}+\varepsilon^*_{jV}),$$

As a result, we obtain an expression for the rate of the bimolecular reaction

$$U_{ij} = K_{ij}\exp(-\beta\varepsilon_{ij})\Sigma_n\Sigma_\sigma\alpha_{ij}(n\sigma)\theta_{ij}(n\sigma)\exp\{\beta[n_1\delta\varepsilon_{iA}+(z-3-n_1)\times$$
$$\times\delta\varepsilon_{iV}+n_2\delta\varepsilon_{jA}+(z-3-n_2)\delta\varepsilon_{jV}+n_3(\delta\varepsilon_{iA}+\delta\varepsilon_{jA})+(2-n_3)(\delta\varepsilon_{iV}+\delta\varepsilon_{jV})]\}.$$

In this case, the contributions from the sites, which are simultaneously at the distance of the first neighbours from both reagents, are not expressed in terms of the functions S_i. The modified expressions for the factors in the concentration component of the reaction rate have a more complex form – see below in Section 54.

As a result of a similar averaging, without taking into account the mutual influence of the neighbours among themselves in the QCA for the mixture with any number of components, the following expressions are obtained for the velocities of mono and bimolecular elementary processes

$$U_i = K_i\theta_i S_i^z = K_i\theta_i\Lambda_i^*,\ \ S_i = \sum\nolimits_{k=1}^{s} t_{ik}\exp(\beta\delta\varepsilon_{ik}) \qquad (51.9)$$

$$U_{ij} = K_{ij}\exp(-\beta\varepsilon_{ij})\theta_{ij}(S_iS_j)^{z-1}, \qquad (51.10)$$

where $\delta\varepsilon_{ik} = \varepsilon_{ik}^* - \varepsilon_{ik}$ is ε_{ik}^* the interaction parameter between the activated complex (the particle i located at the site f in the transition state) of the elementary process α with the neighbouring particle k at the neighbouring site g; where $t_{ij} = \theta_{ij}/\theta_i$ is the conditional probability of finding the particle j near the particle i, calculated in the quasi-chemical approximation. In the derivation of expressions (51.9) and (51.10), it was assumed that the rate constants K_i and K_{ij} weakly depend on the density.

In the function, S_i summation is carried out over all sorts of neighbouring particles. The functions S_i are the factors of the non-ideality function of the reaction system $\Lambda_i^* = S_i^z$, which reflect the influence of each neighbouring particle on the height of the activation barrier of the reaction in (51.9).

A similar situation occurs in the case of a bimolecular reaction (51.10). The structure of the functions S_i is not related to the type of potential lateral interaction functions and the radius of the potential, it is determined by using the quasi-chemical approximation of the account of interactions. The number of functions S_i involved in the reaction rate in the form of factors is determined by the size of the coordination sphere z. For bimolecular reactions, this number of neighbours around two reagents equal to $2(z-1)$ practically doubles. The difference between the expression (51.10) and the formula (51.9) is that the influence of intermolecular interactions is manifested not only through the non-ideality functions, but also through the change in the probability of the encounter of reagents described by the function θ_{ij}. If the potential of interparticle interaction exceeds the nearest coordination sphere, contributions from the subsequent coordination spheres enter into expressions for U_i and U_{ij} in the form of additional factors of functions S_i belonging to different distances [12–17].

52. Self-consistency of the rates of two-site stages with an equilibrium distribution of molecules

The proof of self-consistency of the rates of the two-site stages with the equilibrium distribution of molecules for the two-site reaction is somewhat more complicated than for single-site reactions. This is due to the need to consider the influence of neighbouring molecules (via the non-ideality function) and to connect the probability of encountering reagents with the law of mass action through concentrations.

Let the products $C + D$ be obtained in the direct reaction $A + B$. Using the example of stage $A + C \leftrightarrow B + D$, we verify the self-consistency of the description of the kinetics and equilibrium of non-ideal reaction systems. Here it is necessary to follow the progress of the elementary process and indicate which of the products is obtained from the initial reagent, i.e. indicate that from A we receive C and B produces D. This is due to the structure of the activated complex and its properties should not depend on the reaction path. As above we denote $\varepsilon^*_{Aj} = \varepsilon^*_{Cj} \equiv \varepsilon^*_{(AC)j}$ and $\varepsilon^*_{Bj} = \varepsilon^*_{Dj} \equiv \varepsilon^*_{(BD)j}$, the sort of neighbours can be any $1 \le j \le s$. For a two-site stage, the rate is expressed as $U_{AB}(\alpha) = K_{AB}(\alpha)\exp(-\beta\varepsilon_{AB})V_{AB}(\alpha)$, where $K_{AB}(\alpha)$ is the pre-exponent of the rate constant and $V_{AB}(\alpha)$ is the concentration component of the two-site reaction rate, ε_{AB} is the energy of the lateral interaction of the reagents. It follows from the equality $U_{AC}(\alpha) = U_{BD}(\alpha)$ that the equilibrium constant of the reaction does not depend on the parameters of the activated complex.

$$K_2 = \frac{K_{AC}(\alpha)}{K_{BD}(\alpha)} = \frac{V_{BD}(\alpha)}{V_{AC}(\alpha)} = \frac{\theta_{BD}\exp(-\beta\varepsilon_{BD})[S_B(\alpha)S_D(\alpha)]^{z-1}}{\theta_{AC}\exp(-\beta\varepsilon_{AC})[S_A(\alpha)S_C(\alpha)]^{z-1}} \tag{52.1}$$
$$= \frac{\theta_{BD}\exp(-\beta\varepsilon_{BD})}{\theta_{AC}\exp(-\beta\varepsilon_{AC})}[S_B(A)S_D(C)]^{z-1}$$

where the identity (50.6) or (50.7) is used in the last equality, as above for the relations of non-ideality functions in single-site stages. This follows from the properties of the activated complex, $\varepsilon^*_{Aj} = \varepsilon^*_{Cj} \ne \varepsilon^*_{Bj} = \varepsilon^*_{Dj}$, $1 \le j \le s$. Consider in the detailed record a single factor for one bond from (z–1) neighbours around each reagent inside the central dimeric particle – the ratio of the concentration factors gives

$$\frac{\sum_{j=1}^{s} t_{Aj} \exp\left[\beta(\varepsilon^{*}_{(AC)j} - \varepsilon_{Aj})\right] \sum_{j=1}^{s} t_{Bj} \exp\left[\beta(\varepsilon^{*}_{(BD)j} - \varepsilon_{Bj})\right]}{\sum_{j=1}^{s} t_{Cj} \exp\left[\beta(\varepsilon^{*}_{(AC)j} - \varepsilon_{Cj})\right] \sum_{j=1}^{s} t_{Dj} \exp\left[\beta(\varepsilon^{*}_{(BD)j} - \varepsilon_{Dj})\right]} = \tag{52.2}$$
$$= \sum_{j=1}^{s} t_{Aj} \exp\left[\beta(\varepsilon_{Cj} - \varepsilon_{Aj})\right] \sum_{j=1}^{s} t_{Bj} \exp\left[\beta(\varepsilon_{Dj} - \varepsilon_{Bj})\right],$$

which excludes the presence of the properties of the activated complex in this respect, as well as for single-site reactions.

To transform the pair functions in (52.1), we multiply equation (50.3) on the right and left by $\theta_A\theta_B$, we write it in the form

$$\theta_{AC} = \theta_{BC}\theta_A \exp\left[\beta(\varepsilon_{AC} - \varepsilon_{BC})\right] / (\theta_B S_B(A)) = \theta_{BC}\varphi(A,B,C) \tag{52.3}$$

Taking into account that $\theta_{nm} = \theta_{mn}$, we represent the required relation with the form (we consider the general case of the function θ_{nm} and θ_{kd})

$$\theta_{kd} = \theta_{mn}\varphi(k,B,d) / \left[\varphi(n,d,B)\varphi(m,B,n)\right], \tag{52.4}$$

and since $\varepsilon_{nm} = \varepsilon_{mn}$, once again using the identity (50.6) to exclude the sort B, we obtain

$$\theta_{kd} = \theta_{mn} \frac{\theta_k\theta_d}{\theta_m\theta_n} \exp\left[\beta(\varepsilon_{kd} - \varepsilon_{mn})\right] \sum_{j=1}^{s} t_{dj} \times$$
$$\times \exp\left[\beta(\varepsilon_{nj} - \varepsilon_{dj})\right] \sum_{j=1}^{s} t_{kj} \exp\left[\beta(\varepsilon_{mj} - \varepsilon_{kj})\right]. \tag{52.5}$$

As a result, the ratio of the rate constants of the forward and reverse reactions is expressed in terms of (52.2) and (52.5)

$$K = \frac{\theta_{AB} \exp(-\beta\varepsilon_{AB})}{\theta_{CD} \exp(-\beta\varepsilon_{CD})} \times$$
$$\times \left[\frac{\sum_{j=1}^{s} t_{Aj} \exp\left[\beta(\varepsilon^{*}_{(AC)j} - \varepsilon_{Aj})\right] \sum_{j=1}^{s} t_{Bj} \exp\left[\beta(\varepsilon^{*}_{(BD)j} - \varepsilon_{Bj})\right]}{\sum_{j=1}^{s} t_{Cj} \exp\left[\beta(\varepsilon^{*}_{(AC)j} - \varepsilon_{Cj})\right] \sum_{j=1}^{s} t_{Dj} \exp\left[\beta(\varepsilon^{*}_{(BD)j} - \varepsilon_{Dj})\right]} \right]^{z-1} = \tag{52.6}$$
$$= \frac{\theta_A\theta_B}{\theta_C\theta_D} \left[\sum_{j=1}^{s} t_{Aj} \exp\left[\beta(\varepsilon_{Cj} - \varepsilon_{Aj})\right] \sum_{j=1}^{s} t_{Bj} \exp\left[\beta(\varepsilon_{Dj} - \varepsilon_{Bj})\right] \right]^{z}$$

Formally, taking into account the redefinition of the meaning of the equilibrium constant K, one can consider the flow of a two-site reaction as two independent single-site processes at different sites: $A \leftrightarrow C$ and $B \leftrightarrow D$. The expression (52.6) has the form of a product of two independent processes, each of which is described by its equilibrium (50.1). Therefore, the equilibrium condition of the reaction coincides with the condition for equilibrium of the medium as a whole.

Single-particle approximations. This type of approximation closes the system of equations for equilibrium and kinetic processes through unary DFs (i.e., only through the concentration of components). In this case, all pair functions $\theta_{ij} = \theta_i\theta_j$ are expressed as they do not need to have additional equations (equilibrium and kinetic). The same is true for higher-dimensional DFs. However, by construction, the rate of bimolecular reaction rates is expressed as before: $U_{AB}(\alpha) = K_{AB}(\alpha)\exp(-\beta\varepsilon_{AB})V_{AB}(\alpha)$, which must be substituted $\theta_{ij} = \theta_i\theta_j$. This leads to the following relations (see (51.1))

$$K_2 = \frac{K_{AC}(\alpha)}{K_{BD}(\alpha)} = \frac{V_{BD}(\alpha)}{V_{AC}(\alpha)} = \frac{\theta_B\theta_D\exp(-\beta\varepsilon_{BD})}{\theta_A\theta_c\exp(-\beta\varepsilon_{AC})}[S_B(A)S_D(C)]^{z-1} \tag{52.7}$$

where the non-ideality functions $S_B(A)S_D(C)$ do not contain the interaction parameters of the activated complex, but the right-hand side of expression (52.7): 1) does not break up into two independent processes with concentration factors $\frac{\theta_B}{\theta_A}S_B(A)^z$ and $\frac{\theta_D}{\theta_c}S_D(C)^z$, as follows from (50.1) and as would be the case for single-site independent stages, and 2) contains energy factors that change the activation energies of the elementary stages E_{AC} and E_{BD}, which are included in the pre-exponentials of the rate constants K_{AC} and K_{BD}.

Thus, from the analysis of similar expressions for chaotic and mean-field approximation, it follows that they do not provide a self-consistent description of equilibrium and kinetics, since as a result of equating the reaction rates, analogous equations for isotherms in the same approximation are not obtained. Therefore, these approximations can not be used to describe kinetic processes. Even in the special case of migrations by the vacancy mechanism the exponential factor with the parameters $\varepsilon_{iV} = 0$ turn to zero, but this does not compensate for the absence of the exponent $(z-1)$ instead of z. (In addition, if the properties of the system are taken into account more accurately, even with the initial values of the parameters

$\varepsilon_{iV} = 0$, due to the internal motions of the molecules, these parameters become non-zero.)

53. Correlation effects in stage velocities

As a result of averaging over local environments around the reagents, the following expressions are obtained for the velocities of mono- and bimolecular elementary processes

$$U_i = K_i\theta_i S_i^z,\ U_{ij} = K_{ij}\exp(-\beta\varepsilon_{ij})\theta_{ij}(S_iS_j)^{z-1} \tag{53.1}$$

where $\delta\varepsilon_{ik} = \varepsilon_{ik}^* - \varepsilon_{ik}$, and ε_{ik}^* is the interaction parameter between the activated complex (the particle i located at the site f in the transition state) of the elementary process α with the neighbouring particle k at the neighbouring site g, where $t_{ij} = \theta_{ij}/\theta_i$ is the conditional probability of finding the particle j near the particle i, calculated in the quasi-chemical approximation. In the derivation of expressions (53.1), it was assumed that the rate constants K_i and K_{ij} weakly depend on the density. In the function S_i summation is carried out over all types of neighbouring particles. The functions S_i are the factors of the non-ideality function of the reaction system, which reflect the influence of each neighbouring particle on the height of the activation barrier of the reaction: $S_i = \sum_{k=1}^{s} t_{ik}\exp(\beta\delta\varepsilon_{ik})$ – QCA, $S_i = \sum_{k=1}^{s} \theta_k\exp(\beta\delta\varepsilon_{ik})$ – chaotic approximation, $S_i = \exp\left(\beta\sum_{k=1}^{s}\delta\varepsilon_{ik}\theta_k\right)$ – mean-field approximation. In the absence of intermolecular interactions, the multipliers $S_i = 1$ and the formulas go into the reaction rate for the ideal reaction system (48.2).

Equations (53.1) show that the mutual influence of reagents in the case of non-ideal systems complicates their cooperative behaviour (in contrast to traditional chemical kinetics), and this behaviour changes the 'sensitivity' of the change in velocity from concentration and temperature. Both factors simultaneously influence each other, so the concentration dependences become as determinative as the temperature ones. A similar situation occurs in the case of a bimolecular reaction.

Important in the construction of expressions for functions of non-identity is the inclusion of correlation effects between interacting molecules. It has been shown above that taking into account the correlations between the nearest neighbours ensures the self-consistency of the description of reaction rates and the equilibrium distribution of the components of the reaction mixture. If there are no correlation effects, then there is no such self-consistency, and model parameters, determined from equilibrium and kinetic measurements, do not coincide with each other.

The presence of correlation effects also leads to significant differences in the concentration dependences of the velocities at a fixed temperature. This is illustrated by the concentration dependence of the desorption rate in Fig. 53.1 *a*. The calculation was carried out for fixed parameters of monomolecular desorption. The differences are related only to the accuracy of describing the correlation effects in different approximations. In Fig. 53.1 *b* these same approximations are compared with the experimental data on desorption by a potassium atom from the tungsten surface (the shaded regions reflect the experimental error in the estimation of the parameters).

An even greater difference in the curves due to correlation effects is observed in the course of non-isothermal processes. Figure 53.2 lists the so-called thermodesorption curves, which are obtained by linear heating of the surface according to the law $T = T_0 + bt$, where b is the heating rate (deg/sec), T_0 is the initial temperature for which, at the initial time, θ_0. Those. a differential equation is solved $d\theta/dt = -U_A$, where U_A is the desorption rate, for a given initial filling $\theta(t = 0) = \theta_0$. Heating the surface accelerates the desorption process, and as the substance is depleted, the desorption rate decreases. The calculation is carried out at the same value of all the parameters of the model. The vertical axis represents the amount of matter

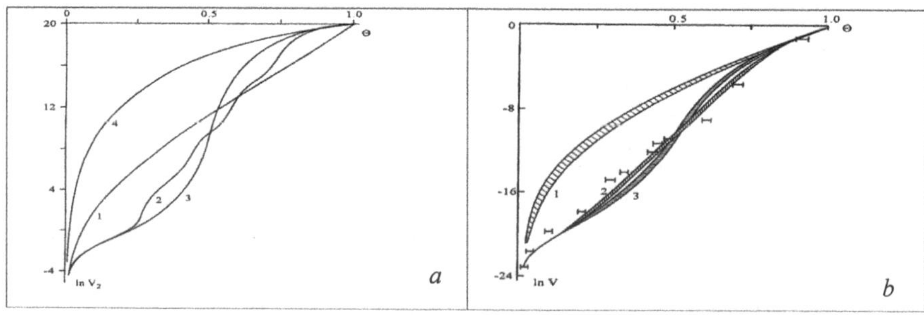

Fig. 53.1. (*a*) Influence of the approximation of accounting for intermolecular interactions on the form of the desorption rate [7]: correlation effects are taken into account in the polynomial (*2*) and quasi-chemical approximations (*3*), and are not taken into account in the molecular field (*1*) and chaotic (*4*) approximation. (*b*) Description of the desorption rate *K* from the surface of W [4,5], calculated in the chaotic (*1*), polynomial (*2*), and quasi-chemical (*3*) approximations. The symbols indicate the experimental desorption rates taking into account the experimental error.

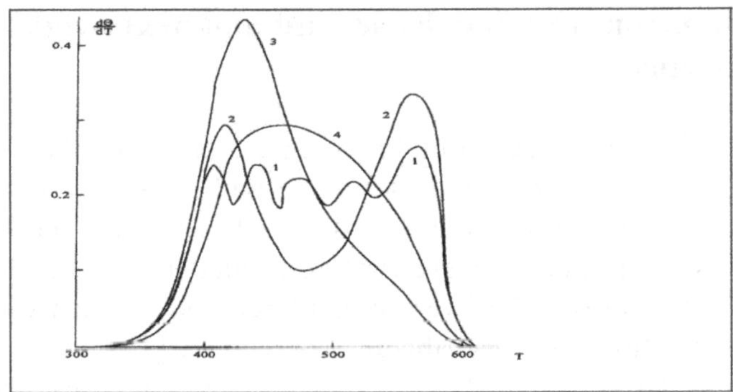

Fig. 53.2. The effect of taking into account the effects of correlation on the form of TDS, calculated in polynomial (*1*), quasi-chemical (*2*), chaotic (*3*) and mean field (*4*) approximations on a uniform surface [19].

that is desorbed at each time point. The chaotic and mean-field approximations do not take into account the correlations, while the QCA and the polynomial approximation preserve correlation effects. In the absence of this accounting, both curves have a single maximum, while taking into account correlations leads to splitting of the curve into several peaks. Thus, the differences are related only to the methods of calculating the non-ideality function. Correlation effects between repulsive particles lead to the splitting of thermal desorption curves on a uniform surface [14–17,19]. It is obtained that the splitting is due to the correlation effects (curves *1* and *2*). Neglecting the correlations for curves *3* and *4* for the same values of the interaction parameters does not split the thermal desorption curves.

For single-particle approximation, it is seen that when self-consistency is absent, then the course of the concentration dependences of the stage rates varies sufficiently with respect to correlations in which self-consistency is conserved. A similar situation in elementary processes for the model of an non-uniform structure – after averaging the velocity in the forward and backward directions can be mismatched. This depends on the ratio of the distribution functions of different types of centres and on the rate of redistribution of reagents between them between them. If surface diffusion is affected, self-consistency will not be ensured. To describe the general multi-stage processes, the relaxation times for each dynamic variable of the reaction system are important.

54. Accounting for the second and next neighbours (uniform systems)

Single-site stages. Let the single-site elementary process $A \to C$ [12-17,20] be realized, where A and C are the components of the lattice solution containing s sorts of particles. The kinetic equations for such a reaction were first constructed in [20]. They use the same approximations that were considered above when closing the system of equations for equilibrium correlators [18].

In QCA, the rates of single-site reactions are expressed as $U_A(\alpha) = K_A(\alpha)V_A(\alpha)$, all values are defined above; only the concentration component of the reaction rate $V_{AB}(\alpha)$ changes, which in this case is written in the form (here and up to the end of Chapter 6 $R = R_{lat} > 1$)

$$V_i(\alpha)=\theta_i\prod\nolimits_{r=1}^{R} S_i\left(r\,|\,\alpha\right)^{z_r}, S_i\left(r\,|\,\alpha\right)=\sum_{j=1}^{s} t_{Bj}(r)\exp\left[\beta(\varepsilon_{Bj}^{*}(r)-\varepsilon_{Bj}(r))\right], \quad (54.1)$$

Hence the ratio of the factors in the non-ideality functions gives

$$S_B\left(r\,|\,\alpha\right)/\,S_A(r\,|\,\alpha)=\sum_{j=1}^{s} t_{Bj}(r)\exp\left[\beta(\varepsilon_{Bj}^{*}(r)-\varepsilon_{Bj}(r))\right]/ \qquad (54.2)$$
$$/\sum_{j=1}^{s} t_{Aj}(r)\exp\left[\beta(\varepsilon_{Aj}^{*}(r)-\varepsilon_{Aj}(r))\right]=S_B(A\,|\,r),$$

that, as above, excludes the properties of the activated complex and leads to isotherms in the QCA approximation when taking into account the interparticle interaction within R c.s.:

$$K_1=\frac{K_A(\alpha)}{K_B(\alpha)}=\frac{V_B(\alpha)}{V_A(\alpha)}=\theta_B\prod\nolimits_{r=1}^{R} S_B(A\,|\,r)^{z_r}\,/\,\theta_A,$$
$$S_B(A\,|\,r)=\sum_{j=1}^{s} t_{Bj}\exp\left[\beta(\varepsilon_{Aj}-\varepsilon_{Bj})\right]. \qquad (54.3)$$

Thus, we obtain an analogue of the expression (50.2) in the form (for more details, see [14, 20])

$$\theta_A=\theta_B\exp[\beta(\nu_B-\nu_A)]\prod\nolimits_{r=1}^{R} S_B(A\,|\,r)^{z_r}. \qquad (54.4)$$

Two-site reaction A + B → C + D. The activated reaction complex occupies two adjacent lattice sites (denoted by n and k) and interacts with neighbouring particles in the r-th coordination sphere, $1 \le r \le R$, both with respect to the site n with particle A, so and with respect to site k with particle B.

We introduce the concept of a single r-th coordination sphere of a dimer molecule AB as a set of sites located at the same distance r from the closest of sites A or B. The coordination spheres of isolated n and k sites overlap, so the single r-th coordination sphere contains a part of the sites related only to r coordinate sphere of one of the sites n or k, and a part of the sites related to the $(r + \lambda)$ coordination sphere of the neighbouring site k or n. The quantity λ depends on the size of the lattice and the number of nearest neighbours.

We divide the set of sites of the uniform r-th coordination sphere into equivalent groups of sites, which are given by their orientation. The orientation of the sites is determined by the angle formed by the straight line connecting the central sites and the straight line between the site in the r-th coordination sphere of the dimeric molecule and the midpoint between the central sites.

We denote the number of different orientations in the unique r-th coordination sphere by π_r, and the number of sites with a given orientation ω_r in terms of $\kappa_{\omega r}$ $(1 \le \omega_r \le \kappa_{\omega r})$. Figure 54.1 shows the distribution of the sites of the plane lattices for $R = 2$.

The interaction potential of the activated complex with the particle j located at the site of the r-th unified coordination sphere of a dimer molecule AB with orientation ω_r, will be denoted by $\varepsilon^*_{ABj}(\omega_r)$. Values $\varepsilon^*_{ABj}(\omega_r)$ are expressed in terms of paired $\varepsilon^*_{ij}(r)$ potentials as follows: $\varepsilon^*_{ABj}(\omega_r) = \varepsilon^*_{Aj}(r_{nm}) + \varepsilon^*_{Bj}(r_{km})$, where m is the number of the site containing the particle j. For example, for a flat lattice $z = 4$ and $R = 2$, we have

$$\varepsilon^*_{ABj}(\omega_1 = 1) = \varepsilon^*_{Aj}(1), \varepsilon^*_{ABj}(\omega_1 = 2) = \varepsilon^*_{Aj}(1) + \varepsilon^*_{Bj}(2), \;\; \varepsilon^*_{ABj}(\omega_1 = 3) = \varepsilon^*_{Aj}(2) + \varepsilon^*_{Bj}(1)$$

$$\varepsilon^*_{ABj}(\omega_1 = 4) = \varepsilon^*_{Bj}(1), \; \varepsilon^*_{ABj}(\omega_2 = 1) = \varepsilon^*_{Aj}(2), \;\; \varepsilon^*_{ABj}(\omega_2 = 2) = \varepsilon^*_{Bj}(2)$$

Similarly, the values $\varepsilon^*_{ABj}(\omega_r)$ are constructed for other z and R. If $A = B$, then the symmetric positions of the sites of the single r-th coordination sphere with respect to the plane passing through the middle of the straight line connecting the central sites and perpendicular to this line correspond to the same values $\varepsilon^*_{AAj}(\omega_r)$, $1 \le \omega_r \le \pi_r/2$, which corresponds to twice the values of $\kappa_{\omega r}$.

The expressions for the velocities of the two-site stages and the proof of their self-consistency when taking into account the

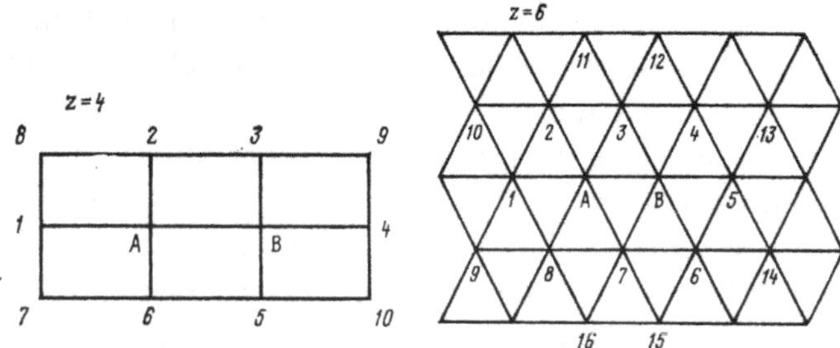

Fig. 54.1 Distributions of sites of planar lattices for $z = 4$ and 6 by equivalent groups of the first two coordination spheres. For $z = 4$, the first c.s. includes sites 1–6 ($\pi_1 = 4$), the second c.s. 7–10 ($\pi_r = 4$): $\omega_1 = 1$ corresponds to site 1, $\omega_1 = 2$ to sites 2 and 6, $\omega_1 = 3$ – sites 3 and 5, $\omega_4 = 4$ – site 4; $\omega_2 = 1$ correspond to sites 7 and 8, $\omega_2 = 2$ – sites 9 and 10. For $z = 6$ the first c.s. ($\pi_1 = 5$), the second – the sites 9 – 16 ($\pi_r = 4$): $\omega_l = 1$ corresponds to site 1, $\omega_l = 2$ corresponds to sites 2 and 8, $\omega_1 = 3$ – sites 3 and 7, $\omega_4 = 4$ to sites 4 and 6, $\omega_l = 5$ corresponds to site 5; $\omega_2 = 1$ correspond to sites 9 and 10, $\omega_2 = 2$ – sites 11 and 16, $\omega_2 = 3$ – sites 12 and 15, $\omega_2 = 4$ – sites 13 and 14.

interaction of neighbours at distances greater than nearest neighbours are given below immediately for an non-uniform lattice.

55. Non-ideal non-uniform systems

In a ‘distributed’ model for a non-uniform lattice, each site is considered a separate type of the region of localization of the molecule – this is the most common type of non-uniform lattice [14,21–23]. Let us consider the question of the self-consistency of the expressions for the rates of elementary reactions (stages) occurring at one and two sites and for the equilibrium state of the reaction system in the QCA, taking into account any radius of the interaction potential between the neighbours R on the distributed model of the non-uniform lattice system. To this end, we will operate with specific individual sites, i.e. before the averaging procedure with different distribution functions for the composition and structure of the non-uniform surface. The averaged models are obtained from the expressions for the distributed model by weighting using the functions discussed in [14,21–23].

Single-site reaction. First, consider a single-site reaction between the adsorbed particle A and the gas particle B: $ZA + B \leftrightarrow ZC + D$, flowing at a site of type q with the number f. The reaction proceeds in the forward and reverse directions. The equation for the velocity of a single-site stage has the form

$$U_{AB} = \frac{1}{M}\sum_{f=1}^{M} U_f^{AB},\ U_f^{AB} = \hat{K}_f^{AB}\theta_f^A S_f^A,$$
$$S_f^i = \prod_{r=1}^{R}\prod_{h\in z_f(r)} S_{fh}^i(r),\ i = A;$$
$$S_{fh}^i(r) = \sum_{j=1}^{s}\theta_{fh}^{ij}(r)\exp\left[\beta\delta\varepsilon_{fh}^{ij}(r)\right]/\theta_f^i = 1 + \sum_{j=1}^{s-1}\theta_{fh}^{ij}(r)x_{fh}^{ij}(r)/\theta_f^i \qquad (55.1)$$
$$x_{fh}^{ij}(r) = \exp\left[\beta\delta\varepsilon_{fh}^{ij}(r)\right] - 1,\ \delta\varepsilon_{fh}^{ij}(r) = \varepsilon_{fh}^{ij*}(r) - \varepsilon_{fh}^{ij}(r)$$

Here M is the number of sites of the distributed system, $\hat{K}_f^{AB}$ is the rate constant of the elementary reaction between the adsorbed particle A and the gas particle B, including the product with the partial pressure of component B in the gas phase of the P_B; the upper symbol '^' means that a pressure factor is applied to the velocity expression for a particle from the gas phase (which is not reflected in the symbol of site f); $\varepsilon_{fh}^{ij*}(r)$ is the interaction parameter of the neighbouring particle j, in the ground state at a distance r, with the activated complex of the reacting molecule i; $\varepsilon_{fh}^{ij}(r)$ is the interaction parameter of two neighbouring particles at a distance r in the ground state.

In the equilibrium condition, the equations for the reaction rates in the forward and backward directions $U_f^{AB} = U_f^{CD}$ give the following relation

$$\hat{K}_f^1 = \hat{K}_f^{AB}/\hat{K}_f^{CD} = \theta_f^C\Psi_f^{CA}/\theta_f^A,\ \Psi_f^{CA} = \prod_{r=1}^{R}\prod_{h\in z_f(r)} S_{fh}^C(r)/S_{fh}^A(r) \qquad (55.2)$$

In the absence of interaction, the function $\Psi_f^{CA} = 1$ and $\hat{K}_f^1$ represents the effective equilibrium constant for the site f of the ideal adsorption system; $\hat{K}_f^1 = K_f^1 P_{\hat{A}}/P_D$, where K_f^1 is the true equilibrium constant of the single-site stage $A \leftrightarrow C$.

For an non-uniform lattice in a QCA, instead of (50.4) we have the following relations for the pair distribution functions

$$\theta_f^i\theta_{fh}^{jk}(r) = \theta_f^j\theta_{fh}^{ik}(r)\exp[\beta\{\varepsilon_{fh}^{jk}(r) - \varepsilon_{fh}^{ik}(r)\}]/S_{fh}^i(j|r)$$
$$S_{fh}^i(j|r) = \sum_{k=1}^{s}\theta_{fh}^{ik}(r)\exp[\beta(\varepsilon_{fh}^{jk}(r) - \varepsilon_{fh}^{ik}(r))] \qquad (55.3)$$

Where $S_{fh}^i(j|r) = \sum_{k=1}^{s}\theta_{fh}^{ik}(r)\exp[\beta(\varepsilon_{fh}^{jk}(r) - \varepsilon_{fh}^{ik}(r))]$

$$\text{Then } \theta_{fh}^{kl}(r) = \theta_{fh}^{mn}(r)\exp[\beta\{\varepsilon_{fh}^{kl}(r) - \varepsilon_{fh}^{mn}(r)\}]\times \\ \times S_{fg}^{k}(m\,|\,r)S_{fg}^{l}(n\,|\,r)\theta_f^k\theta_h^l / (\theta_f^m\theta_h^n) \tag{55.4}$$

Using the fact that the properties of the activated complex do not depend on the direction of the reaction ($\varepsilon_{fh}^{Aj*}(r) = \varepsilon_{fh}^{Cj*}(r)$ for any index j), it follows from (55.1) and the equations for the QCA (55.3) of the constraints that

$$S_{fh}^{C}(r) / S_{fh}^{A}(r) = \left[S_{fh}^{A}(C\,|\,r)\right]^{-1}, \tag{55.5}$$

and therefore, $\Psi_f^{CA}(r) = [S_{fh}^{A}(C\,|\,r)]^{-1}$ that is, the right side of expression (55.2) does not depend on the interaction of AC with its neighbours, and the effective equilibrium constant is expressed only through the parameters of the interaction of particles in the ground state (and not in the transient state) and the equilibrium particle concentrations. This result is completely consistent with the concept of the equilibrium particle distribution, and the resulting expression (55.2) is an equilibrium constant on an non-uniform system. In other words, the equations for the equilibrium distribution are obtained independently of the method of their construction: from the kinetic analysis or directly for the equilibrium state of the molecules of the mixture.

Two-site reaction. A similar conclusion can be obtained from the relations (55.3)–(55.5) if we equate the rates of the forward and backward directions of the two-site reaction stage ($ZA + ZB + C \leftrightarrow ZE + ZD + F$) occurring at two neighbouring sites f and g [1–3]. The reaction rate data are recorded as

$$U_{fg}^{ABC}(1) = \hat{K}_{fg}^{ABC}(fg)\theta_{fg}^{AB}(1)\exp[-\beta\varepsilon_{fg}^{AB}(1)]\prod_{r=1}^{R}\prod_{\omega_r=1}^{\pi(r|qp)}\prod_{h\in z(\omega_r|qp)} S_{fgh}^{AB}(\omega_r) \tag{55.6}$$

$$S_{fgh}^{AB}(\omega_r) = \sum_{j=1}^{s} t_{fgh}^{ABj}(\omega_r)\exp[\beta\delta\varepsilon_{fgh}^{ABj}(\omega_r)],\quad \delta\varepsilon_{fgh}^{ABj}(\omega_r) = \varepsilon_{fgh}^{ABj*}(\omega_r) - \varepsilon_{fgh}^{ABj}(\omega_r)$$

$$t_{fgh}^{ABj}(\omega_r) = \theta_{fh}^{Aj}(r_1)\theta_{gh}^{Bj}(r_2) / \theta_f^A\theta_g^B\theta_h^j,\quad \varepsilon_{fgh}^{ABj}(\omega_r) = \varepsilon_{fh}^{Aj}(r_1) + \varepsilon_{gh}^{Bj}(r_2)$$

And for large distances $\theta_{fh}^{ij}(r > R) = \theta_f^j\theta_h^j$, which consequently gives $S_{fgh}^{AB}(\omega_r)\,|_{r_1>R} = S_{gh}^{B}(r_2)$.

To prove the self-consistency condition, it is necessary to prove the following relation

$$S_{fgh}^{ED}(\omega_r) / S_{fgh}^{AB}(\omega_r) = S_{fh}^{E}(r_1)S_{gh}^{D}(r_2) / [S_{fh}^{A}(r_1)S_{gh}^{B}(r_2)], \qquad (55.7)$$

where the values of ω_r are uniquely related to r_1 and r_2. Then the equilibrium distribution of the molecules will not depend on the method of achieving it through the kinetics of processes in the forward and backward directions or in direct consideration of only the equilibrium configurations. In order to prove equality (55.7), we must remember that $\varepsilon_{fh}^{Aj*}(r_1) = \varepsilon_{fh}^{Ej*}(r_1)$ and $\varepsilon_{gh}^{Bj*}(r_2) = \varepsilon_{gh}^{Dj*}(r_2)$ due to the independence of the properties of the activated complex from the direction of the process. We introduce the notation $\tilde{\theta}_{fh}^{ij}(r) = \theta_{fh}^{ij}(r)\exp[-\beta\varepsilon_{fh}^{ij}(r)]$ and rewrite the left-hand side of (55.7) as

$$\theta_f^A\theta_g^B\sum_{j=1}^{s} F_j\tilde{\theta}_{fh}^{Ej}(r_1)\tilde{\theta}_{gh}^{Dj}(r_2) / \{\theta_f^A\theta_g^B\sum_{j=1}^{s} F_j\tilde{\theta}_{fh}^{Aj}(r_1)\tilde{\theta}_{gh}^{Bj}(r_2)\}$$

Where $F_j = \exp[\beta\{\varepsilon_{fh}^{Aj*}(r_1) - \varepsilon_{gh}^{Bj*}(r_2)\}] / \theta_h^j$, and it must be shown that the ratio $N / L = \tilde{\theta}_{fh}^{Ej}(r_1)\tilde{\theta}_{gh}^{Dj}(r_2) / \left[\tilde{\theta}_{fh}^{Aj}(r_1)\tilde{\theta}_{gh}^{Bj}(r_2)\right]$ does not depend on the index j.

To do this, we express the functions $\tilde{\theta}_{fh}^{Ej}(r_1)$ and $\tilde{\theta}_{gh}^{Dj}(r_2)$ through the functions $\tilde{\theta}_{fh}^{Aj}(r_1)$ and $\tilde{\theta}_{gh}^{Bj}(r_2)$, respectively, using the general QCA for the pair functions (for example, $\tilde{\theta}_{fh}^{Ej}(r_1) = \tilde{\theta}_{fh}^{Aj}(r_1)\tilde{\theta}_{fh}^{EA}(r_1) / \tilde{\theta}_{fh}^{AA}(r_1)$ and so on). This allows you to write $\mathrm{N} / \mathrm{L} = \Lambda_{fh}^{EA}(r_1)\Lambda_{gh}^{DB}(r_2)$, where $\Lambda_{fh}^{ik}(r) = \{\tilde{\theta}_{fh}^{ii}(r)\tilde{\theta}_{fh}^{ik}(r) / [\tilde{\theta}_{fh}^{ki}(r)\tilde{\theta}_{fh}^{kk}(r)]\}^{1/2}$.

Then, considering the obvious equality: $\sum_{j=1}^{s} F_h^j N / \sum_{j=1}^{s} F_h^j L = N / L$ we can see that the left side of the relation (55.7) is equal to $\theta_f^A\theta_g^B\Lambda_{fh}^{EA}(r_1)\Lambda_{gh}^{DB}(r_2) / \{\theta_f^E\theta_g^D\}$.

On the other hand, the formulas $S_{fh}^{E}(r_1) / S_{fh}^{A}(r_1) = \Lambda_{fh}^{EA}(r_1)\theta_f^A / \theta_f^E$ and $S_{gh}^{D}(r_2) / S_{gh}^{B}(r_2) = \Lambda_{gh}^{DB}(r_2)\theta_g^B / \theta_g^D$ can be proved in the same way, therefore, summing, we have the proof of equality (55.7).

The use of approximations that do not take into account the effects of spatial correlation, such as in the mean-field approximation or the random approximation, does not meet the self-consistency conditions, as explained in Section 54, in particular, this leads to renormalization of the activation energy of the process by a value ε_{ij}^{fg} (1).

56. The velocity of thermal motion of molecules

The need for simultaneous description of the transport of molecules in multiphase systems, including within adsorbents, in rarefied gas and dense liquid phases, makes it inconvenient to use the concept of the mean free path of molecules varying from $10^4\lambda$ for a gas to λ for a liquid. In the lattice model, as in the kinetic theory of liquids, the basic concept is the probability of hopping (or displacement) of the molecule $W(\chi)$ by the considered distance χ, rather than the mean free path. This concept is also the main one in Chapter 5 when discussing the question of the averaging of various properties in obtaining a single description of transport processes in three aggregate states.

The key to calculating the hopping probability of a molecule $W(\chi)$ is the concept of a vacancy region through which a molecule moves from one cell to another. As an illustration, Fig. 56.1 shows the scheme for hopping a molecule from site f to the nearest free site g, as well as all neighbouring sites in the first two coordination spheres of the lattice structure with $z = 6$, which simultaneously influence the probability of such displacement. All neighbours located at sites 1 through 25 and in site ξ simultaneously interact with a moving molecule from site f to neighbour site g (hopping length $\chi = 1$). In the general case $\chi > 1$, a vacancy trajectory is necessary to move inside the condensed phase, through which the molecule moves with the thermal velocity $W(\chi)$. Such a trajectory is created in a fluctuation manner. The longer the trajectory χ, the more neighbours participate around it in the formation of the environment which affects the probability of a jump. The task of the theory is to calculate the probability of its formation, so that the molecule can move, and take into account the influence of neighbours on this displacement. For this, it is required to calculate in a self-consistent manner the probabilities of the many-particle configurations forming the vacancy trajectory.

The formulation of the transition state model for non-ideal reaction systems, when the barrier to be overcome is created by the potentials of neighbouring particles and the surface of a solid, is given in [24]. As follows from the ratio of the dimensions, the average thermal velocity of the molecules is expressed in terms of the hopping velocity of the molecule $U_{fg}(\chi)$ between the sites f and g by the distance χ in the form

$$W_{fg} = \chi U_{fg}(\chi) / \theta_f, \tag{56.1}$$

In the state of thermal equilibrium, the probability of molecular hopping between sites is described by formulas

$$U_{fg}^{AV}(\chi) = K_{fg}^{AV}(\chi)V_{fg}^{AV}(\chi), \tag{56.2}$$

where $K_{fg}^{AV}(\chi)$ is the rate constant of the molecule hopping from the site f to the free site g at a distance χ and $V_{fg}(\chi)$ is the concentration dependence of the hopping rate of the molecule;

$$K_{fg}^{AV}(\chi) = K_{fg}^{*AV}(\chi)\exp\left[-\beta E_{fg}^{AV}(\chi)\right], \tag{56.3}$$

$K_{fg}^{*AV}(\chi)$ is the pre-exponent of the rate constant $E_{fg}^{AV}(\chi)$ is the activation energy of the hop determined by the potential of the wall surfaces (the influence of interparticle interactions on the value of the activation jump is determined by the non-ideality function $\Lambda_{fg}^{AV}(\chi)$);

$$V_{fg}^{AV}(\chi) = \theta_{fg}^{AV}(\chi)\Lambda_{fg}^{AV}(\chi), \theta_{fg}^{AV}(\chi) = \theta_{fg_1}^{AV}(1)\prod_{k=1}^{\chi-1} t_{g_k g_{k+1}}^{VV}(1). \tag{56.4}$$

The concentration dependence of the migration rate of the molecule $V_{fg}(\chi)$ is expressed in terms of the product of two factors: 1) $\theta_{fg}^{AV}(\chi)$ is the probability of the vacancy trajectory from χ free sites from the cell f of length χ through the sequence of cells $g(1)$, $g(2)$ and so further to the cell $g \equiv g(\chi)$, for $\chi = 1$ the cell $g(1)$ is finite; 2) $\Lambda_{fg}^{AV}(\chi)$ is a non-ideality function that takes into account the influence of the interactions of molecules around the given trajectory on the probability of a jump inside it:

$$\Lambda_{\xi fg}^{AV}(\chi) = \prod_{r=1}^{R}\prod_{\omega_r=1}^{\pi_r}\prod_{h\in m(\omega_r)}\sum_{k=1}^{V} t_{fgh}^{AVk}(\omega_r \mid \chi)E_{fgh}^{AVk}(\omega_r \mid \chi),$$
$$t_{fgh}^{AVk}(\omega_r \mid \chi) = \frac{\theta_{fh}^{Ak}(r_1)\theta_{gh}^{Vk}(r_2)}{\theta_f^A\theta_g^V\theta_h^k}. \tag{56.5}$$

The $t_{fgh}^{AVk}\left(\omega_r \middle| \chi\right)$ multiplier describes the probability of finding a neighbouring particle j at a site h at a distance r_1 from reagent A and r_2 from vacancy V. The particles A and V themselves are at a distance χ. The environment of the particles around the activated complex is conveniently numbered with the help of the number of sites $m(\omega_r/\chi)$ in the orientations $\omega_r(1 \le \omega_r \le \pi_r)$, π_r is the number of orientations in the r-th unified coordination sphere around the dimer AV on sites of central f and g at a distance χ (a single coordination sphere of radius r, $1 \le r \le R$, is defined as the set of sites located at a distance r from either site f or site g); R is the radius of the

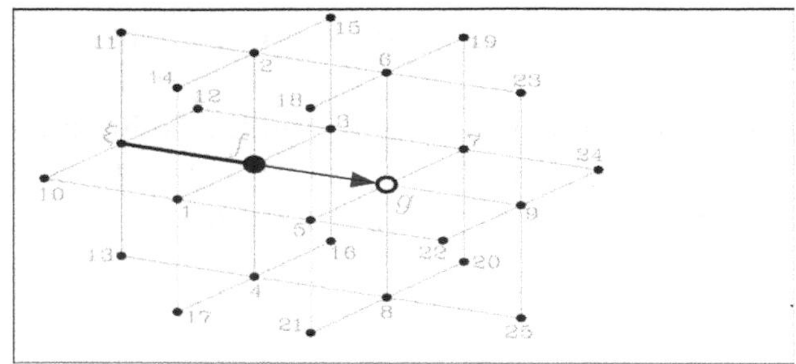

Fig. 56.1. The scheme for hopping a molecule from site f to a free site g and neighbouring sites in the first two coordination spheres of a lattice structure with $z = 6$.

interaction potential. Orientations are measured from the centre of the dimer AV: the point of intersection of the line connecting the central sites fg, and the line connecting the site h to the centre of the dimer AV.

In the process of hopping (in the middle of the fg coupling, see Fig. 56.1), the moving molecule A experiences the influence of neighbouring particles – the energy of this interaction is denoted by the parameters $\varepsilon^{*Aj}_{fh}(r)$ which differ from the analogous energy parameters for particles in the ground state $\varepsilon^{AVi}_{fgh}(\omega_r)$ (i.e., at the lattice sites). The multiplier $E^{AVi}_{fgh}(\omega_r)$ reflects the difference in the interaction of the activated complex with its neighbours through the difference of pair interactions $(\varepsilon^{*ij}_{fg}(r) - \varepsilon^{ij}_{fg}(r))$.

$$E^{AVi}_{fgh}(\omega_r \mid \chi) = \exp\beta\left[\delta\varepsilon^{Ai}_{fh}(r_1) + \delta\varepsilon^{Vi}_{gh}(r_2)\right],$$
$$\delta\varepsilon^{ij}_{fg}(r) = \varepsilon^{*ij}_{fg}(r) - \delta\varepsilon^{ij}_{fg}(r). \qquad (56.6)$$

Traditionally, the pre-exponent of the rate constant $K^{*AV}_{fg}(\chi)$ is expressed in terms of the ratios of the statistical sums of the reactants and the activated complex as [1–3] $K^{*AV}_{fg}(\chi) = \dfrac{kT}{h}\dfrac{F^{*AV}_{fg}}{F^A_f F^V_g}$ where F^{*AV}_{fg} and F^A_f are the statistical sums of the molecule in the transition and ground states for the vacancy $F_V = 1$, h is the Planck constant. In the absence of lateral interactions, formula (56.2) can be written in the form $U_{fg}(\chi) = K_{fg}\theta_f(1-\theta_g)^{\chi}$.

Far from the surfaces of the pore walls, the ratio F_A/F^*_{AV} is the translational degree of freedom in the direction of particle motion and is equal to $(2\pi m\beta^{-1})^{1/2}\chi/h$ then $K_\chi = [(2\pi m\beta)^{1/2}\,\chi]^{-1}$ or $K_\chi = w/4\chi$ where $w = (8/\pi m\beta)^{1/2}$ [25].

Knowing the hopping rates $U_{fg}(\chi)$, we can calculate self-diffusion coefficients and other transport coefficients [24]. In sections 48–54, the elementary process occurred only between reagents at the nearest sites. In this example, it is shown that this approach can be extended to any lengths of distances between reagents (in this case it is related to the length of trajectories) and taking into account the interaction of any radius, the interaction potential between the components of the system. Similarly, it can be used for any type of collision between particles and types of chemical transformations.

References

1. Eyring, H.J., Chem. Phys. 1935. V. 3. P. 107.
2. Temkin M.I., Zh. fiz. khimii. 1938. V. 11. P. 169.
3. Glasston S., Laidler K.J., Eyring H., Theory of absolute reaction rates. Moscow, IL, 1948 [Princeton Univ. Press, New York, London, 1941]
4. Tovbin Yu.K., Thesis. Moscow, Karpov Institute of Physical Chemistry, 1974.
5. Tovbin Yu.K., Fedyanin V.K., Fiz. Tverd. Tela. 1975. V. 17. P. 1511.
6. Tovbin Yu.K., Fedyanin V.K., Kinetika i kataliz. 1978. V. 19. No. 4. P. 989.
7. Tovbin Yu.K., Fedyanin V.K., Kinetika i kataliz. 1978. T. 19. No. 5. With. 1202.
8. Tovbin Yu.K., Fedyanin V.K., Fiz. Tverd. Tela. 1980. V. 22. No. 5. P. 1599.
9. Tovbin Yu.K., Fedyanin V.K., Zh. fiz. khimii. 1980. T. 54. No. 12. P. 3127, 3132.
10. Tovbin Yu.K., Zh. fiz. khimii. 1981. V. 55. No. 2. P. 284.
11. Tovbin Yu.K., Dokl. AN SSSR. 1982. V. 267. No. 6. P. 1415.
12. Tovbin Yu.K., Dokl. AN SSSR. 1984. P. 277. No. 4. P. 917.
13. Tovbin Yu.K., Poverkhnost'. Fizika. Khimiya. Mekanika. 1989. No. 5. P. 5.
14. Tovbin Yu. K., Theory of physical and chemical processes at the gas-solid interface. – Moscow: Nauka, 1990. – 288 p. [CRC, Boca Raton, Florida, 1991].
15. Tovbin Yu.K., Progress in Surface Sci. 1990. V. 34, No. 1-4, P. 1-236.
16. Tovbin Yu.K., Theories of adsorption–desorption kinetics on flat non-uniform surfaces., Dynamics of Gas Adsorption on Non-uniform Solid Surfaces / Eds. by W.Rudzinski, W.A. Steele, G. Zgrablich. Elsevier, Amsterdam, 1996. P. 240–325.
17. Tovbin Yu. K., Thin Films and Nanostructures. Vol. 34. Physico-Chemical Phenomena in Thin Films and at Solid Surface. Eds. L.I. Trakhtenberg, S. H. H. Lin, and O. J. Ilegbusi, Elsevier, Amsterdam, 2007. P. 347.
18. Tovbin Yu.K., Zh. fiz. khimii. 1981. V. 55. No. 2. P. 273.
19. Tovbin Yu.K. Kinetika i kataliz. 1979. V. 20. No. 5. P. 1226.
20. Tovbin Yu.K., Kinetika i kataliz. 1982. V. 23. No. 4. P. 813, 821; No. 5. P. 1231.
21. Tovbin Yu.K., Kinetika i kataliz. 1983. V. 24. No. 2. P. 308, 317.
22. Tovbin Yu.K., Zh. fiz. khimii. 1990. V. 64. No. 4. P. 865. [Russ. J. Phys. Chem. 1990. V. 64. No. 4. P. 461.
23. Tovbin Yu.K., Zh. fiz. khimii. 1992. V. 66. No. 5. P. 1395. [Russ. J. Phys. Chem., 1992 V. 66, No. 5, P. 741.
24. Tovbin Yu.K., The Molecular Theory of Adsorption in Porous Solids, Moscow, Fizmatlit, 2012. [CRC Press, Taylor & Francis Group, 2017.
25. Hirschfelder J. O., Curtiss Ch. F., Bird R. B., Molecular theory of gases and liquids. – Moscow: Inostr. Lit., 1961. – 929 p. [Wiley, New York, 1954].

7

Analysis of thermodynamic interpretations

The material of this chapter combines one common property: all the interpretations discussed are related to the incorrect implicit use of relaxation times. This refers to the priority of establishing a mechanical equilibrium over the chemical equilibrium for the Kelvin equation and the so-called metastable drops, to the equilibrium interpretations of metastable states when changing the thermodynamic parameters of the system and to introducing the concept of the activity coefficient for activated complexes in the theory of absolute reaction rates in describing the rates of elementary stages.

57. The Yang–Lee theory and the Kelvin equation

The Kelvin equation (KE) $P/P_\infty = \exp\{2\sigma\, v_{liq}\beta/R\}$ [1] was discussed in detail in Section 7: two different ways of establishing particular equilibria give two different solutions for complete equilibrium. Chapter 3 shows the existence of equilibrium drops, which are absent in thermodynamics. Analysis of the relaxation times (Section 43) not only gives two states of complete equilibrium, but also explains the reason for this difference, which is related to different properties of the boundary. In this section, these different states of two-phase systems are discussed for their correspondence to the Yang–Lee theory [2–5] together with the states of the metastable drop from the Kelvin equation.

Figure 57.1 shows the isotherm curve (T = const) with a stratification loop in the two-phase region for the volume phase (curve *1*) and the secant curve (curve *2*), constructed according to Maxwell's rule, at the saturated vapour pressure $P_s(T)$ for a given temperature T [6,7]. It relates the coexisting densities of the fluid θ_f and the vapour θ_v. In addition to Fig. 18.1 the densities of the transition zone layers are plotted on the secant *2* in the chemical potential–density coordinates: dots indicate the values of the concentration profile at the vapour–liquid interface $\{\theta_q\}$ in different monolayers, as in Fig. 23.1 *a* (these densities refer to the intermediate region between θ_f and θ_v). Figure 57.1 also shows curve *3* that relates the states of the liquid inside the drop and within the part of its boundary that have the same chemical potential values. This 'node' 3 is shifted with respect to the secant 2 by the magnitude of the chemical potential associated with the jump in the Laplace pressure (P_x) with respect to the pressure P_s for a drop of some radius R. Because of the unique relationship between the pressure P_x and the chemical potential $\mu(P_x)$ the pressure jump automatically leads to a jump in chemical potential!

The point (x) of the intersection of the node 3 at the pressure P_x inside the drop with the isotherm curve refers to the figurative points of the equilibrium states of the system with density θ_x. Continuation of node 3 (absent in this figure) towards the vapour spinodal branch of the isotherm before the intersection with curve 1 is indicated by the symbol (o). The point (o) does not belong to the figurative points of the phase diagram. The density θ_o corresponding to the pressure P_x corresponds to the unstable state of the metastable system and refers to the *macroscopic planar* interface, i.e. quantity θ_0 is not related to any drops.

The equilibrium drop corresponds to the density of the coexisting liquid and vapour phases θ_f and θ_v at a pressure P_s for a given temperature T; here $\mu_{liq} = \mu_{vap}$ and $P_{liq} = P_{vap}$, which satisfies the Yang–Lee theory [2–5]. Equilibrium drops are, in fact, the limiting

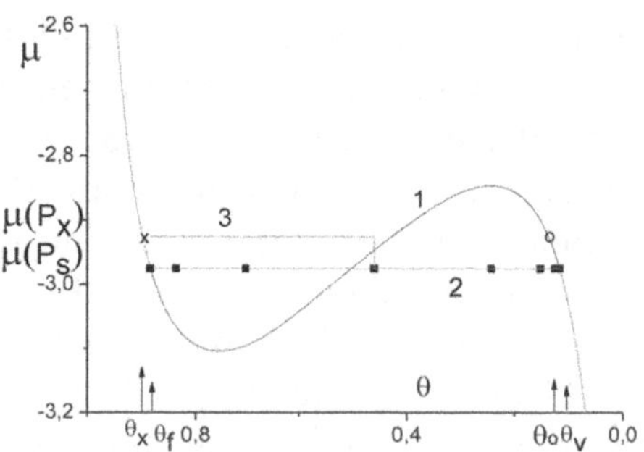

Fig. 57.1. An isotherm with a van der Waals loop (1), a Maxwell secant (2) and a 'node' (3), corresponding to the radius of the drop R.

case of usual stratifying of macroscopical coexisting phases, when the fraction of the new phase is extremely small, and it is necessary to take into account the effect of the interface on the state of the complete system. The metastable drop in the theory corresponds to the densities of the liquid and vapour phases θ_x and θ_v at pressures P_x for the drop and P_s for the vapour. Here $\mu_{liq} \neq \mu_{vap}$ and $P_{liq} \neq P_{vap}$, because the chemical potential and pressure jump *simultaneously* on the dividing surface. These states also satisfy the Yang–Lee theory, since both states of vapour and liquid do not enter the binodal.

For a metastable drop in the Kelvin equation it is assumed that $P_{liq} \neq P_{vap}$, but $\mu_{liq} = \mu_{vap}$. The Yang–Lee theory excludes such states: there are equations of state, both in a drop and in a vapour, connecting their internal densities and pressures. Change of the pressure in any phase, without changing the chemical potential, it is impossible. The density of the point θ_o is on the continuation of single-phase isotherms and falls inside the curve of the binodal. This means, according to the Yang–Lee theory, that the equilibrium state of the system does not correspond to it.

Thus, the molecular theory, even admitting the existence of metastable drops, shows their principal difference from metastable drops in the Kelvin equation. Metastable drops in theory reflect the introduction of a perturbation associated with the curvature of the surface into the very conditions of stratification of coexisting phases due to taking into account the surface tension. This is done through the use of the Laplace equation under the condition that the surface tension is calculated from the concentration profile of the molecules in the transition region.

A metastable drop in the Kelvin equation is constructed purely thermodynamically: first, a condition of mechanical balance between the vapour and the drop is found, and then the equality of their chemical potentials is *declared or allowed*. We note that mechanics, like thermodynamics, does not determine the equations of state themselves in different phases, but the condition of the mechanical equilibrium of the interface allows one to relate the stress at the interface between the phases without the connection with the state of chemical equilibrium of the neighbouring phases. In Gibbs' approach, the mechanical equilibrium is linked by analogy with a film or membrane (see page 229 in [8]), which is another material, and with which the molecules of vapour and drops can not be in chemical equilibrium. The subsequent use of the chemical equilibrium between the vapour and drop molecules is realized in the presence of a film that is constant in time and can not be adjusted by the chemical equilibrium condition (in particular, the exchange of molecules between the vapour and the drop can be excluded). This is confirmed by the analysis in section 45.

In any experiments with flexible membranes, mechanical equilibrium is determined by their properties, and not only by the properties of the phases on both sides of it. For mechanical equilibrium, the surface tension value is introduced without the chemical equilibrium condition on both sides of the surface. Therefore, the surface tension for mechanical equilibrium in the presence of a film (as a foreign body) between phases on both sides of the surface and surface tension for coexisting phases in direct contact are different concepts, and they can not be confused! Thus, both ways of the establishment of complete equilibrium in Section 7 operate with different concepts of surface tension.

It is in the *analogy*, and not in *identity*, of the state of the surface under mechanical equilibrium and the subsequent introduction of the postulate of chemical equilibrium with the determination of the magnitude of the surface tension in the coexisting phases that is the fundamental logical inaccuracy in obtaining the Kelvin equation. This refers to the implicit use of the concept of non-equilibrium or dynamic surface tension. The latter was introduced much later for a particular case of the creation of a freshly formed (without relaxation) surface of a liquid [9–11]. The non-equilibrium surface tension is specified in the derivation of the Kelvin equation by an incorrect relationship between the relaxation times between mechanical and chemical equilibrium (Sections 12 and 45). This conclusion fully applies to the metastable drop in the theory into which the Laplace equation is introduced reflecting the influence of mechanical equilibrium on the chemical equilibrium (Appendix 1 and Section 59). The incorrectness of the Kelvin equation is due to the confusion of the concepts of surface tension in the non-equilibrium and equilibrium states. Moreover, the non-equilibrium state of the system itself is *artificially* introduced in the derivation of the Kelvin equation through the establishment of mechanical equilibrium with an undefined degree of chemical equilibrium, whereas the known relationships between the relaxation times of pressure and the chemical potential require adjustment of the pressure to the evolution of the chemical potential.

As a result, only at the cost of distorting the meaning of the molecular property of the system (which is σ) in thermodynamics can the size of the drop R be introduced. The answer to the question of Section 7 on the essence of the linear size of the drop radius is that it is the result of confusing concepts from thermodynamics and mechanics: parameter R is not an intensive parameter of thermodynamics – it is taken from mechanics. This example shows that for a 'model-free' classical thermodynamics even a curved boundary is a complex object, and the initial statements of thermodynamics are more general concepts than a particular case of curved surfaces. Therefore, any deviation from the three stages of thermodynamics should be controlled by the methods of statistical physics. For more than a century, different correlation ratios have been built on the basis of the erroneous Kelvin equation. Only recently have we started to get rid of the use of the Kelvin equation in the problems of adsorption porosimetry [12] and condensation processes [13], where it possessed the maximum prevalence.

58. Small systems by J.W. Gibbs

In the works of J.W. Gibbs there are two types of additions to the postulates of thermodynamics for bulk phases, which are used in obtaining widely known results on surface phenomena.

The first type of additions is the use by J.W. Gibbs of the same sequence of establishment of equilibria (first mechanical equilibrium, then chemical equilibrium) when considering the thermodynamics of a curved boundary, as in the Kelvin equation. In particular, this sequence is emphasized in the analysis of the stability of curved boundaries (pp. 220–230 in [8]) and in the treatment of other problems in the thermodynamics of small bodies. (The original texts of Gibbs are not repeated: they refer to the relevant pages.) This led to the fact that the thermodynamic criterion of stability $dP/dV < 0$, referring to macroscopic systems, significantly overstates the range of real parameters corresponding to complete equilibrium for small bodies under the condition of the equality of all three quantities in different phases independent of each other: P, T and μ (corresponding, respectively, to mechanical, thermal and chemical equilibria). During the following time this served as a justification for introducing the concepts of metastable states, and the possibility of their interpretation using thermodynamic equations.

The second type of assumption is specially introduced for small bodies, for which Gibbs rejects the same own definitions introduced for phases and phase equilibria (pp. 252–253 in [8]). In particular, he believes that for small bodies it is possible to preserve previously introduced concepts and mathematical relationships, although it emphasizes – in its small bodies there are no inner uniform regions and the properties of a small body are variable inside it. This assumption immediately excludes the concept of 'phase' and 'surface tension' introduced by Gibbs for heterogeneous systems (i.e., the phase approximation does not occur – Section 6). Nevertheless, to this object all the thermodynamic equations are applied, including the concept of the work of formation of a small body W, and the Laplace equation for the connection of surface tension and the pressure difference inside the small body and outside it. As a result, the famous formula for the work of the formation of a new phase $W = \sigma A/3$, associated with the formation of the phase interface and the change in the internal state of the new phase, is obtained. Gibbs used the same rejection of his own definitions for two-phase states and surface tension when considering two neighbouring small drops from immiscible phases (p. 261 and 295 in [8]).

If we use the concept of two-phase systems with internally uniform properties (according to Gibbs), then the work of the formation of a new phase in equilibrium rather than artificially introduced metastable (more precisely, non-equilibrium) conditions will be $W = \sigma A$. This difference causes three times the large numerical divergence when using this expression in the equations of condensation kinetics or the like. since it is included in the exponent of the rate of the nucleation process [14].

When going over to analogous constructions for solid micro-crystals, one should once again pay attention to the convention of applying thermodynamic functions to a small system because of the absence of internal uniform regions associated with the concept of phase. From the standpoint of the mechanics of a continuous medium, it is well known that under such conditions a small body is in a state of internal stresses, variable in magnitude along the radius of the body [15]. These include Lamé's work [16] on the distribution of stresses in a medium with internal and external pressures. Recall that the work of Lamé became classic, and after his work the basic mechanical moduli of compression and shear in the theory of elasticity became established. These same questions are also dealt with in modern mechanics courses (see, for example, [17]).

Gibbs, as the founder of the tensor calculus (see the section on page 263 in [8]), was aware of these results. Therefore, when he constructed the thermodynamics of small crystals, he could not use the same arguments as for liquid drops. For a solid, Gibbs introduced two concepts of surface tension: one for a deformed state when a new surface is created in the course of mechanical stresses γ, and a second concept for the equilibrium process of creating a new surface by crystallization and dissolution σ. Therefore, it was no longer possible to use the notion of deformations equally, and for small solid bodies Gibbs gave another justification for the formula $W = \sigma A/3$. Instead of resorting to the presence of a deformed small body, he uses opposing arguments. Gibbs introduces two non-thermodynamic concepts (pp. 255–257–258 in [8]): the assumption of a linear relationship between the size and surface tension of the crystal, and the condition for preserving the similarity of the crystal to itself when its size changes.

Recall that the fundamentals of the thermodynamics of the Gibbs' macrosystems do not allow the introduction of any additional ideas about the structure or body states that are not related to the energy characteristics of the First and Second Laws.

All subsequent work on the thermodynamics of solids was conducted in terms of calculating each of these two contributions. However, the statistical theory leads to the conclusion about their simultaneous participation in all processes that change the state of a solid body, since the surface always changes the state of deformation of the near-surface region with respect to the volume state (i.e., these concepts can not in principle be divided, although for macroscopic bodies this effect is negligible) [17].

At the same time, it should be noted that Gibbs allowed for small microcrystals the possibility of mutual influence of the state of different faces on each other. In subsequent works this aspect of Gibbs' work was omitted and was no longer discussed. In particular, in the formulation of Wulff's rule, such macroscopic systems are considered in which such mutual influence of different faces is absent.

After Gibbs' work, the order of first mechanical and then chemical equilibrium (according to Kelvin), when considering the curved boundaries of phases and small bodies, became a thermodynamic axiom. It can not be ruled out that precisely because of the need to introduce these additional assumptions for small bodies, and also because of the question of the inconsistency of the mechanical and chemical equilibrium conditions for the curved interfaces (there was no concept of 'chemical potential' before Gibbs), the need for more a detailed description of the thermodynamic characteristics of systems, which ultimately led Gibbs in 1902 to the development of a statistical approach [8].

59. Molecular theory of metastable spherical drops

The molecular theory of equilibrium drops is presented in Chapter 3. To pass to metastable drops, it is necessary to add to it the Laplace equation, which relates the internal pressure in the drop to the pressure of the surrounding phase (Appendix 1):

$$\pi_1 - \pi_\kappa = 2\sigma / (R + \rho_r - 1), \tag{59.1}$$

where R is the radius of the drop without the contribution of the monolayers of the transition region, ρ_r is the number of the reference layer or the surface of tension on which the pressure jump in the metastable system between the vapour π_κ and the liquid π_1 in the drop is realized, $\pi_1 \neq \pi_\kappa$. On the same layer ρ_r, a jump in the chemical potential (fixed through the external pressure of the thermostat aP) is realized: up to $q = \rho_r$ we have $\pi_q = \pi_1$, and from $q = \rho_r+1$ to $q = \kappa$ we have $\pi_q = \pi_\kappa$. For equilibrium bulk coexisting phases, $\pi_1 = \pi_\kappa$. As a result, we obtain a closed system of equations (24.8) and (59.1), in which a dividing surface with a pressure rupture on the so-called surface of tension is introduced. The equation of the local isotherm for the monolayer q relates the probability of filling the cell θ_q and the chemical potential of the molecule μ_q in the layer q, $2 \le q \le \kappa-1$. Equation (59.1) assumes knowledge of the surface tension σ, which is obtained only after determining the concentration profile and fixing the position of the dividing surface.

The expressions for surface tension σ through local pressures are indicated in Appendix 1 (see also [11, 19]). For the metastable drops existing at metastable pressure P_{met} during the construction of the concentration profile, it is necessary to determine the position of the reference phase interface ρ_r on which the jump between the internal vapour pressure π_κ and the liquid π_1, described by the Laplace equation (19), is realized. These provisions are defined as: a) the equimolecular reference surface ρ_e [11,19,20] – the fixation of the coordinate of the dividing surface through the material balance at the equimolecular interface

$$\sigma_e = \frac{1}{F_{\rho_e}} \left[\sum_{q \le \rho_e} F_q(\pi_1 - \pi_q^T) + \sum_{q > \rho_e} F_q(\pi_\kappa - \pi_q^T) \right] \tag{59.2}$$

The quantity $\rho_r = \rho_e$ is found from condition

$$\sum_{q=1}^{\rho_e} F_q(\theta_q - \theta_1) + \sum_{q=\rho_e+1}^{\kappa} F_q(\theta_q - \theta_\kappa) = 0 \tag{59.3}$$

The quantities F_q are defined in (24.4) and, for the sake of simplicity of notation, they omit the symbol R.

b) the reference surface ρ_s through the equilibration of forces and moments of forces on the dividing surface [11, 19, 21]

$$\sigma_s = \frac{1}{R_{\rho_e}} \left[\sum_{q \le \rho_e} R_q(\pi_1 - \pi_q^T) + \sum_{q > \rho_e} R_q(\pi_\kappa - \pi_q^T) \right]. \tag{59.4}$$

The quantity $\rho_r = \rho_s$ is found from the condition of equilibration of the moments of forces

$$\sum_{q \le \rho_s} R_q(R_q - R_{\rho_s})(\pi_1 - \pi_q^T) + \sum_{q > \rho_s} R_q(R_q - R_{\rho_s})(\pi_\kappa - \pi_q^T) = 0; \tag{59.5}$$

c) the reference surface ρ_G by Gibbs (this definition of surface tension is purely thermodynamic [8])

$$\sigma_G = \frac{1}{F_{\rho_G}} \left[\sum_{q \le \rho_G} F_q(\pi_1 - \pi_q^T) + \sum_{q > \rho_G} F_q(\pi_\kappa - \pi_q^T) \right]; \tag{59.6}$$

The quantity $\rho_r = \rho_G$ is found from the condition $\min(\sigma_G)$ of the surface tension when the position ρ_G is varied inside the transition region and when the concentration profile $\{\theta_q^i\}$ is recalculated for each value of the reference surface, $1 < \rho_G < \kappa$. The states of the coexisting phases (but not the transition region) are considered fixed. The quantities F_q are defined (24.4).

d) the reference surface ρ_K by Kondo [11,19,22,23] is also determined purely by the thermodynamic method, but its calculation occurs with a fixed concentration profile

$$\sigma_K = \frac{1}{F_{\rho_K}} \left[\sum_{q \le \rho_K} F_q(\pi_1 - \pi_q^T) + \sum_{q > \rho_K} F_q(\pi_\kappa - \pi_q^T) \right]. \tag{59.7}$$

The quantity $\rho_r = \rho_K$ is found from the condition $\min(\sigma_K)$ of the surface tension when the position ρ_K, $1 < \rho_K < \kappa$, for a fixed concentration profile $\{\theta_q^i\}$ is varied. Here the displacement of the position ρ_K reflects a purely mathematical procedure.

Properties of metastable drops. As a first example, let us consider the influence of the position of the surface of tension according to Gibbs (59.6). The pressure jump occurs on the ρ_r-th spherical monolayer within the transition region, $1 \leq \rho_G \leq \kappa$. The influence of the position of the dividing surface on the profiles of the transition region is demonstrated by the curves in Fig. 59.1 [24]. Shown are the changes in the profiles at the extreme positions of the reference surface: in the second ($\rho_r = 2$) and the penultimate ($\rho_r = \kappa - 1$) monolayers for the three values of the drop radii $R = 40$ (1), 100 (2), 250 (5) for $\tau = 0.7$, $\alpha = 1$. The change in the number of the layer q at which the pressure jump occurs necessarily changes the nature of the distribution of molecules within the transition region. The degree of this change in the values of σ can be different, but the state of the transition layer will change with any of the molecular parameters. The curves show the most important difference between molecular models [6.25] from thermodynamic postulates: a change in the value of ρ_r changes the nature of the distribution of molecules within the transition region (or different concentration profiles are obtained). According to the molecular theory, changing any of the molecular parameters changes the properties of the interphase layer.

The curves in Fig. 59.1 are typical examples of concentration profiles, and they show that the width of the transition region κ in the drop is always less than κ for a plane boundary. In this case, the differences in the values of κ in the calculation of metastable and equilibrium drops rarely exceed one monolayer, with the exception of the critical temperature region.

The dependence of the width of the transition region $\kappa(R)$ on the drop radius R over a wide temperature range is shown in Fig. 59.2. A stepwise change in the curves is a consequence of the discrete nature of $\kappa(R)$ as the number of monolayers in the transition region.

Figure 59.3 shows the dependences of the normalized surface tension σ/σ_{bulk} on the radius of the drop R for $\tau = 0.69$ (*a*) and 0.82 (*b*). The variants of the calculations of $z_{q,p}(R)$ are given by the formulas of Section 24 for $\alpha = 0$ and $\alpha = 1$ and for two ways of choosing the number of the monolayer ρ_r that defines the reference surface: $\rho_r = \rho_e$ defines an equimolecular surface (59.2), and $\rho_r = \rho_m$ corresponds to the extremum σ. All curves start from small values of surface tension and $\sigma \to \sigma_{bulk}$ at $R \to \infty$. Naturally, the curves corresponding to $\rho_r = \rho_m$ are located above the curves corresponding to $\rho_r = \rho_e$.

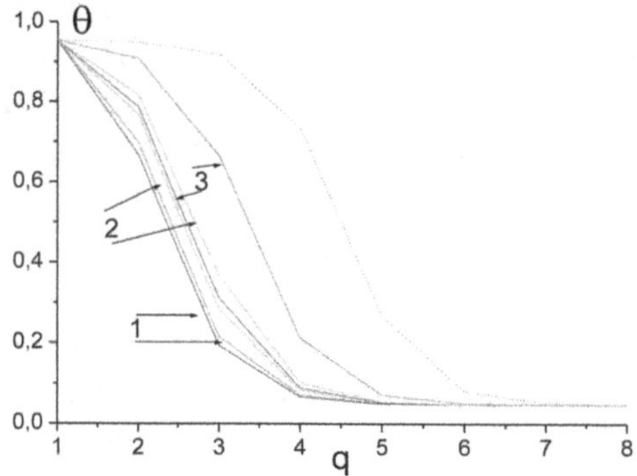

Fig. 59.1. The range of profile changes at the positions of the reference surface in the second ($\rho = 2$) and penultimate ($\rho = \kappa{-}1$) monolayers for the three values of the drop radii $R = 40$ (*1*), 100 (*2*), 250 (*3*) is indicated by arrows related to one size drop. Calculation for $\tau = 0.7$, $\alpha = 1$. As above, the points are the profile in the bulk phase.

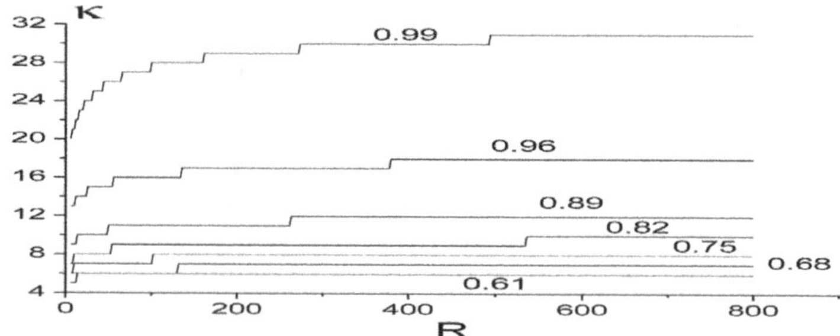

Fig. 59.2. Dependence of the width of the transition region of the liquid–vapour boundary κ on the radius of the spherical drop R for different temperatures $\tau = T/T_c$. The values of τ are shown on the curves.

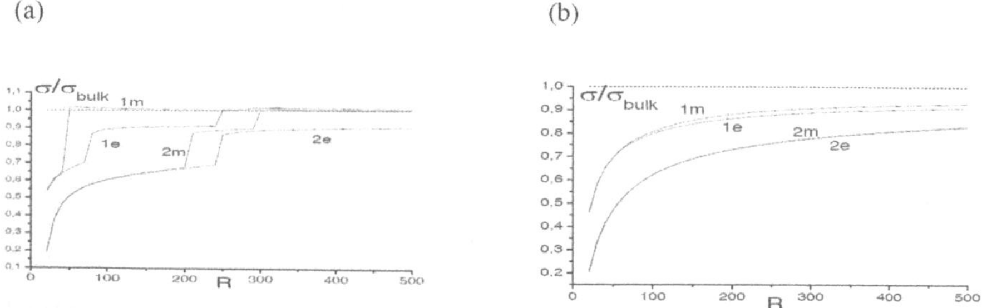

Fig. 59.3. Dependence of the normalized values of the surface tension σ/σ_{bulk} on the radius of the drop R. Results are given for two methods of calculating the structural characteristics: 1 – by formulas with $\alpha = 0$, 2 – by formulas with $\alpha = 1$, and for two methods of choosing ρ_r: e – $\rho_r = \rho_e$ and m – $\rho_r = \rho_m$. The results are given for two values $\tau = T/T_c$: $\tau = 0.68$ (a), 0.82 (b).

The curves in Fig. 59.3 can be broken down into three types. The first type includes increasing curves without jumps (Fig. 59.3 *b*). These curves are realized at relatively elevated temperatures ($\tau > 0.75$). The second type includes curves that contain a jump in the surface tension, as a result of which they reach the value $\sigma > \sigma_{bulk}$ and then $\sigma \to \sigma_{bulk}$ at $R \to \infty$, remaining $\sigma > \sigma_{bulk}$. Curves of this type are observed only at low temperatures, near the temperature of the triple point $\tau = 0.55 \div 0.62$ (curves 1m, 2m and 1e in Fig. 59.3 *a*). The third, intermediate type includes curves that, like curves of the second jump type of σ, but, like curves of the first type, do not reach the value $\sigma = \sigma_{bulk}$. They are realized in the intermediate temperature range.

Kondo drops. The thermodynamic definition of σ by Gibbs [8] was refined in Kondo's paper [22]. To calculate the surface tension: instead of real displacement of the position of the surface of tension in the transition region, it was assumed that the density profile in the transition region was fixed, and only the position of the surface of tension was displaced. Molecular theory allowed us to analyze this refinement and compare it with the equimolecular definition [26].

The essence of the analysis consists in investigating the fulfillment of the thermodynamic conditions determining $\sigma(q)$ at a given temperature: 1) fixing the supersaturation of the vapour $\Delta\mu$; 2) the fulfillment of the Laplace equation $\Delta_L = 0$, where $\Delta_L = \pi_1 - \pi_\kappa - 2\sigma/R$, for $\sigma > 0$; and 3) determination of the position of the dividing surface q_r, $2 \le q_r \le \kappa-1$. The layer number q_r is fixed to the minimum of $\sigma(q)$ in the inner part of the transition region between the vapour and the liquid. The value of q corresponding to the minimum of $\sigma(q)$ determines the position of the surface of tension q_m, and the value of $\sigma(q_m)$ itself is the required surface tension. This minimum $\sigma(q_m)$ and the drop size $\rho = R + q_m - 1$ are used in the Laplace equation.

For comparison, an equimolecular surface is also used. In this case, the profile θ_q, in contrast to $\sigma(q)$, does not depend on q_m, therefore the position of q_e is uniquely determined from the supersaturation of the vapour $\Delta\mu$ and R (T = const).

For Kondo's definition (59.7), it was found that for each fixed value of R, there are two values corresponding to the maximum P_{max} and the minimum P_{min} supersaturation that are realized in metastable systems. Figure 59.4 shows the regions of admissible values in the coordinates (R, P_{met}/P_s) corresponding to the equilibrium state between metastable vapour and liquid, for medium and high temperatures τ = 0.86 (*1*), 0.79 (*2*), 0.73 (*3*), 0.66 (*4*). The lower the temperature, the

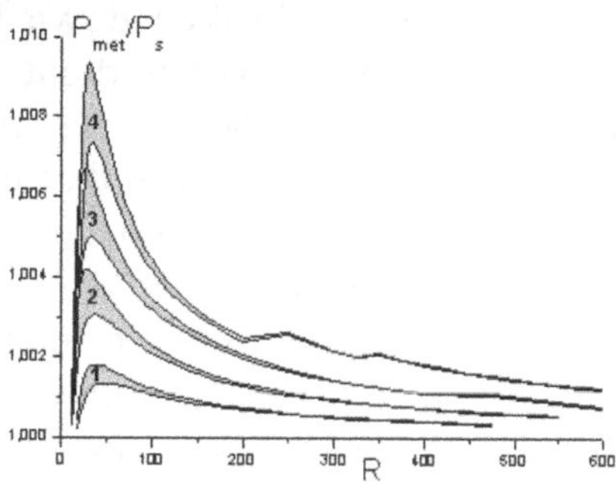

Fig. 59.4. The regions of existence of solutions with respect to R and P_{met}/P_s, which correspond to the metastable states of the vapour–liquid system, for $\tau = 0.86$ (1), 0.79 (2), 0.73 (3), 0.66 (4).

greater the difference between P_{met}/P_s and higher is the corresponding two-valued region. The presence of this area indicates the multiplicity of solutions connecting the parameters of the state of the system. The value P_{max} corresponds to the corresponding value of R_m and the maximum value of the internal pressure in the drop $\pi_1(R_m)$ at a given temperature.

The essence of the polysemy is explained as follows. Since density profile θ_q is uniquely determined for each R and P_{met} in the range of admissible values, then changing the position of the reference surface $\rho = R + q_r - 1$, we determine for this profile the minimum value of $\sigma(q_r)$ and the corresponding value of q_m. In addition, we determine the position of the reference surface $\rho = R + q_L - 1$, on which the Laplace equation is satisfied. The values of q_L as a function of $(P_{met} - P_{min})/P_s$ at $\tau = 0.79$ for $R = 50$ (1), 100 (2), 200 (3) are shown in Fig. 59.5 *a*. On the same field, the symbols show the values of q_m. It can be seen from the figure that for the same pressure the values of q_L differ from q_m by not more than one. In Fig. 59.5 *b* the values of $\sigma(q)$ and $1 < q < \kappa$ are given for integer values of q_L, that is, for $q_L = q_m$. The symbols show the values of $\sigma(q)$ for $q_L = q_m = q_e$, i.e. on an equimolecular surface.

Figure 59.5 shows the fundamental difference between the two methods of selecting a dividing surface. For each radius there is a set of values of P_{met} and for each of them a minimum surface tension from the layer number of the transition region is realized (Fig. 59.5 *b*). The highest value of vapour supersaturation at $R = 50$ on the field (59.5 *b*) corresponds to $P_{met} = P_{max}$ – the upper curve corresponds to it. At a minimum, the value of σ is shifted from the right edge to the left when moving from P_{min} to P_{max}. All the curves refer to the fixed value of the liquid part of the drop R. This figure demonstrates that at R = const the quantities P_{met} (or $\Delta\mu$) and q_m are interrelated, thanks to the Laplace equation, and the entire set of solutions satisfies the three thermodynamic conditions listed above. At the same time, the condition for choosing an equimolecular dividing surface leads to a single value of q_e.

Figure 59.6 *a* shows at a temperature of $\tau = 0.79$ the intersection of the line $P_{met}/P_s = 1.003$ with the region of existence of solutions defining the region of admissible values of R at a given pressure. This range of values of R varies from 20 to 65. In Fig. 59.6 *b* the q_m values corresponding to the surface of tension for the specified range of R are given. First, the q_m values increase until they reach the value corresponding to R_m, then remain constant for some time, and then decrease to $q_m = 2$.

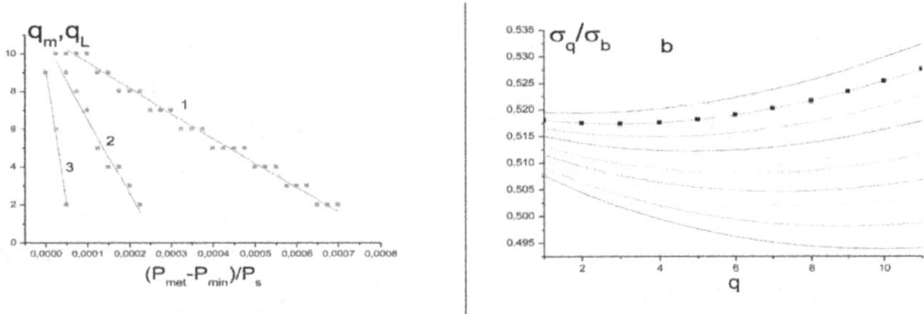

Fig. 59.5. (a) Dependences of q_m and q_L on the pressure P_{met} in a metastable drop at $\tau = 0.79$ for $R = 50$ (1), 100 (2), 200 (3); lines refer to q_L, symbols to integer values q_m. (b) Dependences of the surface tension value σ on the layer number q at pressure values corresponding to integer values of q_L, i.e. at $q_L = q_m$ for $R = 50$. The symbols show the σ-profile at a pressure for which $q_L = q_m = q_e$.

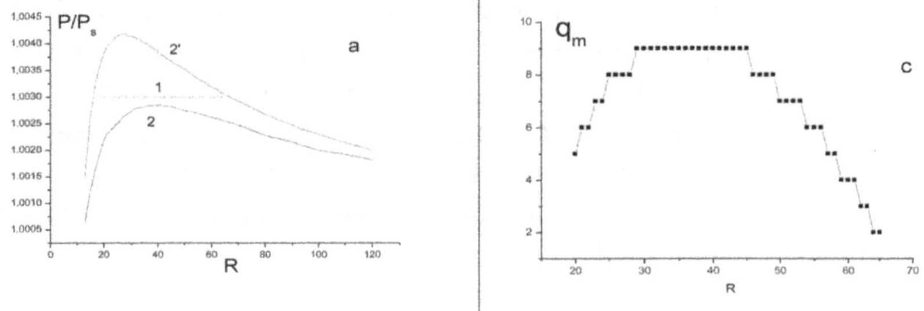

Fig. 59.6. (a) The intersection of the line $P_{met}/P_s = 1.003$ with the region of existence of solutions at a temperature $\tau = 0.79$, which determines the region of permissible values of R at a given pressure. (b) the values of ρ_m as a function of R in the range of acceptable values of R.

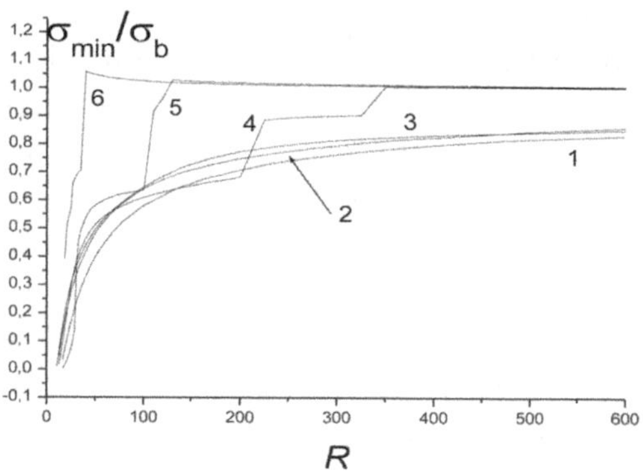

Fig. 59.7. The dependence of surface tension $\sigma_{\min}$ on the radius of the liquid part of the drop R at τ = 0.86 (1), 0.79 (2), 0.73 (3), 0.66 (4), 0.60 (5), 0.53 (6).

The examples in Figs. 59.4–59.6 demonstrate the inadequacy of the thermodynamic conditions [22] for a unique relationship between the degree of supersaturation and the drop radius.

To select a single value of q_m, an additional condition is necessary, which is absent in thermodynamic conditions [22]. For example, if R = const is chosen to be the minimum of all the values of $\sigma(q_m)$ for the admissible values of P_{met}/P_s, then such a dependence of $\sigma_{\min}(R)$ is shown in Fig. 59.7 for a wide range of drop sizes. This dependence largely coincides with the dependence obtained in [24, 27]: at low temperatures the value of the surface tension abruptly exceeds the volume value of σ_b and gradually decreases to it, and at high temperatures, σ_b gradually approaches. Another way to select a single value from the range of admissible P_{met} values is to choose a pressure for which $q_L = q_m = q_e$, where q_e is the equimolecular position of the reference surface. In this case the values of $\sigma(q_e)$ differ somewhat from the values of $\sigma(q_m)$, but the discrepancies are small and commensurate with other methods of calculating σ [24,27].

The analysis of Kondo drops [22] showed that their characteristics are qualitatively similar to those of other metastable drops. In particular, they give the largest values of drop radii, which correspond to the applicability of thermodynamics: the minimum values of the radii of R_{t2} are in the range from 80 to 150λ, which fully agrees with the estimates in Chapter 4.

60. Comparison of the properties of equilibrium and metastable drops

The systems of equations for the concentration profile for equilibrium and metastable drops differ by the presence of an additional Laplace equation for metastable drops. Equations (24.8) for the concentration profile remain unchanged, so it is natural to expect that the properties of metastable and equilibrium drops will not differ very much. Because $P_{\text{met}} > P_{eq}$, then, by virtue of the equation of state for the interior of the drop, $\theta_1^* > \theta_1$ and therefore $\sigma_{\text{met}}(R|T) > \sigma_{eq}(R|T)$. For both types of drops, the theory gives the dependence of the surface tension on the radius. For $R \to \infty$, both the quantities $\sigma_{\text{met}}(R|T)$, $\sigma_{\text{eq}}(R|T) \to \sigma_{\text{bulk}}(T)$, where $\sigma_{\text{bulk}}(T)$ is the volume value of the surface tension on a flat boundary at a given temperature T. With decreasing radii $\sigma_{\text{met}}(R|T)$ and $\sigma_{\text{eq}}(R|T)$ decrease to zero. The values of the radii R_0 corresponding to the equality $\sigma(R_0) = 0$ represent the lower boundary of existence of the phase. The minimum values of the metastable drops R_0^* according to the molecular theory correspond to the minimum phase formation size for the equilibrium drops $R_0^* = R_0$, because $P_\alpha = P_\beta$, which follows directly from the Laplace equation for $\sigma_{\text{met}} = 0$ and $R > 0$.

Below are given examples of comparisons of the properties of equilibrium and metastable drops [24]. The following properties are compared: local densities, pressure, surface tension, total free energy and total mass of drops of the same size, the difference between chemical potentials inside metastable and equilibrium drops of the same size at a given temperature. A comparison of the properties of different drops was carried out for a wide range of temperatures $\tau = T/T_c$, where T_c is the critical point in the bulk phase, from the triple point ($\tau - 0.55$) to the near-critical region $\tau < 0.9$. When comparing the traditional definitions of the surface tension of metastable drops with respect to three types of a dividing surface are used: an equimolecular surface, a surface chosen from the balance of moments of forces, and a dividing surface selected by the surface tension extremum. For an equilibrium drop, the surface tension is determined on an equimolecular surface. Let's start by comparing the concentration profiles, through which all of the listed properties are calculated.

Figure 60.1 shows the results of calculating concentration profiles in the transition region for a drop of radius $R = 100$ over a wide temperature range from a relatively low temperature of $\tau = 0.68$ to a near-critical temperature ($\tau = 0.96$). The following profiles are compared: equilibrium (N) and metastable (L) for the equimolecular dividing surface $\rho_r = \rho_e$. For comparison, the corresponding profiles for a planar lattice are indicated by a prime. The profiles for

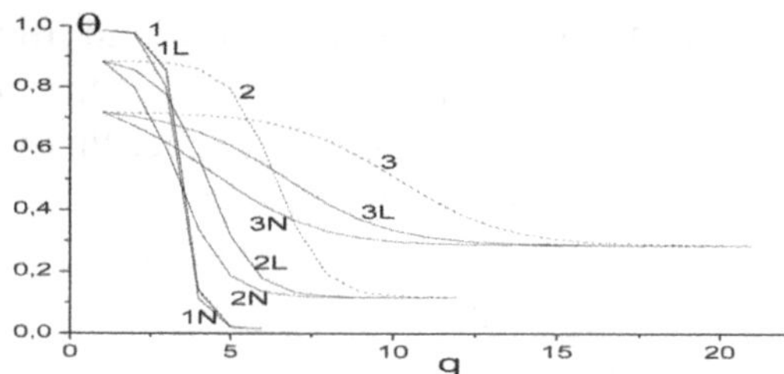

Fig. 60.1. Profiles of the density of the transition region from liquid to vapour at temperatures τ = 0.68 (1), 0.82 (2), 0.96 (3). For a drop of radius R = 100. The notation on the curves: ((*N*) – without taking into account the Laplace equation, (*L*) – taking into account the Laplace equation (8) and $\rho_r = \rho_e$. For comparison, the corresponding profiles for the planar lattice are given by curves 2 and 3.

metastable states are closer to the profiles for a planar lattice than in the case of complete equilibrium of the system [24].

The general results can be formulated as follows: firstly, the profiles for the metastable states (*L*) are closer to the profiles for the planar lattice than in the case of complete equilibrium of the system (*N*). For the lowest temperature (τ = 0.54), all variants of calculations practically coincide, and the width of the transition region is only about 2–4 monolayers. The profiles for different drop radii differ slightly from each other. These calculations are qualitatively correlated with calculations of the profiles obtained by the density functional method [28, 29], although the numerical values differ. In many respects this depends on the differences in the densities of the coexisting vapour and liquid phases obtained by different methods. At the same time, the width of the transition region obtained in this work and in the calculations [28–31] agrees satisfactorily. For high temperatures, the differences between the profiles of the two types of drops and a flat lattice increase.

Figure 60.2 shows the relative properties of metastable drops as a function of their radius R (the radius of the liquid part of the drop is understood as the radius of R) [32]. The region of the vapour corresponds to the radius $R + \kappa$. The value of $R + \rho$ is between R and $R + \kappa$. Figure 60.2 *a* shows the ratio of the densities of the liquid part of the metastable and equilibrium drops, Fig. 60.2 *b* shows the ratio of their internal pressures in the liquid, and Fig. 60.2 *c* gives the difference of the chemical potential $\Delta\mu = \mu_1 - \mu_\kappa$ on the same dividing surface ρ of the metastable drop, to which the surface tension relates. Figure 60.2 *d* shows the dependence of the relative surface tension σ/σ_b on the radius of the drop R. Here

$\sigma_b = \sigma_\infty$ is the value of the surface tension at $R \to \infty$, or, which is the same, for a planar lattice.

The curves in Fig. 60.2 *a*–60.2 *c* for the equimolecular surface and the balance-of-forces surface have the same form: first, the decrease in radius increases the considered value to a certain maximum, after which it decreases to zero. This behaviour corresponds to the presence of maximum supersaturation. This general trend of the presence of a single maximum for the properties under discussion may have differences (local jumps at low temperatures and small radii are possible), because of the discreteness of the change in the number of layers of the transition region $\kappa(T)$ with decreasing temperature [24]. But the general nature of the decrease in the characteristics under discussion with decreasing drop radius remains, and it demonstrates the presence of maximum supersaturation.

The free Helmholtz energy pertaining to the transition region in the QCA and normalized to one site is calculated by the formulas (24.9).

Figure 60.3 shows the logarithmic dependences of the total energy ratios of a metastable and an equilibrium drop of the same size (a)

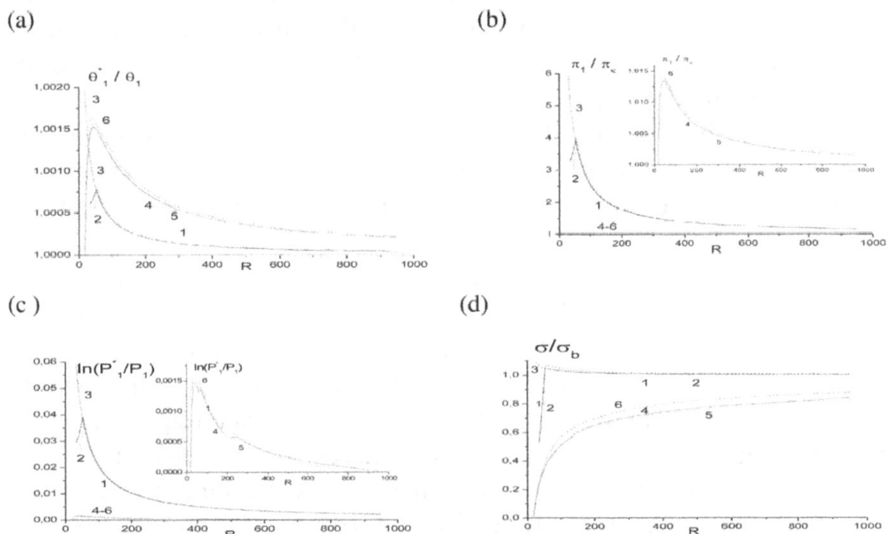

Fig. 60.2. The properties of metastable drops as a function of their radius R: the ratio of the densities of the liquid part of the metastable and equilibrium drops θ_1^*/θ_1(a), the ratio of their internal pressures in the liquid π_1^*/π_1 (b), the difference of these pressures (c), the relative surface tension σ/σ_b (d). The results are given for two values of the reduced temperature $\tau = 0.55$ (1, 2, 3) and 0.89 (4, 5, 6) for three methods of selecting the dividing surface: equimolecular (1.4), balance of moments of forces (2.5) and surface tension (3,6), $\alpha = 1$.

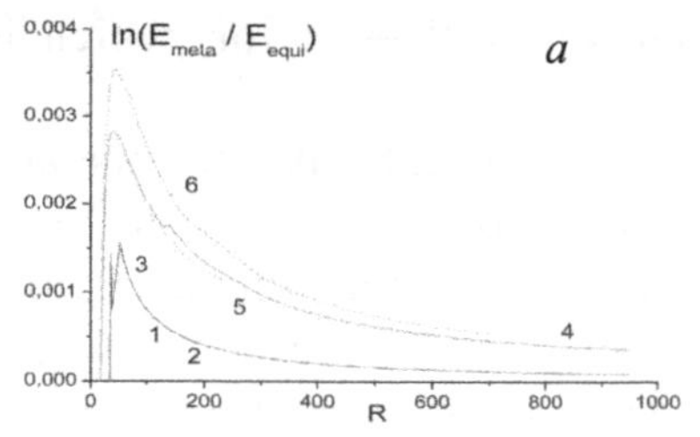

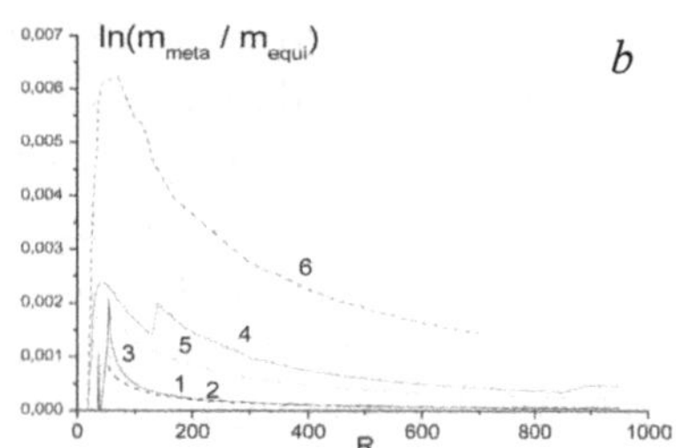

Fig. 60.3. Relative difference between metastable and equilibrium drops as a function of their radius R for temperatures $\tau = 0.55$ and 0.89 for the total free energy E_{meta}/E_{equi} (a) and the drop mass m_{meta}/m_{equi} (b). The results are given for three methods of selecting the dividing surface – the designations of the curves are the same as in Fig. 60.2.

and the ratio of the mass of the same drops (b) to their size R. The subscript indicates the type of drop.

In view of the non-uniformity of the drop, its mass m is defined as the product of local densities by the local volumes of different regions of the drop $m = V_{liq}\theta_1 + \Sigma_{q=2}^{\kappa-1} 4\pi(R+\rho)^2 \theta_q$ where the drop volume $V = V_{liq} + V_\kappa$ consists of the liquid volume $V_{liq} = 4\pi R_{liq}^3/3$ and the volume of the transition layer from liquid to vapour of width κ ($V_\kappa = \Sigma_{q=2}^{\kappa-1} 4\pi(R+\rho)^2$), q is the layer number of the transition region $2 \le q \le \kappa-1$. The value $q = 1$ refers to the liquid part of the drop, $q = \kappa$ refers to the vapour phase.

Accordingly, the total free energy of the drop E is expressed as $E = V_{liq}E_1 + \Sigma_{q=2}^{\kappa-1} 4\pi(R+\rho)^2 E_q$, where E_1 is the free energy inside the drop per one fluid site, and E_q is the analogous value for the site in the layer q of the transition region from liquid to vapour.

Calculations based on the molecular model of a spherical drop located in the vapour phase showed fairly close values of the internal properties of metastable drops of different sizes for all three traditional methods for calculating the properties of a metastable drop, differing in the position of the dividing surface (equimolecular, by the balance of the torques and on the surface of tension).

Methodically, the most important value is the magnitude of the discrepancy between the total free energy of the metastable and equilibrium drops, and also between the total mass of the substance contained in these drops. It was found in a wide temperature range from the triple point to the near-critical region that the total free energy and the total mass of the substance in such drops differ by not more than 0.3% by using an equimolecular dividing surface and a surface chosen by the balance of the moments of forces.

A similar difference between an equilibrium drop bounded by an equimolecular surface and a metastable drop bounded by a surface of tension showed a divergence of 0.6% at high temperatures. At low temperatures ($\tau < 0.7$), such a comparison is impossible because of the absence of a surface of tension in the metastable drop, which automatically leads also to the absence of maximum supersaturation [24, 27]. The absence at the low temperatures of the concept of the surface of tension and maximum supersaturation indicates that the conditions for thermodynamic constructions are not fulfilled for all possible parameters of the state of the system.

The small difference between the total free energy and the total mass of the substance between the metastable and equilibrium drops of equal size indicates that: 1) under modern experimental methods of investigation, it is difficult to distinguish the differences in the contributions for the drops of these two types in the thermodynamic characteristics of the system, and 2) the presence in the system drops that can be in different thermodynamic states, requires the use of more perfect kinetic models to describe the condensation process than traditional approaches, taking into account only the metastable drops. The very fact found in the works [24, 27] of the existence of equilibrium drops indicates a fundamental inadequacy of the use of thermodynamic description in the process of vapour condensation.

Comparison of the properties of metastable drops. The presence of metastable drops in the molecular theory makes it possible to compare their properties with metastable drops in thermodynamics. Here the differences between the properties of the metastable drops are more significant than the differences between the equilibrium and metastable drops in the molecular theory. The qualitative difference between the two types of metastable drops (associated with the presence of state equations in the theory and its absence in thermodynamics) is differently related to the states of these drops with the Yang–Lee theory. Further:

1) The metastable drop in thermodynamics corresponds to a single value of the surface tension σ_{bulk}, measured in an experiment for macrosystems. The theory, in contrast to thermodynamics, gives the size dependences of the surface tension $\sigma_{met}(R)$.

2) The size dependence $\sigma_{met}(R)$ significantly reduces the difference in pressure in the liquid and vapour (supersaturation) and leads to the existence of a maximum supersaturation for small drop radii in comparison with the unlimited growth of the internal pressure in

a metastable drop from thermodynamics according to the Kelvin equation.

3) From the mathematical point of view, the solution for metastable drops in the theory is less stable than for equilibrium drops, and at low temperatures finite supersaturations are not always realized for them, as in the metastable drops of thermodynamics.

4) It is natural to raise the question of whether it is possible to pass to the Kelvin equation from the metastable drops in theory. The molecular theory operates a multilayer model, regardless of the type of boundary: flat or curved. An analysis of the system of coupled equations on the molecular densities (24.8) showed that the Kelvin equation can be obtained [7] if we confine ourselves to taking into account only one intermediate monolayer between two phases and exclude the equation of state for it. In a multilayer model, the equations for the concentration density profile have a sufficient number of degrees of freedom to describe a gradual transition from one phase to another, whereas in the thermodynamic approach there is only one variable (pressure). As a result, the effect of the change in the density profile on it with respect to the profile of the plane boundary is attributed to a change in this thermodynamic variable. This change in the real profile can be attributed with the same success to a change in the other phase parameter. For example, the liquid density θ_1, as discussed in [7, 33], was exactly the same. Therefore, any one-layer models contradict the molecular pattern of the multilayer interphase region.

5) In thermodynamics there are no expressions for either the size of the metastable drop, except for the intense parameter R introduced because of the confusion of the concepts of mechanics and thermodynamics (Section 57), nor on the minimum size of the embryo. The well-known Frenkel's interpretation [14] about the maximum of the Gibbs potential corresponding to the unstable state of the embryo is a reformulation of the well-known Gibbs result $W = \sigma A/3$ [8] corresponding to the work of the formation of a new phase (the method of obtaining which was indicated in Section 58). Interpretation of Fig. 50 in [14] has a purely methodological meaning to show the relationship between the internal and surface contributions for a new phase under the conditions of a metastable system for a *predetermined* drop size. Thus, any value can be used as a given size of the embryo.

From the Gibbs' two non-thermodynamic concepts (pp. 255-258 [8]) on the assumption of a linear relationship between the size and surface tension of the crystal and the condition for the similarity of the crystal to itself when its dimensions change, it follows unambiguously that for $R \to 0$ the contribution from the surface tension should vanish. In this limit, the value of the surface contribution σA vanishes not only because of the decrease in the surface area $A = 4\pi/3R^2$ for $\sigma(R{\to}0) > 0$, but also because of the decrease in the surface tension $\sigma \to 0|_{R\to 0}$ (for $A > 0$). Gibbs' requirement can only be met by the value of the minimum size of phase R_0 (Sections 25–28). It should be understood as the term of the thermodynamic limit $R \to 0$, because it is uniquely related to the very concept of the phase. The quantity R_0 also follows from the theory for metastable drops, and it is the only value at a given temperature which coincides with the value of R_0 for the equilibrium drops.

61. Quasi-thermodynamics

In this section, we summarize the use of the scheme for calculating the thermodynamic characteristics of Kelvin and Gibbs surface tension of curved surfaces in small systems. These topics include: 1) modifications of thermodynamic equations for refinements of Gibbs constructions, 2) constructions that are intermediate between classical and statistical thermodynamics – approaches to which molecular parameters are introduced; 3) statistical theories that use concepts of the surface of tension by analogy with thermodynamics [11, 19].

In Appendix 1 and Sections 59 and 60, examples of the first direction are given. They regard the interface of phases not as a mathematical plane, but following the ideas of the van der Waals molecular theory [34, 35] (see also [19]) as a region of finite width. In addition, the term 'quasi-thermodynamics' is the theory of surface layers based on the local formulation of thermodynamics. This version of the quasi-thermodynamic theory is presented in [22], it is close to the version originally developed by Tolman [36], and serves as a bridge between the macroscopic thermodynamics and the molecular theory based on statistical mechanics.

The standard position of quasi-thermodynamics is the preservation of the kind of functional relationships established in thermodynamics for the thermodynamic characteristics and in the mechanics of continuous media for mechanical deformations, for each local region with the coordinate r, where r is located inside the transition region. When developing a quasi-thermodynamic theory, ideas of the width of the transition region were not yet known, and this assumption was assumed to be natural. However, later the molecular statistical calculations showed [11,19] that the surface layer is only a few molecular sizes in thickness, and the main postulate of the quasi-thermodynamic theory has been violated. Briefly recall the main provisions of the quasi-thermodynamic theory assuming that, at least near the critical point, when the width of the transition region is much larger than the size of the molecules, these bases are correct. On the other hand, this example clearly shows the qualitative shortcomings of thermodynamics in comparison with molecular approaches.

Quasi-thermodynamics itself only fixes the necessity for the existence of a non-uniform density profile $\theta(z)$ in the transition region (z is the coordinate along the normal inside the transition region), but does not determine the equation for the concentration profile and width of the transition region. The basic postulate of quasi-thermodynamics modifies the concepts of macroscopic thermodynamics in that any intense thermodynamic quantity is uniquely determined at each point as a function of the temperature and density of the number of molecules at the same point z. We denote by $c(z)$, $v(z)$ and $f(z)$, respectively, the number of molecules per unit volume, the volume per molecule, and the free energy per molecule (by definition, we have $v(z) = c(z)^{-1}$), which corresponds to the condition (22.16). From an increase in the area A by an amount δA by reversible isothermal displacement of the vessel sidewall, we obtain a relationship between the local specific volume and the tangential pressure component inside the plane transition region in the form

$$\frac{\partial f(z)}{\partial v(z)} = -P_T(z) \tag{61.1}$$

where $f(z,T) = f(c(z,T);\, T)$, i.e., at a given temperature, the free energy F of the system as a whole is a functional of $c(z)$. Therefore, the equilibrium form $c(z)$ must be determined from the condition of a minimum of F for a given temperature and volume.

The quasi-thermodynamic condition of local equilibrium (see [22, 36]) is written as

$$f(z) + P_T(z)/c(z) = \mu. \tag{61.2}$$

where μ is the total chemical potential for the bulk phases α and β. Inside the bulk phases $P_T(z)$ is reduced to the usual hydrostatic pressure $P_{\alpha,\beta} = P$. Equation (61.2) is one of the fundamental relations of quasi-thermodynamics. It can be shown from this that $s(z) = -\partial f(z)/\partial T$ is the entropy per molecule at point z. Consequently, the internal energy per molecule $u(z)$ at the point z is given by the equality $u(z) = f(z) + Ts(z)$. Equation (61.2) also implies (comparing with (22.7)) the expression for the surface tension

$$\sigma = \int_{-\infty}^{\infty} (P - P_T(z))dz, \tag{61.3}$$

identical to (22.20), derived from the mechanical definition of surface tension for a plane boundary. It follows that the expression (22.20) can be derived from the quasi-thermodynamic equilibrium condition (61.2).

We will not dwell on the case of spherical drops described in the quasi-thermodynamic approach, since their concrete interpretation in a discrete description is given in Appendix 1. Any real expression of quasi-thermodynamic equations for a profile is given only by statistical mechanics, the solution of which allows us to find the equilibrium density profile $\theta(r)$ inside the transition region. A rather detailed critique of quasi-thermodynamics is given in Section 2.5 [19].

The presence of an interface also imposes limitations on the molecular models involved in describing the transition region. An important role is played by the equations of state used – they must satisfy the actual experiment. Otherwise, formal thermodynamic constructions for finding the surface tension do not allow them to calculate. It is well known that the van der Waals equation is extremely important from a methodological point of view. But it has not found its practical application for describing vapour–liquid systems; instead, the virial equation of state was actively used [37, 38]. Similar problems arise when using the van der Waals equation as the basic equation of state for the bulk phase when considering the interface of phases. In [39, 40], according to the quasi-thermodynamic calculations of the transition region, Hill replaced this equation because of its roughness by the Tonks equation [41]. Such a replacement made it possible to obtain a reasonable form of the concentration profile of the vapour–liquid interface.

We note that quasi–thermodynamic constructions began long ago: this includes the theory of the thermodynamic melting point of Pavlov [42, 43] on the account of size effects in a first-order phase transition; all the thermodynamic work described above for different positions for dividing surfaces; many molecular approaches use the thermodynamic position of metastable drops [11, 19], and even use the properties of metastable nuclei [28–30].

Even the very presence of the crystal lattice in crystals is a definite compromise between the absence of the structure of matter in 'model-free' thermodynamics and the constancy of the crystal's properties in terms of volume, because of its isotropy at the level of elementary cells (although the difference from this isotropy inside the cells is well known from the optical branches of the vibrational spectra) [44]. It is for this reason that by introducing the concept of phase, in thermodynamics all phase transitions are treated identically regardless of the aggregate state of the substance, whereas in molecular approaches a specific consideration of the symmetry of the substance is necessary for the correct description of phase transitions [45].

The Tolman model. For quasi-thermodynamic constructions it is necessary to classify different works on taking into account the dependence of the surface tension on the drop radius, which leads to a modification of the Kelvin equation. To do this, a derivative $d\sigma/dR$ (or correction in another form of record) is introduced, which formally reflects the change in the magnitude of the surface tension with the size of the drop. The term 'formally' means the absence of any means of determining this dependence within the framework of thermodynamics itself, with the exception of experimental measurements, which is practically impossible. Various attempts have been made to thermodynamically construct the dependence $\sigma(R)$ [23].

All problems of small bodies rest on the way to determine 'surface tension'. The introduction into thermodynamic approaches of any molecular specificity distorts their essence and leads to hybrids that can not give satisfactory results. This can be seen from the example of the well-known Tolman equation [11, 46].

From the consideration of two types of dividing surfaces (equimolecular and surfaces of tension), the author succeeded in expressing the dependence of the surface tension of the drop on the curvature of the dividing surface in the form $\ln(\sigma(R)/\sigma_\infty) = \int_\infty^R A(R)dR/(1+A(R))$ where $A(R)=2\xi(1+\xi+\xi^2/3)$, $\xi = \delta/R$, δ = const is a quantity of the order of the molecule size. Calculation of this formula leads to the fact that at the minimum drop size $\delta = R$ and $\xi = 1$, corresponding to the molecular associate, consisting of an atom with its nearest neighbours, the ratio $\sigma(R)/$

σ_∞=0.28 is obtained. Thus, the 'associate' has a noticeable surface tension, which contradicts the physical meaning of such a concept.

Approximate calculations based on statistical mechanics for liquid argon showed that δ should be of the order of 3 *A*, and in the case of a liquid drop it should be positive. Thus, taking the value $\delta = 3$ *A*, we can expect that the surface tension on a spherical boundary with a radius of 100 *A* will be 6% less for a drop (larger for a bubble) than for a flat boundary. Thus, the formulas obtained give a purely qualitative estimate indicating a decrease in the surface tension with decreasing drop radius.

Asymptotic molecular theory. Problems with the use of molecular models arise not only for rough state equations, but also for models with an incomplete interaction potential: the asymptotic theory [47,48] operates with the long-range part of the potential, but excludes the short-range (repulsive) part. In [49], expressions for the normal and tangential components of the pressure tensor were obtained using the asymptotic theory of accounting for the interaction between bodies in the liquid phase. This truncated potential yielded the result: the tangential component of the pressure tensor is observed in the transition region at a distance of about 80λ, while the width of the transition region itself is only ~4λ [23,49]. The microscopic theory shows a unique relationship between the width of the transition region and the difference in the components of the pressure tensor [6,25,27] and (Appendix 1), so in the transition region of the pressure tensor component can not differ from the pressure in the bulk phase. The authors [23, 49] obtained that between the phases there is a transition region with a variable average pressure different from the pressure in these phases, which contradicts the mechanical stability of the system.

The use of estimates for the width of the transition region and its separation into the liquid and vapour parts in this theory together with the equations of thermodynamics led in the limit $R \to 0$ for small drop sizes to the existence of two different pressures in the vapour and inside the drop [23]. This fact can be understood by comparing the value of the thermodynamic limit $R \to 0$ with the value of R_t (Section 26), when the limitation on the applicability of the equations of thermodynamics comes. Indeed, in a metastable drop there is a difference between the internal pressure and the pressure in the vapour, and at a radius R_t these values will differ. At the same time, we note that the separation of the transition region into two subregions provided the limitation of the internal pressure in the drop instead of its unlimited increase in the Kelvin equation.

We also note that asymptotic molecular theory can not be used to analyze the deformation states of any phase due to the lack of taking into account the repulsive potential. Consequently, the molecular models involved in describing the properties of interfaces must be sufficiently correct.

Statistical theories and thermodynamics. The equations of statistical physics for curved interfaces are considered in monographs [11, 19]. All theories took as their basis the assumption of thermodynamics about the presence of a surface of tension, despite the ambiguity of the existing definitions of surface tension. They *a priori* laid the existence of a pressure jump between the vapour and the liquid phases, and further work was done on the construction of surface tension bonds with equations for the distribution of unary and pair functions. Doubts in Section 4.8 of [19] concerning the correctness of quasi-thermodynamic constructions do not, however, affect the very fact of the *a priori* introduction of mechanical equilibrium, as in the Kelvin equation.

The resulting complex system of equations for unary and pair distribution functions was not solved in full integral form, and it was used in the form of a density functional theory that does not take into account correlation effects (the latter excludes its use in problems of kinetics). The mathematical complexity of the problem did not allow earlier: neither to check the correspondence of metastable drops with the Yang–Lee theory [2–5] nor to detect equilibrium drops. In discrete form, this continuum material from the statistical physics of curved interfaces is set forth in Appendix 1 and in Section 24.

Concluding the discussion of metastable drops (Sections 57–60), we note that the molecular theory [6.25] allowed us to investigate all possible ways of determining σ and verify the correctness of thermodynamic hypotheses. The difference between the molecular theory and thermodynamics is that, in fact, there are two parameters in the theory that characterize the state of the drop for a given supersaturation degree P_{met}/P_s: this is the radius of the drop R and the width of the transition region κ. However, in fact, we are talking about a much larger number of degrees of freedom related to the profile within the transition region $\theta(r)$, through which the value of κ is determined. Such a difference in the degrees of freedom leads to more diverse variants of molecular distributions. As a consequence, it was found that different definitions lead to different concentration profiles and values of σ. The analysis showed a significant difference in molecular distributions from the assumptions on the thermodynamic hypotheses that were mentioned above.

We note the quasi-thermodynamic determination of the surface tension through the equilibration of forces and moments of forces on the dividing surface [21], which qualitatively differs from the methods of introducing σ previously considered on the basis of purely thermodynamic hypotheses [11,19,22]. This difference is well known [11]: the conditions for the equality of forces and moments of forces with respect to the dividing surface are sufficient to introduce local pressures inside the transition region, and to obtain equations that ensure the internal mechanical stability of the interface. In this case, the local stability conditions imply the macroscopic Laplace equation, and it should not be used *a priori* as a postulate, with respect to the molecular theory. Here it emphasizes that microscopic molecular equations independently describe all properties without attracting additional connections from thermodynamics or mechanics of continuous media.

It was found in the calculations that at all temperatures the criterion for the applicability of thermodynamics exceeds the width of the transition region $R_t > \kappa$. This indicates the inappropriateness of the application of thermodynamic methods in describing the properties of the transition region along the normal to the interface of any phase. An exception is the critical temperature range. Thus, the molecular theory not only defined the lower bound of the applicability of the thermodynamics R_t in problems with small systems, but also showed that the finding of σ can not also be associated with the use of thermodynamic constructions.

62. Relaxation times of metastable drops to equilibrium states

Let us consider the results of a direct molecular calculation of the relaxation time of metastable drops to their equilibrium state at saturated vapour pressures, depending on the size of the drops. In terms of physical meaning, both types of drops – equilibrium and metastable – are determined for asymptotically large times, which, in principle, can exceed the characteristic times of existence of real local states in a non-equilibrium system during nucleation.

The existence of a metastable drop follows from the condition of mechanical equilibrium in the absence of chemical equilibrium. Such a state can not exist for a long time, since the condition for the minimum free energy of the local subregion necessary for true equilibrium is not satisfied for it [14, 50]. When analyzing various macroprocesses, an important role is played by the characteristic times of processes in local volumes. One of such possible local processes is the transition between the metastable and the equilibrium states of the drops [51].

We use the equations papers [6, 25] (Sections 24 and 59) describing the state of both types of drops at the molecular level: the boundary is a region non-uniform in density separating uniform regions of vapour and liquid. During the relaxation process of the metastable drop, its density, internal pressure, free energy and surface tension for drops of a fixed radius R change. There is no inhibition on the drop surface – the equilibrium between the layers is established more rapidly than inside the volume of the drop. (This follows directly from the notion of 'non-autonomy' of the liquid–vapour boundary, and also from the analysis of Section 44). The density variation inside the drop is due to the diffusion of molecules to the interface and the outflow of molecules from the surface. The total relaxation time of a drop with the establishment of an equilibrium density inside it depends on the size of the drop R. We will consider this process without being connected with the mass transfer processes in the vapour phase – i.e. we estimate the minimum relaxation time of density from an increased value to a smaller value when the size of the drop changes during its partial desorption, which is not complicated by the outflow of molecules around the drop.

The process of mass transfer inside the drop is described by the diffusion equation, as indicated in Section 36. The calculation was carried out for argon atoms with the Lennard-Jones potential [38] for five temperatures $\tau = T/T_c$ ($T_c = 151$ K). Figure 62.1 shows the dependences of the density inside metastable drops calculated from the equimolecular dividing surface, $z = 12$, taking into account only the interactions of the nearest neighbors. All of them have the same form: as the radius of the drop R increases, first the density $\theta_1(R)$ sharply increases for small radii, and then decreases monotonically. The equilibrium densities $\theta_1^{eq}(\tau)$ at the same temperatures are 0.9760 ($\tau = 0.55$), 0.94852 ($\tau = 0.65$), 0.9051 ($\tau = 0.75$), 0.83566 ($\tau = 0.85$), and 0.70551 ($\tau = 0.95$). These quantities in equilibrium drops are determined by the densities of the binodal curve and remain constant for all radii.

To determine the time to establish the equilibrium value of the density within the entire drop one must know the mass transfer coefficients D. Most literature data on the coefficients refer to self-diffusion coefficients (or label transfer) D^*. For argon drops, experimental data were used for the bulk liquid phase [52]. The values of the mass transfer coefficients in order to obtain the necessary estimates of the relaxation times from Eq. (36.2) depend on the specific conditions of the organization of the process, so different approaches are possible for its evaluation (see Section 36).

We specify the methods for calculating the thermodynamic factor $D_e = g_e D^*$ [53–55], where the thermodynamic factor g_e is related to the chemical potential of the component μ_A as $d\mu_A/d\theta_A$. Or $g_e = 1 + \partial\ln\gamma_i/\partial\ln\theta_i$, where the activity coefficient in the quasi-chemical approximation used for taking into account the lateral interactions [56,57] is equal $\gamma_A = (t_{AA}/\theta_A)^{z/2}$, here $t_{AA}(\theta_A)$ is the probability of finding the particle A next to another particle A: $t_{AA} = 2\theta_A/(\delta + b)$, $\delta = 1 + x(1 - 2\theta_A)$, $b = (\delta^2 + 4x\theta_A^2)^{1/2}$, $x = \exp(-\beta\varepsilon_{AA}) - 1$. When calculating g_e in binary alloys [53–55], the contribution of vacancies (due to their smallness) is neglected in the thermodynamic characteristics. In dense one-component substances, the fraction of vacancies is also small, but the vacancies themselves can not be considered equilibrium in the conditions of mass flow. In this case, we should use another expression for the function g_k, which relates the self-diffusion coefficient and the mass transfer coefficient, which follows from the kinetic analysis (formula (32.6) in [56]): $D_k = g_k D^*$, where $g_k = 1 + \frac{\partial t_{AA}}{\partial\theta_A}\left\{\frac{1}{1 - t_{AA}} + \frac{(z-1)x}{1 + x t_{AA}}\right\}$ is the «kinetic» factor in the same quasi-chemical approximation, connecting D_k and $d(\mu_A - \mu_V)/d\theta_A$. We also consider model estimates of the value of D [54] if one considers that the differences in the activation energies of the self-diffusion coefficients E^* and the mass transfer E_m are due to the formation energy of the vacancy H_V: $E^* = E_m + H_V$. Then $D_V = g_V D^*$, where $g_V = (1 - \theta)^{-1}$.

For the calculation of the functions g_e and g_k, independent experimental data on the heats of sublimation, $H_{\text{subl}} = 1840$ cal/mole [58], were used to determine the value $\varepsilon_{AA} = H_{\text{subl}}/z$ at $z = 12$. The mass transfer coefficients in the listed approaches with functions g_e, g_k and g_V are given in Table 62.1. The calculations yield qualitative estimates, so the experimental data for D^* and H_{subl} and the phase diagram used to calculate the density of the coexisting vapour–liquid phases were taken from different sources and not self-consistent

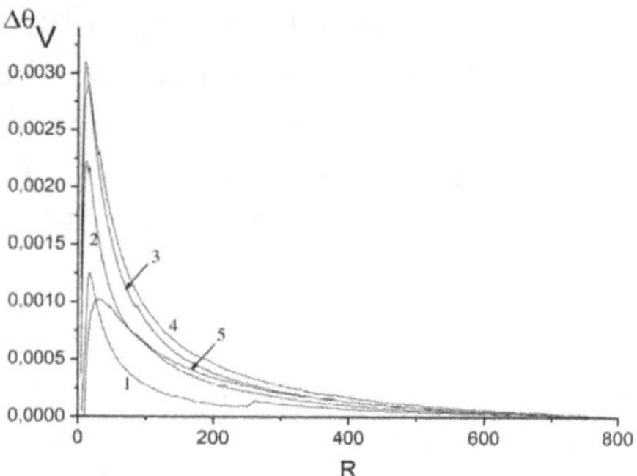

Fig. 62.1. The deviation of the values of the vacancy density (θ_V=1–θ) of a metastable drop as a function of its radius R (calculation with an equimolecular surface) from the equilibrium value $\Delta\theta_V = \theta_V(R) - \theta_V^{eq}$ of the vacancy densities for different reduced temperatures τ: τ = 0.555 (1), 0.65 (2), 0.75 (3), 0.85 (4), 0.95 (5).

Table 62.1. The self-diffusion coefficients (D^*) [52] and the mass transfer in liquid argon at different temperatures on the stratification curve calculated from the equilibrium (g_e) and kinetic (g_k) transition functions to the mass flux and from the model estimate (g_V).

τ	D^*, m²/s	g_e	D_e	g_k	D_k	g_V	D_V
0.55	$1.68\cdot10^{-9}$	0.62	$1.042\cdot10^{-9}$	26.22	$4.40\cdot10^{-8}$	41.67	$7.00\cdot10^{-8}$
0.65	$3.60\cdot10^{-9}$	0.41	$1.476\cdot10^{-9}$	8.45	$3.04\cdot10^{-8}$	19.61	$7.06\cdot10^{-8}$
0.75	$5.30\cdot10^{-9}$	0.20	$1.06\cdot10^{-9}$	2.37	$1.26\cdot10^{-8}$	10.53	$5.58\cdot10^{-8}$
0.85	$7.35\cdot10^{-9}$	–0.02	$-1.47\cdot10^{-10}$	–0.16	$-1.18\cdot10^{-9}$	6.10	$4.48\cdot10^{-8}$
0.95	$9.10\cdot10^{-9}$	–0.25	$-2.275\cdot10^{-9}$	–1.31	$-1.19\cdot10^{-8}$	3.39	$3.08\cdot10^{-8}$

with one another. This led to negative values of the mass transfer coefficients D_e and D_k as they decrease for high temperatures. The decrease in the mass transfer coefficients in the near-critical region is a well-known fact [59], as well as the appearance of negative values of this coefficient describing the so-called 'upward' diffusion associated with the process of phase formation [56], and not with the spreading of an ensemble of non-interacting particles [53, 54]. The model estimate does not work near g_V in the near-critical region, because it reflects only the tendency to change the vacancy density.

Typical results of calculating the relaxation times of metastable drops in a wide range of radii are shown in Fig. 62.2. Here, relaxation times are given only for positive values of the mass transfer coefficients. The solid curves correspond to the calculations with the mass transfer coefficient from the functions g_V, the dashed curves correspond to the mass transfer coefficient from the functions

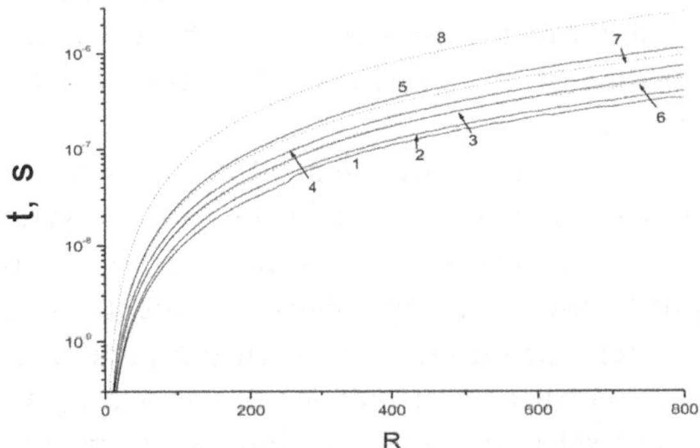

Fig. 62.2. Dependence of the equilibrium establishment time on the radius of the metastable drop at the saturation line at various reduced temperatures τ: 1 and 6 – $\tau = 0.55$, 2 and 7 – 0.65, 3 and 8 – 0.75, 4 – 0.85, 5 – 0.95. The solid lines correspond to the calculations of the mass transfer coefficient D from the model estimate with g_{γ}, the dashed lines correspond to the kinetic factor g_k.

g_k. The unit of measurement of length is the number of monolayers (monolayer size δ).

All curves increase with the size of the drops. For small R, the curves start from nanosecond range fractions and reach a range of up to several microseconds. In both cases, the relaxation time is greatest for the lowest temperature and with increasing temperature the relaxation time decreases, but this change fits within one order of magnitude in time. (If we look at the near-critical temperature region with the thermodynamic factor g_e at $\tau = 0.85$, then the relaxation times increase by about two orders of magnitude.)

For small drops with a radius $R = 6$ of the molecule diameters (or of the order of 2 nm), the relaxation times are ~1 nanosecond. This estimate is obtained without taking into account the fluctuation corrections to the drop size, which increase the lower values of their stable sizes, but this can not increase substantially the range of the characteristic relaxation times of small drops. Therefore, we can conclude that the lower range of values of the relaxation times does not exceed the nanosecond range, and the process of internal relaxation of the states of the metastable drop during the times of several collisions of molecules in the gas phase takes it to its equilibrium state.

For large drops with a radius of the order of $R = 800\delta$ (~0.3 μm), the relaxation times increase by 3–4 orders of magnitude. However, this range is much smaller than the often allowed range of millisecond time resolution in the modeling of nucleation processes

[60]. At a fixed drop radius ($6 \leq R \leq 800$), the range of relaxation time variation from the melting point to the critical temperature does not exceed one order: the lower the temperature, the slower the relaxation process.

Thus, the relaxation times of metastable drops show that with a decrease in their size, a rapid change in the internal state of the drop occurs. This process is much faster than previously thought. The presence of equilibrium drops broadens the spectrum of physically realizable drop states during nucleation. In the general case, in order to calculate the dynamics of the nucleation process, it is necessary to take into account all types of drops that can form within a certain local region of the vapour during diffusion inhibition of its mixing.

63. Metastable states

One of the most problematic questions of physical and chemical analysis is the question of the nature of metastable states and the methods of its description. Classical thermodynamics included metastable states in the number of concepts related to small perturbations to equilibrium [8]. The basis for such inclusion was experimental data on chemical processes in which there were no reactions without a catalyst and mechanical processes associated with friction of sliding of solids (Section 42). At the same time, metastable systems included systems with the formation of a new phase (supersaturated or supercooled vapour, supercooled liquid, etc.), which were described by the equations of state of single-phase systems.

As shown above, in vapour–liquid systems the concept of metastable states is associated with the use of the Kelvin equation, which artificially introduces non-equilibrium through the mixing of the concepts of dynamic and equilibrium surface tension. Therefore, for simple vapour–liquid systems, such an interpretation has a conventional meaning. In more complex situations: amorphous systems, liquid crystals, polymer solutions, etc. Representations of metastable states are often introduced as frozen states at low temperatures [14, 61, 62]. An even greater variety of states is realized in solid-phase systems – for them there is a very wide spectrum of relaxation times of internal rearrangements, especially for different deformation states (Section 43), thus the metastable states, like frozen non-equilibrium states, are very widespread.

In all textbooks, when talking about the possible states of physico-chemical systems in a metastable state (see for example [63, 64]), Fig. 62.1 *a*. "The curve represents the dependence of the Gibbs energy G on the parameter x under the conditions of the existence of the system. The equilibrium condition $dG = 0$ is satisfied at all points of the maximum and minimum. The difference between them is determined by the value of the second derivative, which must be positive at the minimum points ($d^2G > 0$) and negative at the maximum points ($d^2G < 0$). The minimum points (A', A'' and A''') correspond to *stable equilibrium*, and the maximum points to *unstable equilibrium*. The different levels of stability of the equilibrium state correspond to different levels of the minima position on the curve of Fig. 62.1 *a*. (A‴) will be thermodynamically more stable in comparison with the states to which higher positions of the minimum (A′ and A″) correspond, and the transition from the first position to the other requires work. The state A‴ is the most stable for the given conditions of existence of the system. The states (A′ and A″) corresponding to small relative stability are called metastable". This text explains the difference between metastable states and the most stable (equilibrium) state.

The field in Fig. 62.1 *b* shows the graph of the dependence of the Gibbs thermodynamic potential on the molar volume v [65]. "At the critical values of the molar volume v_i ($i = 1, 2, ..., 6$), the thermodynamic potential $G(v_i,\alpha)$ undergoes local minima reflecting local non-equilibrium potentials in zones of hydrostatic stretching of various scales. The critical values of v_i correspond to the following states in a deformable solid: v_0 is an equilibrium crystal; v_1 - zones of stress microconcentrators, in which nuclei of dislocations are nucleated; v_2, v_3 – zones of meso- and macroconcentrators of stresses, in which local structural-phase transitions occur with formation of meso- and macrobands of localized plastic deformation, respectively; v_4 corresponds to the intersection of the curve $G(v,\alpha)$ with the abscissa; with a further increase in the local molar volume, the change in the Gibbs thermodynamic potential occurs under the conditions $G(v) > 0$ and the system becomes unstable: various types of material destruction develop in it; at $v > v_4$, two phases can exist: the atom–vacancy phase (for $v = v_5$) and the local vacuum (for $v \sim v_6$) in the form of micropores, cracks and discontinuities". The curve in Fig. 62.1 *b* from [65] is similar in shape to the curve on the field in Fig. 62.1 *a*, but here the situation refers to very strongly non-equilibrium states of a solid, and, nevertheless, it is characterized by the free Gibbs energy, as in the equilibrium state. The reason for including a strongly deformed state in the metastable state was a reference to Leontovich [66], who allowed the representation of non-equilibrium states as equilibrium states in some external field. Such an interpretation is permissible in general thermodynamic constructions [67] when discussing conditions for the minimality of the non-equilibrium free energy as a function of the external field.

It is obvious that the non-equilibrium distribution of particles corresponds to their non-uniform distribution in space even in the bulk phase. However, the converse statement about the non-equilibrium of the system in the case of inhomogeneity is incorrect. At the interface of coexisting phases in the equilibrium state, a non-uniform distribution of particles within the transition region is always realized. Therefore, the construction of A.M. Leontovich [66], who introduced non-equilibrium states with the help of an external field, is not correct. It is not accidental, he was forced in the molecular-statistical substantiation of his statement for thermodynamics to introduce an additional restriction on infinitesimal deviations from equilibrium [68]. Section 40 shows that for small deviations from equilibrium the number of dynamic variables does not increase in comparison with the number of thermodynamic parameters in the equilibrium state. Additional dynamic variables appear in strongly non-equilibrium states, but they can not be represented as an effect of external fields [66].

Recall that the strictly equilibrium state of a solid corresponds to a state in which the internal stresses are completely absent [16]. This condition is certainly not satisfied in examples in Fig. 62.1 *a* and *b*, with the exception of points related to a strictly equilibrium state – the point with the deepest minima (point A‴ in the field in Fig. 62.1 *a*). In the case of internal stresses, the system is not considered to be in equilibrium – these are deformed states. The degree of deviation of the deformed state from the equilibrium state can vary within very wide limits. Thus, in the fields in Fig. 62.1 *a* and *b*, different variants of the energy curves representing the properties of solids are represented along the ordinate axis. They characterize the average value of the energy characteristic of a system relating to an arbitrary large time interval. For all curves of the type shown in Fig. 62.1 there is usually no information on the time range under consideration. The central question is the meaning of the energy curves presented, on which a certain parameter of the

a

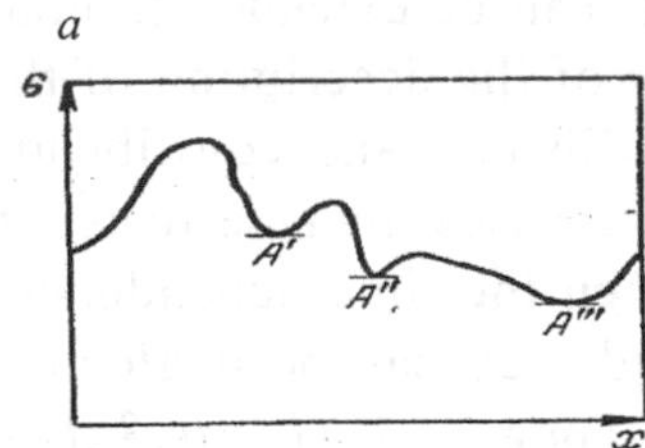

b

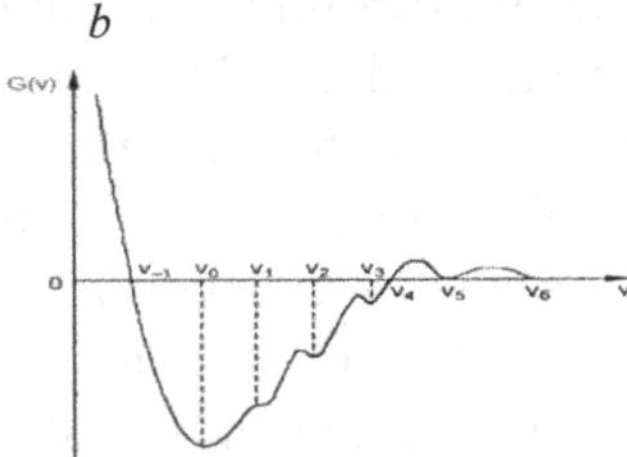

Fig. 62.1. a – Change in *Gibbs energy* $G(x)$ in an arbitrary physicochemical process, x – state parameter [63]. b – dependence of the Gibbs thermodynamic potential $G(v)$ on the molar volume v, taking into account the local zones of stress concentrators of different scale [65].

system (coordinates, concentrations, specific volume, etc.) is present on the abscissa axis, but not time.

The answer to this main question of the interpretation of the experiments is given by the formulated method for calculating non-equilibrium thermodynamic potentials.

In solid bodies, two types of motion are basic: oscillatory and translational in diffusion, which are separated by characteristic times by about ten or more orders of magnitude. (In the general case of intense mechanical perturbations of a solid, it is necessary to additionally include equations for the dynamics of the deformation of the sample [69].) On the ordinate axis, for any system under discussion, the energy characteristic should be a non-equilibrium analog of the Gibbs free energy, and not a potential or internal energy. Averaging over a large time interval inevitably involves vibrational motions that, due to their rapid relaxation, are in locally equilibrium states, depending on the particle configurations in the coordinate space, and their free energy consists, as usual, of the potential energy and entropy of the vibrational motion.

According to the rules of statistical mechanics [4,5,8,67], in order to calculate any equilibrium function $G(x)$, it is necessary that the free energy for a given state parameter x has a minimum: $\partial G/\partial x = 0$. This condition allows us to find x^*, which forms the required dependence $G(x^*)$. Therefore, the true minimum value of free energies, as a function of x, is not directly related to the total curves in Fig. 62.1.

An equilibrium state is a state that differs by an infinitesimal amount, from a state of strict equilibrium that arises from natural mean square fluctuations. The answer to the question of the extent to which the equilibrium potentials *G*, *F*, *U* can be extended to non-equilibrium states depends on the accuracy of the description of the state of the system by means of non-equilibrium and equilibrium distribution functions. If all elementary processes manage to reach their limit values on a given time scale, then the time dependences are insignificant in their contributions, and they can be neglected. From the mathematical point of view, the answer is: $G_{\text{real}}(\theta_i,\theta_{ij},t) = G_{\text{equi}}(\theta_i) + \delta G(\theta_i,\theta_{ij},t)$, where $\delta G(\theta_i,\theta_{ij},t) < \varepsilon$, here ε is the preassigned accuracy of the description of functions; The deviation $\delta G(\theta_i,\theta_{ij},t)$ of the non-equilibrium G_{real} function (θ_i,θ_{ij},t) from the equilibrium value $G_{equil}(\theta_i)$ depends on the concentration θ_i and on the average number of neighbor pairs θ_{ij} between the different components *ij*. In practical situations, taking into account the transition from δG to $\delta\theta_i$, the error in local concentrations should not exceed one or a fraction of percent of the characteristic values of equilibrium values. For a solid, these values are less than 10^{-3}–10^{-6} (depending on the temperature) [70]. All other states of the system are non-equilibrium, and their evolution should be described by kinetic equations.

If the kinetic theory and the equilibrium equations are self-consistent (Chapter 6), then for a long time, when an equilibrium state can be reached at elevated temperatures in solids, this non-equilibrium free energy becomes the equilibrium function of the 'ordinary' Gibbs thermodynamic potential. Only this state corresponds to the only real point of the minimum value of the potential indicated in the fields of Fig. 62.1 *a, b* (this point is described by equations for the equilibrium distribution). All other points on the curves in Fig. 62.1 *a, b* are not equilibrium, and they should be described by the kinetic equations (38.6), (38.7). Or the discussed drawings are schemes that are in no way connected with equilibrium thermodynamics, and sufficiently small deviations of the state parameters require a transition to dynamic models.

Usually it is considered that as an argument (on the abscissa axis) any values can be chosen, as listed above: parameters of state, concentration, coordinates, time, etc. But, this is an inaccurate wording. For relaxation processes, only the 'time' is the natural coordinate, and as a functional connection between the thermodynamic potential and, for example, the concentration, the kinetic equation describing the change in concentration over time is used. In the general case, as an argument along the abscissa axis, a vector of dynamic variables describing the evolution of the system should be deposited. It is important to emphasize that for non-equilibrium processes the fixation of only the state parameters used for the equilibrium state is insufficient to describe the dynamic state of a solid. When calculating the dynamics, it is necessary, as a minimum, to take into account the changes in all pair distribution functions that are directly related to the structural characteristics. As a result, the whole set of conditions leads to the need to explicitly take into account the time factor of the process. Thus, we are talking about the connections $G(t)$ and $v(t)$, but these connections are usually absent.

To the metastable states one can not apply equations of strictly equilibrium systems. The fact that the rigorous Yang–Lee mathematical theory separates metastable states from the equilibrium state of matter means that for any physically realizable metastable state, the non-equilibrium parameter should be clearly formulated. As a rule, this parameter is associated with the existence in the system of diffusion restrictions on the mixing of molecules in its volume or with the existence of high activation barriers for the realization of the chemical transformation.

A correct description of the dynamics of the vapour–liquid equilibrium has long been based on kinetic equations for the distribution function of a new phase in terms of sizes [71–73], and porous systems based on molecular models [12]. The experiment indicates the need to use kinetic equations for the distribution functions of phases in terms of size and for all alloys [74], which have different degrees of dispersion. In describing them, analogous kinetic equations [75, 76] are being introduced (see also [77] – the diffusion processes inevitably require the consideration of the time factor). For the physicochemical analysis of solids, kinetic approaches (usually diffusion type) have not found wide application. The resulting equations are based only on intermolecular interaction potentials of particles and do not contain additional thermodynamic representations.

64. Incorrect use of the coefficient of activity in kinetics

One of the main problems in calculating the rates (liquid-phase, solid-phase, surface) reactions in condensed phases is to take into account the influence of the medium. In the condensed phase, the molecules of the reagents are constantly in the field of action of neighbouring molecules. The change in the internal states of the reagents during the formation of the activated complex (AC) of the elementary process can cause a response in the environment, which in turn affects the course of this process.

The equilibrium properties of non-ideal systems in thermodynamics are actively described by the concept of the activity coefficient (10.7) . The relationship between this coefficient and the molecular theory is discussed in Appendix 3 using the example of associated solutions. The importance of taking into account the real properties of potential functions and the need to take into account correlation effects between interacting molecules in solution is demonstrated. We consider how the equations for the non-ideal mixture change in the presence of associates, when the volume of associates exceeds the volume of the initial components of the mixture, and the accuracy of the traditional methods for accounting for residual contributions in the 'theory of associated solutions'.

Equation (10.8) demonstrates the way in which the reaction system, which was formed at the end of the thirties, was taken into account in the framework of Eyring's TARR [78]. For the activated complex, a similar activity coefficient was introduced, as for the ordinary component of the solution. It was believed that this allowed for taking into account the transition from the gas phase process to the process in the condensed phase. Alternative was the view of M.I. Temkin [79.80] who believed that it is necessary to directly take into account interparticle interactions and their effect on the rate of the elementary stage without using the concept of an activated stage complex (for more details on the history of accounting for the non-ideality of reaction systems, see the review [81]). Temkin demonstrated this with the example of non-uniform reactions [79] at ordinary pressures, and was particularly vividly demonstrated in the description of the process of ammonia synthesis at elevated pressures (more than 300 atm) [80]. For high pressures of the gas phase, the pressures themselves must be replaced by volatility, and the rate constants present in the rates of elementary reactions must be modified by taking into account the effect of the displacement of the adsorption equilibrium on the catalyst surfaces. In [80] this was done through an indirect account of the change in the equation of state of the adsorbed layer, but without introducing the concept of an activity coefficient.

The interpretation of the Eyring school has become widespread in the calculations of the rates of liquid-phase reactions based on the assumption of an equilibrium distribution of the components of the medium around the reagents [82–84]. Equilibrium adjustment of the medium during the formation of AC is possible only with a slow process of activation of the initial reagents, i.e. slow motion of the reaction subsystem from its energy states during the thermofluctuation activation process.

Both points of view existed in different processes. In the theory of surface processes at the gas–solid interface, practically all the models are based on the assumption that the reagents surrounding the reagent in the course of the elementary reaction [56, 81] are in the invariance states; the elementary act is instantaneous and the surrounding molecules, without changing their state, create only an 'external field' in which this process proceeds (although the electronic subsystem of neighboring molecules can respond to the reaction process [85, 86]). Both situations can arise during the kinetic course of the reaction, when there is no diffusion inhibition and the components of the solution are distributed in equilibrium.

A comparison of the two types of non-ideality accounting was first carried out in Ref. [88]. Expressions of the rates of elementary stages for fast reactions with representations of the activated complex as an intermediate state of the reaction subsystem, but without changing the configurations of neighboring molecules are given in Chapter 6, and it is shown that they are self-consistent with a description of the equilibrium state of the entire system.

The reaction rate at an equilibrium state of the environment. In the case of a slow reaction, the characteristic time of the elementary process is much greater than the time for the reorganization of the surrounding molecules. The rapid mobility of the environment leads to an equilibrium distribution of neighbouring molecules around the initial reagents, which corresponds to Eyring's hypothesis [78] about the existence of an activity coefficient. Introducing the notion of an activity coefficient for AC α_i^* (or α_{ij}^*), we mean averaging over all possible equilibrium states of the environment. However, the change of the nearest neighbours of the reagents always changes the potential relief of the reaction and, in principle, another channel for realizing the reaction between the initial reagents becomes possible. Together, this makes the use of thermodynamic concepts contradictory to the dynamics of reaction systems.

In the case of a slow reaction, one can construct an expression for the rate of the reaction stage, in which AC is regarded as an analog of a real molecule (according to [78]) and around which an equilibrium distribution of neighbouring molecules is realized. This was done in [88]: $U_i = K_i^*\theta_i$, $U_i = K_i^*\theta_i$, where $K_i^* = K_i\exp(-\beta\delta F_i)$, the exponential factor K_i^* in the rate constant takes into account the change in free energy during the elementary reaction δF_i of the molecule i (the bimolecular stage was also considered there)

Using the QCA of accounting between the components of the solution, the following equation was obtained for the rate of the monomolecular reaction $A \rightarrow B$

$$U_i = K_i\theta_i(S_i^*)^z,\ S_i^* = \exp\sum_{j=1}^{s}\left\{t_{ij}\ln\left[t_{ij}\exp(-\beta\varepsilon_{ij})\right] - t_{ij}^*\ln\left[t_{ij}^*\exp(-\beta\varepsilon_{ij}^*)\right]\right\} \quad (64.1)$$

where $t_{ij} = \theta_{ij}/\theta_i$ and $t_{ij}^* = t_{ij}\exp(\beta\delta\varepsilon_{ij})/\sum_{k=1}^{s} t_{ik}\exp(\beta\delta\varepsilon_{ik})$, and the energy difference $\delta\varepsilon_{ij}$ is determined above in the formulas (49.4), (51.10). As a result, the structure of formula (49.7), (51.9) is preserved, but in it the function S_i, which takes into account the imperfection of the reaction system, changes.

The formulas show [87] that the rates of elementary processes $\delta\varepsilon_{fg}^{A\lambda}$ depend on the difference between the intermolecular interaction of AC and the initial reagents with the environment, which form the values of the activation barriers for each specific local composition. If $\varepsilon_{fg}^{*ij} = \varepsilon_{fg}^{ij}$, then $S_{hg}^{*A} = 1$, and formula (64.1) for U_A has the same form as for ideal reaction systems $U_A^{id} = k_A\theta_A$ (the law of mass action).

The role of the relaxation of the medium can be naturally traced by considering the ratio of velocities in the case of slow (sl) and rapid (ra) reactions: $\eta_A = I_A(\mathrm{sl})/U_A(\mathrm{ra})$. The structure of the relations η can be analytically studied only for an isotropic particle distribution for $\mathbf{s}$ = 2. In this case, $t_{AA} = 1-2(1-\theta)/(1+b)$, $b = \{1 + 4\theta(1-\theta)[\exp(\beta\varepsilon_{AA})-1]\}^{1/2}$ (here $\varepsilon_{AR} = \varepsilon_{RR} = 0$, where the lower symbol R refers to the solvent).

For small densities, we obtain $\eta_A = \{\exp[-c\theta\ln\theta]/(1+c\theta)\}^z$, where $c = \exp(\beta\varepsilon_{AA}^*)-\exp(\beta\varepsilon_{AA})$, and for large densities $\eta_A = \exp[-z\beta\varepsilon_{AA}(1-\theta)]$. Hence it follows that in the region of small fillings, if $\varepsilon_{AA}^* > \varepsilon_{AA}$ ($c > 0$), then the relaxation of the medium accelerates the reaction rate and this acceleration increases with increasing component A concentration. If $\varepsilon_{AA}^* < \varepsilon_{AA}$ ($c < 0$), then the relaxation of the medium reduces the reaction rate compared to its absence, and with an increase in the concentration of component A this slowing increases. At high densities, the situation is reversed: if $\varepsilon_{AA}^* > \varepsilon_{AA}$, the relaxation of the medium slows down the reaction rate, and if $\varepsilon_{AA}^* < \varepsilon_{AA}$, then – accelerates the reaction rate. Thus, the relaxation of the medium can have a different effect on the reaction rate for small and large reagent densities.

Figure 64.1 shows the calculated rates of fast and slow monomolecular reaction and the value of $-\ln(\eta_A)$ for the whole range of the change in the density of reagent A (for $s = 2$, the medium is represented by other molecules of A and the relaxation consists in the redistribution of molecules around the AC during their migration). These calculations correspond to the rate of non-dissociative desorption on a square lattice ($z = 4$).

Note that $-\ln(\eta_A) = -\ln\left(U_A(\mathrm{sl}) / U_A(\mathrm{ra})\right) = \beta\left[\Delta E_A^{\mathrm{ef}}(\mathrm{sl}) - \Delta E_A^{\mathrm{ef}}(\mathrm{ra})\right]$, where $\beta\Delta E_A^{\mathrm{ef}} = -\ln(U_A / U_A^{\mathrm{id}})$. Here, the effective activation energies characterize the degree of deviation of the reaction rate in the non-ideal reaction system compared to the ideal one. Therefore, instead of the dependences $\eta(\theta)$, it is convenient to consider the dependences $\ln(\eta(\theta))$. The inset in Fig. 64.1 shows the concentration dependences of the ratios $T_{AA} = t^*_{AA}/t_{AA}$ that characterize local changes in the distribution of components A due to their migration under the influence of AC for a slow reaction. This ratio tends to unity at $\theta \Rightarrow 1$, for small θ the maximum effect of AC: $T_{AA} = \theta \exp(\beta\delta\varepsilon_{AA})$. For a quick reaction T_{AA}=1 for all 0.

Figure 64.1 also illustrates the qualitative difference between the concentration dependences of the reaction rates for various relaxation of the medium: in the absence of relaxation, $\ln U_A(\mathrm{ra})$ changes practically linearly with increasing θ, and for equilibrium

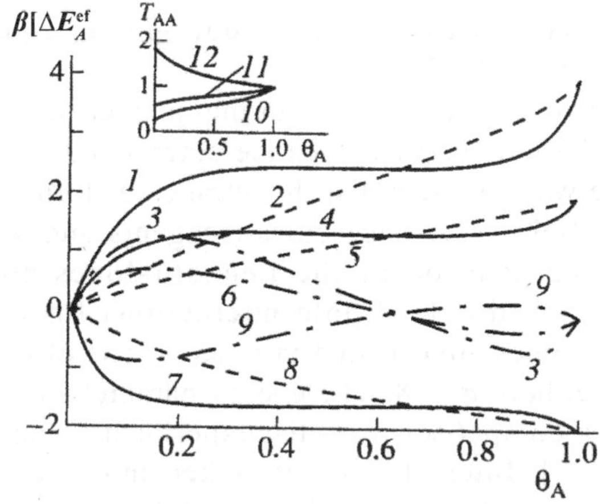

Fig. 64.1. Effective activation energies for slow (1.4.7) and rapid (2.5.8) monomolecular reactions, their difference $\beta[\Delta E_A^{\mathrm{ef}}(\mathrm{sl}) - \Delta E_A^{\mathrm{ef}}(\mathrm{ra})]$ (3.6.9) and function $T_{AA} = t^*_{AA}/t_{AA}$ (10–12) (insert); $s = 2$, $\beta\varepsilon_{AA} = 1$, $\varepsilon^*_{AA}/\varepsilon_{AA} = 0$ (1–3,10); 0.5 (4–6, 11); 1.5 (7–9,12) [87].

relaxation of the medium $\ln U_A(\text{sl})$ changes sharply in the regions $\theta < 0.2$ and $\theta > 0.8$ and remains practically constant at $0.2 < \theta < 0.8$.

This general property of the influence of the character of the relaxation of the medium is also preserved for other situations: mono- and bimolecular stages in solutions, which were investigated in [89, 90] (see also below). It can be used as a basis for analyzing the experimental concentration dependences of the logarithms of the reaction rates over a wide range of concentrations.

Dipole system. Both ways of taking into account the effect of the non-ideality of the reaction system in the dipole solvent on the rate of the stages were considered in [90] within the framework of the molecular theory (lattice gas model) [12, 56]. Potential functions of the solvent and reagents take into account short-range Lennard-Jones contributions, which stabilize the system with a dipole interaction, and dipole–dipole contributions, which depend on the orientation of the molecules. As a lattice structure for water simulation (for brevity, we will call so the dipole solvent), we often use the tetrahedral structure of diamond [91, 92], in which the mean-field approximation was used. In [93] this approach was generalized to the case of taking into account spatial correlations between interacting molecules in the quasi-chemical approximation. The latter made it possible to proceed to the calculation of the rates of elementary reactions, which was impossible because of the discrepancy between the description of the equilibrium distribution of molecules and reaction rates with their participation in the mean-field approximation, as in the original model for the dipole solvent [91, 92]. Note that lattice models are relatively often used to calculate the vapour–liquid equilibrium of non-aqueous systems [94,95].

The crystal structure of diamond with the number of neighbors z = 4 was chosen as the lattice. Such a lattice corresponds well to short-range order in the water system and the tetrahedral form of hydrogen bonds. The force field of all molecules (reagents and solvent) was modeled by a superposition of the Lennard-Jones potential for non-electrostatic and dipole–dipole interactions for electrostatic interactions. The dipoles are oriented in space in one of eight discrete directions [93]; i.e. here $\sigma_j = 8$. The energy parameters of the model were found from a comparison with the experimental curve of water stratification [96,97]. Interactions were taken into account up to 4 c.s., details of the calculation are given in [90].

The potential parameters for reagents A, B, C, D were chosen so that the rate of interaction of molecule A with water would be 2 times less than the water–water interaction at the same molecular arrangement, *B* with water 3 times less, *C* with water 4 times less, *D* with water 5 times less. To ensure that the presence of reagents in the solution had little effect through lateral interactions on the course of the reaction, the concentrations of the components $i = A, B, C, D$ were chosen to be $\theta_i = 0.001$. The calculations were carried out for the case of a low concentration of reagents in order to study the role of dipole solvent molecules. The concentration of water varies from 0 to 0.998 in the case of monomolecular and up to 0.996 in the case of a bimolecular reaction. Thus, the main contribution of the influence of the environment on the reaction rate is exerted by solvent molecules.

The main parameter of the model is the energy of the transition state ε^*_{ij}, which is assumed to be proportional to the binding energy of the molecule with the ground state ε_{ij}: $\varepsilon^*_{ij} = \alpha\varepsilon_{ij}$. The dependences of the reaction rate ratio *U* on the solvent concentration for different values of the energy parameter $\alpha_{ij} = \varepsilon^*_{ij}/\varepsilon_{ij}$ were calculated, where ε^*_{ij} is the energy parameter of the interaction of the AC formed from the reagent *i* with the neighbouring particle of the type *j*. For simplicity, the same α was adopted for all components of the solution. If $\alpha = 1$, the formation of AC does not significantly affect the course of the reaction and the process proceeds as in the case of weakly interacting reaction components (an ideal reaction system). A value of $\alpha = 0$ means that in the transition state, the AC does not interact with the environment, which is an almost unrealistic omission for liquid-phase reactions.

Monomolecular reaction. Figure 64.2 shows how the rate of the monomolecular reaction depends on the concentration of the solvent. The rate U_A of the monomolecular reaction is plotted in relative units as a function of the solvent concentration θ_{H_2O} for fast (solid line) and slow (dots) reactions at 300 K and the values of the parameter $\alpha = 0.25$ (*1*), 0.5 (*2*), 0.75 (*3*). For a fast reaction, the change in the rate as a function of the concentration of the solvent is monotonic and reaches its lowest value when the water concentration reaches its maximum value in the liquid phase. With a slow reaction, an increase in the water fraction leads to a rapid decrease in the reaction rate in the region of low densities and when θ_{H_2O} reaches a value of ~0.2, practical constancy is observed in which the dotted curves intersect solid curves in the region of large θ_{H_2O}. The reason for the difference in the course of the curves for fast and slow reactions is that for these two approximations the weights of the configurations

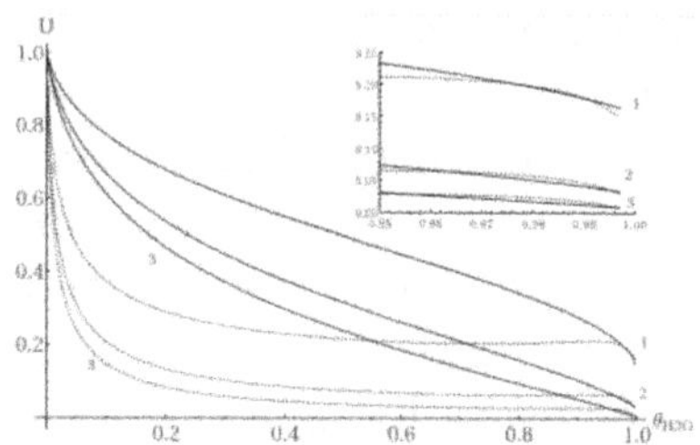

Fig. 64.2. The rate U_A of the monomolecular reaction (in arbitrary units) as a function of the concentration of the solvent θ_{H_2O}; $T = 300$ K, $\alpha = 0.25$ (*1*), 0.5 (*2*), 0.75 (*3*). On the inset – the region of large densities.

of the solvate shells of AC differ in the expression for the reaction rate differently. The limiting values of both expressions for $\theta_{H2O} \to 0$ and $\theta_{H2O} \to 1$ coincide.

The results of the calculation confirm the previous conclusion [87] that there is a significant difference in the concentration dependences of the rates, depending on the nature of the relaxation of the environment. Recall that the difference in reaction rates can reach up to 4–5 times in the region of large and / or small fillings of the solvent. The displacement of the position of θ_{H_2O} for the points of intersection of the curves from the value $\theta_{H2O} \sim 0.5$ [87] to high densities is related to the nature of the dipole–dipole interaction: depending on the mutual orientations of neighbouring molecules, they can both be attracted and repelled. (Whereas in the case of a spherically symmetric potential, the nature of the interaction between the molecules does not depend on their orientations.) The position of the point of intersection is also affected by the interaction energies of the solvent molecules with the reagents.

Bimolecular reaction. Figure 64.3 shows how the rate of the bimolecular reaction depends on the concentration of the solvent. This figure shows the total velocity of U_{AC}, which is the sum $\sigma_A\sigma_C = 64$ of different partial velocities from different locations of reagents A and C. The calculation in Fig. 64.3 is carried out for $\alpha = 0.5$ (1), 0.75 (2) and 1.5 (3) . The results of comparisons of the normalized values of the rates of the bimolecular reaction U are given at $T = 300$ K for two limiting cases of environment relaxation: fast (solid lines) and slow (points) reactions.

As in the case of a monomolecular reaction, in the case of a fast bimolecular reaction we observe a practically monotonic change in the value of the rate, depending on the concentration of the solvent, which reaches its lowest value when the solvent concentration reaches its maximum value. In this case, the solvent inhibits the course of the reaction, making it difficult to meet reagents and increasing the value of the effective activation energy of this stage due to lateral solvent–reagent interactions. For a slow reaction, as the fraction of the solvent increases, the reaction rate decreases much more rapidly (curve 2), then there is a region of a slight change in the velocity, but when θ_{H2O} reaches ~0.8, an increase in the reaction rate occurs, passing through a maximum, reaches a common point at $\theta_{H2O} \to 1$. In the region of large densities, curves intersect for two types of relaxation of the environment. As above, the reason for the difference in the course of these two types of curves is that for these two approximations the weights of the configurations of the solvate shells of AC differ in the expression for the reaction rate differently.

To illustrate the existence of a point of intersection of curves for fast and slow reactions, a calculation is given for the situation when $\alpha = 1.5$ (curve *3*). In this case, an increase in the concentration of the solvent increases the reaction rate in both cases, however, even here, at high densities, there is an obligatory intersection of the curves. (Both curves 3 are given on a scale reduced by three orders of magnitude to put all the curves in one field.)

The obtained results testify to the generality of the previous conclusions [87] about the significant difference in the concentration dependences of the reaction rates depending on the nature of the relaxation of the environment and the mandatory intersection of curves corresponding to different types of relaxation of the nearest environment. The position of the point of intersection θ_{H2O} is due to the nature of the dipole–dipole interaction and the coupling of reagents with the solvent. According to the general analysis of the lattice gas model (LGM) [81,98], the presence of attraction between the laterally interacting particles can lead to a non-linear

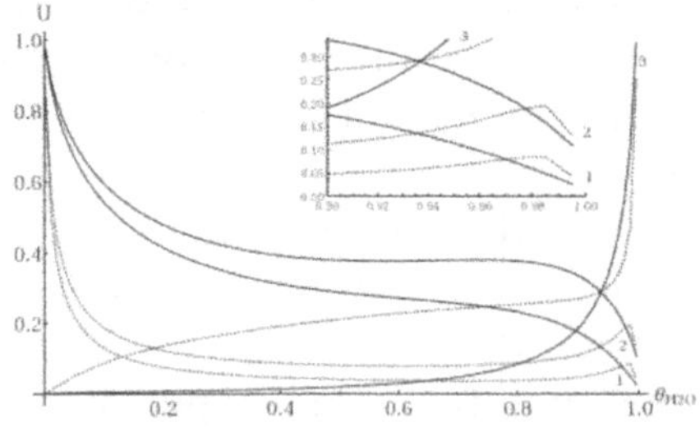

Fig. 64.3. The rate U_A of the bimolecular reaction (in relative units) as a function of the concentration of the solvent θ_{H2O}; T = 300 K, α = 0.25 (*1*), 0.5 (*2*), 1.5 (*3*). The inset – the region of large densities.

concentration dependence of the rate of the elementary reaction. The average values of the rate shown in Fig. 64.3 are obtained by weighting over all orientations of neighbouring molecules. The nature of the non-linearity of curves *1* and *2* for a fast reaction is small. In contrast, the course of the corresponding curves for the slow relaxation of the environment is much more nonlinear, and gives maxima in the region of large θ_{H2O}. This type of non-linearity is a consequence of phase transitions that are absent in the given system (and it is not present when considering the reaction as a fast process) [81,98]. This indicates that the assumption of an equilibrium distribution of surrounding molecules leads to an non-physical result, i.e. to the distortion of the properties of a real system.

Self-diffusion coefficient. The process of self-diffusion can be formally written in the form of a bimolecular reaction $(H_2O)_f + V_g = V_f + (H_2O)_g$, in which the indexes of adjacent sites f and g appear in the form of indices, and the symbol V_g denotes a free site with the number g. Thus, the hopping of the water molecule is a particular case of the bimolecular reaction proceeding at the sites f and g. The theory makes it possible to find the value of the coefficient of self-diffusion of water and to study its dependence on temperature (Fig. 64.4) [90].

The calculation is carried out under the condition that the considered hopping of a molecule into an adjacent vacancy occurs much faster than a local restructuring of the whole environment. The transition to the processes of thermal motion of molecules is associated with a graphic illustration of the correctness of the hypothesis of a weak change in the state of neighbours during the jump of the molecule under consideration, according to the Frenkel hypothesis [14]. The assumption [78] about the equilibrium distribution of neighbours would be in contradiction with the logic of the process: that the selected molecule should be moved so that similar neighbours quickly adapt to the transition state of this stage. That is, the set of neighbour jumps is necessary to make one hop of the molecule in question.

For the energy parameters of water molecules found in [93], the optimal value of the parameter $\alpha = 0.65$ is determined from the comparison of the constructed theoretical curve with the experiment [99]. (It should be noted that the same experiment can also be satisfactorily approximated by the molecular dynamics method [100].) The good agreement obtained indicates the adequacy of the lattice model and the validity of its assumptions. The found value

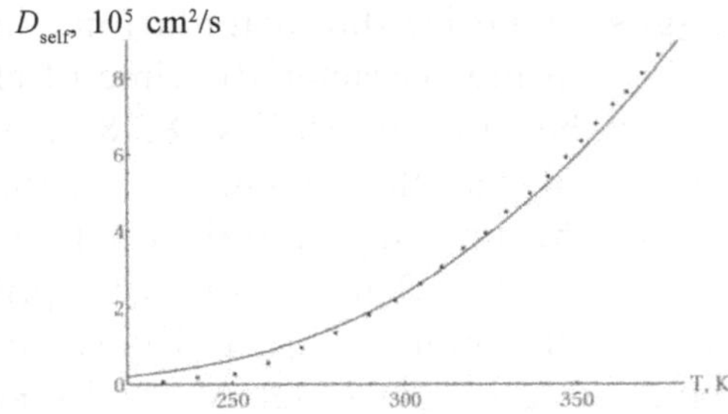

Fig. 64.4. Comparison of the temperature dependence of the coefficient of self-diffusion of water molecules: experiment (points) [99] and calculation of this work (line).

of α indicates a relatively small activation energy, which must be overcome by the water molecule during the jump in the course of this 'reaction'.

Thus, the comparison of the rates of relaxation of the medium to the concentration dependences of the rates of mono- and bimolecular reactions (in the kinetic regime of their course) leads to differences in reaction rates up to 4–5 times and they have a different course of concentration dependences. The considered equations of reaction rates refer to two limiting cases: $\tau_{rel} \ll \tau_{reaction}$ (slow reaction) and $\tau_{rel} \gg \tau_{reactions}$ (fast reaction), which are determined independently by the characteristic migration times of τ_{migr} and rotations τ_{rot} of the solvent molecules in the pure solvent $\tau_{rel} = \max(\tau_{migr}, \tau_{rot})$ and reactions $\tau_{reaction}$ in an inert solvent (or vacuum). Analysis of the concentration dependences of the reaction rates in a wide range of solvent concentrations θ_R ($0.3–0.7 < \theta_R < 1$) allows one to discriminate which of the two situations is realized in the experiment. If there are less strict ratios $\tau_{rel} < \tau_{reaction}$ and $\tau_{rel} > \tau_{reaction}$, then two situations are possible, caused by the imperfection of the reaction system. The first situation is that when the change in ΔE^{ef} preserves these relations. Then, for the same characteristic concentration dependences of reaction rates on θ_R, these relations can be discriminated. The second situation is such that when the times τ_{rel} and $\tau_{reactions}$ become commensurable, and to take this fact into account it is necessary to use more exact equations [56, 88], which take into account the equilibrium with respect to the degrees of freedom and the disequilibrium with respect to the remaining degrees of freedom.

The traditional idea of the existence of the chemical potential of AC in dense phases is in contradiction with the real times of molecular motions of molecules and with the procedure of statistical averaging of molecular distributions. Typical times of translational displacements and changes in molecular orientations in dense phases are, as a rule, larger than the characteristic time of chemical transformation, estimated on the order of 10^{-13} s [82,83]. Actually, the chemical reaction associated with the passage of the top of the activation barrier occurs in the unchanged field of neighboring molecules that form a local barrier. The notions of equilibrium relaxation of the molecular environment are unlikely, since the relaxation rate would have to be on the order of 10^{-14} s and less, which is physically unrealistic. Analysis of the molecular model does not exclude the possibility of realizing the equilibrium distribution of atomic and molecular polarization, which has characteristic relaxation times of less than 10^{-13} s. In this case, the model must be supplemented by taking into account the polarization contributions, which are discussed in [101, 102].

Concluding the analysis of thermodynamic interpretations by discussing the question of the inadmissibility of introducing parameters into any thermodynamic equations, it should be emphasized that the plausibility of the expression for the reaction rate constant through the internal states of the reagents and the transition state in the theory of absolute reaction rates (k_0') is associated with the influence of environments through activity coefficients. For equilibrium characteristics, activity coefficients are a formal way of expressing experimental data. But for a transitional state such an interpretation gives a qualitatively wrong result. The relaxation times of the transition state are characterized by τ_{temp}. The calculation of the activity coefficient for an activated complex implies the realization of averagings over all configurations with a characteristic relaxation time τ_{den}. To use α_M^* we required the time τ when the activated complex repeatedly terminates the reaction, $\tau > \tau_{den} \gg \tau_{temp}$. This leads to qualitative distortions in the magnitude of the reaction rate since the introduction of γ^{**} violates the meaning of the activated complex as a transition state [78,82,83]. Calculations [87,89,90] illustrate this discrepancy. This factor is principally important for the averaging of the kinetic equations in obtaining the mean values $< S_{fg}^{(r,n)} \theta_{fg} >$ of the transported moments that enter into the expressions for the kinetic equations of Chapter 5.

References

1. Thomson W.T., Phil. Mag. 1971, V. 42, P. 448.
2. Yang C.N., Lee T.D., Phys. Rev. 1952. V. 87. P. 404.
3. Lee T.D., Yang C.N., Phys. Rev. 1952. V. 87. P. 410.

4. Hill T.L., Thermodynamics of Small Systems. Part 1. New York Amsterdam: W. A. Benjamin, Inc., Publ., 1963. Part 2. 1964.
5. Huang K.. Statistical Mechanics. Moscow, Mir, 1966. 520 p.
6. Tovbin Y.K., Zh, fiz. khimii. 2010. V. 84. No. 10. P. 1882. [Rus. J. Phys. Chem. A, 2010, V. 84. No. 10. C. 1717].
7. Tovbin Y.K., Fiz.-khim. poverkhnosti i zashchita materialov. 2010. V. 46. No. 3. P. 261. [Protection of Metals and Physical Chemistry of Surfaces, 2010, Vol. 46, No. 3, P. 309].
8. Gibbs J.W., Thermodynamics. Statistical Mechanics. Moscow, Nauka, 1982.
9. Rice O.K., J. Phys. Chem. 1927. V. 31. P. 207.
10. Prigogine I., Defay R., J. Chem. Phys. 1949. V. 46, p. 367.
11. Ono S., Kondo C. Molecular theory of surface tension. Moscow, IL, 1963. [Handbuch der Physik, Vol X (Springer) 1960].
12. Tovbin Yu.K., The Molecular Theory of Adsorption in Porous Solids, Moscow, Fizmatlit, 2012. [CRC Press, Taylor & Francis Group, 2017].
13. Anisimov M.P., Usp. khimii. 2003. V. 72. No. 7. P. 664.
14. Frenkel J., Kinetic theory of liquids., Publishing House of the USSR Academy of Sciences, 1945.
15. Timoshenko S.P., Goodier J., Theory of elasticity. Moscow, Nauka, 1979.
16. Lame G., Lecons sur la Theorie ... d el'Elasticite, Paris, 1852.
17. Landau L.D., Lifshits E.M. Theory of elasticity. T.7. (page 11, problem 2).
18. Tovbin Yu.K., Zh. fiz. khimii. 2017. V. 91. No. 9. P. 1453. [Rus. J. Phys. Chem. A, 2017, V. 91, No. 9, P. 1621].
19. Rowlinson, J., Widom B., Molecular theory of capillarity. Moscow, Mir, 1986. p. [Oxford: ClarendonPress, 1982].
20. Adamson A. W., The Physical Chemistry of Surfaces, Mir, Moscow, 1979. [Wiley, New York, 1976].
21. Buff F.P., J. Chem. Phys. 1955. V. 23. P. 419.
22. Kondo S.J., Chem. Phys. 1956, V. 25, p. 662.
23. 23. Rusanov A.I.. Phase equilibria and surface pnehomena. Leningrad, Khimiya, 1967.
24. Tovbin Yu.K., Rabinovich A.B., Izv. AN, Ser. khim., 2009. No. 11. P. 2127. [Russ. Chem. Bull. 58, 2193 (2009)].
25. Tovbin Yu.K., Zh. fiz. khimii. 2010. V. 84. No. 2. P. 231. [Rus. J. Phys. Chem. A, 2010, V. 84. No. 2. P. 180].
26. Tovbin Yu.K., Rabinovich A.B., Zh. fiz. khimii. 2013. V. 87. No. 2. P. 337. [Russ. J. Phys. Chem. A, 2013, V. 87, No. 2, P. 329].
27. Tovbin Yu.K., Rabinovich A.B., Izv. AN. Ser. khim. 2010. No. 4. P. 663. [Rus. Chem. Bull. 2010. T. 59. No. 4. P. 677].
28. Bykov T.V., Shchekin A.K., Kolloid. zh. 1999, V. 61, P. 164.
29. Bykov T.V., Shchekin A.K., Neorgan. mater, 1999, V. 35, P. 759.
30. Bykov T.V., Zeng X.C., J. Chem. Phys., 1999, V. 111, P. 3705.
31. Bykov T.V., Zeng X.C., J. Chem. Phys., 1999, V. 111, P. 10602.
32. Tovbin Yu.K., Rabinovich A.B., Izv. AN. Ser. khim. 2010. No. 4. P. 839. [Russ. Chem. Bull. 59, 857 (2010)].
33. Hill T.L., Thermodynamics of Small Systems. Part 2. New York Amsterdam: W. A. Benjamin, Inc., Publ., 1964.
34. Van der-Waals I.D., Konstamm F., Courses in thermostatics. Moscow, ONTI, 1936.
35. Bakker G., Kapillaritat und Oberflachenspannung, Handbuch der Experim-ental physik, Bd. VI, Leipzig, 1928.

36. Tolman R.C., Journ. Chem. Phys. 1948, V. 16, P. 758.
37. Mason E.A., Spurling T.H., The Virial Equation of State. Moscow, Mir, 1972. [The International Encyclopedia of Physical Chemistry and Chemical Physics. Topic 10. The Fluid state. V. 2].
38. Hirschfelder J. O., Curtiss Ch. F., Bird R. B., Molecular theory of gases and liquids. – Moscow: Inostr. Lit., 1961. – 929 p. [Wiley, New York, 1954].
39. Hill T.L.. J. Chem. Phys. 1951. V. 19. P. 261.
40. Hill T.L.. J. Chem. Phys. 1952. V. 20, P. 141.
41. Tonks L., Phys. Rev. 1936, V. 50, P. 955.
42. Pawlow P. Z., Phys. Chem. (Munich) 1909, V. 65, P. 548.
43. Couchman P.R., Jesser W.A., Nature. 1977, V. 236, P. 481.
44. Sirotin Yu.I., Shaskol'skaya M.P., Fundamentals of crystal physics. Moscow, Nauka, 1979.
45. Landau L.D., Zh. Eksp. Teor. Fiz. 1937. V. 5. P.627.
46. Tolman R.C., J. Chem. Phys. 1949, V. 17, P. 333.
47. Kuni F.M., Vestnik LGU. 1964. N. 22. P. 3.
48. Kuni F.M., Phys. Lett. A. 1968, V. 26, p. 305.
49. Kuni F.M., Rusanov A.I., Dokl. AN SSSR. 1967. V. 174. No. 2. P. 406.
50. Landau L.D., Lifshitz E.M., Statistical Physics. V. 5. Moscow, Nauka, 1964.
51. Tovbin Yu.K., Komarov V.N., Zaitseva E.S., Zh. fiz. khim. 2016. V. 90. No. 10. P. 1570. [Rus. J. Phys. Chem. A, 2016, V. 90, No. 10, P. 2096].
52. Naghizadeh J., Rice S.A. J. Chem. Phys. 1962, V. 36, P. 2710.
53. Borovskii, I.B., Gurov, K.P., Machukova, I.D., Ugaste, Yu.E., Processes of mutual diffusion in alloys. Moscow, Nauka, 1973.
54. Bokshtein, B.S., Bokshtein, S.Z., Zhukhovitskii, A.A., Thermodynamic and Kinetic of Diffusion in Solids, Moscow: Metallurgiya, 1974.
55. Gurov B.A., Kartashkin B.A., Ugaste Yu.E., Mutual diffusion in a multiphase metallic system. Moscow, Nauka, 1981.
56. Tovbin Yu. K., Theory of physical and chemical processes at the gas-solid interface. – Moscow: Nauka, 1990. – 288 p. [CRC, Boca Raton, Florida, 1991].
57. Hill T.L., Statistical Mechanics. Principles and Selected Applications. – Moscow: Izd. Inostr. lit., 1960. – 486 p. [N.Y.: McGraw–Hill Book Comp. Inc., 1956].
58. Properties of elements. Directory, ed.. M.E. Drits. Moscow, Metallurgiya, 1985.
59. 59. Fisher I.Z., The problem of many bodies and plasma physics. Moscow, Nauka, 1967.
60. Raist P. Aerosols. Introduction to the theory. Moscow, Mir, 1987.
61. Chandrasekar S., Liquid crystals. Moscow, Mir, 1980.
62. Tager A.A.. Polymer chemistry. Moscow, Khimiya, 1978. 544 p.
63. Kireev V.A., Physical chemistry course. Moscow, Khimiya, 1975.
64. Kubo R. Thermodynamics. Moscow, Mir, 1970.
65. Panin V.E., Egorushkin B.E. Fiz. mezomekhanika. 2008. V. 11. No. 2. P. 9.
66. Leontovich A.M., Zh. Eksper. Teor. Fiz., 1938. V. 8. V. 844.
67. Bazarov I.P., Thermodynamics. Moscow, Vysshaya shkola. 1991.
68. Leontovich A.M. Introduction to thermodynamics. Statistical Physics. Moscow, Nauka, 1983.
69. Ionov V.N., Selivanov V.V. Dynamics of destruction of a deformed body. Moscow, Mashinostroenie, 1987.
70. Tovbin Yu.K., Titov S.V., Komarov V.N., Fiz. Tverd. Tela. 2015. V. 57, No. 2. P. 342. [Physics Solid State, 2015, V. 57, No. 2, P. 360].
71. 71. Fuchs N.A. Mechanics of Aerosols. Moscow, Khimiya, 1959. 500 p.

72. Lushnikov A.A., Sytygin A.G., Usp. khimii. 1976. V. 45. P. 385.
73. Lushnikov A.A., Phys. Rev. E. 2007. V. 76. P. 011120.
74. New materials. ed. Yu.S. Karabasov. Moscow, MISA, 2002.
75. Zhilyaev A.P., Pshenichnyuk A.I., Superplasticity and grain boundaries in ultrafine-grained materials. Moscow, Fizmatlit, 2008.
76. Chuvil'deev V.N. Non-equilibrium grain boundaries in metals. Theory and applica¬tions. Moscow, Fizmatlit, 2004.
77. Aaronson H.I., et al., Mechanisms of Diffusional Phase Transformations in Metals and Alloys. CRC Press, Taylor & Francis Group, Boca Raton, FL. 2010.
78. Glasston S., Laidler K.J., Eyring H., Theory of absolute reaction rates. Moscow, IL, 1948 [Princeton Univ. Press, New York, London, 1941].
79. Temkin M.I., Zh. fiz. khim. 1941. V. 15. P. 296.
80. Temkin M.I., Zh. fiz. khim. 1950. V. 24. P. 1312.
81. Tovbin Yu.K., Progress in Surface Sci. 1990. V. 34. No. 1-4. P. 1-236.
82. Entelis S.G., Tiger R.L. Kinetics of reactions in the liquid phase. Moscow, Khimiya, 1973.
83. Melvin-Hughes E.A., Equilibrium and kinetics of reactions in solutions. Moscow, Khimiya, 1975.
84. Marcus R.A., Ann. Rev. Phys. Chem. 1964. V. 15. P. 1.
85. Levich V.G., Itogi nauki. Elektrokhimiya. Moscow: VINITI, 1967. P. 5.
86. Dogonadze R.R., Kuznetsov A.M., Itogi nauki i tekhniki. Kinetika i kataliz. V. 5. Moscow, VINITI, 1978. P. 5.
87. Tovbin Yu.K., Votyakov E.V., Zh. fiz. khimii. 1997. V. 71. No.1. P. 271. [Russ. J. Phys. Chem. 1997. V. 71. No. 2. P. 214].
88. Tovbin Yu.K., Zh. fiz. khim. 1996. V. 70. No. 10. P. 1783. [Russ. J. Phys. Chem. 1996. V.70. No. 10, C. 1655].
89. Tovbin Yu.K., Titov S.V., Sverkhkriticheskie flyuida: teoriya i praktika. 2011, V. 6, No. 2. P. 35. [Rus. J. Phys. Chem. C, 2011. V. 5. No. 7. P. 1135].
90. Tovbin Yu.K., Titov S.V., Zh. fiz. khim. 2013. V. 87. No. 2. P. 205. [Rus. J. Phys. Chem. A, 2013, V. 87, No. 2, P. 185].
91. Bell G.M.. J. Phys. S. 1972. V. 5. No. 9. P. 889.
92. Bell G.M., Salt D.W., J. Chem. Soc: Faraday Trans. Pt. 2. 1976. V. 72. No. 1. P. 76.
93. Titov S.V., Tovbin Yu.K., Izv. AN. Ser. khimii. 2011. No. 1. P. 12. [Rus. Chem. Bull. 2011. V. 60, No. 1, P. 11].
94. Smirnova N.. Molecular theory of solutions. Leningrad. Khimiya, 1987.
95. Prausnitz J.M., et al., Molecular thermodynamics of fluid-phase equlibria. 2nd Ed. New Jersey: Prentice-Hall Inc. Englewood Cliffs, 1986.
96. Water. A comprehensive treatise. Ed. F. Franks. New York - London: Plenum, 1972. V. 1.
97. Eisenberg, D., Kautsman V. Structure and properties of water. Leningrad, Gidrometeoizdat, 1975.
98. Tovbin Yu.K., Dynamics of gas adsorption on non-uniform solid surfaces. Eds. W.Rudzinski, W.A. Steele, G. Zgrablich. Amsterdam: Elsevier, 1996. P. 240.
99. Angell C.A., Water: A Comprehensive Treatise, F. Franks, ed. 1978. V. 7. P. 23.
100. Malenkov G.G., et al., J. Molec. Liquids. 2003. V. 106. No. 2–3. P. 179.
101. Misurkin I.A., Titov S.V., Zh. fiz. khimii. 2008. V. 81. P. 1781. [Russ. J. Phys. Chem. A 82, 1672 (2008)].
102. Tovbin Yu.K., Zh. fiz. khim. 2014. V. 88. No. 11. P. 1752. [Rus. J. Phys. Chem. A, 2014, V. 88, No. 11, P. 1932].

Conclusion

Statistical thermodynamics answered the questions posed in the Preface about the limitations of the use of the equations of macroscopic thermodynamics in the description of small systems. The molecular theory singled out three characteristic sizes of the radius of spherical drops (as the simplest example of a small body), corresponding to:

(1) the beginning of the process of appearance of a dense phase with a surface tension $\sigma(R_0) \geq 0$, $R_0 \sim 10\lambda$, where λ is the average distance between adjacent molecules of the drop. The minimum phase size R_0 allowed to unambiguously separate the notions of clusters and droplets (as small phases). The constructed theory of small systems made it possible to introduce a strict statistical definition of the 'embryo' of the new phase (by the condition $\sigma = 0$), which is a criterion distinguishing between the concepts 'cluster/associate' or 'small phase'.

(2) the region of applicability of the thermodynamic description of the surface tension of a droplet, when for $R > R_{t2} \sim 90\lambda$ the discreteness of the matter and the contributions of spontaneous fluctuations can be completely neglected. When $R_{t1} < 41\lambda$, density fluctuations must be taken into account. The region of thermodynamic description for bulk phases without an allowance for boundary contributions is given by analogous quantities $R_{t1}^{(V)} = 17\lambda$ and $R_{t2}^{(V)} = 29\lambda$. The theory has shown that thermodynamics can not be applied to small systems and to the calculation of the surface tension of any interfaces. Although statistical mechanics limits the applicability of thermodynamics beyond the phase approximation (for $R < R_t$), but extends this approximation to small systems for all $R > R_0$.

(3) large droplet size regions, in which surface tension values are close to the bulk value, $R_b \sim 10^2 \div 10^3\lambda$; at $R > R_b$, the dimensional dependence of the surface tension $\sigma(R) = \sigma_{\text{bulk}}$ can be neglected.

Statistical thermodynamics clarifies and restricts various thermodynamic constructions in the field of non-equilibrium processes. From the standpoint of the non-equilibrium theory, the artificiality of the concept of 'passive forces' laid down by Gibbs in the thermodynamics for bulk phases, which reflects the level of knowledge of those years with respect to the chemical kinetics and properties of solids, but not the violation of any laws of thermodynamics, was shown. Also, the principle of deriving criteria for estimating the yield of system states from the condition of local equilibrium was developed. This makes it possible to correctly relate the description of equilibrium and non-equilibrium processes, which is especially important for ensembles of small solids, which have been actively investigated recently (and which currently have virtually no thermodynamic interpretation). The principal importance of taking into account the relaxation times of various properties and the incorrectness of using the concept of the coefficient of activity of the activated complex in the theory of absolute reaction rates are noted.

The formulation of the list of constraints completes the construction of classical thermodynamics, since it specifies the field of its application and limits its use to small systems.

We should separately dwell on the problem of the Kelvin equation (KE). The analysis showed that this equation can not be derived either from the equilibrium or from the kinetic molecular theory. Historically, it has become widespread in all problems with a curved interface: it is used for both equilibrium and dynamic conditions. When discussing the conditions for its application, one can only speak of deviations between the calculated values of the KE and the molecular theory under static conditions. Today, KE is used for all droplet sizes, ranging from the micron range to the nanometer range. The natural criterion for the use of KE is the difference between $P(R)/P_s$ from a similar value in the molecular theory. Such a criterion is the dimensionless parameter $\psi = 2\beta\sigma V_0/R$. The smaller the deviation of the parameter ψ from zero, the more 'justified' is the use of KE the less it distorts the molecular characteristics.

Here it is necessary to separate two situations: the problem of forming an isolated new phase in which the main role is played by intermolecular interactions and the problem of the form of a liquid film near the surface of a solid body in which the surface potential plays the main role, like any other external field (Gibbs considered the action of the gravitational field on surface tension).

In the first problem, the deviation from the KE appears only due to the size dependence of the surface tension $\sigma(R)$ (provided there are no external fields).

In the second problem, the distribution of the film and the formation of its meniscus is the result of the combined action of the surface and intermolecular potentials, and the difference in the shape of the meniscus from the circular one, embedded in the KE, largely depends on the shape of the surface. In the case of cylindrical pores, the form of the vapour–liquid interface varies greatly from the pore size: at small nanometer diameters, the meniscus can be approximated by a spherical shape, whereas as the diameter increases, the meniscus flattens and tends at the centre of the pore to a flat shape, partially retaining a curved shape (convex or concave, depending on how the interaction potential of the liquid molecules with the walls is) only near the walls (reference [12] of Chapter 7). Therefore, the analysis of capillary condensation in the pores with the help of KE is conditional – for the use of KE the shape of the meniscus should remain spherical.

Parameter ψ corresponds to both situations. The increase in R equally flattens the phase boundary of an isolated small body in the bulk phase and, as indicated above, the meniscus in the pores. In the first case, deviations of $P(R)/P_s$ in terms of KE due to differences in surface tension correspond to a large range of sizes from 20 nm at temperatures near the melting point to 50 nm at elevated temperatures, but outside the critical region.

In the second case, the specific results of the comparisons of the KE and the theory in terms of the $P(R)/P_s$ values for the simplest gases (nitrogen, argon and others, which obey the so-called law of the corresponding states) show: at a diameter of 50 nm, the deviations from the molecular theory are 1%, at 12 nm ~ 10%, and at 4.5 nm the difference is ~100% (*and in general is not applicable* at sizes less than 4.1 nm!) (Ref. [12] Chapters 7). Thus, the application of the KE here is limited from below by values of the order of 20 nm.

The above numbers refer to weakly interacting molecules. As σ increases, the numerator of the parameter y increases accordingly, and similar deviations refer to large values of the radius R. So, to compare the characteristic values, it can be pointed out that the above characteristic values of the deviations of the KE and the theory for argon should be attributed to the large linear size of the aluminum droplets in 2.6, in 3.8, mercury in 7.9 times. Thus, the lower limit of the applicability of the KE for atoms with a strong attraction can exceed 100 nm and enter the submicron range. In these qualitative estimates, of course, temperature effects affecting the values of the parameter ψ and the magnitude of the surface tension depending on T and R are not taken into account, of course. Therefore, the value of the parameter, permissible for using the approximation with the help of KE, must be specified in concrete conditions.

The analysis of the foundations of classical thermodynamics from the point of view of statistical thermodynamics has often confirmed two opinions in the literature on the place of thermodynamics among other disciplines. Most often quoted are the words of A. Einstein, 1949, "A theory is the more impressive the greater the simplicity of its premises, the more different kinds of things it relates, and the more extended its area of applicability. Therefore the deep impression that classical thermodynamics made upon me. It is the only physical theory of universal content which I am convinced will never be overthrown, within the framework of applicability of its basic concepts"

The analysis showed that the first three stages (out of four) of thermodynamics introduced in the Preface are not subject to doubt – when moving from small bodies to macroscopic dimensions, all the positions introduced by Gibbs in the formulation of the phase approximation for describing non-uniform systems are completely satisfied. However, the fourth stage of classical thermodynamics, which pertains to small systems, can not be attributed to its basic propositions.

Also popular is the opinion of J.W. Gibbs, 1902: "But although, as a matter of history, statistical mechanics owes its origin to investigations in thermodynamics, it seems eminently worthy of an independent development, both an account of the elegance and simplicity of its principles, and because it yields new results and places old truths in a new light in departments quite outside of thermodynamics". The words about the presentation of 'old truths in a new light' proved to be prophetic not only in relation to the numerous applications of statistical physics at the present time, but also in relation to the analysis of the very foundations of thermodynamics for small systems from the positions developed by Gibbs to the methods of statistical description of the thermodynamic properties of systems. The fourth stage of thermodynamics, connected with the approach to small systems, turned out to be incorrect. The problem of the thermodynamic description of curved surfaces, which began with the Kelvin equation (1871), arose from the formal transfer of mechanical representations to thermodynamics. This is a purely historical circumstance, because then mechanical ideas prevailed about the curved boundaries of the phase separation, and they were in no way connected with the notion of chemical potential, which was not yet introduced. During the development of Gibbs' thermodynamics (1878), it was impossible to foresee those mathematically rigorous results of statistical mechanics that would indicate the incorrectness of the thermodynamic approach to small systems. We are talking about the Yang–Lee condensation theory (1952) and the existence of stable solutions to the droplet concentration profile without taking into account, and taking into account the influence of the Laplace equation (2010) (both types of solutions satisfy the Yang–Lee theories). This radically changes the situation with the treatment of small systems – their description is possible only within the framework of statistical thermodynamics.

The molecular theory made it possible to reveal the main reason for the difference between the molecular statistical and thermodynamic approaches in describing small systems. The reason for the incorrectness of the application of thermodynamics is based on neglecting the consideration of the real properties of systems connected under isothermal conditions with the relaxation times of impulse and mass transfer. This is a non-trivial reason, caused by the essence of thermodynamics itself: its bases are based solely on experimental data. If they are violated, then thermodynamics can not give an adequate description of the properties of systems. Equilibrium thermodynamics ignores the well-known relationships between the relaxation times of impulse transfer and mass, assuming that for large times this does not play a role. However, the molecular theory shows that the trajectory of the system's output to the limiting states (at large times) during phase transformations is ambiguous, and the difference in the relaxation times for impulse and mass transfer plays a fundamental role, since it leads to two different final states.

The experiment shows that the impulse relaxation times are, as a rule, much less than the relaxation times of the mass. This automatically excludes metastable states from the phase equilibrium of the vapour–liquid system, which, in contradiction to the Yang–Lee condensation theory, is present in the Kelvin equation, and leads to the appearance of equilibrium drops. This once again confirmed the basic position of thermodynamics that the violation of the conditions relating to the properties of experimental systems should lead to a violation of the conditions for the applicability of thermodynamic relations. Formally, for macroscopic systems, ignoring the role of relaxation times in the transfer of impulse and mass, does not affect the conditions for complete phase equilibrium at asymptotically large times. Here the term 'asymptotically' large times plays a key role. (For simple liquid systems, they are realized, and for solid ones, not always.) The relaxation times of the dynamic variables of the system form a hierarchy of relaxation times of different dynamic variables with respect to a given time scale in the experimental system under study. 'Automatically' for this time scale, the size scale of the subsystem and the type of dynamic variables that corresponds to the condition of internal complete equilibrium for these dynamic variables are formed, and the sizes of subsystems in which complete equilibrium is not realized for each of the remaining dynamic variables of the entire system. Therefore, the concept of complete equilibrium must always be strictly defined in relation to a fixed time scale. The discussed choice of the time scale was omitted by Gibbs (1875) from an analysis of the conditions for generalizing the principle of mechanical equilibrium to thermodynamic variables. Small systems were considered by Gibbs later in 1878 without revision of the previously formulated provisions.

At one time the development of classical thermodynamics served as a catalyst for the development of science. Classical thermodynamics remains today the foundation for all natural sciences, in which it is necessary to take into account the conservation of energy and the study of thermal effects. The internal energy of matter is associated with different types of interactions on different spatial scales and questions related to thermodynamics will arise in the case of macroscopic sizes of the systems under study. The properties of particles themselves can be related to their quantum nature and through intermolecular interactions in ensembles lead to the corresponding dependences of thermodynamic functions on their quantum nature. With a decrease in the size of small systems classical thermodynamics loses its validity and it is necessary to take into account the size effects indicated in this book. The found size criteria are characteristic values that reflect the properties of short-range potentials. The specifics of Coulomb charges were not discussed in the book. According to the most rough estimates, all the found size criteria will be increased approximately two times, because in systems with charge components they are neutral particles.

Thermodynamics is not an interpolation tool because of its model-free field of science. The introduction of any molecular specificity into it distorts its essence and leads to quasi-thermodynamic approaches, more precisely to 'hybrids', creating the illusion of scientific consideration. The term thermodynamic model is incorrect because the thermodynamics reflects the experiment to the full extent of simultaneous manifestation in it of all molecular effects. Any molecular interpretation is possible only on the basis of a molecular model, and its applicability should be tested by comparison with experiment and have a sufficient correct basis (that is, do not be too crude) to allow some extrapolation of results outside the parameter range (used in the comparison with experiment).

Today, the important role of thermodynamics for the organization of the experiment and its interpretation remains: it is control over the correctness of the use of the law of conservation of energy in real systems and the 'storage' of fundamental information on thermodynamic characteristics and the organization of experimental measurements to measure new and refine old thermodynamic characteristics.

For many solid-phase systems, the information extracted from the experiment today is approximate, the degree of approximation is difficult to control – there is not always evidence of achieving complete equilibrium. Achieving equilibrium should reflect the dynamic nature of processes that ensure the constancy of the values of the parameters in time and the independence of the values of the characteristics from the path of the transition. At low T, the system necessarily goes into a solid state, and for monitoring in them, measurements of the relaxation times of thermophysical characteristics, which are very few, are needed. The binding to solid state states follows from Nernst's heat theorem, which establishes a common reference point for all thermodynamic characteristics.

The recent development of the statistical mechanics of 'three-aggregate' state equations and their boundaries has made it possible to reach the same level of generality in describing three aggregate states and different phases, both in equilibrium and in non-equilibrium conditions, which were formerly characteristic only of thermodynamics. This opens the possibility of transition to a correct description of processes involving small systems, providing their interpretation of the thermophysical and thermodynamic characteristics, provided it is supplemented by modern statistical thermodynamics.

Appendix 1

Metastable drops

Chapter 3 shows that the molecular theory leads to the existence of equilibrium drops, which are not found in thermodynamics: they do not have a pressure jump inside the transition region and the Laplace equation does not need to be calculated for the concentration profile of the vapour–liquid interface. The theory makes it possible to obtain the equation of state both within the coexisting phases and the transition region between the phases. This eliminates the need to draw information about the state of the boundary from an independent condition of mechanical equilibrium (not related to the current chemical potential values). The mechanical equilibrium at the boundary of the equilibrium drop is ensured by the equality of the chemical potential and the equation of state in each transition monolayer. This kind of detailed information is impossible in thermodynamics.

Today, all the work on drops uses the Laplace equation, as in the derivation of the Kelvin equation, so the goal of this application is to show that the theory allows us to include in our consideration the traditional condition of mechanical equilibrium (via the Laplace equation). This allows us to compare the thermodynamic functions of equilibrium and metastable drops. This appendix gives an account of the traditional thermodynamics of metastable drops and its microscopic version in a discrete approach (including the 4th method of determining the surface of tension), which allows one to simultaneously discuss the differences between continual and discrete descriptions. In Chapter 7, both types of drops are compared with thermodynamic metastable drops, which follow from the Kelvin equation.

First, we consider the traditional thermodynamic description of a drop. In order not to complicate the formulas, we follow the exposition [1, 2], which is oriented only to a spherical boundary.

Determination of surface tension. A spherical surface (bubble or drop) is the only stable in the absence of an external field. We consider the situation with curved surfaces by the example of a drop of liquid in the surrounding vapour. We consider a closed system consisting of two phases α (internal) and β (external), separated by a spherical dividing layer, that is, the case of a transition zone of constant curvature. Consider a limited part of the system enclosed in a conical vessel, such as shown in Fig. A1.1, where ω and r denote the solid angle of the cone and the radial coordinate, measured from the vertex of the cone 0 [1], respectively.

The vessel consists of the part of the cone, enclosed between $r = R_\alpha$ and $r = R_\beta$, where $R_\alpha < R_\beta$. We assume that the system consists of (s–1) independent components, there are no external fields, and the centre of curvature of the spherical surface layer lies at the point 0. Then any intensive property of the system can depend only on r. We choose the sphere $r = a$ (dotted line in Fig. A1.1) as the dividing surface. It will divide the total volume V into two volumes V_α and V_β, equal to, respectively

$$V_\alpha = \omega\left[a^3 - (R_\alpha)^3\right]/3, \quad V_\beta = \omega\left[(R_\beta)^3 - a^3\right]/3, \tag{A1.1}$$

where the area of the dividing surface will be equal to: $A = \omega a^2$.

It is assumed that while the system is in mechanical equilibrium the pressure P_α in the internal phase α can not equal the pressure P_β in the outer phase β. Inside the bulk phases α and β, the pressures P_α and P_β are constant. However, the method can not be applied to the case of spherical drops so small that it is impossible to achieve uniform properties even at the centre of the phase α.

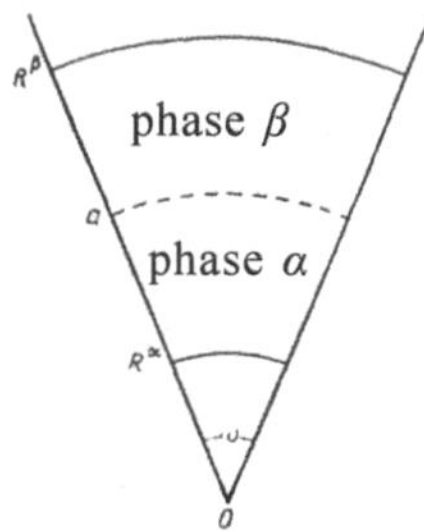

Fig. A1.1. Scheme for determining the surface tension of a drop.

Imagine that the solid angle ω increases by $d\omega$ by reversible isothermal displacement of the side wall, with all other variables remaining unchanged. The work done by the system in this process is proportional to $d\omega$ and can be written as $-\eta d\omega$. This work can be considered as consisting of two parts: the work $P_\alpha dV_\alpha + P_\beta dV_\beta$, connected with the change in the volumes of two uniform bulk phases, and the excess work, which we denote by $-\sigma dA$. Thus, it can be written that

$$dW = -\eta d\omega = P_\alpha dV_\alpha + P_\beta dV_\beta - \sigma dA. \tag{A1.2}$$

From a purely mathematical point of view, the term $-\sigma dA$ is simply the correction term introduced for the sum to give the complete work dW, which is a completely determined quantity containing (in addition to contributions from uniform parts in the phases α and β) the complex effects arising as a result of the presence of the surface layer. The coefficient σ can be considered, however, as a work done on the system associated with a unit increment of the area of the dividing surface for a given curvature. This allows us to consider the real system as if it consisted of two uniform phases, α and β, separated by a spherical film of zero thickness, with a radius a and a uniform tension σ in all directions.

Then it is natural to take σ, determined according to (A1.2), for the surface tension of the spherical boundary.

Using (A1.1), we obtain from (A1.2) [3]

$$\sigma = \frac{1}{3}a(P_\alpha - P_\beta) + \frac{K}{a^2}, \quad K = \eta + \frac{1}{3}(R_\beta)^3 - \frac{1}{3}(R_\alpha)^3, \tag{A1.3}$$

The quantity K does not depend on the choice of the dividing surface. Therefore, according to (A1.3), the surface tension determined by (A1.2), unlike the planar case, depends on the choice of the dividing surface.

Taking into account the definition of K and the expressions (A1.1), equality (A1.3) can be rewritten in an alternative form:

$$\eta\omega = \sigma A - P_\alpha V_\alpha - P_\beta V_\beta. \tag{A1.4}$$

Differentiating (A1.3) with respect to a, we obtain [3]

$$\sigma + \frac{1}{2}a\left[\frac{\partial\sigma}{\partial a}\right] = \frac{1}{2}a(P_\alpha - P_\beta). \quad \text{(A1.5)}$$

Here $[\partial\sigma/\partial a]da$ is the change in the surface tension value associated with the mathematical displacement of the dividing surface by da, provided that all physical quantities within the system and external conditions remain unchanged (the notation $[\partial\sigma/\partial a]$ should be distinguished from the derivative by the real dependence on the radius, which will be recorded as $\partial\sigma/\partial a$). This change in the magnitude of surface tension is due simply to the arbitrariness in the choice of the dividing surface, and it should not be confused with the variation of σ from the increase in the radius of the physical surface of the discontinuity. The derivative $[\partial\sigma/\partial a]$ plays an important role in the thermodynamics of a spherical interface.

Since the work done by the system associated with the change in dR_α and dR_β is $-P_\alpha\omega\,(R_\alpha)^2dR_\alpha + P_\beta\omega\,(R_\beta)^2dR_\beta$, the general expression for the elementary work dW instead of (A1.2) should be written in the form

$$dW = -\eta\omega + P_\beta\omega(R_\beta)^2 dR_\beta - P_\alpha\omega(R_\alpha)^2 dR_\alpha. \quad \text{(A1.6)}$$

Each of the terms on the right-hand side of (A1.6) does not depend on the choice of the dividing surface. From the formulas (A1.1) we obtain

$$dV_\alpha = \frac{d\omega}{3}[a^3 - (R_\alpha)^3] + \omega a^2 da - \omega(R_\alpha)^3 dR_\alpha \quad \text{(A1.7)}$$

$$dV_\beta = \frac{d\omega}{3}[(R_\beta)^3 - a^3] + \omega(R_\beta)^3 dR_\beta - \omega a^2 da \quad \text{(A1.8)}$$

$$dA = a^2 d\omega + 2\omega a da. \quad \text{(A1.8)}$$

Then, solving these equations for dR_α, dR_β and $d\omega$, substitute the result in (A1.6) and using (A1.1) and (A1.4), we obtain

$$dW = P_\beta dV_\beta + P_\alpha dV_\alpha - \sigma dA - \{P_\alpha - P_\beta - 2\sigma / a\}Ada. \quad \text{(A1.9)}$$

The last term on the right-hand side of this equation determines the differences in different ways of describing the thermodynamics of a spherical interface.

Using (A1.5), we can rewrite (A.1.9) in an equivalent form:

$$dW = P_\beta dV_\beta + P_\alpha dV_\alpha - \sigma dA - \left[\frac{\partial\sigma}{\partial a}\right] Ada \qquad \text{(A1.10)}$$

It is not difficult to show that (A1.10) remains unchanged with an infinitesimal unphysical change

$$dV_\alpha = -dV_\beta = Ada, \quad dA = 2Ada / a, \qquad \text{(A1.11)}$$

This is due to a purely mathematical displacement of the position of the dividing surface.

We now consider a change not only of ω, R_α and R_β, but also of other state variables. Equation (A1.10) remains valid even in this general case. Therefore, if we denote by dQ the heat received by the system, and by dU the change in its internal energy during this process, then the expression for the First Law of thermodynamics takes the following form:

$$dU = dQ - P_\beta dV_\beta - P_\alpha dV_\alpha + \sigma dA + \left\{P_\alpha - P_\beta - 2\sigma / a\right\} Ada. \qquad \text{(A1.12)}$$

It should be noted that this equation is also valid for systems that are not in thermodynamic equilibrium if the values of P_α and P_β are completely determined for them.

Surface tension. The mechanical equilibrium condition for a system consisting of two phases α and β separated by a spherical membrane of radius a and having a uniform tension σ leads to the well-known relation

$$P_\alpha - P_\beta = 2\sigma / a, \qquad \text{(A1.13)}$$

where $P_\alpha > P_\beta$, the pressure at the liquid boundary from the inside differs from the pressure outside.

On the other hand, we have the relation (A1.5), which is applicable to any system with a spherical interphase boundary. We now choose the dividing surface in such a way that the derivative $[\partial\sigma/\partial a]$ disappears, noting the quantities associated with this choice by the index s. The radius of this dividing surface a_s is determined from the relation [3]

$$\left[\frac{\partial\sigma}{\partial a}\right]_{a=a_s} = 0. \qquad \text{(A1.14)}$$

Thus, the chosen dividing surface turns out to be one of the most important, since only for such a surface from (A1.5) follows a simple form of the relation (A1.13):

$$P_\alpha - P_\beta = 2\sigma_s / a_s, \tag{A1.15}$$

According to Gibbs [4], this particular dividing surface is called the *surface of tension.* The agreement between (A.1.15) and the relation (A1.13) means that the *mechanical action* of a real spherical boundary having a complex structure can be replaced by the action of a simple flexible film whose location coincides with the position of the surface of tension having zero thickness and tension σ, the same in all directions [4,5].

The substitution of (A1.3) into (A.1.15) leads to the relation

$$a_s = \left(\frac{6K}{P_\alpha - P_\beta} \right)^{1/3} \tag{A1.16}$$

Using it, we can exclude K from (A1.3) and obtain [3]

$$\sigma = \frac{a_s^2 \sigma_s}{3a^2} + \frac{2\sigma_s a}{3a_s} \tag{A1.17}$$

Since the surface tension values are always positive, it is easy to see from (A1.17) that σ reaches its minimum value σ_s for $a = a_s$. Therefore, we can conclude that as long as a_s is large in comparison with the thickness of the transition layer, all the values of σ related to those dividing surfaces that are located inside or near this layer (we can expect that there is also a surface of tension in it), are almost equal to the minimum value of σ_s [3] (more precisely $\sigma = \sigma_s \left\{ 1 + \frac{(a-a_s)^2}{a_s^2} + o\left(\frac{(a-a_s)}{a_s} \right)^3 \right\}$). This means that from a macroscopic point of view, the surface of tension can be considered as *practically independent* of the position of the dividing surface, as long as the latter lies inside the surface layer (except for the case of drops and bubbles so small that their radii are comparable to the thickness of the surface layer).

Thus, the physical surface tension in the case of a spherical boundary can be completely determined by means of such σ. Finally, we must emphasize that the relations obtained in this subsection are also valid for systems that are not in equilibrium.

The considered difference of spherical surfaces in comparison with flat surfaces allows to repeat word for word and rewrite expressions for fundamental thermodynamic equations and expressions between excess values. As an example, we give only the expression for the excess free energy

$$F_b = \sum_{i=1}^{s-1} \mu_i N_b^i + \sigma A. \tag{A1.18}$$

If the dividing surface is chosen so that the sum $\sum_{i=1}^{s-1} \mu_i N_b^i$ vanishes, then (A1.18) reduces to equality

$$F_b = \sigma A. \tag{A1.19}$$

It follows that the surface tension on the spherical boundary is equal to the excess free energy per unit area of the dividing surface, if the latter is chosen so as $\sum_{i=1}^{s-1} \mu_i N_b^i$ vanishes. However, in the case of any other dividing surface, σ is no longer equal to excess free energy.

Discrete version of the theory [6]. We use the developments of [3,7–9], in which we explicitly describe how the surface tension depends on the position of the dividing surface to which it relates, and give a discrete exposition of the mechanical determination of the surface tension to show that it is fully equivalent continuum description, and to show what the refinements are when accounting for fluctuations in small systems. The symmetry of the spherical interface is considered in Section 24.

The presence of normal stresses in deformed cells in the form of curved monolayers should cause the appearance of tangential stresses. This situation is analogous to the relationship between normal and tangential stresses in macrosystems [10]. If we denote by S_q the surface area in the monolayer q ($S_q = 4\pi q^2$), then the mechanical equilibrium condition is formally written as

$$\left(\pi^N S\right)_q = \left(\pi^{N+1} S\right)_{q+1} + \left(\pi^T \Delta S\right)_q, \tag{A1.20}$$

where the subscript of the layer q refers to both factors in the parentheses. Here π_q^N is the normal component of the pressure tensor in the layer q, related to the surface of this monolayer S_q. The difference in the normal components of the pressure tensor in the layers q and $q + 1$ is balanced by the tangential component of the pressure tensor π^T_q in the layer q. The component π^T_q in a curved coordinate system is directed at an angle to the radius of curvature of the local section along which the normal component acts. The slope of the component π^T_q depends on the curvature of the surface under consideration. Therefore, the action of the tangential component consists in stretching the lower surface of the monolayer q from $S_{q-0.5}$ to its upper surface $S_{q+0.5}$ (or vice versa in compression from the surface $S_{q+0.5}$ to the surface $S_{q-0.5}$).

Equation (A1.20), constructed on a discrete set of monolayers of the boundary of a drop, can be written in two ways, using finite differences [11,12]: (a) to expand the increments of $(\pi^N S)_q$ and go to the difference derivatives for π^N_q and S_q, or (b) consider the difference derivative of the product $(\pi^N S)_q$. The difference increment of the function α between the layers q and $q + 1$ is denoted by $\Delta(\alpha)_q = \alpha_{q+1} - \alpha_q$.

(a) In the first case we rewrite (A1.20) as

$$\pi_q^N S_q = \pi_q^{N+1} S_{q+1} + \pi_q^T (S_{q-0.5} - S_{q+0.5}). \tag{A1.21}$$

Representing π^{N+1}_q in the form of the difference derivative as $\pi^{N+1}_q = \pi^N_q + \Delta\pi^N_q/\Delta q$ it follows that $\Delta\pi^N_q/\Delta q = \pi^N_q(S_q - S_{q+1})/S_{q+1} - \pi^T_q(S_{q-0.5} - S_{q+0.5})/S_{q+1}$. Neglecting the small difference between $(S_{q-0.5} - S_{q+0.5})$ and $(S_q - S_{q+1})$, which is equal to $1/q^2$, we obtain: $\Delta\pi_q^N / \Delta q = (\pi_q^T - \pi_q^N)(1 - S_q / S_{q+1})$. In the last bracket, the area ratio can be represented in the form $S_q/S_{q+1} = (q/q+1)^m = 1 - m/q$, where $m = 2$ for the sphere and $m = 1$ for the cylinder, which gives

$$\Delta\pi_q^N / \Delta q = m(\pi_q^T - \pi_q^N) / q. \tag{A1.22}$$

The discrete equation (A1.22) is equivalent to the first form of writing a continual differential equation – it is a microscopic analog of the macroscopic equation for the normal component of the pressure tensor. For comparison, we write out the equivalent forms of writing equations for the variation along the radius of the normal pressure component [1,9]: $\dfrac{dP_N(r)}{dr} = \dfrac{2[P_T(r) - P_N(r)]}{r}$, (1st form), $\dfrac{d(r^3 P_N(r))}{dr} = r^2[2P_T(r) + P_N(r)]$, (2nd form), and $\dfrac{d(r^2 P_N(r))}{d(r^2)} = P_T(r)$ (3rd form).

(b) In the second case, equation (A1.20) is written as

$$\Delta(\pi^N S)_q = \pi_q^T \Delta S_q, \tag{A1.23}$$

Here, the difference of the order of $1/q^2$ between the areas $(S_{q-0.5}-S_{q+0.5})$ and (S_q-S_{q+1}) is also neglected. To the right is a discrete increment of the surface area $\Delta S_q = 2q\lambda$, which is proportional to $dr^2 = 2rdr$ ($dr = \lambda$ and $r = q$) in the continuum calculus. Thus, the discrete equation (A1.23) is equivalent to the third form of the continuous record of the mechanical equilibrium equation in the drop.

The boundary conditions of both discrete equations are the pressure valurd in the liquid drop at $q = 1$ and in the vapour phase at $q = \kappa$. The quantity $\kappa = q_{\text{vap}} - q_{\text{liq}}$, $r \gg \kappa$, here q_{vap} and q_{liq} are the numbers of monolayers that determine the radii of the vapour and liquid phases, bounding the transition region.

We emphasize that the discrete analogues (A1.22) and (A1.23) of the first and third equivalent forms of equations [9] follow from the same discrete equation (A1.20). The difference in these equations consists in the method of expressing the difference between the pressures (the first form) or in the difference of the product of pressure on the area of the drop (the third form). Formally, the differences consist only in the replacement of the difference derivative by the differential derivative [11, 12]. However, there is a fundamental difference between the expressions for the components of the pressure tensor themselves. At the macroscopic level, we are dealing with stresses in local volumes containing a large number of particles. Here, the monolayer size is equal to the diameter of the molecule and the expressions for π_q^T and π_q^N reflect the curvature on the characteristic cell size $\Delta q = \lambda$.

In principle, the microscopic theory allows us to abandon the use of the concept of surface tension for the calculation of quasi-equilibrium distributions of molecules at the vapour–liquid interface in metastable conditions. For this, using the first form of equations for the normal components of the pressure tensor, an alternative description of the properties of the transition region of the boundary can be obtained. The difference equation (A1.22) determines the way the normal component π_q^N varies from layer to layer. For each layer q there exists a unique connection between π_q^N and the local section of the profile $\{\theta_q\}$ and $\pi_q^T(\{\theta_q\})$, which constitutes the system of equations on $\{\theta_q\}$. The equation as a whole from one boundary (with a given vapour pressure) to the other (up to the inside of the drop) determines the pressure inside the drop. This description, in principle, is suitable for any curved surface. After determining the profile $\{\theta_q\}$, local values $(aP)_q$ – the variable profile of the chemical potential in the metastable system, can be calculated.

The first mechanical determination of surface tension. The microscopic theory makes it possible to obtain a molecular interpretation of surface tension. To this end, consider equation (A1.23) and obtain by summation over the layers an integral relationship between the vapour pressure and the liquid, which describes the hydrostatic equilibrium of the drop as a whole. By transforming this connection, by analogy with the derivation of [1,7,9], we can construct a discrete analogue of the Laplace equation, from which the first definition of the surface tension σ is introduced.

Summation over the layers on the left side of the expression (A1.23) leads to the integral form of the notation $\Sigma_{1\kappa} = S_{\kappa+1}\pi_{\kappa+1} - S_0\pi_0$, where $S_{\kappa+1}$ and S_0 are the areas of the boundary circles of the drop from the vapour side and the liquid, $\pi_{\kappa+1}$ and π_0 are the pressures from the side of the vapour and the liquid. The sum on the right side (A1.23) gives $\sum_{1\kappa} = \sum_{q=1}^{\kappa} \pi_q^T \Delta S_q = \sum_{q=1+R}^{\kappa+R} \pi_q^T 2q\Delta q = 2\lambda \sum_{q=1+R}^{\kappa+R} \pi_q^T q$.

Thus, $S_{\kappa+1}\pi_{\kappa+1} - S_0\pi_0 = 2\lambda \sum_{q=1+R}^{\kappa+R} \pi_q^T\ q$, or taking into account the definitions of S_q and ΔS_q indicated above, we have

$$R_{\kappa+1}^2 \pi_{\text{vap}} - R_0^2\, \pi_{\text{liq}} = 2\lambda \sum_{q=1+R}^{\kappa+R} \pi_q^T q \qquad \text{(A1.24)}$$

Equation (A1.24) is identical to the continual expression for hydrostatic pressure around a drop of radius R. The derivation of the equation for the first mechanical determination of σ is largely identical to the derivation of σ for the continuum calculus. The essence of the derivation is the transition from the formula (A1.25) to the equation for mechanical equilibrium with respect to the reference dividing surface ρ_s, relative to which the surface tension σ_s determined from the condition of mechanical equilibrium of the moments of forces in the transition region is determined. We note that without determining the reference surface, it is impossible to uniquely introduce the surface tension and it is necessary to use the first form of the equations for the normal components of the pressure tensor, as indicated above.

We take the area in the form of a sectorial strip in the yz plane rotating around the axis of the cone with the centre coinciding with the centre of the drop and having the shape of a sector with an angle $d\theta$ and bounded by the circles $q = R_\alpha \equiv R_0$ and $q = R_\beta \equiv R_{\kappa+1}$ (Fig. A1.2). In our case, for a discrete number of monolayers of the interface, the end of the sector strip with the angle $d\theta$ has a discrete broken line passing along the faces of the cells (instead of a strict line along the radius). To obtain a mechanical definition of the surface tension for a spherical boundary, let us imagine a hypothetical system consisting of the phases α and β, uniform up to a spherical film of radius ρ_s separating them [1]. The values of ρ_s and σ_s are determined from the condition that the hypothetical system is mechanically equivalent to the real system both by the resultant force, and by the resultant moment acting on the selected sectoral area. The subscript s reflects the position of the dividing surface $\rho = \rho_s$, referring to the hypothetical phase boundary having zero thickness, and the uniform tension σ_s related to the Laplace equation.

The resultant stress $d\sum_x$, acting in the x-direction normal to the selected area in the real system, is given by

$$d\sum_x = -d\theta \sum_{q=R+1}^{R+\kappa} \pi_q^T q\,dq, \quad (dq = \Delta q = \lambda). \tag{A1.25}$$

On the other hand, in a hypothetical system, the stress $d\sum^{\alpha\beta}$, acting in the x-direction on our sectoral strip can be represented in the form

$$d\sum^{\alpha\beta} = -d\theta \sum_{q=R+1}^{R+\kappa} \pi_q^{\alpha\beta} q\,dq + \sigma_s \rho_s d\theta, \tag{A1.26}$$

where we use the notation

$$\pi_q^{\alpha\beta} = \pi_q^{\alpha} \text{ for } 1 \le q \le \rho_s \text{ and } \pi_q^{\alpha\beta} = \pi_q^{\beta} \text{ for } \rho_s < q \le \kappa. \tag{A1.27}$$

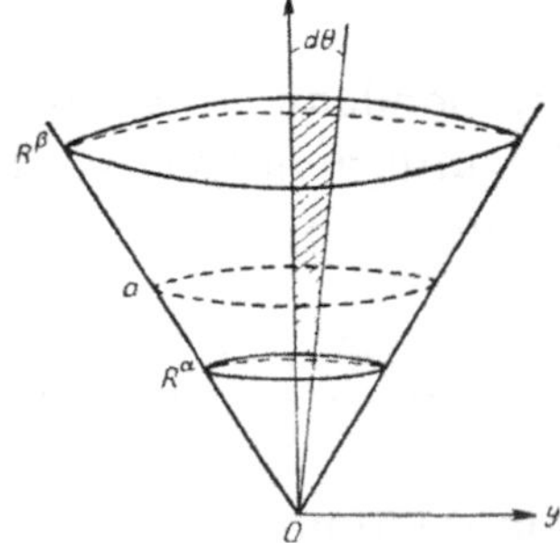

Fig. A1.2. The shaded part represents an area in the form of a sectorial strip. A rectangular coordinate system x, y, z with the origin at the vertex of the cone 0 and the z axis directed along the axis of the cone is introduced.

Equating (A1.25) and (A1.26), we obtain a formula that defines the surface tension for a spherical interface:

$$\sigma_s = \sum_{q=R+1}^{R+\kappa} (\pi_q^{\alpha\beta} - \pi_q^T) q \, dq / \rho_s, \tag{A1.28}$$

where the value of ρ_s is determined by the equilibrium condition with respect to the resultant moment of forces inside the transition region of the boundary:

$$\sum_{q=R+1}^{R+\kappa} (\pi_q^{\alpha\beta} - \pi_q^T) q (q - \rho_s) = 0 \tag{A1.29}$$

where it is taken into account that $dq = \lambda$ is the same for all layers q and it can be shortened. In the expression (A1.29) there are both discrete (number of layers) and continual values (π_q), therefore the quantity ρ_s has in the general case a continual value, which is taken into account by the relations between q and ρ_s in (A1.27). For the vapour region ($q > R + \kappa$) and drops ($q < R + 1$), the pressures and fluid densities remain constant, and in principle the summation limits in formulas (A1.28) and (A1.29) can be formally extended to a larger (infinite) range of values of q, as is done in the continuum calculus, in order to emphasize the independence of the introduced definitions from the width of the transition region. However, in the above forms of writing equations (A1.28) and (A1.29), the inclusion of the boundedness of the quantity κ is visibly reflected, which is of fundamental importance in the interpretation of the temperature dependences of the surface tension [13]. In the case of $R \gg \kappa$, the expressions (A1.28) and (A1.29) automatically become expressions for the surface tension and the positions of the dividing surface of the flat boundary, which correspond to the conditions of mechanical equilibrium.

Let us verify the correspondence of expression (A1.28) and the condition of mechanical equilibrium (A1.24). To do this, we rewrite the last formula in the form

$$\rho_s^2 (\pi_{\text{vap}} - \pi_{\text{liq}}) = 2\lambda \sum_{q=1+R}^{\kappa+R} (\pi_q^{\alpha\beta} - \pi_q^T) q, \tag{A1.30}$$

where ρ_s is introduced via relation

$$2\lambda \sum_{q+1+R}^{\kappa+R} \pi_q^{\alpha\beta} q = R_{\kappa+1}^2 \pi_{\text{vap}} - R_0^2 \pi_{\text{liq}} - \rho_s^2 (\pi_{\text{vap}} - \pi_{\text{liq}}), \tag{A1.31}$$

which is also obtained from (A1.24) by substituting in it a hypothetical pressure profile $\pi_q^{\alpha\beta}$ given by the expression (A1.27) instead of the real profile for the tangential components of the pressure tensor π_q^T.

As a result, we have the Laplace equation for a spherical drop in the lattice gas model

$$\pi_{\mathrm{vap}} - \pi_{\mathrm{liq}} = \left(2\lambda / \rho_s^2\right) \Sigma_{q=1+R}^{\kappa+R} (\pi_q^{\alpha\beta} - \pi_q^T) q = 2\sigma_s / \rho_s, \qquad \text{(A1.32)}$$

in which σ_s is given by the expression (A1.28), which proves the self-consistency of the definition of both σ_s and ρ_s by the formula (A1.29). For any violation of equality (A1.29), the moment of forces is violated in the transition region, and the value of ρ_s remains uncertain, since the transformation procedure (A1.29) providing the recalculation of the mechanical equilibrium in the transition region to the reference dividing surface is not defined.

Mechanical equilibrium in the continuous and discrete calculus. Before proceeding to the second mechanical definition of surface tension, we discuss the relationship between the three forms of mechanical equilibrium recording in the continuous and discrete calculi.

Equations for a drop in the continuous calculus are equivalent in three forms of writing. The discrete analogues of the first (A1.22) and the third (A1.23) form, as shown above, follow directly from one equation (A1.20). The difference in these equations consists in the way of expressing the difference between the pressures (the first form) or in the difference in the product of the pressure on the area of the drop (the third form). From the point of view of continuous analysis, the transition from the third form to the first is carried out by simple differentiation of r^2 in the denominator and in the form of a factor in the numerator.

However, the final formulas for the discrete notation for the third form $\Delta(\pi^N S)q = \pi_q^T \Delta S_q$ (A1.23) and for the first form $\Delta\pi_q^N / \Delta q = m\left(\pi_q^T - \pi_q^N\right)/q$ (A1.22) differ from each other due to differences in the rules for calculating difference derivatives. For the first form, the increment in the layer $\Delta\pi_q^N \Delta q = m\lambda\left(\pi_q^T - \pi_q^N\right)/q$ leads to the coincidence with the differential form of the record $dP_N(r) = m\lambda(P_T(r) - P_N(r))/r$, where $\lambda = dr$.

The third forms of writing derivatives in discrete and continuous analysis of the product of functions lead to similar expressions:

discrete calculus: $\Delta(q^2\pi_q^N)_q / \Delta(q^2) = \left[(q+\lambda)^2 \pi_{q+1}^N - q^2\pi_q^N\right] / \Delta(q^2) = = q^2\Delta(\pi_q^N) / \Delta(q^2) + (2q+\lambda)\pi_{q+1}^N / (2q+\lambda) = q^2\Delta(\pi_q^N) / \Delta(q^2) + \pi_{q+1}^N$.

continuous analysis: $d\left(r^2\pi_q^N\right)_q / d(r^2) = r^2 d\pi_q^N / d(r^2) + \pi_q^N$.

That is, the differences in the second summand consist in the value of the function π_{q+1}^N at another point $q+1$ instead of π_q^N. On the macroscale at $\Delta r = \lambda \to 0$ both methods of recording the third form of the equations of mechanical equilibrium coincide.

For the second differential continuum form of recording, the corresponding discrete record is constructed as

$$\Delta(q^3\pi_q^N)_q / \Delta q = \left[(q+\lambda)^3 \pi_{q+1}^N - q^3\pi_q^N\right] / \lambda = q^2\left[2\pi_q^T + \pi_q^N\right]. \quad \text{(A1.33)}$$

The differences form both the additional terms $(3r\lambda + \lambda^2)\pi_{q+1}^N$, and the values of the function π_{q+1}^N at the other point $q + 1$. The transition from the formula (A1.33) to the first form of writing leads to the following expression

$$\Delta\pi_q^N / \Delta q = q^{-1}\left\{2\pi_q^T + \pi_q^N - q^2\pi_{q+1}^T - (3q\lambda + \lambda^2)\pi_{q+1}^N\right\},$$

which differs from both types of records, both in continuous and discrete analysis. However, for $\Delta r = \lambda \to 0$, both ways of recording the second form of the equations of mechanical equilibrium coincide, and the form two goes into the first form. For a finite value of λ this is not so.

This difference creates the first problem using the second form of the record. Thus the transition to the first discrete form of writing from the second discrete form of writing leads to another equation with respect to $\Delta\pi_q^N/\Delta q$ (additional terms and coordinates are obtained). This situation is not unexpected. Differences between discrete and continuous calculi are well known [12]. At the same time, the use of the lattice gas model (LGM) in contrast to the general problems of discrete calculus [12] consists in the fact that for all functions (here, first of all, this refers to $\pi_q^{T,N}$) no approximation by polynomials is necessary, since the functions themselves in the LGM. This simplifies their practical use.

In the second continuous record form on the left side there is a derivative with respect to the radius of the product of the normal component of the tensor by volume, which reflects the derivative of the work associated with deformation of the layer along the normal. Its right-hand side can be represented as $3q^2\pi_q$, where $\pi_q = (2\pi_q^T + \pi_q^N)/3$ is the average pressure in the cell of the layer q. That is, unlike traditional thermodynamic constructions [1–4], which use only tangential components of the pressure tensor to construct the surface tension, the second form of mechanical equilibrium leads to a different relationship between the work and the components of the pressure tensor. This defines the second problem with the second form of recording directly related to the second mechanical definition of surface tension.

Second mechanical determination of surface tension. The second mechanical definition is based on the displacement of the surface element of the conical wall of a two-phase system with a spherical boundary, bounding the same cone-shaped sector as for the first determination. The force acting across the surface element of the conical wall, enclosed between r and $r + dr$ and between φ and $\varphi + d\varphi$, can be written as $\pi^T(r)r \sin\theta \, dr \, d\varphi$. Let us give the solid angle of the cone an infinitesimal increment $d\omega$, then the work on moving the surface element of the conical wall will be expressed as

$$dW = d\omega \sum_{q=R+1}^{R+\kappa} \pi_q^T q^2 dq, \tag{A1.34}$$

so that according to the general thermodynamic relation $dW = \pi^\alpha dV^\alpha + \pi^\beta dV^\beta - \sigma dA$ [1–4], we have

$$d\omega \sum_{q=R+1}^{R+\kappa} \pi_q^T q^2 dq = \pi^\alpha dV^\alpha + \pi^\beta dV^\beta - \sigma dA, \tag{A1.35}$$

where $dA = \rho^2 d\omega$ is the change in the surface area of the interface, for a given displacement. Since $dV^\alpha = d\omega \sum_{q=R+1}^{\rho} q^2 dq$ and $dV^\beta = d\omega \sum_{q=\rho}^{R+\kappa} q^2 dq$, then taking into account (A1.27) we obtain the second mechanical definition of the surface tension

$$\sigma_2 = \sum_{q=R+1}^{R+\kappa} (\pi^{\alpha\beta} - \pi_q^T) d^2 dq / \rho^2. \tag{A1.36}$$

which must be satisfied for any positions of the dividing surface ρ.

The formal differentiation of the given expression σ_2 by ρ yields the formula

$$\left[\frac{d\sigma}{d\rho}\right] = -\frac{2}{\rho^3} \sum_{q=R+1}^{R+\kappa} \left(\pi^{\alpha,\beta} - \pi_q^T\right) q^2 dq + (\pi^\alpha - \pi^\beta), \tag{A1.37}$$

coinciding with the traditional thermodynamic definition, which allows a change in the value of surface tension σ_2 associated with a mathematical change in the position of the dividing surface by an amount $d\rho$ (here, in the discrete calculus of $\Delta\rho$), provided that all physical quantities within the transition region and , in particular, the physical radius of the drop R, remained unchanged [1–4].

Thermodynamic treatments lead to different definitions of surface tension. There are two thermodynamic approaches in determining σ through the mathematical (Ono) and physical (Gibbs) displacements of the dividing surface and the two above-mentioned mechanical approaches [1–4]. It is believed that both thermodynamic approaches and the second mechanical definition lead to equivalent formulations in which the position of the dividing surface is found from the minimum (in the general case of the extremum) of the function $\sigma(\rho)$. This procedure involves fixing the state of the transition layer when ρ changes. This hypothesis is necessary for thermodynamic constructions, since it is necessary to determine the position of the dividing surface on which the surface tension is introduced. The state of the system at the microlevel is described by a much larger number of variables than at the macrolevel, so the requirement of the constancy of any physical property within the transition region with a change in the value of ρ is some *a priori* assumption (hypothesis), which need not necessarily be satisfied at the molecular level. This was clearly demonstrated by the analysis of Kondo drops [3] (see [14] and Section 59).

References

1. Ono S., Kondo S., Molecular theory of surface tension. Moscow: IL, 1963. [Springer, Berlin, Gottinhen, Heidelberg, 1960].
2. Rowlinson G., Widom B. Molecular theory of capillarity. Moscow, Mir, 1986. [Clarendon, Oxford, 1982].
3. Kondo S., Journ. Chem. Phys., 25, 662 (1956).
4. Gibbs J.W., Thermodynamics. Statistical mechanics. Moscow, Nauka, 1982.
5. Tolman R.C., J. Chem. Phys., 1948. V. 16. P. 758.
6. Tovbin Yu.K., Zh. fiz. khimii. 2010. V. 84. No. 2. P. 231. [Russ. J. Phys. Chem. A. 2010. V. 84. No. 2. P. 180].
7. Kirkwood J. G., Buff F. P., Journ. Chem. Phys. 1949. V. 17. P. 338.
8. Hill T.L., J. Phys. Chem. 1952. V. 56. P. 526.
9. Buff F.P., J. Chem. Phys. 1955. V. 23. P. 419.
10. Timoshenko S.P., Goodier J.N., Theory of Elasticity. Moscow, Nauka, 1979.
11. Godunov S.K., Ryaben'kii V.S., Difference schemes (introduction to theory). Moscow, Nauka, 1973.
12. Gelfond A.O., Calculus of finite differences. Moscow, Nauka. 1967.

13. 13. Tovbin Yu.K., Rabinovich A.B., Izv. AN, ser. khim. 2009. No. 11. P. 2127. [Russ. Chem. Bull. 58, 2193 (2009)]
14. 14. Tovbin Yu.K., Rabinovich A.B., Zh. fiz. khimii. 2013. V. 87. No. 2. P. 337. [Russ. J. Phys. Chem. A 87, 329 (2013)]

Appendix 2

Transfer equations and dissipative coefficients

The following subjects are discussed in this appendix: what are the transfer equations for the properties contained in the system (38.6) and (38.7), the principle of constructing the dissipative coefficients and applying the first equation from the system of equations (38.7) to describe the dynamics of solid-phase processes within the two-level model.

Transfer equations. Let us write for specificity the structure of equations (38.6) and (38.7) of a one-component system in the absence of contributions from external fields. The transport equations reflect the evolution of the mass, impulse and energy of the system. Substituting S_f ($S_f = m_f$, $m_f v_{fi}$, $mv_f^2/2$) into equation (38.6), we obtain an analog of the well-known system of hydrodynamic equations [1–4]:

The continuity equation (only molecular mechanisms of molecular transport are considered)

$$\frac{\partial\langle\theta_f\rangle}{\partial t}+\sum_{j=1}^{3}\frac{\partial\langle\theta_f v_{fj}\rangle}{\partial r_{fj}}=I_f^{(m)}=0, \qquad \text{(A2.1)}$$

since there are no direct (collisionless) displacements of molecules over long distances, and their collisions do not lead to mass transfer (however, in this case $I_f^{(ik)} \neq 0$ and $I_f^{(e)} \neq 0$).

The equation of the viscous flow

$$\theta_f\left(\frac{\partial u_{fk}}{\partial t}+\sum_{j=1}^{3}u_{fj}\frac{\partial u_{fk}}{\partial r_{fj}}\right)+\sum_{j=1}^{3}\left[\frac{\partial\left[P_f^{ne}\delta_{jk}-\Pi_{ik}^{ne}(f)\right]}{\partial r_{fj}}\right]=I_f^{(ik)}. \qquad \text{(A2.2)}$$

where u_{fk} is the microhydrodynamic velocity of molecules at site f in the direction $k = x, y, z$; P_f^{ne} is the non-equilibrium pressure at the given local parameters of the system at site f (see Section 43) [5]. The principle of constructing the components of the impulse flux density tensor $\Pi_{ik}^{ne}(f)$ is discussed below. A characteristic feature of this approach is that all characteristics (P_f^{ne} and $\Pi_{ik}^{ne}(f)$) are expressed through unary and pairwise distribution functions (DF) with identical functional connections for any degree of non-equilibrium. The symbol of nonequilibrium (ne) means the use of non-equilibrium paired DFs. In the case of local equilibrium, formula (A2.2) becomes the hydrodynamic equation. In them, there are no exchange terms, $I_f^{(ik)} = 0$, and the quantities P_f^{ne} and $\Pi_{ik}^{ne}(f)$ are calculated by means of equilibrium paired DFs.

The energy transfer equation

$$\frac{\partial\left(\theta_f(m_f u_f^2/2 + U_f^{full})\right)}{\partial t} + \\ + \sum_{i=1}^{3} \frac{\partial}{\partial r_{fi}} \left\{ \left(\theta_f u_f^2 u_{fi}/2\right) + \left[\theta_f U_f^{full} + P_f^{ne}\right] u_{fi} \right\} + \\ + \sum_{i=1}^{3} \frac{\partial}{\partial r_{fi}} \left[L_{fi}^{ne} - \sum_k u_{fk} \Pi_{fi,fk}^{ne} \right] = I_f^{(e)}. \tag{A2.3}$$

where U_f^{full} is the total internal energy per unit volume, L_{fi}^{ne} is the vector of the heat flux density under non-equilibrium conditions [5]. This expression generalizes the well-known expression for energy transfer under the condition that local equilibrium is fulfilled for the case of local disequilibrium. The rearrangement $\partial P_f^{ne} u_{fi} / \partial r_{fi}$ of the term to the internal energy $\theta_f U_f^{full} u_{fi}$ reflects the enthalpy flow for the unit volume of the system.

Equations of transfer of paired properties. The first group for transferring the pair properties of the equations (38.7) consists of 5 functions $\left\langle m_f S_\Psi^m \right\rangle$ (and vice versa $\left\langle S_f^m m_\Psi \right\rangle$) for different S_ψ^m (below the mass in all terms are canceled).

For $S_\psi^m = m_\psi$ and $S_{f\psi}^{m,m} = m_f m_\psi$ we have an equation for the evolution of paired DFs, taking into account the separation of spatial and concentration variables,

$$\frac{\partial\langle\theta_{f\psi}\rangle}{\partial t}+\sum_{j=1}^{3}\left\{\frac{\partial\langle\theta_{f\psi}v_{fj}(\psi)\rangle}{\partial r_{fj}}+\frac{\partial\langle\theta_{f\psi}v_{\psi j}(f)\rangle}{\partial r_{\psi j}}\right\}=$$
$$=\frac{\partial\theta_{f\psi}}{\partial t}+\sum_{j=1}^{3}\left\{\frac{\partial\theta_{f\psi}u_{fj}(\psi)}{\partial r_{fj}}+\frac{\partial\theta_{f\psi}u_{\psi j}(f)}{\partial r_{\psi j}}\right\}=I_{f\psi}^{(m,m)} \tag{A2.4}$$

In the second term, new unknown functions of the type $\theta_{f\psi}\langle v_{fj}(\psi)\rangle$ appeared. The symbol (ψ) indicates that the value is determined for a fixed occupied state of the neighbouring site ψ by the molecule A.

Kinetic equations for new variables $\theta_{f\psi}u_{fi}(\psi)$ are written out as equations for the conservation of the mass and impulse of particles of the pair $f\psi$: $S_{f\psi}^{m,ik}=m_f i_{\psi k}(f)$, k = 1–3 (here $i_{\psi k}(f)$ is the total impulse of the molecule, reflecting the translational motion and the influence of the potential fields of neighbours on this motion [5]):

$$\frac{\partial\theta_{f\psi}u_{\psi k}(f)}{\partial t}+\sum_{j=1}^{3}\left\{\frac{\partial\theta_{\psi f}\langle v_{fj}(\psi)i_{\psi k}(f)\rangle}{\partial r_{fj}}+\right.$$
$$\left.+\left[\frac{\partial\theta_{f\psi}u_{\psi j}(f)u_{\psi k}(f)}{\partial r_{\psi j}}+\frac{\partial t_{\psi f}[P_{\psi}^{ne}(f)\delta_{jk}-\Pi_{ik}^{ne}(\psi f)]}{\partial r_{\psi j}}\right]\right\}=I_{f\psi}^{(m,ik)}, \tag{A2.5}$$

Note that in the square bracket there is a conditional probability $t_{\psi f}$ instead of a pair DF, associated with the method of calculating the components of the impulse transfer tensor [5]. Here the term in the second part of the formula $\Pi_{ik}^{ne}(\psi f)$ in the square brackets is a modified expression for the dissipative coefficient constructed on the site ψ – here the second index f points to the site containing the fixed particle *A*. In equation (A2.5), the first summand of the sum is a new unknown $\theta_{f\psi}\langle v_{fj}(\psi)i_{\psi k}(f)\rangle$ dynamic variable, to describe the evolution of which we need a microscopic analogue of the first equation of the Keller–Friedman chain [6] inside the cell f, ($\psi \in V(f)$).

To describe the evolution $S_{f\psi}^{m,e}=m_f e_{\psi}(f)$ (mass$_f$ – energy$_\psi$) we have an equation for $\theta_{f\psi}\langle e_{\psi}(f)\rangle$ (here $e_{\psi}(f)$ is the total energy of the particle, taking into account the kinetic and potential contributions [5]), in which taking into account the fixation of the neighbouring molecule at the site f, we obtain

$$\frac{\partial\theta_{f\psi}\langle e_\psi(f)\rangle}{\partial t}+\sum_{j=1}^{3}\left\{\frac{\partial\theta_{f\psi}\langle v_{fj}(\psi)e_\psi(f)\rangle}{\partial r_{fj}}+\right.$$

$$\left.+\sum_{i=1}^{3}\frac{\partial}{\partial r_{\psi i}}\theta_{f\psi}\left\{u_\psi^2(f)u_{\psi i}(f)/2+U_\psi^{full}(f)u_{\psi i}(f)\right\}\right\}+ \qquad \text{(A2.6)}$$

$$+\left\{\sum_{i=1}^{3}\frac{\partial}{\partial r_{\psi i}}\left\{t_{\psi f}[P_\psi^{ne}(f)u_{\psi i}(f)+L_{\psi i}^{ne}(f)-\sum\nolimits_k u_{\psi k}(f)\Pi_{ik}^{ne}(\psi f)]\right\}\right\}=I_{f\psi}^{(m,e)}$$

In the second term (A2.6), a new unknown function appeared $\theta_{f\psi}\langle v_{fi}(\psi)e_\psi(f)\rangle$, built on two internal sites f and ψ of the region $V(f)$, requiring its new kinetic equation. The third term takes into account the flux of internal energy $U_\psi^{full}(f)$ pertaining to one site ψ, but in the presence of a particle at site f. In the fourth term, the modified expressions for the pressure and dissipative coefficients of the heat $L_{\psi i}^{ne}(f)$ and impulse $\Pi_{ik}^{ne}(\psi f)$ appear.

The evolution of the pair property impulse$_{\text{fn}}$–impulse$_{\psi k}$ $S_{f\psi}^{(in,ik)}=i_{fn}(\psi)i_{\psi k}(f)$, n, k = 1–3 is described by the following equation

$$\frac{\partial\theta_{f\psi}\langle i_{fn}(\psi)i_{\psi k}(f)\rangle}{\partial t}+$$

$$+\sum_{j=1}^{3}\left\{\frac{\partial\theta_{f\psi}\langle v_{fj}(\psi)i_{fn}(\psi)i_{\psi k}(f)\rangle}{\partial r_{fj}}+\right.$$

$$\left.+\frac{\partial\theta_{f\psi}\left\langle\dfrac{\partial\theta_{f\psi}<i_{fn}(\psi)i_{\psi k}(f)v_{\psi j}(f)>}{\partial r_{\psi j}}\right\rangle}{\partial r_{\psi j}}\right\}=I_{f\psi}^{(in,ik)}, \qquad \text{(A2.7)}$$

Under the sign of the time derivative (A2.7), a correlator $\theta_{f\psi}\langle i_{fn}(\psi)i_{\psi k}(f)\rangle$ is associated with the new dynamical variable that appeared in (A2.5) $\theta_{f\psi}\langle v_{fj}(\psi)i_{\psi k}(f)\rangle$.

In the flux terms there are the same triple correlators of hydrodynamic velocities. The most consistent procedure from the statistical point of view of the closure of triple correlators is the Kirkwood superposition approximation [7] (see also [8]). In each of the subspaces, the triple correlators are expressed in terms of unary and paired DFs:

$$\left\langle v_{fj} v_{fk} v_{\psi i} \right\rangle = \frac{\left\langle v_{fj} v_{fk} \right\rangle \left\langle v_{fj} v_{\psi i} \right\rangle \left\langle v_{fk} v_{\psi i} \right\rangle}{\left\langle v_{fj} \right\rangle \left\langle v_{fk} \right\rangle \left\langle v_{\psi i} \right\rangle},$$
$$\theta_{f\psi h}^{ikj}(r_f r_\psi r_h) = \frac{\theta_{fh}^{ij}(r_f r_h)\theta_{f\psi}^{ik}(r_f r_\psi)\theta_{\psi h}^{kj}(r_\psi r_h)}{\theta_f^i \theta_\psi^k \theta_h^j}. \tag{A2.8}$$

This approximation was first introduced to describe the spatial equilibrium distribution of triplets of molecules $\theta_{f\psi h}^{ikj}(r_f r_\psi r_h)$ [7]. Concentration triple correlators are present in all exchange terms (38.7).

All subsequent equations of the complete system after formula (A2.7) contain third correlators with respect to hydrodynamic velocities, and expressions (A2.8) should be used to close them. The remaining equations of system (38.7) related to energy transfer are written out in the form: three functions 'impulse$_{fk}$–energy$_\psi$' (k = 1–3)

$$\frac{\partial \theta_{f\psi} \left\langle i_{fk}(\psi) e_\psi(f) \right\rangle}{\partial t} + \sum_{j=1}^{3} \left\{ \frac{\partial \theta_{f\psi} \left\langle v_{fj}(\psi) i_{fk}(\psi) e_\psi(f) \right\rangle}{\partial r_{fj}} + \right.$$
$$\left. + \frac{\partial \theta_{f\psi} \left\langle v_{\psi j}(f) i_{fk}(\psi) e_\psi(f) \right\rangle}{\partial r_{\psi j}} \right\} = I_{f\psi}^{(ik,e)}, \tag{A2.9}$$

one function 'energy$_f$–energy$_\psi$'

$$\frac{\partial \theta_{f\psi} \left\langle e_f(\psi) e_\psi(f) \right\rangle}{\partial t} + \sum_{j=1}^{3} \left\{ \frac{\partial \theta_{f\psi} \left\langle v_{fj}(\psi) e_f(\psi) e_\psi(f) \right\rangle}{\partial r_{fj}} + \right.$$
$$\left. + \frac{\partial \theta_{f\psi} \left\langle e_f(\psi) e_\psi(f) v_{\psi j}(f) \right\rangle}{\partial r_{\psi j}} \right\} = I_{f\psi}^{(e,e)}. \tag{A2.10}$$

The use of closure (A2.8) leads to the appearance of a non-linearity for pair correlators in equations (A2.7), (A2.9) and (A2.10), describing the mean-square fluctuations of the pair properties 'impulse–energy' (by the construction of means $u_{fi} = u_{\psi i}$ within the domain f).

Thus, although the general structure of the transport equations for the paired properties is that for each unknown function new unknown functions of higher dimension appear at the hydrodynamic velocities: either for both sites of the pair due to the displacement of this property by the hydrodynamic flow, or for one of the sites of the pair. However, the closure of the system of equations (A2.8) reduces all the flow terms to the mean of the thermal velocities. The above modified dissipative coefficients differ from the usual dissipative coefficients by the presence of one of the neighbours for which the averaging is not carried out.

To work with the constructed system of equations, it is necessary to introduce a procedure for constructing dissipative coefficients and take into account the dimensions of the obtained dynamic variables at the microscopic and hydrodynamic levels.

Closure of hydrodynamic flows. To close the transport equations in hydrodynamics it is necessary to have the dissipative coefficients. The flows of the properties (the impulse flux density tensor $\Pi_{ik}(f)$ and the energy flux density vector L_{fi}) in the presence of local equilibrium are closed within the framework of the phenomenological laws of non-equilibrium thermodynamics and for each dynamic variable the property flow is considered with respect to a particular variable [3]. It follows from (38.6) and (38.7) that in the case of local equilibrium $U_{fg}^{VA} = U_{fg}^{AV}$ and $U_{fg}^{AA^*} = U_{fg}^{A^*A}$, and the right-hand sides of the transfer equations vanish. They go over into the well-known equations of hydrodynamics [1–4]; the system establishes a dynamic equilibrium of the transport flows of molecules in the forward and backward directions.

The general idea of constructing dissipative coefficients is connected with calculating the amount of transfer of a property through an isolated plane used earlier in microscopic hydrodynamics [9,10]. It does not change when three-aggregate states are considered. Here we shall only outline the principle of constructing dissipative coefficients by the example of spherically symmetric particles (more precisely, a monatomic fluid) in order to focus our attention only on the concentration dependences of the self-diffusion transfer coefficients (D^*), shear (η) and bulk (ξ) viscosities, and also the thermal conductivity (κ) to exclude the effect of internal degrees of freedom on the coefficients of bulk viscosity and thermal conductivity. It follows from the kinetic theory [1, 2] that the transfer coefficients characterize the flows for small deviations of the state of the system from the equilibrium state. We will characterize the state of the fluid far from the pore walls by the concentration θ and the temperature T. We will calculate the equilibrium particle distribution with respect to each other in a quasi-chemical approximation taking into account direct correlations between the interacting particles.

To calculate the dissipative coefficients, we select in space some plane 0 and consider the particle fluxes and the impulses and energy transferred by them. We will use the notion of the mean velocity of moving particles w. We draw two planes parallel to the plane 0 (with $x = 0$) at distances $x = \pm\rho$, where ρ is the mean free path of the particle, then the properties of the particles in these planes are written as $S(x = \rho) = S(x = 0) \pm\rho dS/dx$, where the symbol S denotes the concentration, the impulse (in the direction of γ, for example) or the energy of the particles moving along the X axis. The flow of the quantity S through the plane 0 is composed of two oppositely directed motions of the particles from the planes $x = \pm\rho$.

Two channels of impulse and energy transfer are realized in dense fluids (Fig. A2.1). The first is connected with the displacement of particles, as in the rarefied phase, and the second is determined by collisions between the particles. The particle under consideration can not intersect this plane 0 if its trajectory has been blocked by other particles located at the sites up to the plane 0 or by a particle located in the immediate vicinity of the other side of the plane 0 and preventing its crossing of the given plane. Both these cases are not taken into account by the elementary kinetic theory in gas, and it is necessary to use the kinetic theory of condensed systems [10–12].

The transfer of the property S through the distinguished plane, where S is 1) the number of molecules – to calculate the self-diffusion coefficient D_i^* and the mass transfer coefficients D_{ij}, 2) the impulse amount – to calculate the shear coefficients η and the bulk viscosity ξ, and 3) the amount of energy – to calculate coefficient of thermal conductivity λ. There are two channels for transferring the property S: $\eta_{fg} = \eta_{fg}(1) + \eta_{fg}(2)$, $\lambda_{fg} = \lambda_{fg}(1)+\lambda_{fg}(2)$, where the number (1) means the transfer of molecules through the selected plane – the calculation of the coefficients D_i^*, D_{ij}, $\eta_{fg}(1)$ and $\lambda_{fg}(1)$; and (2) the transfer of the impulse and energy properties through collisions – the calculation of the coefficients $\eta_{fg}(2)$, ξ and $\lambda_{fg}(2)$.

This principle of constructing molecular transport models in a uniform bulk phase has been generalized to non-uniform systems at the phase boundary, taking into account the influence of the surface potential [10].

In the absence of local equilibrium, the flow of properties of any dynamic variable $\Pi_{ik}^{ne*}(f)$, L_{fi}^{ne*} will be a function of all dynamic variables of the system. When varying any property, the derivatives

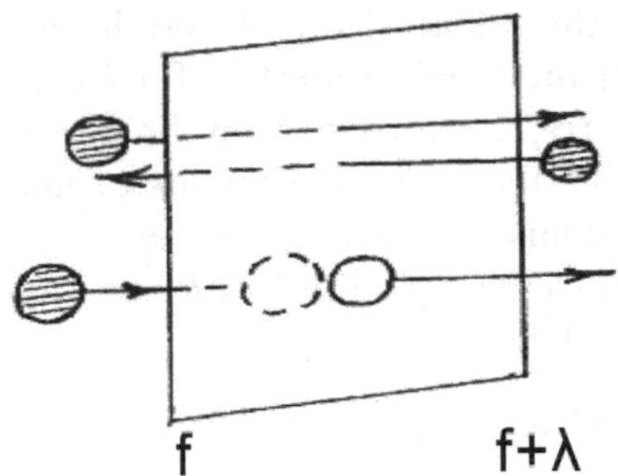

Fig. A2.1. A scheme for transferring properties across a selected plane.

must be used for all independent dynamic variables $\frac{\partial J_f}{\partial x} = \sum_{i=1}^{b} \frac{\partial J_f}{\partial y_i} \frac{\partial y_i}{\partial x}$, where $b = b_1 + b_2$ is the list of dynamic variables y_i. Obviously, in the case of local equilibrium, the number $b = b_1$. Far from equilibrium, the list of independent variables is further extended by $b_2 = 15$ variables on the scale considered associated with the transfer of paired properties [13, 14].

Consider the local flow of properties $J_f^{(r)} = \langle \theta_f v_{fj} S_f^{(r)} \rangle$, where $r \in b_1$, and $J_{f\psi}^{(r,n)} = \langle \theta_{f\psi} v_{fj} S_{f\psi}^{(r,n)} \rangle$, r, $n \in b_2$. We select the plane through which the flow of the property passes $S_f^{(r)}$, $S_{f\psi}^{(r,n)}$, situated between the neighbouring sites f and ψ with the coordinates $x = x_0 - l/2$ and $x = x_0 + l/2$, l is the distance between the sites f and ψ inside the region f. In the continual approximation, the expressions for the $J_f^{(r)}$ fluxes can be written in terms of the thermal velocities $U_{f\psi}$ of the molecules moving by two molecular mechanisms (direct jumps and their collisions) between the sites f and ψ, and likewise $J_{f\psi}^{(r,n)}$ – the thermal velocities of the molecules $U_{(hf)\psi}$ moving between the sites f and h, the state of the site ψ does not change in the course of this process, as [10]

$$J_f^{(r)} = U_{f\psi} S_f^{(r)}\big|_{x=x_0-l/2} - U_{f\psi} S_f^{(r)}\big|_{x=x_0+l/2} = \sum_{j=1}^{b} K_{f\psi}^{rj} \frac{\partial y_j}{\partial x}, \quad r \in b_1, \qquad \text{(A2.11)}$$

$$J_{f\psi}^{(r,n)} = U_{(hf)\psi} S_{f\psi}^{(r,n)}\big|_{x=x_0-l/2} - U_{(hf)\psi} S_{f\psi}^{(r,n)}\big|_{x=x_0+l/2} = \sum_{j=1}^{b} K_{(hf)\psi}^{r,n,j} \frac{\partial y_j}{\partial x}, \quad r,n \in b_2, \qquad \text{(A2.12)}$$

The coefficients of the expansion of the flows with respect to the spatial variations of the dynamic variables are dissipative coefficients that characterize the effect of the change in the dynamic variable y_j on the transfer of the local property r for $j \in b_1$ and on the transfer of the local property r, n for $j \in b_2$ in the domain f. The coefficients are calculated from the current solutions of the kinetic equations to the evolution of unary and pair properties.

Imagine $S_f^{(r)}$ as $\langle S_f^{(r)}\rangle = d_r y_r$, for $j=r\in b_1$, and similarly $\langle S_{f\psi}^{(r,n)}\rangle = d_{r,n} y_{r,n}$ for $j = r$, $n \in b_2$, then

$$K_{f\psi}^{rj} = \partial\left(U_{f\psi} S_f^{(r)}\right) / \partial y_j = \\ = \partial\left(U_{f\psi} S_f^{(r)}\right) / \partial y_j = K_{f\psi}^{r=j} \delta_{rj} + K_{f\psi}^{r*}, \\ K_{(hf)\psi}^{r,n,j} = \partial(U_{(hf)\psi} S_{f\psi}^{(r,n)}) / \partial y_j = K_{(hf)\psi}^{r,n=j} \delta_{(r,n)j} + K_{(hf)\psi}^{r,n*j}, \tag{A2.13}$$

where δ_{rj} is the delta function and

$$K_{f\psi}^{r=j} = U_{f\psi} d_r, \quad K_{f\psi}^{r*j} = y_r \partial(U_{f\psi} d_r) / \partial y_j, \\ K_{(hf)\psi}^{r,n=j} = U_{(hf)\psi} d_{r,n} \quad K_{(hf)\psi}^{r,n*j} = y_{r,n} \partial(U_{(hf)\psi} d_{r,n}) / \partial y_j. \tag{A2.14}$$

Equations (A2.13) show that among the complete set of dynamic variables j there is a situation when the symbol r corresponds to the symbol j, which is the 'ordinary' dissipation coefficient when this property is transferred ($r = j$) under the condition of local equilibrium. In the absence of local equilibrium, this coefficient $K_{f\psi}^{r=j}$ forms only part of the flow of property r. The coefficient $K_{f\psi}^{r*j}$ forms an additional part of the flow for the same dynamic variable. The remaining coefficients of the type $K_{f\psi}^{r*j}$, for $r \neq j$, correspond to cross contributions from the influence of other dynamic variables. All the same applies to the coefficients for the transfer of pair properties (r, $n = j$).

Taking into account (A2.13), the expressions for the flows (A2.11) and (A2.12) will be rewritten as

$$J_f^{(r)} = K_{f\psi}^{r=j} \frac{\partial y_r}{\partial x} + \sum_{j=1}^{b} K_{f\psi}^{r*j} \frac{\partial y_j}{\partial x}, \quad r \in b_1, \tag{A2.15}$$

$$J_{f\psi}^{(r,n)} = K_{(hf)\psi}^{r,n=j} \frac{\partial y_{r,n}}{\partial x} + \sum_{j=1}^{b} K_{(hf)\psi}^{r,n*j} \frac{\partial y_j}{\partial x}, \quad r \in b_2, \tag{A2.16}$$

In the case of local equilibrium, by virtue of the equilibrium relationship between unary $K_{f\psi}^{rj}$ and paired DFs, only one nonzero coefficient remains among the coefficients $K_{f\psi}^{r=j}$, and all the $K_{(hf)\psi}^{r,n,j}$ coefficients vanish.

In the general case of the absence of local equilibrium, the system (38.6), (38.7) can not be divided into subsystems and a complete system of equations must be solved in order to obtain all the current values of the dynamic variables. The solution of the system of equations (38.6), (38.7) allows one to obtain the average values of five dynamic variables $< S_f^{(r)} >$ and their 15 pair combinations $< S_{f\psi}^{(r,n)} >$ as a function of time. Introducing in the usual way [15], the deviations of dynamic variables $\Delta S_f^{(r)} = S_f^{(r)} - < S_f^{(r)} >$, and their root-mean-square deviations as $\Delta S_{f\psi}^{(r,n)} =< S_{f\psi}^{(r,n)} > - < S_f^{(r)} >< S_\psi^n >$ as a result of the solution of this system of equations, the time dependences of root-mean-square fluctuations of dynamic variables will be obtained $S_f^{(r)}$. In particular, this refers to a function $<u_{fi}u_{\psi k}>$ that is directly obtained from the solution of the system (38.6), (38.7) (recall that $u_{fi} = u_{\psi i}$ inside the region *f*) and the expressions (A2.15) and (A2.16) determine all the dissipative coefficients in the absence of local equilibrium.

Unlike the higher moments used in the analysis of turbulent flows, which can be introduced for one or two times [3,4,6,8], all the new 15 dynamic variables $< S_{f\psi}^{(r,n)} >$ refer to one point in time, for which defines the variables $< S_f^{(r)} >$.

As an example, we give the expressions for the coefficients of shear (η_{zx}^{turb}) and bulk (ξ_{zz}^{tur}) viscosities and thermal conductivity (κ_{fi}^{ne}) of the turbulent flow.

$$\eta_{zx}^{turb} = u_y \sum_{j=1}^{b} \frac{\partial(U_{f\psi} d_z)}{\partial y_j} \frac{\partial y_j}{\partial u_z}. \tag{A2.17}$$

$$\xi_{zz}^{tur} = u_z \sum_{j=1}^{b} \frac{\partial(U_{f\psi} d_z)}{\partial y_j} \frac{\partial y_j}{\partial u_z} \tag{A2.18}$$

where all the quantities refer to the coordinate *f*. In (A2.17), the first derivative under the summation sign is ordinary, while the second derivative is a functional derivative (otherwise it would be zero). This functional derivative is also taken over the complete system of equations of dimension *b*. It follows from the structure of (A2.17) that η_{yx}^{turb} is a function of the local flow velocity and all cross dissipative coefficients.

$$\kappa_z^{turb} = T\sum_{j=1}^{b}\frac{\partial(U_{f\psi}d_{ez})}{\partial y_j}\frac{\partial y_j}{\partial T}. \qquad \text{(A2.19)}$$

where κ_{fi}^{ne} is the local coefficient of thermal conductivity in the direction i (here $i = z$) in the framework of the Kolmogorov equations [8] for the transfer $S_f^{(e)} = \langle e_f \rangle = C_v(f)T_f$, T_f is the local temperature of the region $V(f)$ and $C_v(f)$ is its heat capacity at constant volume per molecule; that is $d_e = C_v$, which depends on paired DFs. The expressions for $C_v(f)$ in terms of unary and paired DFs are discussed in detail in [10]. Expression (A2.19) is written out using the same functional derivative.

Two-level models. The constructed equations have a wide potential range of applications. We confine ourselves to discussing only the range of problems associated with small solid particles. Hydrodynamic variables are usually determined in elementary macroscopic volumes containing a large number of molecules [3]. In the microscopic theory, a microscopic volume of the order of the volume of one molecule is considered as an elementary volume (cell) [10]. This is the minimum size at which one can talk about the existence of average flow characteristics (section 37). Small particles form the second level of the size of the system. An important role in non-uniform systems is played by ensembles of small particles, the size of which does not fit into the Gibbs phase approximation, since each of them can no longer be considered macroscopic. This specificity should be taken into account in solid-phase systems consisting of ensembles of small bodies, with the details of models used for modelling.

As one of the most important areas in the field of applications of new equations, one should point to the theoretical justification of the 'discrete element method' approach, widely known in mechanics [16–20], which is a family of numerical methods for calculating the motion of a large number of particles, such as molecules or grains of sand. Such situations are very common in technologies: in the sintering processes [21], in the flow of bi-dispersed bulk material [22,23], if the granular flow is similar to gas or liquid, and many others. The behaviour of individual particles is described in terms of a continuous medium using computational fluid dynamics. Considering in an explicit form the forces of different nature acting between the particles, it should be noted that the forces can have a macroscopic character (such as friction, bounce, etc.) along with the Coulomb and van der Waals molecular forces. In this connection, the treatment of the discrete element method as a generalization of the finite element method, on the one hand, and as an analog of the MD method, although the particles are not molecules, on the other hand, points to its intermediate position and the predominance of the practical value of the received qualitative information over the rigorous statistical substantiation of results. The question of reconciling both points of view in such situations is still open [24].

The works on investigation of the shear flow of suspensions, as well as the results of computer simulation of the properties of disperse systems under dynamic conditions [25–27], which show the determining role of the microstructure of disperse systems in the formation of their volume–rheological properties, are similar in style and the causes and conditions for the appearance of a macroscopic non-uniformity – the discontinuity of continuity in highly concentrated systems. Here we have the same problem of correct determination of the forces having a complex nature and depending on the state of the medium, the velocity of the particles and the state of the surface.

To simulate such processes, it is necessary to describe the system at different levels: as a molecular level problem, in order to take into account the specific local non-equilibrium distributions of reagents for the creation of specific nanoparticles (microcrystals) and as a task to create in such a macroscopic system such flows that provide for the spatial displacement of nanoparticles. These estimates for the minimum size of the field of application of thermodynamic approaches make it possible to divide the use of theoretical methods for local processes and for supermolecular flows. These estimates apply to all methods of a continuous medium. In fact, they determine the boundary of the application of the very equations of thermodynamics of irreversible processes and the calculation of dissipative coefficients for them.

Dense solid-phase systems. The systems specified in Section 1 are generally complex non-uniform systems. In the general case, four types of characteristic time-of-flight relaxation time scales can be distinguished, corresponding to the complete and partial equilibrium distribution of the components [28]. A fully equilibrium state refers to the state of all components of the system with commensurate mobilities of the components. First, we single out two relaxation time scales associated with (1) the transfer of vacancies with a weak relative redistribution of the components of the mixture, and (2) and with the relaxation of the relative redistribution of the components of the mixture for quasi-equilibrium vacancies. For large differences in the mobilities of the components, the total number s can be divided into 'fast' (mobile) and 'slow' (partially frozen): $s = s_1 + s_2$, s_1 – the number of fast components, and s_2 – the number of slow components. (3) On the third time scale, the transfer of slow components is practically absent, and the migration of mobile components occurs by the vacancy mechanism in the field of action of the potentials of the slow components, and their relaxation is studied in partial equilibrium. (4) On the fourth time scale, the fast components are distributed in equilibrium, and the migration of slow components by the vacancy mechanism occurs in the field of action of the fast components.

The last two situations correspond to the processes of adsorption, absorption, desorption, drying and wetting, etc., when slow components form a 'fixed' matrix, but the latter is usually rearranged, deformed and relaxed under the influence of mobile vapour and liquid components (in existing models this factor is taken into account roughly). Usually, such matrices are ensembles of small phases connected in a single whole macrobody. In these bodies, there can exist a developed surface, whose role increases with the decrease in the size of non-uniform phases – changing to micro-non-uniform phases uses different terms: microcrystals, grains, granules, globules, etc. Below, the term 'grain' will be used.

In practice, in mechanics of continuous media, there is a reliably established upper bound for the grain size in macroscopic systems [29]. It follows from the fact that 1 cm^3 contains several million grains. This gives the size of the side of the cube L of one grain about ~10^5 nm (and provides good statistics on the macroscale at $L \sim 10^7$ nm). It is natural to take this value of $L \sim 10^5$ nm as the upper value of the mesoscale. On it, each macrograin should be regarded as a separate body with its boundary. When moving to micro-non-uniform systems, the average size L decreases. It is advisable to introduce the following classification of grain sizes, based on the results obtained earlier on the size dependences of the surface tension of vapour–liquid in droplets [30, 31]. Note that the sizes of spherical droplets and cubic grains have close values: the ratio between the volume of the cube and the sphere is connected as $L = 2^{-1/3}D \sim 0.8D$, where L is the side of the cube, D is the diameter of the sphere, and the surface-to-volume ratio for the sphere ($S/V = 6/D$) and for the cube ($S/V = 6/L$) is the same. We denote $R_{mac} \sim L \sim 10^5$ nm.

In Section 27, three characteristic droplet sizes R were singled out: $R_0 \sim 10\lambda$, $R_t \sim 60\lambda$, and $R_b \sim 10^2 \div 10^3\lambda$. Section 34 lists the grain size ranges for which thermodynamics can be used. Thus, the general range of grain sizes to which it is possible to correctly use the thermodynamic description of the diffusion process within the framework of non-equilibrium thermodynamics goes from R_{t2} (36 nm) to R_{mac} (10^5 nm). However, moving from R_{mac} (one can neglect the role of grain surfaces in thermodynamic functions) to micro-non-uniform systems with a characteristic grain size $R < 10^2 \div 10^4$ nm, the question arises of the commensurability of R and R_{t2} and the need for different averaging procedures for substance flows. This range is naturally divided by the value of R_b found by the dependence of the surface tension on the size of the small droplet on two ranges: 1) $R_{t2} < R < R_b$ (the surface tension depends on the grain radius) and 2) $R_b < R < R_{\text{mac}}$ (the surface tension does not depend on the grain radius). In the second range, the grain size has practically no effect on the surface tension, whereas in the first range the grain sizes affect the surface (equilibrium and dynamic) tension values, which can also be manifested in diffusion processes.

Types of grains and boundaries. The main property of micro-non-uniform systems is the developed surface of grains with 'interphase' boundaries. In micro-non-uniform solids, it is necessary to isolate the regions inside the grains and their near-surface transition regions, and also to differentiate the state of the *intermediate regions* between neighbouring grains. Transitional sections are analogues of the interface between phases, which are characterized in thermodynamics by surface tension between coexisting or immiscible phases. Intermediate areas exceed the size of the transition areas. The size of the intermediate regions can be commensurable with the grain size, then they also need to allocate their *transition areas*. These boundaries differ in the nature of the transition regions between grains according to the type of coherence of the crystal structure [32] and the type of intermediate regions [10,33].

The grain boundaries are coherent, partially coherent and incoherent. The coupling of phases with different crystal lattices should provide for mutual accommodation of these lattices due to elastic displacements of atoms from their equilibrium positions (coherent boundaries), and also due to inelastic displacements associated with discontinuities in the continuity of the material, caused by misfit dislocations and vacancies condensing at the boundaries coherent boundaries). The existence of completely incoherent conjugation is defined as the absence of tangential shear stresses when a particle of a new phase is inserted into the corresponding cavity in the matrix in which there is no friction between the surface of this particle and the inner surface of the cavity. Under these conditions, the boundaries can freely slip relative to each other. This type of boundaries, however, is interpreted somewhat simplistically, both because of the impossibility of creating an ideally spherical boundary for the new phase (which eliminates the possibility of slippage) and because of the appearance of intermediate transition regions of amorphous and/or altered crystal structure.

An important role in solids is played by the voids that arise in various processes of redistribution of atoms in the course of mutual diffusion or mechanical disturbances. In this case, in addition to the various crystal conjugations of neighbouring grains, the character of the distribution of voids (isolated or forming a porous subsystem of a solid body) plays an important role. The type of incoherent boundaries should be classified into three classes of contacts (considering the possibility of free slippage as an idealized exception). The first class is the presence of intermediate regions and regions with an amorphous and/or distorted crystalline structure with a predominant normal stress when the majority of the bonds between the grains are ruptured. Such regions take on a mechanical load from outside the body and participate in the transfer of impulse throughout the volume of the solid. The second class is areas with an amorphous and/or distorted crystalline structure outside the zones of mechanical stress. By mechanical behaviour the microporous region also refers to them – opposite walls of micropores influence each other through direct potential interaction, but mechanical contact (as an elastic body) is absent in the absence of a mobile phase and is strongly attenuated when the micropores completely fill the mobile phase. The third class is the regions with a meso- and macroporous structure. They do not directly interact with opposite walls of the pores in the absence of the mobile phase, but there is indirect influence through the mobile phase with complete filling of the pores. It should be noted that the deformation component is taken into account when classifying boundaries. A subsystem of slow components and a subsystem of fast components are considered together.

Equations of diffusion transfer. Equations [34] take into account the existence of a non-uniform system in which internal local redistributions of components along different sublattices occur in the volume dV. These equations reflect these two-level non-uniform grain distributions and their boundaries. We denote by x the coordinate of the system along which diffusion is considered. The processes on different sublattices have different characteristic times, which are determined by the constants of elementary velocities of component jumps in the right-hand sides of the kinetic equations. Important is the non-uniformity of the spatial structure of the system, given by the volume shares and the method of locating sites of different types [35–39].

The diffusion description of the transport process is associated with the rejection of the site-by-site description of the concentrations in different sites of the system. Since the hopping velocities (Section 55) depend on local concentrations and on pair functions, in the general case of the absence of local equilibrium, the diffusion equations must take into account the evolution of these independent variables. Therefore, migration processes are characterized by four types of transfer coefficients of components and their pairs. For the flow along the x-axis, the equations [34] were obtained in the form

$$\begin{aligned}
\dot{\theta}_q^i(x) &= \sum_p \sum_k \frac{\partial}{\partial x} D_{qp}^{ik} \frac{\partial \theta_q^k}{\partial x} + \sum_{p\eta} \sum_{kl} \frac{\partial}{\partial x} D_{q(p\eta)}^{i(kl)} \frac{\partial \theta_{p\eta}^{kl}}{\partial x} \\
\dot{\theta}_{q\xi}^{ij}(x) &= \sum_p \sum_k \frac{\partial}{\partial x} D_{q\xi(p)}^{ij(k)}(r) \frac{\partial \theta_q^k}{\partial x} + \sum_{p\eta} \sum_{kl} \frac{\partial}{\partial x} D_{q\xi(p\eta)}^{ij(kl)} \frac{\partial \theta_{p\eta}^{kl}}{\partial x}
\end{aligned} \tag{A2.20}$$

The coefficients D_{qp}^{ik}, and $D_{qp\eta}^{i(kl)}$, $D_{q\xi(p)}^{ij(k)}$ and $D_{q\xi(p\eta)}^{ij(kl)}$ characterize the diffusion fluxes of the component i from the sites of type q and the pairs of components ij on pairs of sites $q\xi$. The second terms characterize the flows under the influence of the gradients of the pairs. Summation over the pair functions in (A2.20) is carried out over linearly independent pairs kl. The choice of the latter is related to the normalization conditions. All the coefficients D are obtained by expanding in series in gradθ_p^k and grad$\theta_{p\eta}^{kl}$ the flows of particles and their pairs through jumps $J_{mm+1}^i(qp)$, where m is the number of the layer perpendicular to the x axis, between the layers m and $m + 1$ there is a secant plane with coordinate x. In a one-step approximation in time, the flow between adjacent layers is written as $J_{mm+1}^i(qp) = U_{mm+1}^{Vi}(qp) - U_{mm+1}^{iV}(qp)$, where $U_{mm+1}^{iV}(qp)$ is the migration rate of particle i from site q of layer m to site p of layer $m + 1$; V is the vacancy symbol. The total flow of pairs of particles ij is composed of the corresponding migration contributions by the vacancy mechanism of both components of the pair i and j. In the second equation (A2.20), the hopping rates of particles i appear in the presence of neighbours j: $U_{pq\xi}^{(Vi)j} = U_{qp}^{iV}\psi_{q\xi}^{ij}$, where $\psi_{q\xi}^{ij} = t_{q\xi}^{ij}\exp\left[\beta(\varepsilon_{ij}^{*} - \varepsilon_{ij})\right] / S_{q\xi}^{i}$.

Equations (A2.20) are an example of equation (A2.1) for a strongly non-equilibrium solid-phase system. Its reduction with the transition to the ordinary diffusion equations in the case of local equilibrium is discussed in [34], and the transition to the problem of estimating the relaxation times near the local equilibrium for a two-scale grain structure (the supra-atomic structure is modeled with the grains of a dispersed body of characteristic size $L \gg \lambda$) are discussed in [28].

Other situations discussed in Section 41, in particular analogues of turbulence in solids, require the analysis of the following equations of the system (38.7), except (A2.1). For them, the coefficients (A2.17)–(A2.19) obtained above must be specified with the properties of the two-level model under discussion.

References

1. Huang K. Statistical mechanics. Moscow, Mir, 1966. [Wiley, New York, 1963].
2. Rumer Yu.B., Ryvkin M.Sh., Thermodynamics, statistical physics and kinetics. Moscow, Nauka, 1977.
3. Landau L.D., Lifshitz E.M. Theoretical physics.VI. Hydrodynamics. Moscow, Nauka, 1986. 734 p.[Pergamon, New York, 1987].
4. Bird R. B., Stuart W. E., Lightfoot E. N., Transport phenomena. Moscow, Khimiya, 1974. [Wiley, New York, London, 1965].
5. Tovbin Yu.K., Zh. Fiz. Khimi. 2017. Vol. 91. No. 3. P. 381. [J. Phys. Chem. A 91, 403 (2017)].
6. Keller L., Fridman A.D., Proc. 1st Intern. Congr. Appl. Mech., Delft, 1924. P. 395.
7. Kirkwood, J., J. Chem. Phys. 1935. V. 3. P. 300.
8. Monin A.S., Yaglom A.M. Statistical hydrodynamics. Moscow: Nauka, 1965 Part 1. 639 p; 1967 Part 2. 720 pp.
9. Tovbin Yu.K., Zh. Fiz. Khimii. 1998. V. 72. No. 8. P.1446. [Russ. J. Phys. Chem. 1998. V. 72. No. 8. P. 1298].
10. Tovbin Yu.K., The Molecular Theory of Adsorption in Porous Solids, Moscow, Fizmatlit, 2012. [CRC Press, Taylor & Francis Group, 2017].
11. Tovbin Yu.K., Khim. Fiz., 2002. V .21. No. 1. P. 83.
12. Tovbin Yu.K., Zh. Fiz. Khimii. 2002. T. 74. No. 1. C. 76. [Russ. J. Phys. Chem. 2002. V. 76. № 1. P. 64].
13. Tovbin Yu.K., Zh. Fiz. Khimii. 2014. T. 88. No. 2. P. 261. [Russ. J. Phys. Chem. A 88, 213 (2014)].
14. Tovbin Yu.K., Teoret. Fundamentals of chemical. technol. 2013. V. 47. No. 6. P. 734.
15. Landau L.D., Lifshitz E.M. Theoretical physics. V. Statistical Physics. Moscow: Nauka, 1964.

16. Bicanic N., Discrete Element Methods, in Encyclopedia of Computational Mechanics, Ed. by E. Stein, R. de Borst, and T. J. R. Hughes (Wiley, New York, 2004), Vol. 1.
17. 2nd International Conference on Discrete Element Methods, Editors Williams, J.R. and Mustoe, G.G.W., IESL Press. 1992.
18. Williams, J. R., O'Connor R., Discrete Element Simulation and the Contact Problem, Archives of Computational Methods in Engineering. 1999. V. 6. No. 4. P. 279.
19. Herrmann H.J.P., Statistical Physics. Invited Papers from STATPHYS 20. P. 51 / Reprinted from Physica A. 263 No. 1-4. Elsevier. North-Holland 1999.
20. Munjiza A., The Combined Finite-Discrete Element Method Wiley. 2004.
21. Galkin V.A., Smoluchowski's equation. Moscow, Fizmatlit. 2001.
22. Dorofeenko S.O., Polyanchik E.V., Manelis G.B., DAN. 2008. V. 422. P. 1.
23. Dorofeenko S.O., Teoret. Fundamentals of chemical. technol. 2007. V. 41. P. 1.
24. Tovbin Yu.K., Ross. nanotekhnologii. 2010, Vol. 5, No. 11–12, P. 715.
25. Uryev N.B., Potanin A.A., Flowability of suspensions and powders. Moscow, Khimiya. 1992.
26. Potanin A.A., Muller V.M., Kolloid. Zh. 1995. V. 57. No. 4. P. 533.
27. Uryev N.B., Kuchin I.V., Usp. khimii. 2006. V. 75. P. 36.
28. Tovbin Yu.K., Zh. Fiz. Khimii. 2017. V. 91. No. 8. P. 1243. [Rus. J. Phys. Chem. A, 91, No. 8, P. 1357 (2017)].
29. Timoshenko S.P., Goodier, J. Theory of Elasticity. Moscow, Nauka, 1979.
30. Tovbin Yu.K., Zh. Fiz. Khimii. 2010. V. 84. No. 4. P. 797. [Russ. J. Phys. Chem. A 84 (4), 705 (2010)].
31. Tovbin Yu.K., Zh. Fiz. Khimii. 2010. V. 84. No. 10. P. 1882. [Russ. J. Phys. Chem. A 84 (10) 1717 (2010)].
32. Khachaturyan A.G., The theory of phase transitions and the structure of solids. Moscow: Nauka, 1974.
33. Tovbin Yu.K., Izv. AN. Ser. khim. 2003. No. 4. P. 827. [Russ. Chem. Bull. 2003. V. 52. № 4. P. 869].
34. Tovbin Yu.K., Dokl. AN SSSR. 1988. V. 302. No. 2. P. 385.
35. Ovchinnikov A.A., Timashev S.F., Belyi A.A., Kinetics of diffusion-controlled chemical processes. Moscow, Khimiya. 1986.
36. Chalykh A.E., Diffusion in polymer systems. Moscow, Khimiya. 1987.
37. Timashev S.F., Physicochemistry of membrane processes. Moscow, Khimiya.1988.
38. Tovbin Yu.K., Usp. khimii. 1988. Vol. 57. P. 929.
39. Tovbin Yu.K., Zh. Fiz. Khimii. 1997. 71. No. 8. P. 1454. [Russ. J. Phys. Chem. 1997. V. 71. No. 8. P. 1304]

Appendix 3

Coefficients of activity in associated solutions

The notion of 'associated solution' reflects the existence in the mixture of new stable particles – associates, which consist of the original components of the system. Depending on the number of initial components of different sorts that are included in the associates, their appearance may differ greatly in their composition and structure. In this case, the principal property of associates is the possibility of their decay under the conditions (temperature and pressure) considered for the initial components, so that the chemical equilibrium between the associates and the original components is established in the system. The same condition allows the possibility of the transition of the original components between different associates, than a wide range of associates is formed. For liquid solutions, the concept of associated solutions focuses on the relationship between internal and external molecular interactions for all molecules present in the mixture.

The use of activity coefficients simplifies the formal recording of the thermodynamic functions of real systems by analogy with ideal systems. In the concept of associated solutions, the dominance of the energy of internal bonds between the constituent components of associates over the lateral interactions with the remaining components of solutions is fundamental.

Intermolecular interactions in condensed phases are approximated by different potential functions [1–4]. Most often, intermolecular potential functions are modeled by the Lennard-Jones potential. Even in such simple systems, the interaction of the nearest neighbouring molecules is about 70% of the total binding energy, which is formed as a result of taking into account all neighbours at different distances. In polar media in the absence of charged ions, it is necessary to take into account dipole–dipole interactions, which depend on the mutual orientation of the molecules. Such interactions are formed by local charges on atoms and groups of atoms belonging to different molecules. This type of interaction is characterized by its length over distances of the order of up to ten molecular diameters. It is also possible for the induction effect that occurs when molecules are deformed under the influence of the electrostatic forces of neighbouring molecules, which grabs a sufficiently extended region around the components under consideration. This effect is of greatest importance in the interaction between molecules with a large dipole moment and easily polarizable molecules. Thus, in real solutions, intermolecular interactions are of considerable length and in the analysis of the thermodynamic characteristics it is necessary to take this circumstance into account.

The original idea of associated solutions was reduced to the formulation of ideal associated solutions [5]. Physically, it is quite difficult to imagine strict compensation of local charges on different molecules, so that the associates practically do not interact with the other components of the solution and with each other, so later the notions of non-ideal associated solutions appeared.

In [6] the concept of associated solutions is analyzed from the point of view of correspondence to real potential functions, which are used in describing the equilibrium distribution of solution components. For the analysis, thc molecular theory of solutions within the framework of the lattice gas model (LGM) [3,4,7,8] is used, which makes it possible to calculate molecular distributions and thermodynamic functions from intermolecular forces. In [9,10], the main limitations on lattice models were removed: the lattice stiffness, the internal motions of the centres of mass of molecules inside the cells, the possibility of their rotation and orientation interactions. This allowed us to consider a wide range of potentials of intermolecular interactions that simultaneously take into account all the intermolecular contributions to the overall interaction potential that reach this r-th coordination sphere:

$$\xi_{ij}(r) = u_{\text{rep}}(ij\,|\,r) + u_{\text{el}}(ij\,|\,r) + u_{\text{disp}}(ij\,|\,r) + u_{\text{ch.tr.}}(ij\,|\,r) \tag{A3.1}$$

where there are contributions to the total interaction between a pair of molecules ij: 1) exchange (repulsive) interaction (u_{rep}); 2) direct electrostatic interaction (u_{el}); 3) dispersion interaction (u_{disp}); 4) the interaction due to charge transfer ($u_{ch.tr.}$). As the distance r between the molecules i and j increases, the terms on the right-hand side of equation (A3.1) decrease or may be zero. (An inductive interaction leading to multiparticle interactions is excluded from the 'usual' list [1–4] in (A3.1)).

To specify the orientations of polyatomic molecules of commensurable sizes, we represent a vector from the centre of the molecule mass directed to one of its atoms, and we will characterize the orientation of the rigid molecule by the angles between the given vector and the external (fictitious) field vector. In a spherical coordinate system, the orientation is given by the angle η_x, $0 \leq \eta_x \leq 2\pi$, which determines the projection of the vector onto the X0Y plane reckoned from the X axis (the external field vector coincides with the z axis) and the angle η_z, $0 \leq \eta_z \leq 2\pi$, between the vector and the Z axis (for linear molecules), and also by the additional angle η, $0 \leq \eta \leq 2\pi$, which describes the rotation with respect to a given vector for non-linear molecules. For brevity, we shall denote the set of values of the angles η_x, η_z, η of the molecule i by the single symbol ϕ_i.

By analogy with the transition from a continuous set of values of the coordinates of the centre of mass of molecules to the discrete one used in lattice models, we introduce a set of discrete molecular orientations relative to their mass centres instead of continuous orientations. In this case, the values of ϕ_i correspond to the discrete values of the angles η_x, η_z, η [11]. The number of different orientations of the molecule i will be denoted by σ_i. This model made it possible to explain the existence of the upper and lower critical points on the phase diagrams of the stratification of binary solutions [11]. In later works [12–15], the number of orientations was considered as a parameter of the model, not related to the lattice structure z.

To describe the molecular orientations, we introduce the random variable γ_i^ϕ, which characterizes the orientation ϕ_i of the molecule i: $\gamma_i^\phi = 1$ if the molecule i has the orientation ϕ_i, and $\gamma_i^\phi = 0$ otherwise. For the value γ_i^ϕ, the equality $\sum_{\varphi_i=1}^{\Phi_i} \gamma_i^\varphi = 1$ is fulfilled. For vacancies, there are no orientations, $\sigma_s = 1$. Then we can introduce the values $\gamma_i^\phi(f) = \gamma_f^i / \gamma_i^\phi$ characterizing the complex event: at site f there is a molecule i and it has the orientation ϕ_i. This allows us to introduce a new notion of the kind of particle ℓ at site f if the sort of molecule i and its orientation ϕ_i are related to one index ℓ ($\ell \leftrightarrow i$, ϕ_i), then $\gamma_i^\ell \equiv \gamma_i^\phi(f)$, where the values of the index ℓ vary from unity to $\Phi = \Sigma_{i=1}^s \sigma_i$. (Let us agree to refer the last index Φ to vacancies.)

It should also be taken into account that for fixed indices r, ϕ_i and ψ_i, the change in the position of the neighbouring molecule (variation of g_r) changes the values of the energy parameters. This is the main difference of the potentials that reflects the directivity of the bonds from the spherical potentials. Although it is obvious that the sequence of indices in itself does not change the values of the energy parameters $\varepsilon_{ij}^{\phi\psi}(f,g|r) = \varepsilon_{ji}^{\psi\phi}(g,f|r)$. Equations for the molecular distribution are written as

$$\theta_f^\Phi = \theta_f^l \exp\beta(v_f^l - v_f^\Phi)\prod_{r=1}^{R}\prod_{g\in z_f(r)} S_{fg}^l(r), \quad \sum_{l=1}^{\Phi}\theta_f^l = 1, \tag{A3.2}$$

$$\begin{gathered} S_{fg}^l(r) = 1 + \sum_{\lambda=1}^{\Phi-1} t_{fg}^{l\lambda}(r) x_{fg}^{l\lambda}(r), \quad x_{fg}^{lm}(r) = \exp[-\beta\varepsilon_{fg}^{lm}(r)] - 1, \\ \theta_{fg}^{lm}(r)\theta_{fg}^{\Phi\xi}(r) = \theta_{fg}^{l\xi}(r)\theta_{fg}^{\Phi m}(r)\exp[\beta\omega_{fg}^{lm\Phi\xi}(r)], \\ \sum_{\lambda=1}^{\Phi}\theta_{fg}^{l\lambda}(r) = \theta_f^l, \quad \sum_{l=1}^{\Phi}\theta_{fg}^{l\lambda}(r) = \theta_g^\lambda, \\ \omega_{fg}^{lmb\Phi}(r) = \varepsilon_{fg}^{lm}(r) + \varepsilon_{fg}^{b\Phi}(r) - \varepsilon_{fg}^{l\Phi}(r) - \varepsilon_{fg}^{bm}(r). \end{gathered} \tag{A3.3}$$

Equations (A3.2) are the equations for the molar fractions of the particles l $\left(\theta_l = \left\langle \gamma_f^1 \right\rangle\right)$. They describe the vapour–liquid equilibrium when substituted $v_f^l = v_l$ into the expression for the values of chemical potentials v_l of molecules in the gas phase. In the theory of liquid solutions, it is usually assumed that the mole fraction θ_l of the component i is known, $\theta_i = \sum_{\varphi_i=1}^{\Phi_i}\theta_i^\varphi$. In this case, equations (A3.2) allow us to determine the molar fraction of molecules i with a specific orientation ϕ_i, and also allow us to use experimental data on vapour–liquid equilibrium or other thermodynamic characteristics of solutions for determining the parameters of intermolecular interaction. For molecules in the dense phase, the values v_f^l take into account the internal motions of molecules, modified by the influence of neighbouring molecules [10].

Equations (A3.3) is a system of equations for paired distribution functions $\theta_{fg}^{l\lambda}(r) = \langle \gamma^l_f \gamma^\lambda_{g_r} \rangle$ characterizing the probability of finding particles l and λ (i.e., molecules i with orientation ϕ_i at site f and molecule j with orientation ψ_j at site g_r) at a relative distance r. The size of the algebraic system of equations (A3.3) can also be reduced by introducing new variables, as explained in [7].

Equations (A3.2), (A3.3) are a generalization of the equations given in [11–15], used earlier to study the phase diagrams of binary solutions of molecules with specific interactions. They reflect the contributions of all possible molecular orientations of each variety. 'Selection' of orientations is carried out through specific values of energy parameters, which in the case of specific interactions 'cut out' the locally oriented locations of neighbouring molecules, taking into account the tendency of molecules to organize specific solvation at which the internal motions of molecules change, which changes the values of the parameters v_l (for more details, [10,16]).

Equations (A3.2) and (A3.3) follow a modified expression for the chemical potential of the molecule i, which is an orientation-averaged expression for the 'quasiparticles' k_i, $\theta_i = \sum_{k_i=1}^{\sigma_i} \theta_f^{k_i} / \sigma_i$ also containing contributions from different coordination spheres

$$\mu_i = v_i + kT\ln(\theta_f^i) + \frac{kT}{2}\sum_{r=1}^{R} z(r) \sum_{\ell=1}^{\Phi} t_{fg}^{k_i\ell}(r) \ln \frac{\hat{\theta}_{fg}^{k_i\ell_j}(r)}{(\theta_f^i)^2 \sigma_j}, \tag{A3.4}$$

where $\hat{\theta}_{fg}^{k_i\ell}(r) = \theta_{fg}^{k_i\ell}(r)\exp(-\beta\varepsilon_{fg}^{k_i\ell})$, here the type of contact k_i of the molecule i at its fixed position depends on the site number g in the r-th coordination sphere $z(r)$ around the site f. In (A3.4) for simplicity of recording it is assumed that each contact is different from the others and it is possible not to introduce a weighting factor. The co-factor σ_j provides the transition (A3.4) to the formula for point particles in the case of a spherical potential. Here $\mu_i^{res} = \mu_i - \mu_i^{id} = kT\sum_{r=1}^{R} \ln \gamma_i(r)$ consists of contributions from different coordination spheres, $\mu_i^{id} = v_i + kT\ln\theta_i$, where $\ln\gamma_i(r) = \frac{z(r)}{2}\sum_{\ell=1}^{\Phi} t_{fg}^{k_i\ell}(r)\ln\frac{\hat{\theta}_{fg}^{k_i\ell_j}(r)}{(\theta_f^i)^2\sigma_j}$.

For point particles from formulas (A3.4), an expression for the chemical potential follows $\mu_i = v_i + kT\ln\theta_i + \frac{kT}{2}\sum_{r=1}^{R} z(r)\ln\frac{\hat{\theta}_{ii}}{\theta_i^2}$.

In the particular case of $R = 1$, this formula becomes the well-known expression [4,17]. Here and below, we use equations that explicitly contain paired functions that more clearly reflect the physical meaning. Thus, taking into account the extent of potential functions leads to the fact that the residual contribution to the activity coefficient consists of the sum of the terms of contributions from individual coordination spheres: $\mu_i^{res} = \frac{kT}{2}\sum_{r=1}^{R} z(r)\ln\frac{\hat{\theta}_{ii}(r)}{\theta_i^2}$. We note that the expansion of the excess chemical potential over different distances r in the formula $\mu_i^{res} = kT\ln\gamma_i^{res} = kT\sum_{r=1}^{R}\ln\gamma_i(r)$ refers to a spherically symmetric potential of radius $R > 1$. To relate to the expressions given below, we give an equivalent notation for $\gamma_i(r)$:

$$\ln\gamma_i(r) = \frac{z(r)}{2}\ln\frac{\hat{\theta}_{ii}(r)}{\theta_i^2} = \frac{z(r)}{2}\sum_{j=1}^{s} t_{ij}(r)\ln\frac{\hat{\theta}_{ij}(r)}{\theta_i^2}. \tag{A3.5}$$

If we use more accurate methods for describing the distribution of neighboring molecules than the quasi-chemical approximation, then the only term remains μ_i^{res} in the expression for, and the expression itself γ_i^{res} becomes more cumbersome [18].

Mean-field approximation (MFA). In the early works, the so-called mean-field approximation was actively used, with the help of which the presence of the first term in the formal expansion of Margules for the chemical potential of the components of the binary mixture was explained [5.19] (see below). In the framework of this approximation, the effect of direct correlations, which is present in the quasi-chemical approximation, is neglected. As a result, equations

$$\begin{aligned} a_i P_i &= \frac{\theta_i}{\theta_s}\exp[\sum_{r=1}^{R} z(r)\beta\sum_{j=1}^{s}(\varepsilon_{sj}(r) - \varepsilon_{ij}(r))\theta_j] = \\ &\frac{\theta_i}{\theta_s}\exp(z\beta\sum_{j=1}^{s}(\hat{\varepsilon}_{sj} - \hat{\varepsilon}_{ij})\theta_j), \end{aligned} \tag{A3.6}$$

where new effective interaction parameters are introduced $\hat{\varepsilon}_{ij} = \sum_{r=1}^{R} z(r)\varepsilon_{ij}(r)/z$, which include contributions from extended potential regions in all coordination spheres [20]. The main conclusion from the expression (A3.6) is that taking into account the long-term contributions of potential functions in the MFA does not lead to a change in the concentration dependence in comparison with taking into account only the nearest neighbours, therefore any variants of potential functions are reduced only to renaming the short-range parameters $\varepsilon_{ij} \equiv \varepsilon_{ij}(r=1)$. Therefore, in the expression for the chemical potential (for the residual contribution), there are only α_{ij} terms like for $R = 1$. As an example, we give a well-known expression for the chemical potential in a three-component strictly regular solution [5] $\mu_1 = kT\ln\theta_1 + \alpha_{12}\theta_2^2 + \alpha_{13}\theta_3^2 + (\alpha_{12} - \alpha_{23} + \alpha_{13})\theta_2\theta_3$, (expressions for μ_2 and μ_3 are obtained by cyclic replacement of the indices 1, 2 and 3) here, which excludes $\alpha_{ij} = z\varepsilon_{ij}/2$ the possibility of a difference in the concentration dependences of the activity coefficient for interaction potentials of different lengths.

Similar effective short-range interaction parameters are obtained by using the potential (A3.1) if, in addition to averaging over different distances r, averaging over different orientations of neighboring molecules is performed ϕ_i, which illustrates the inapplicability of this approximation to the molecular interpretation of thermodynamic characteristics.

Associates. The transition to associates in non-ideal mixtures within the molecular theory is carried out in a similar way, as for ideal systems. Specific intermolecular interactions are intermediate between van der Waals interactions and a chemical bond, and in many respects they are associated with a partial charge redistribution, which increases the energy of interaction of neighbouring components in comparison with non-specific interactions. The formation of any associate increases the size of the original component, and as the degree of association grows, larger molecules grow in size.

Assume that in the associated solution of their original components A and B there are complexes A_i formed from i monomolecules A, complexes B_j and complexes A_iB_j formed as a result of the association of i molecules A and j molecules B [5]. The complexes are in equilibrium with each other and with monomers A and B. Possible reactions between these particles can be represented by the equations $A_i = iA_i$; $B_j = jB_j$, $A_iB_j = iA_1 + jB_1$. If the total numbers of moles A and B in the solution are equal to n_A and n_B, and the number of moles of the various complexes present in the solution are $n_{A(i)}$, $n_{B(j)}$, $n_{A(i)B(j)}$, then

$$n_A = \sum_i i n_{A(i)} + \sum_i \sum_j i n_{A(i)B(j)}$$
$$n_B = \sum_j j n_{B(j)} + \sum_j \sum_i j n_{A(i)B(j)} \tag{A3.7}$$

Denoting the chemical potentials of complexes in a solution through $\mu_{A(i)}$, $\mu_{B(j)}$, $\mu_{A(i)B(j)}$, we have [5] that $\mu_{A(i)} = i\mu_{A(1)}$; $\mu_{B(i)} = i\mu_{B(1)}$; $\mu_{A(i)B(j)} = i\mu_{A(1)} + j\mu_{B(1)}$, where the macroscopic chemical potentials of the components A and B are denoted by μ_A, μ_B and $\mu_A = \mu_{A(1)}$ and $\mu_B = \mu_{B(1)}$. Thus, the macroscopic chemical potentials μ_A and μ_B are equal to the chemical potentials of monomeric molecules.

Taking the non-ideality of the solution into account retains the balance expressions (A3.7) unchanged. The above formulas for chemical potentials (A3.6) could be used if we neglect the differences in the sizes of the associates and the original components. However, it is not. The question of the role of differences in the size of the components of mixtures on lattice structures has been devoted to a large number of papers [3–5,21–37]. A certain number of places are assigned to the molecule in the lattice, the arrangement of which agrees with the shape of the molecule and its flexibility. Much attention was paid to chain molecules, which are represented by a linear sequence of segments, each of which occupies one site. The forces of attraction are taken into account in the same way as in theories for small spherical molecules. All available works are limited to taking into account only the lateral interactions of the nearest neighbors. This is reflected in the Barker–Guggenheim model, which takes into account both the size of the associate and the presence of a multitude of different functional groups on the polyatomic molecule, which are approximated in terms of the 'contact model'. According to this model, a certain characteristic size of the surface area of a polyatomic molecule with its energy parameter is singled out in the system.

The expression for the activity coefficient of component i is expressed as two terms

$$\ln\gamma_i = \ln\gamma_i^{comb} + \ln\gamma_i^{res}, \tag{A3.8}$$

where the first term γ_i^{comb} term reflects the size (volume) and shape of the large particle i, due to the contribution of the surface contacts of the given molecule to the total energy of the system. The second term γ_i^{res} reflects the lateral interactions of a large molecule with its neighbours.

The appearance of two terms in the formulas for $\ln \gamma_i$ is connected with the choice as the point of reference of an ideal mixture consisting of molecules with approximately the same size: $\mu_i^{id} = \nu_i + kT\ln\theta_i$. If the sample for an ideal system were calculated taking into account the different sizes γ_i of the components, then only the non-ideal part of the contributions, associated γ_i^{res} with lateral interactions, should be present in the expression. In turn, it can be represented as one γ_i^{res} or two terms reflecting the structure of molecules or free volume when using group energy parameters (see, for example, [4]).

The above equations for large particles on the basis of the molecular theory for non-ideal systems were generalized: in [21-37] molecules were considered that were in the form of three-dimensional parallelepipeds and their simple modifications. As a result, the presence of associates leads only to an increase in the number of components in the solution and additionally it is necessary to take into account the differences in the sizes of the associates and the initial components in the molecular theory.

Since our task is to discuss the relationship between the above molecular equations that reflect the extent of intermolecular interactions and the structure of the expression for the activity coefficient, then taking into account the different sizes of the associates, we have the formula (A3.8), in γ_i^{res} which

$$\ln\gamma_i^{\mathrm{res}} = \frac{1}{2}\sum_{r=1}^{R}\sum_{h\in S_i(f)} z_h(r)\sum_{\ell=1}^{\Phi} t_{hg}^{k_i\ell}(r)\ln\frac{\hat{\theta}_{hg}^{k_i\ell_j}(r)}{(\theta_h^i)^2\sigma_j} \qquad \text{(A3.9)}$$

where intermolecular interactions are taken into account in the quasi-chemical approximation. The formula (A3.9) differs from the expression (A3.4) in that it contains sets of numbers $z_h(r)$ that are constructed with respect to the contact number h (the energy properties of the contact h are fixed by the symbol k) on the surface of the large molecule instead of the coordination spheres with respect to central site f containing the center of mass of the molecule i. Here the list of contacts h refers to the entire surface $S_i(f)$ of a molecule i having a centre of mass at site f. The coordinate recalculation from the center of the molecule f to the contact h is given by the geometry of the molecule, which is considered known. As a consequence, the dependence of the sets of numbers $z_h(r)$ on the presence of neighboring particles inside the region of the interaction radius R appears in the formula (9), in contrast to the formula (A3.4).

The generalization of the system of equations (A3.2) and (A3.3) for associates leads to a system that is more cumbersome and larger in dimension, but the physical $\mu_i^{res} = kT\sum_{r=1}^{R} \ln \gamma_i(r)$ meaning of each contribution is completely preserved, including the expansion in terms of contributions from different distanccs. Notc that the formula (A3.9) refers only to the disorganized mutual arrangement of molecules of different sizes. In the case of the appearance of ordered structures that are characteristic of liquid-crystal substances, this expression should reflect the disparity of the distributions of neighboring molecules.

Molecular interpretation. The obtained results indicate the importance of taking into account the real properties of potential functions and the need to take into account the correlation between interacting molecules in the solution. Neglecting the effects of correlation in the mean-field approximation makes molecular models devoid of physical meaning. For the molecular interpretation of experimental data, the quality of the molecular models used and the accuracy of the description of the actual factors involved are important. It is well known that any description of experimental data by selecting model parameters is a poorly-specified problem that allows an ambiguous solution. Therefore, depending on the physical validity of the functional relationships of the equations used, the reliability of the molecular interpretation of the experiment largely depends.

In this connection, it is necessary to emphasize the qualitative difference between the thermodynamic approach and molecular models in the processing of experimental data. The thermodynamic problem consists in the development of functional connections only between the measured parameters of the state of the system, the number of which is always small. If we neglect the influence of external fields, then the parameters of the solution state are: temperature (T), pressure (P) and concentration of the components of mixtures θ_i (molar fractions). The mixing of the thermodynamic characteristics and molecular properties (the size and orientation of the molecule, intermolecular potentials, the structure of the solution, etc.) is unacceptable. Molecular properties can be used only in molecular models that operate with a large (in general, macroscopic) number of variables describing the states of all molecules of the mixture. When constructing models, naturally, the number of variables is reduced by the simplifications introduced. To take into account the local structure of the solution and its effect on the thermodynamic characteristics, molecular models use, in addition to concentrations, additional pairs of higher and higher order distribution functions.

The introduction of the concepts of chemical potential and the coefficient of activity closes classical thermodynamics and does not allow any molecular interpretation of these characteristics in them without involving molecular models. Additional binding to the concept of ideal solutions when separating excess contributions to thermodynamic characteristics makes it possible to characterize the degree of non-ideality of the system and make a number of conclusions from the form of the concentration dependences of excess functions only about the intensity of deviations from the behaviour of ideal systems.

In the case of significant deviations from the properties of ideal systems, in practice, mathematical methods for describing the redundant functions are introduced. For the first time such constructions were introduced by Margules [5,19], and now they are used in different versions, the so-called thermodynamic models. Thus, the logarithm $\ln\gamma_1 = \sum_{k\geq 2}\alpha_k x_2$ of the activity coefficient in a binary mixture can be represented as a series with integer exponents. If, for example $\ln\gamma_1 = \alpha_2 x_2^2 + \alpha_3 x_2^3 + \alpha_4 x_2^4$, then, taking into account the Gibbs-Dugem equation, the activity coefficient of the second component is defined as $\ln\gamma_2 = (\alpha_2 + 3\alpha_3/2 + 2\alpha_4)x_1^2 - (\alpha_3 + 8\alpha_4/3)x_1^3 + \alpha_4 x_1^4$. The same constructions create formal possibilities for the quantitative description of the experiment also in the case of a larger number of variables, although the written out series for ln γ_1 is a conventional mathematical method of expanding a series of functions of one variable that is devoid of physical premises.

The molecular theory provides 'decoding' the coefficient of activity through molecular parameters. When deriving equations for the thermodynamic characteristics by the methods of statistical thermodynamics, the simplifications and the field of their application are always formulated. Therefore, all molecular parameters are always limited in magnitude by the physical meaning and/or the conditions of mathematical transformations, as well as the domain of applicability of the constructed equations. This accompanying information obtained in the construction of molecular models and the derivation of expressions for the chemical potential can not be obtained in any 'thermodynamic models', and therefore the latter create the illusion of interpretation. Thus, thermodynamic models are not controlled by the concentration range of application, as well as by the number of phases and by the concentration regions of existence of the phases.

In essence, the thermodynamic description is always oriented to 'accurate' experimental information, and not to give 'model' equations for its approximate description. As noted above, there is always a multiplicity (i.e. arbitrariness) of variants of thermodynamic models, therefore, in principle, there can not be 'model thermodynamic' theories. This position illustrates the connections in the binary mixtures ln $\gamma_1 = ax_2^2$ and ln $\gamma_2 = ax_1^2$, which can improve the quantitative agreement with the experimental curve, but from the physical point of view the coefficient α does not have a strictly molecular meaning. The same arbitrariness is represented by proposals to represent the residual contribution of ln γ_i^{res}, caused by intermolecular interactions, in the form of terms from different types of potential functions for specific and non-specific constituents [38]. It was shown above that this procedure is impossible in principle by the very nature of statistical averaging procedures relating to different spatial coordinates. From the additivity of the dispersion and dipole potential contributions to the total energy of the system (A3.1), it is not possible to represent these contributions in the activity coefficients as separate terms. Different potential contributions act simultaneously in space and can not be separated as independent. From a physical point of view, it is the joint action of different potentials that is important for the formation of associates, and also allows us to approach the molecular interpretation of the dielectric permittivity of polar liquids that is absent at present, including water, whose solutions are often interpreted from the position of associates formation.

The molecular theory shows that in the general case the structure of the expression for the activity coefficient (A3.8) is approximate. The first term γ_i^{comb} is a consequence of the historical choice for determining redundant functions through expressions for ideal mixtures of the approximately commensurable size. Large differences in the size of the components of the solution with this choice of 'excess properties' lead to the appearance of a contribution γ_i^{comb} even in the absence of intermolecular interactions. Depending on the accuracy of the molecular model used for describing the differences in the size of components in γ_i^{comb} there may be one or two terms [4], therefore, as explained above, one can not use γ_i^{comb} in 'thermodynamic models'.

Intermolecular interactions between all components of the mixture, including associates as their components, are taken into account by the second term in (8). The number of terms in γ_i^{res} depends on the accuracy of the correlation effects and does not depend on the method of constructing the potential functions: for the molecule as a whole or in the contact approximation in the framework of atom–atom potentials [1–4] or quantum chemical methods [39]. The structure of the expression γ_i^{res} is determined by the extent of the potential functions. When the accuracy of the quasi-chemical approximation is sufficient to consider the experiment, the maximum of the number of terms in γ_i^{res} is equal to the radius of the interaction potential, measured in the coordination spheres, which is determined by the total action of all the potentials, rather than individual contributions of different potential functions. With increasing accuracy of accounting for lateral interactions ln γ_i^{res} consists of a single term for any number of contributions in the formula (A3.1), but each contribution affects the concentration behaviour ln γ_i^{res}.

The increase in the accuracy of the molecular model inevitably links the configurations of large molecules and their intermolecular interactions. This leads to the impossibility of separation of ln γ_i according to the formula (A3.8). It is necessary to calculate ln γ_i as a single characteristic of the mixture – this is the essence why the thermodynamic models can not exist, as noted above.

References

1. Hirschfelder J.O., Curtiss C., Bird R.B. Molecular theory of gases and liquids. Moscow: IL, 1961. 930 p (Wiley, New York, 1954).
2. Kaplan I.G., Introduction to the Theory of Intermolecular Interactions. Moscow, Nauka, 1982. 312 p.
3. Prigogine I.P., The Molecular Theory of Solutions. Amsterdam; New York, Amster¬dams Interscience Publishers Inc., 1957.
4. Smirnova N.A., Molecular theory of solutions. Leningrad, Khimiya, 1987.
5. Prigogine I., Defay R., Chemical Thermodynamics. Novosibirsk, Nauka, 1966. [Longmans Green, London, 1954].
6. Tovbin Yu.K., Zh. Fiz. Khimii. 2012. V. 86. No. 8. P. 1355.
7. Hill T.L., Statistical Mechanics, New York, McGraw-Hill, 1956.
8. Tovbin Yu.K., Theory of physico-chemical processes at the gas–solid interface, Moscow, Nauka, 1990. [CRC, Boca Raton, Florida, 1991].
9. Tovbin Yu.K., Zh. Fiz. Khimii. 1995. V. 69. No. 2. P. 214.
10. Tovbin Yu.K., Zh. Fiz. Khimii. 1995. V.69. No. 2. P. 220.

11. Barker J.A., Fock W., Disc. Farad. Soc. 1953. No. 15. P. 188.
12. Anderson G.A., Wheeler, J.C., J. Chem. Phys. 1978. V. 69. P. 2082.
13. Walker J.C., Vause S.A., Phys. Lett. 1980. V. A79. P. 421.
14. Goldstein R.E., Walker J.C., J. Chem. Phys. 1983. V. 78. P. 1492.
15. Walker J.C., Vause S.A., Ibid. 1983. V. 78. P. 2660.
16. Tovbin Yu.K., Zh. Fiz. Khimii. 2015. V. 89. No. 11. P. 1752.
17. Fowler R.H., Guggenheim E.A. Statistical Thermodynamics. Cambridge Univer. Press, 1939.
18. Tovbin Yu.K., Kinetika i kataliz. 1982. V.23. P. 1231.
19. Durov V.A., Ageev E.P., Thermodynamic theory of solutions. Moscow, Editorial URSS. 2002.
20. Tovbin Yu.K., Teoret. eksper. khimiya. 1982. V. 18. No. 4. P. 419.
21. Guggenheim E.A., Mixtures. Oxford Univer. Press, 1952.
22. Miller A.R., Theory of Solutions of High Polymers. Oxford, Clarendon Press, 1948.
23. Barker J.A., J. Chem. Phys. 1952. V.20. P. 1526.
24. Flory R.J., Principles of Polymer chemistry. New York, Cornell Univ. Press, 1953.
25. Shakhparonov M.I., Introduction to the molecular theory of solutions. Moscow: GITTL, 1956. 507 pp.
26. Sanchez I.S., Lacombe R.H., J. Phys. Chem. 1976. V.80. P. 2352.
27. Lacombe R.N., Sanchez I.C., Ibid. P. 2568.
28. Sanchez I.S., Lacombé R.H., Macromolecules. 1978. V. 11. P. 1145.
29. Tovbin Yu.K., Khim. Fizika. 2011. V.30. No.4. P.27. [Russ. J. Phys. Chem. B. 5, 256 (2011).
30. DiMarzio E.A., J. Chem. Phys. 1961. V. 35. P. 658.
31. Mitra S.K., Alnatt A.R., J. Phys. Ser. C. Solid State Phys. 1979. V.12. P. 2261.
32. Tumanyan N.P., Shahatuni A.G., Arm. Khim. Zh. 1982. V. 35. P. 103.
33. Tumanyan N.P. , Sokolova E.P., Zh. fiz. khimii. 1984. V.58. P. 2488.
34. Tovbin Yu.K., Izv. AN SSSR. ser. khim. 1997. No.3. P. 458. [Russ. Chem. Bull. 46, 437 (1997)].
35. Tovbin Yu.K., Khim. Fizika. 1997. V.16. No.6. P. 96.
36. Tovbin Yu.K., Zhidkova L.K., Komarov V.N., Izv. AN SSSR. ser. khim.. 2001. No. 5. P. 752. [Russ. Chem. Bull. 50, 786 (2001)].
37. Tovbin Yu.K., Senyavin M.M., Sokolova E.P., Zh. fiz. khimii. 2003. V. 77. No. 4. P. 714 [Russ. J. Phys. Chem. A 77, 636 (2003)].
38. Durov V.A., Shilov I.Yu., Zh fiz. khimii. 2012. V. 86. № 2. P. 216 [Russ. J. Phys. Chem. A 86, 162 (2012)].
39. Simkin B.Ya., Sheikhet I.I., Quantum-chemical and statistical theory of solutions. Moscow, Khimiya. 1989 [Prentice Hall, Englewood Cliffs, NJ, 1995].

Index

A

C

D

E

N

O

P

Q

R

S

T

V